XX International Meeting on Physical Interpretations of Relativity Theory 2017

Journal of Physics: Conference Series
Volume 1051

Moscow, Russia
3 – 6 July 2017

ISBN: 978-1-5108-6791-8
ISSN: 1742-6588

Printed from e-media with permission by:

Curran Associates, Inc.
57 Morehouse Lane
Red Hook, NY 12571

Some format issues inherent in the e-media version may also appear in this print version.

Printed by Curran Associates, Inc. (2018)

For permission requests, please contact the Institute of Physics
at the address below.

Institute of Physics
Dirac House, Temple Back
Bristol BS1 6BE UK

Phone: 44 1 17 929 7481
Fax: 44 1 17 920 0979

techtracking@iop.org

Additional copies of this publication are available from:

Curran Associates, Inc.
57 Morehouse Lane
Red Hook, NY 12571 USA
Phone: 845-758-0400
Fax: 845-758-2633
Email: curran@proceedings.com
Web: www.proceedings.com

Table of contents

Volume 1051

XX International Meeting "Physical Interpretations of Relativity Theory 2017"

3–6 July 2017, Moscow, Russian Federation

Accepted papers received: 27 June 2018
Published online: 17 July 2018

Preface

Plenary papers

Papers

Poster papers

IOP Conf. Series: Journal of Physics: Conf. Series **1051** (2018) 011001 doi:10.1088/1742-6596/1051/1/011001

PREFACE

In the period from July 3 to July 6, 2017 Bauman Moscow State Technical University held the conference on modern problems of the theory of relativity, cosmology and astrophysics PIRT-2017.

Since 1988 PIRT conference has been held at Imperial College in London. Since 2003 the conference has been held at BMSTU as well. The main organizers are BMSTU, Liverpool University and the Russian Academy of Sciences.

The chairman of the International Organizing Committee is the head of Bauman Moscow State Technical University, Professor Anatoly Alexandrov. Academician Vladislav Pustovoit is the Chairman of the International Academic Committee. The Academic Committee also includes well-known academicians of RAS S.N. Bagaev, N.S. Kardashev, I.B. Fedorov, A.M. Cherepashchuk and etc.

The 2017 conference, which was announced as the Russian-British Year of Education and Science, received support from the Russian Foundation for Basic Research (grant no. 17-02-20289), and a grant from the UK's Science and Innovation Network of the British Embassy in Moscow.

110 reports were delivered at the conference (10 plenary, 69 oral, 31 poster reports). The reports were presented from 54 universities and research centers from 27 countries (183 participants, including 61 foreign participants).

The working languages of the conference were Russian and English, simultaneous interpretation of all reports was carried out. Previously, theses of all the reports were published.

Within 4 days, there were delivered reports on current problems in the field of gravity, cosmology, relativistic electrodynamics, high-energy astrophysics, and the experimental consequences of the theory of relativity.

As a part of the conference, there was held a session on the problem of creating a new generation gravitational antenna. Scientists from Great Britain, USA, France, Australia, as well as Russia, where the idea of interferometric method of recording gravitational waves first appeared (M.E. Gertsenshtein, V.I. Pustovoit, 1962) took part in it.

There was also held a meeting on the organization of international scientific research housed by Bauman University and universities of the United Kingdom with the participation of other countries. The meeting was attended by Dr. Jonathan Brenton, Minister and Deputy Head of the UK Mission in Russia, Vladislav Pustovoit, Academician of the Russian Academy of Sciences.

During the conference the journal Nature Physics in the person of Dr. Bart Verberck (Great Britain) held a competition to select the best poster reports and awarded two prizes. During the conference videotaping of scientific sessions was carried out, outstanding scientists were interviewed, and a short film about PIRT-2017, international cooperation of scientists in the field of gravity, cosmology and the theory of relativity was made. Also for participants there was organized a cultural program which included a visit to the historical places of Moscow.

IOP Conf. Series: Journal of Physics: Conf. Series **1051** (2018) 011002 doi:10.1088/1742-6596/1051/1/011002

International Organizing Committee of PIRT-2017

- **Prof. A.A. Aleksandrov** (Chairman), Bauman University, Moscow, Russia
- **Prof. V.O. Gladyshev** (Co-Chairman), Bauman University, Moscow, Russia
- **Prof. P. Rowlands** (Co-Chairman), University of Liverpool, Liverpool, United Kingdom
- **Prof. Dr. M. R. Adhikari**, University of Calcutta, IMBIC, International Egyptian Engineering Mathematical Society, Institute for Polymath, India
- **Prof. J.B. Almeida**, University of Minho, Physics Department, Braga, Portugal
- **Prof. D. Blair**, University of South Western Australia, Australia
- **Prof. A. Giazotto**, INFN - Sezone di Piza, Pisa, Italia
- **Prof. G. Pizzella**, Physics Department, University of Rome, Rome, Italy
- **Prof. G.M. Webb**, Commercial Space Technologies Ltd., London, Great Britain
- **Prof. N. Stergioulas**, Department of Physics Aristotle University of Thessaloniki 54124 Thessaloniki, Greece
- **Prof. A.P. Fournier-Sikr**, European Space Agency, France
- **Prof. M.C. Duffy**, University of Sunderland, Sunderland, British Society for the Philosophy of Science, United Kingdom
- **Prof. H.P. Mazumdar**, Physics and Earth Sciences Division, Indian Statistical Institute, Kolkata, India
- **Prof. E. Mychelkin**, Fesenkov Astrophysical Institute NAS RK, Almaty, Kazakhstan
- **Prof. G.Yu. Bogoslovsky**, Skobeltsyn Institute of Nuclear Physics, Lomonosov Moscow State University, Russia
- **Prof. Yu.S. Vladimirov**, Lomonosov Moscow State University, Moscow, Russia
- **Prof. G.N. Izmailov**, Moscow Aviation Institute, Moscow, Russia
- **Prof. V.O. Manturov**, Bauman University, Moscow State University, Moscow, Russia
- **Prof. V.N. Melnikov**, Centre for Gravitation and Fundamental Metrology, VNIIMS and Institute of Gravitation and Cosmology, Peoples' Friendship University of Russia, Moscow, Russia
- **Prof. A.N. Morozov**, Bauman University, Moscow, Russia
- **Prof. V.N. Rudenko**, Sternberg State Astronomical Institute, Lomonosov Moscow State University, Moscow, Russia
- **Prof. S.V. Siparov**, Academy of Civil Aviation, S.-Petersburg, Russia
- **Prof. S.V. Chervon**, Ulyanovsk State Pedagogical University, Ulyanovsk, Russia
- **Prof. A.S. Chernikov**, Bauman University, Moscow, Russia
- **Dr. N.D. Tyannikova** (Scientific secretary), Bauman University, Moscow, Russia
- **Dr. Verberck Bart**, Springer Nature, United Kingdom

IOP Conf. Series: Journal of Physics: Conf. Series **1051** (2018) 011003 doi:10.1088/1742-6596/1051/1/011003

Academician Committee of PIRT-2017 (International Advisory Council)

- **Member of RAS, Prof. V.I. Pustovoit** (Chairman), Russian Academy of Science, Bauman University, Russia
- **Prof. A.N. Morozov** (Co-Chairman), Bauman University, Moscow, Russia
- **Member of RAS, Prof. S.N. Bagaev**, Institute of laser physics SB RAS, Novosibirsk, Russia
- **Member of RAS, Prof. N.S. Kardashev**, Astro Space Center of P.N. Lebedev Physical Institute of RAS, Russia
- **Member of RAS, Prof. I.B. Fedorov**, Bauman University, Moscow, Russia
- **Member of RAS, Prof. A.M. Cherepashchuk**, Sternberg Astronomical Institute Moscow Russia
- **Corr.-Member of RAS, Prof. N.N. Kolachevsky**, P.N. Lebedev Physical Institute of RAS, Russia
- **Corr.-Member of RAS, Prof. N.N. Rozanov**, Saint Petersburg National Research University of Information Technologies, Mechanics and Optics; Vavilov State Optical Institute
- **Prof. J.B. Almeida**, University of Minho, Physics Department, Braga, Portugal
- **Prof. A. Giazotto**, INFN - Sezone di Piza, Pisa, Italia
- **Prof. D. Blair**, University of South Western Australia, Australia
- **Prof. C. Bradaschia**, IstitutoNazionale di FisicaNucleare INFN), e Laboratorio di FisicaSperimentale del dipartimento di Fisica, Universita di Pisa, S. Piero a Grado (Pisa) Italy
- **Prof. M. Sakellariadou**, King's College London, University of London, London, Great Britain
- **Prof. P. Rowlands**, University of Liverpool, Liverpool, United Kingdom
- **Prof. R. Triay**, Centre de Physique Theorique CNRS - University de Provence, Marseille, France
- **Prof. J. Hough**, Department of Physics and Astronomy, University of Glasgow, Glasgow, U.K.
- **Prof. R. Weiss**, Department of Physics, Massachusetts Institute of Technology, Cambridge, U.S.A.
- **Prof. S. Whitcomb**, Physics/LIGO laboratory, California Institute of Technology, Pasadena, U.S.A.
- **Prof. N. Deruelle**, Institut des Hautes Etudes Scientifiques, Bures sur Yvette, France
- **Prof. G. Pizzella**, Physics Department, University of Rome, Rome, Italy
- **Prof. C. Romero**, Universidade Federal da Paraiba, Departamento de Fisica, Joao Pessoa, Brazil
- **Prof. H.P. Mazumdar**, Physics and Earth Sciences Division, Indian Statistical Institute, Kolkata, India
- **Prof. E. Mychelkin**, Fesenkov Astrophysical Institute NAS RK, Almaty, Kazakhstan
- **Prof. B.M. Bolotovskii**, Lebedev Physical Institute of RAS, Moscow, Russia
- **Prof. Yu.S. Vladimirov**, Moscow State University, Moscow, Russia
- **Prof. V.O. Gladyshev**, Bauman University, Moscow, Russia
- **Prof. A.D. Dolgov**, Institute of Theoretical and Experimental Physics of RAS, & INFN, Ferrara, Italy
- **Prof. V.N. Lukash**, Astro Space Center of P.N. Lebedev Physical Institute of RAS, Russia
- **Prof. V.N. Melnikov**, Centre for Gravitation and Fundamental Metrology, VNIIMS and Institute of Gravitation and Cosmology, Peoples' Friendship University of Russia, Moscow, Russia
- **Prof. V.N. Rudenko**, Sternberg State Astronomical Institute, Lomonosov Moscow State University, Moscow, Russia

Peer review statement

All papers published in this volume of *Journal of Physics: Conference Series* have been peer reviewed through processes administered by the proceedings Editors. Reviews were conducted by expert referees to the professional and scientific standards expected of a proceedings journal published by IOP Publishing.

IOP Conf. Series: Journal of Physics: Conf. Series **1051** (2018) 012001 doi:10.1088/1742-6596/1051/1/012001

Generation and detection of high frequency gravitational waves at intensive electromagnetic excitation

V S Gorelik[1,2], V I Pustovoit[2,3], V O Gladyshev [2], A N Morozov[2], V L Kauts[1,2], E A Sharandin[2], I V Fomin[2] and D I Portnov[2]

[1]Lebedev Physical Institute of Russian Academy of Sciences, Moscow, Russia
[2]Bauman Moscow State Technical University, Moscow, Russia
[3] Scientific and Technological Center of Unique Instrumentation of Russian Academy of Sciences, Moscow, Russia
E-mail: gorelik@sci.lebedev.ru

Abstract. The ways of generating and detecting of high frequency gravitational wave in free space and in nonlinear dielectric media under intensive electromagnetic emission excitation have been analyzed. It was shown that there is the opportunity to realize the parametric process of gravitational wave generation with high frequency $\omega_g = 2\omega_0$ due to combining of two electromagnetic waves with frequency ω_0. In the dielectric medium synchronism conditions may be reached if the frequency ω_0 of exciting light corresponds to so call unitary polaritons, characterized by unity refraction index. The detection of gravitational waves with frequency $\omega_g = 2\omega_0$ may be examined by means of the third harmonic generation registration in corresponding dielectric media.

1. Introduction

Detection of gravitational waves is one of the most important problems of our time, interesting both from fundamental point of view, and also for many applications [1-7]. The existence of gravitational waves, arising from cosmic objects oscillation and disturbance, was predicted in the classical Einstein's works. The interest in the solution of this problem has increased significantly after recently direct detection of gravitational waves [6]. The experimental results, presented in [6], consist of gravitational waves detection at relatively low frequencies ($10-100$ Hz), emitted by astrophysical bodies. The results, obtained in [6], based on laser interferometry technique with using of Michelson interferometer, first proposed and theoretically justified in [8]. Thus, the scientific and practical importance and actuality of gravitational waves detection and generation is not in doubt. In accordance with the theory, the gravitational wave radiating intensity is proportional to the frequency of the sixth powered degree (see, for example, [7]). In this connection, it is interesting to investigate the possibility of generating and detecting of the high-frequency gravitational waves. The theory [9] predicts the opportunity the registration of emitted by astrophysical objects gravitational waves at high frequency. According to [9], in cosmic space the modulation of high frequency ω_0 electromagnetic star emission by low frequency ω_g gravitational waves should take place, resulting combinational light scattering with frequency $\omega' = \omega_0 \pm 2\omega_g$, convenient for detection in optical range. Besides this, high-frequency gravitational waves emission may be generated due to the parametric processes in cosmic space [8]. In this case, on the first step the gravitational waves with frequency $2\omega_0$ is emerging as the result of two-photon absorption of intense high frequency ω_0 electromagnetic star radiation. The indirect detection of such high frequency gravitational waves may be realized by the third harmonic of electromagnetic emission registration. The third $3\omega_0$ - electromagnetic emission is emerging as the result of $2\omega_0$ - gravitational wave combining with initial ω_0 - electromagnetic wave. The high frequency $3\omega_0$ - electromagnetic emission may be detected by the very sensitive receiver. For effective parametric conversion processes realizing the momentum conservation laws (synchronism conditions) should be satisfied, i.e. the propagation phase velocity of initial ω_0 electromagnetic radiation should be close to the phase velocity of $2\omega_0$ gravitational wave and $3\omega_0$ detected electromagnetic wave. Thus the third harmonic electromagnetic emission generation in cosmic space at long enough distance was predicted [9].

In this paper we propose the new ways for high-frequency gravitational waves generation and detection on the base of excitation of different parametric processes in dielectric media by powerful lasers sources of light.

2. To the theory of gravitational wave emission in the presence of powerful electromagnetic radiation

Gravitational waves correspond to propagation of metric tensor g_{ik} deviation h_{ik} in the space, described as [7];

$$g_{ik} = g_{ik}^{(0)} + h_{ik}; \quad \left(\Delta - \frac{1}{c^2}\frac{\partial^2}{\partial t^2} \right) h_k^i = \frac{8\pi G}{c^4} T_k^i . \tag{1}$$

The whole energy of gravitational emission depends on the quadruple mass momentum $D_{\alpha\beta} = \int \mu \left(3x^\alpha x^\beta - r^2 \delta_{\alpha\beta} \right) dV$ and is given by the following relation, obtained by A. Einstein:

$$-\frac{dE}{dt} = \frac{G}{45c^3}\left(\frac{\partial^3}{\partial t^3} D_{\alpha\beta} \right)^2 . \tag{2}$$

For plane gravitational wave $h_{\alpha\beta} = h_{\alpha\beta}^{(0)} e^{i\omega_g (t-\frac{x^1}{c})} + [h_{\alpha\beta}^{(0)}] * e^{-i\omega_g (t-\frac{x^1}{c})}$ gravitational wave emission intensity is proportional to $(\omega_g)^6$, i.e. sharply increases with frequency ω_g increase.

In the presence of gravitational field Maxwell equations have the shape [9]:

$$\begin{aligned} rot\vec{E} + \frac{1}{c\sqrt{\gamma}}\frac{\partial}{\partial t}\left(\sqrt{\gamma}\vec{B}\right) = 0; \quad \vec{B} = \frac{\vec{H}}{\sqrt{g_{00}}}; \\ rot\vec{H} - \frac{1}{c\sqrt{\gamma}}\frac{\partial}{\partial t}\left(\sqrt{\gamma}\vec{D}\right) = 0; \quad \vec{D} = \frac{\vec{E}}{\sqrt{g_{00}}}. \end{aligned} \tag{3}$$

Here $\vec{E},\vec{H}$ are electric and magnetic fields correspondingly, and $\gamma = \mathrm{Det}|\gamma_{\alpha\beta}| = 1 - h_{22}^2 - h_{23}^2$ for some polarization geometry.

Due to modulation of initial electromagnetic field with frequency ω_0 by gravitational wild with frequency ω_g the secondary emission radian emerges, described by field $\vec{E}$ .Thus we have:

$$\vec{E} = \vec{E}_0 + \vec{E}^{'}; \vec{E}_0 = \vec{e}_0 e^{i\omega_0 (t-\frac{x^1}{c})} + (\vec{e}_0)^* e^{-i\omega_0 (t-\frac{x^1}{c})} .; \vec{E}^{'} = \vec{e}^{'} e^{i\omega_g (t-\frac{x^1}{c})} + (\vec{e}^{'})^* e^{-i\omega_g (t-\frac{x^1}{c})} . \tag{4}$$

From (3) and (4) equations it follows:

$$(rot\,rot + \frac{1}{c^2}\frac{\partial^2}{\partial t^2})\vec{E}^{'} = -\frac{1}{2c^2}\frac{\partial}{\partial t}(\vec{E}_0 \frac{\partial}{\partial t}\gamma) - \frac{1}{2c}[\vec{H}_0, grad\frac{\partial \gamma}{\partial t}]. \tag{5}$$

From relation (5) under condition, that $\omega' = \omega_0 \pm 2\omega_g$, at the distance L of gravitational and electromagnetic waves interaction, in [9] was obtained :

$$\vec{e}(L) = \frac{i\left[\left(h_{22}^{(0)}\right)^2 + \left(h_{23}^{(0)}\right)^2\right]\omega_g L}{2c}\frac{\omega_0}{\omega_0 \pm 2\omega_g}\vec{e}_0. \tag{6}$$

Thus the task of low frequency gravitational waves detection is decided by registration of the high frequency secondary combinational electromagnetic emission, amplitude of which satisfies:

$$|\frac{\vec{e}_\pm}{\vec{e}_0}| = \frac{2\pi G S_L}{\tilde{n}^4 \omega_g}\frac{\omega_0}{\omega_0 \pm \omega_g}. \tag{7}$$

Now let us discuss of the high frequency gravitational waves registration, emerging during parametric processes. According to the known [9] relation for small intensity gravitational waves, it takes place:

$$(\Delta - \frac{1}{c^2}\frac{\partial^2}{\partial t^2})h_{ik} = \frac{16\pi G}{c^4}(T_{ik} - \frac{1}{2}g_{ik}T). \tag{8}$$

Here T_{ik} energy-momentum tensor, $T = T_i^i$. Taking into account the relations (3,9), it was obtained [9]:

$$h_{ik} = \frac{4i\pi G W_0}{c^3\omega_0}\left(\exp\left[2i\omega_0\left(t - \frac{x^1}{c}\right)\right] + \exp\left[-2i\omega_0\left(t - \frac{x^1}{c}\right)\right]\right), \tag{9}$$

where $W_0 = \frac{E_0^2}{4\pi}$ - the energy density of exciting electromagnetic field. Thus at the presence of intensive electromagnetic field may be generated in space the gravitational wave with high frequency $\omega_g = 2\omega_0$, i.e. doubled of exciting light frequency. In other words, in vacuum parametric process of electromagnetic field self-interactions with doubling of exciting light frequency is permitted in spite of inversion symmetry of vacuum taking place. The intensity of such parametric generation of gravitational waves should be proportional to the squared intensity of exciting emission as for the second harmonic generation in nonlinear crystals. Such parametric generation of high frequency $\omega_g = 2\omega_0$, gravitational waves at long enough distance should take place only when the synchronism conditions are satisfied, i.e. momentum conservation law is fulfilled. On the other hand, the gravitational wave with frequency $\omega_g = 2\omega_0$ may combine with electromagnetic wave with frequency ω_0, resulting third harmonic electromagnetic radiation at frequency $\omega' = 3\omega_0$. Thus, for detecting of gravitational waves with high frequency $\omega_g = 2\omega_0$, the electromagnetic emission at frequency $\omega' = 3\omega_0$ may be recorded, if all synchronism conditions are proved.

Another opportunity for high frequency $\left(\frac{\omega_0}{2\pi} >> 100 \text{ Hz}\right)$ gravitational wave generation is at the presence of outer static magnetic field. According to theoretical works [10-11] the following equation for description of the hybrid gravitational and electromagnetic fields may be used:

$$\begin{aligned}(\Delta - \frac{1}{c^2}\frac{\partial^2}{\partial t^2})a &= \frac{2kH_0}{c^2}\sqrt{G}b;\\ (\Delta - \frac{1}{c^2}\frac{\partial^2}{\partial t^2})b &= \frac{2kH_0}{c^2}\sqrt{G}a.\end{aligned} \tag{10}$$

Here a and b characterize electromagnetic and gravitational fields correspondingly, H_0 is the value of static magnetic field, k is corresponding wave vector, G - gravitational constant. At small distance L of propagation of electromagnetic wave in static magnetic field the coefficient α of electromagnetic-gravitation wave's conversion is very small:

$$\alpha = \frac{G H_0^2 L^2}{c^4} << 1. \tag{11}$$

But for long distance L of wave's propagation, with synchronism condition satisfying (the equality of phase velocity of gravitational and electromagnetic waves), coefficient α should increase. It is waited the appearance of such kind phenomenon for pulsars, characterizing of the giant static field presence.

3. Parametric process in nonlinear media under laser excitation

The using of powerful laser radiation permits now to observe [12-21] a number of nonlinear and parametric processes in dielectric media Stimulated Raman Scattering (STRS), Parametric Light Scattering (PLS), Second Harmonic Generation (SHG), Two-Photon Light Absorbance (TPLA), Two-Photon Excited Luminescence (TPEL) and others. The experimental scheme for STRS-spectra recording in crystals is illustrated by Figure 1.

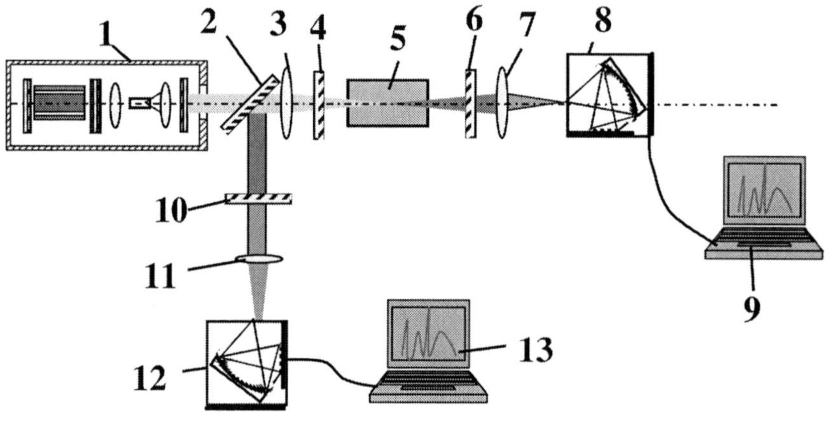

Figure 1. Experimental setup for excitation of STRS in crystals; 1 - YAG:Nd^{3+} laser ($\lambda_0 = 0{,}532$ mkm) with frequency doubling of initial laser emission ($\lambda_L = 1{,}064$ mkm); 2 - semitrasparent mirror; 3,7 - lense; 4,6 - dielectric mirrors; 5 - STRS-active crystal; 8,12 - spectrometer; 9,13 – computer.

For STRS pumping, YAG:Nd^{3+} laser ($\lambda_0 = 0{,}532$ mkm), generating ultra-short (80ps) giant pulses with 100 Hz repetition and about 1 GW/cm^2 light power density, have been used [16-17]. As the result of Raman Scattering excitation the several Stokes and anti-Stokes Stimulated Raman Scattering satellites in this spectrum may be recorded. From Figure 2 we can see that four Stokes and three anti-Stokes are simultaneously excited in calcite ($CaCO_3$) monocrystal. At the first stage in STRS only one Stokes line is emerging. In this case, the conservation law for energy and momentum may be described as:

$$\hbar\omega_0 = \hbar\omega' + \hbar\omega_V ; \quad \hbar\vec{k}_0 = \hbar\vec{k}' + \hbar\vec{k}_V . \tag{12}$$

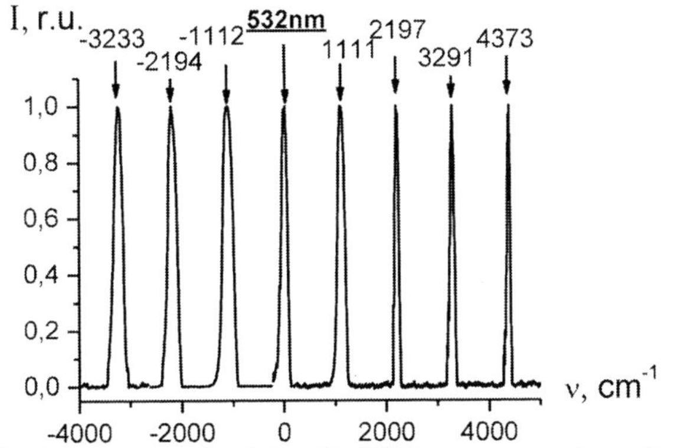

Figure 2. STRS normalized spectrum of calcite ($CaCO_3$) monocrystal, excited by ultra short pulses of the powerful -YAG:Nd^{3+} laser ($\lambda_0 = 0{,}532$ mkm)

With laser pulse power, increase the additional Stokes and anti-Stokes satellites are observed (see Figure 2). In this case four photon parametric processes is realized. The following conservation laws for energy and momentum for four photon parametric processes have the shape:

$$\hbar\omega_{St} + \hbar\omega_{aSt} = 2\hbar\omega_0; \quad \hbar\vec{k}_{St} + \hbar\vec{k}_{aSt} = 2\hbar\vec{k}_0. \tag{13}$$

Thus, the result of Stokes (ω_{St}) and anti-Stokes (ω_{aSt}) parametric processes excitation, the pair of free initial light quanta with frequency ω_0 are burning. Such photons are entangled, because of its simultaneous burning as the result of elemental four photon parametric process. In real crystal there is the possibility of simultaneous absorbance of Stokes (ω_{St}) and anti-Stokes (ω_{aSt}) quanta with the emergence of scalar excitons, symmetry of which coincides with the symmetry of gravitational wave with frequency $\omega_g = 2\omega_0 = \omega_{St} + \omega_{aSt}$. For gravitational waves, refraction index is equal to unity. Thus, such high frequency gravitational wave should be strong enough at long distance only if the corresponding momentum conservation law for such processes is proved. Such situation may be realized if exciting light emission corresponds to so call unitary polaritons [22-23]. For unitary polaritons in dielectric medium refraction index is equal to unity. Correspondingly in this case the phase velocity of gravitational wave ($\omega_g = 2\omega_0$) and laser wave ($\omega_L = \omega_0$) velocity should be the same. As the result of parametric processes of two exciting light wave combining, the generation of coherent gravitational wave at powerful enough laser pumping should take place.

Another way for synchronism conditions satisfying is the use of disordered solid sample: powder or microstructure dielectrics, non destructive at high intensity laser pumping. The experimental scheme for STRS exciting in such type samples is illustrated by Figure 3.

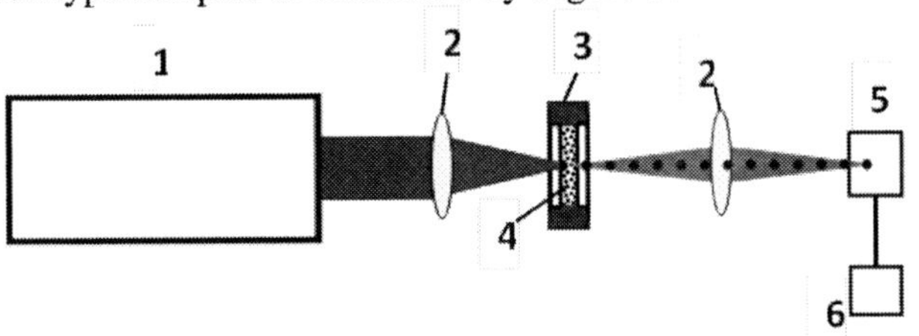

Figure 3. Experimental setup for excitation of STRS in microcrystal powder; 1 - powerful pulsed laser, 2 - lenses, 3 - holder, 4 - STRS-active powder, 5 - spectrometer, 6 - computer. Dark circles correspond to gravitational waves.

Due to micro structured powder disordering, the photon localizing inside of the sample in the region, comparable to wavelength of exciting radiation, should take place. If the intensity of exciting laser emission is large enough, the density of electromagnetic field W_0 should be very high. Thus, in the case, satisfying to relation (9), the gravitational emission with frequency $\omega_g = 2\omega_0 = \omega_{St} + \omega_{aSt}$ should be generated. For detection of such type gravitational waves (dark circles at figure 3). The third harmonics generation signals should be recorded by receiver 5.

Recently [24-27] the mesoporous photonic crystals for investigations of nonlinear processes in dielectric media, embedded into voids of such type crystals, have been used. The important property of such type is the opportunity to essentially decrease of electromagnetic waves group velocity at some spectral range, near the edges of so call photonic band gap. As the result of Purcell effect (photonic density of states increase) at such conditions realizing the efficiency of nonlinear processes may be sharply enhanced [28]. The local second and third harmonic generation in mesoporous photonic crystals have been recorded [25-27] by using of femtosecond pulse laser emission with intensity at the samples up to $1\ \mathrm{TW/cm^2}$. So the mesoporous photonic crystals are useful objects for high frequency gravitational waves generation and detection investigations.

4. Conclusion

Thus we have analyzed the conditions of generating and detecting of high frequency gravitational wave in free space and in nonlinear dielectric media under intensive electromagnetic emission excitation. It was shown that there is the opportunity to realize the parametric process, resulting gravitational wave generation with high frequency $\omega_g = 2\omega_0$ due to combining of two electromagnetic waves with frequency ω_0. In the dielectric medium synchronism conditions satisfying may be realized if the frequency ω_0 of exciting light corresponds to so call unitary polaritons, characterized by unity refraction index. The detection of gravitational waves with frequency $\omega_g = 2\omega_0$ may be on the base of third harmonic generation in corresponding dielectric media.

References

[1] Weber J 1960 *Phys. Rev.* **117** 306
[2] Weber J 1967 *Phys. Rev.Lett.* **18** 498
[3] Weber J 1968 *Phys. Rev. Lett.* **20** 1307
[4] Weber J 1969 *Phys. Rev. Lett.* **22** 1320
[5] Weber J 1970 *Phys. Rev. Lett.* **25** 180
[6] Abbot B P *et al* (LIGO Scientific Collab.) 2016 *Phys Rev. Lett.* **116** 061102
[7] Pustovoit V I 2016 Physics-Uspekhi **59** 1034
[8] Gertsenstein M E and Pustovoit V I 1963 *Sov. Phys. JETP* **16** 433
[9] Pustovoit V I and Chernozatonskii L A 1981 *JETP Letters* **34** 229
[10] Zeldovich J B 1973 *Preprint of Applied Mathematics institute of Russian Academy of Sciences* **38** p 11
[11] Gertsenstein M E 1962 *Sov. Phys. JETP* **14** 84
[12] Gorelik V S 2010 *Eur. Phys. J. Appl. Phys* **49** 33007
[13] Vodchits A I, Voinov Yu P, Gorelik V S, Kudryavtseva A D, Orlovich V A and Tcherniega N V 2013 *Bulletin of the Lebedev Physics Institute* **40** 329
[14] Gorelik V S, Mitin G G and Sushchinskii M M 1975 *Sov. Phys.-JETP* **42** 419
[15] Mitin G G, Gorelik V S, Kulevskii L A, Polivanov Yu N and Sushchinskii M M 1975 *Sov. Phys. JETP* **41** 882
[16] Gorelik V S, Maximov O P, Mitin G G and Sushchinskii M M 1977 *Solid State Communications* **21** 615
[17] Sushchinskii M M and Gorelik V S 1983 *Zhurnal Prycladnoi Spektroscopii* **38** 95
[18] Gorelik V S, Kudryavtseva A D, Tareeva M V and Tcherniega N V 2014 *Inorg. Mat.* **50** 1217
[19] Gorelik V S, Kudryavtseva A D, Tcherniega N V, Vodchits A I and Orlovich V A 2013 *Journal of Russian Laser Research* **34** 50
[20] Gorelik V S, Kudryavtseva A D, Orlovich V A, Sverbil P P, Tcherniega N V, Vodchits A I,
[21] Voinov Yu P and Zlobina L I 2013 *Journal of Russian Laser Research* **34** 523

[22] Almohamed Y, Barille R, Vodchits A I,Voinov Yu P, Gorelik V S, Kudryavtseva A D, Orlovich V A and Tcherniega N V 2015 *JETP Letters* **101** 365
[23] Gorelik V S and Burdanova M G 2014 *Laser Physics* **24** 125001
[24] Gorelik V S and Burdanova M G 2016 *Laser Physics* **26** 035001
[25] Zaytsev K I, Katyba G M, Yakovlev E V, Gorelik V S and Yurchenko S O 2014 *Journal of Applied Physics* **115** 213505
[26] Zaytsev K I, Gorelik V S, Katyba G M and Yurchenko S O 2014 *Journal of Physics: Conference Series* **541** 012072
[27] Voinov Yu P, Gorelik V S, Zaitsev K I, Zlobina L I, Sverbil P P and Yurchenko S O 2015 *Physics of Solid State* **57** 453
[28] Gorelik V S, Zaytsev K I, Moiseenko V N, Yurchenko S O and Aliev I N 2015 *Inorganic Materials* **51** 419
[29] Gorelik V S and Kapaev V V 2016 *JETP* **123** 373

Acknowledgements
This work was supported in part by RFBR grant 16-52-00026 (BRFFR F16R-063).

Dark energy – a critical value

P Rowlands
University of Liverpool, Liverpool, UK
Email: p.rowlands@liverpool.ac.uk

Abstract. Experimental evidence has converged towards a flat or Euclidean geometry for the universe with an Ω parameter or ratio of universe density to critical density of precisely 1. A considerable portion of that density is provided by the so-called dark energy which has added an acceleration term to the Hubble redshift. Measurements by the Planck satellite give values of order 68 % for the dark energy contribution. This value is intriguingly close to $2/3$, a critical value which would make the redshift velocity and acceleration related by the Hubble constant. The same value would also be obtained by applying Mach's principle to explain inertia and using the equivalence principle to relate the inertial and gravitational values of the mass of an object at the centre of a Hubble universe.

1. Introduction

Dark energy is frequently described as unexpected, exotic and mysterious. It may be none of those things. The explanation could be extremely simple. The result that indicates this possibility – the smoking gun, in effect – is the Planck satellite's finding that dark energy is approximately 68 % of the total energy of the universe, rather than the previous value of 74% from supernova measurements. The measured flatness of the universe is also significant.

Cosmology is currently defined, experimentally, by the Hubble redshift and the Hubble parameter, and, theoretically, by the Einstein field equations. The Hubble constant, H_0, the Hubble parameter at the present time, is defined in terms of the Hubble radius, r_H, as c/r_H. At the present time

$$v = H_0 r = \frac{cr}{r_H}.$$

The Einstein field equations of 1915,

$$G_{\mu\nu} \equiv R_{\mu\nu} - \frac{1}{2} R g_{\mu\nu} = \frac{8\pi G}{c^4} T_{\mu\nu}$$

were subsequently modified when Einstein added a cosmological constant Λ to prevent collapse of the universe

$$G_{\mu\nu} \equiv R_{\mu\nu} - \frac{1}{2} R g_{\mu\nu} + \Lambda g_{\mu\nu} = \frac{8\pi G}{c^4} T_{\mu\nu}.$$

In 1922 Alexander Friedmann solved the Einstein field equations for possible universes, which could be classified into three, depending on the universe density: subcritical, critical, supercritical [1]. At the critical density, the universe will be flat or Euclidean, with curvature parameter k equal to zero. According to the first Friedmann equation, in the absence of a cosmological constant Λ,

$$H^2 = \left(\frac{\dot{R}}{R}\right)^2 = \frac{8\pi G}{3}\rho - \frac{kc^2}{R^2}.$$

At the present time, the critical density for zero curvature is given by

$$\rho_{crit} = \frac{3H_0^2}{8\pi G}$$

and the ratio of the actual density ρ to this critical value, the density parameter, $\Omega = \frac{\rho}{\rho_{crit}}$, will determine the universe's evolution and ultimate fate. The value $\Omega = 1$ has a special significance, creating a clear separation of a flat or Euclidean universe from a closed universe with $\Omega > 1$ and spherical geometry, and an open universe with $\Omega < 1$ and hyperbolic geometry. Since 2000, the experimental value has been apparently converging towards the precise value of 1, which is taken as indicating that the universe is flat and possibly infinite, inflationary view of cosmology, against previous consensus [2]. It also means that general relativistic calculations are taken at the Newtonian limit, if we incorporate energy into the mass term. In this case

$$H^2 = \left(\frac{\dot{R}}{R}\right)^2 = \frac{8\pi G}{3}\rho = \frac{kc^2}{R^2}$$

becomes equivalent to

$$\rho_{crit} = \frac{3v^2}{8\rho G r^2},$$

which implies the Newtonian relation $\frac{Gm}{r} = \frac{v^2}{2}$, where m is the mass enclosed within radius r.

The mass density, of course, doesn't have to be solid matter. It can include energy. So, we could replace ρ by $\rho + \frac{3P}{c^2}$, which includes a vacuum energy term with positive (inward) pressure P from, say, radiation, without any significant change to the meaning of Ω. It acts *with* the density; ρ becomes bigger (if we assume $k = 0$). However, the inclusion of a 'cosmological constant' Λ or a vacuum energy term with negative (outward) pressure, as in

$$H^2 = \left(\frac{\dot{R}}{R}\right)^2 = \frac{8\pi G}{3}\rho = \frac{kc^2}{R^2} - \frac{\Lambda c^2}{3}$$

while, not changing the definition of Ω, would have major physical consequences. This acts *against* the density. (Again, we take $k = 0$,)

It seems that a term of this kind was discovered in 1998, through an outward acceleration in the red-shift velocity of distant galaxies [3-4]. This 'dark energy', can be incorporated into Ω as the component Ω_Λ, but it has remained an unexplained phenomenon, and the early values of its magnitude, ranging from 71,4 to 74 per cent of the total energy of the universe, gave no obvious clue as to its origin. However, it may be that, just as with Ω itself, there is a particular critical value of Ω_Λ with a precise physical significance, and new data may suggest that it converges on this value.

2. A new critical parameter

The data from the Planck satellite centre on a figure of 0.68 for Ω_Λ [5-6].

Planck Collaboration XVI		best fit Ω_Λ	0.6825
	68 % confidence limits		0.686 ± 0.020
With lensing		best fit Ω_Λ	0.6964
	68 % confidence limits		0.693 ± 0.019
With WMap		best fit Ω_Λ	0.6817
	68 % confidence limits		$0.685 + 0.018 / - 0.016$

IOP Conf. Series: Journal of Physics: Conf. Series **1051** (2018) 012002 doi:10.1088/1742-6596/1051/1/012002

Planck Collaboration I temperature anisotropies data alone
68 % confidence limits 0.67 + 0.027 /– 0.023

This suggests an intriguing possibility, with the vacuum energy close to two-thirds of the total energy of the universe. If this fraction should turn out to be the preferred value, then a simple calculation suggests some interesting consequences.
If we suppose that

$$\frac{\rho_{vac}}{\rho_{crit}}=\frac{2}{3}$$

then vacuum density becomes

$$\rho_{vac}=\frac{H_0^2}{4\pi G}.$$

This is equivalent to a 'dark' energy density or negative pressure

$$-P=\frac{H_0^2c^2}{4\pi G}$$

and cosmological constant

$$\Lambda=8\pi G\rho_{vac}=2H_0^2.$$

We can incorporate P (or Λ) into Friedmann's first equation, with $k=0$,

$$H^2=\left(\frac{\dot{R}}{R}\right)^2=\frac{8\pi G}{3}\rho=\frac{kc^2}{R^2}-\frac{\Lambda c^2}{3}$$

to give

$$H^2=\left(\frac{\dot{R}}{R}\right)^2=\frac{8\pi G}{3}\left(\rho+\frac{3P}{c^2}\right),$$

where P is negative, or into Friedmann's second equation to give

$$\frac{\ddot{R}}{R}=\frac{4\pi G}{3}\left(\rho+\frac{3P}{c^2}\right).$$

If ρ is the mass density of a uniform and isotropic Hubble universe with mass $m=\frac{4}{3}pG\rho r^3$, within radius r, then we can express this in terms of a force on a unit mass, combining the effect of gravity and dark energy:

$$F=\frac{Gm}{r^2}-H_0^2r=\left(\frac{4}{3}\pi G\rho-H_0^2\right)r.$$

A force per unit mass of this form means that the acceleration responsible for dark energy can be expressed as

$$a=\frac{v^2}{r_H}=H_0^2r.$$

This is a remarkable result, suggesting that the acceleration observed in the red-shift, like the velocity, depends only on Hubble's constant H_0 and the distance, and is thus perhaps an integral component of the same process that produces the red-shift velocity v. Integrating

$$a=v\frac{dv}{dr}=H_0^2r$$

with respect to v and r between the limits 0 and v and 0 and r gives the exact Hubble red-shift law:

$$v=H_0r=\frac{cr}{r_H}.$$

Both acceleration and velocity equations have been obtained solely from the data, with no conjectural or speculative element whatsoever, and no additional theoretical content or modelling (It is also of interest that, starting from $v = H_0 r$, we can produce a whole succession of time derivatives of v, each related to the previous one by the factor H_0).

If the data stands as it is today, the equation is valid to within the experimental confidence limits, and certainly to within a factor $1{,}02 \pm 0{,}02$. It is so close to being exactly true, and of such exceptional physical significance if it is, that we are justified in saying that the test of its exact validity should be one of the aims of future probes. We may recall that the Boomerang collaboration (2000), finding the 95% confidence interval for Ω to be between 0,88 and 1,12, were immediately able to claim, with full justification, that this provided 'evidence for a Euclidean geometry of the Universe'. Even now, $\Omega = 1$ is hardly established to better than about 1%, erring on the side of an increased Ω value, and inferentially an increased Ω_Λ. Planck Collaboration I, including data from lensing, constrains 'departures from spatial flatness at the percent level', that is $\Omega_k = -0.0096$ with 68% confidence limits of $0{,}010 - 0{,}0082$, that is, a total Ω of $1{,}0096 + 0{,}0082 - 0{,}010$ [2].

3. Gravity, inertia and dark energy

The force of gravity involving the bulk of the mass in the universe is an attractive one proportional to $\frac{1}{r^2}$. The force of dark energy is a repulsive one proportional to r. In physics, these are dual laws. One can be replaced by the other by a transformation. And they are the only laws of force which can maintain stable orbits. In the present case, the physical significance of $a = H_0^2 r$ stems from the fact that the velocity term can be derived directly from the acceleration, implying that the acceleration, whatever its origin, is actually *responsible* for the velocity. Clearly, if this is true, there must be a significant impact on possible models of cosmological evolution. There is, however, another equation which we can derive directly from the same starting-points, which, again without any hypothetical or model-dependent input, suggests even further significance. This concerns the origin of the inertial properties of matter ($F = ma$). Why does mass have the dual properties of inertia and gravitation? Would a body have inertial properties if the rest of the universe wasn't present? Ernst Mach thought that inertia is caused only by the presence of the rest of the universe [7].

Now, some aspects of general relativity can be conveniently explained by assuming that the presence of the 'velocity of light' in the mathematics means that we can have a 'magnetic' component of gravity in the same way as we can have one of electricity, and set up equations similar to Maxwell's equations. Any equation in electromagnetism will have a corresponding one in gravitational theory. In the Machian theory of inertia due to Dennis Sciama [8-9] we assume that the 'magnetic' component is responsible for inertia.

We begin by writing the acceleration in the form

$$a = \frac{dv}{dt} = \frac{c^2 r}{r_H^2}$$

and then, following Sciama, consider the possibility of explaining inertia along the lines of Mach's principle using a gravitomagnetic inductive force between two masses with relative acceleration, which could be derived from general relativity. Inertia will be a long-range force that increases with distance.

Now there is a long-range inductive force between charges e_1 and e_2, which can be derived from Faraday's law of induction.

$$F = \frac{G}{c^2 r} e_1 e_2 \sin\theta \frac{dv}{dt}.$$

So, we assume that there is also one between masses m_1 and m_2,

$$F = \frac{G}{c^2 r} m_1 m_2 \sin\theta \frac{dv}{dt}.$$

Sciama considered that, using this force, and assuming that isotropy removes the angular dependence θ, the inertia of a body of mass $m = m_1$ could be attributed to the action of the total mass $m_H = m_2$ within the observable universe, specified by radius r_H, so making the inductive force equation equivalent to the Newtonian inertial equation $F = Kma$, with K a constant and $a = \frac{dv}{dt}$. The inertial force on a unit mass due to the entire mass in the Hubble universe m_H would then be:

$$F = \frac{G}{c^2 r} m_H \sin\theta \frac{dv}{dt}.$$

Inertia provides one standard for defining a unit mass, gravitation provides another, and the connection can be made via the equivalence principle. If we now suppose that mass (Hubble) m_H defines a radial inertial field of constant magnitude from the centre of a local coordinate system, and, at the same time, use the principle of equivalence to equate the magnitude of this to the gravitational field $\left(\frac{Gm_H}{r_H^2}\right)$, which, independently of the local coordinate system, defines a unit of gravitational mass within the same event horizon, we obtain

$$\frac{Gm_H}{c^2 r} \frac{dv}{dt} = \frac{Gm_H}{r_H^2}$$

and this gives us the exact expression for the acceleration which would result if the dark energy constitutes exactly two-thirds of the total energy of the universe:

$$a = \frac{c^2 r}{r_H^2} = H_0^2 r.$$

The calculation suggests that an exact value of two thirds for the dark energy contribution would not only link the dark energy and the Hubble red-shift as aspects of the same phenomenon, but could be of additional interest in connection with Mach's principle and the origin of inertia. Again, there is an indication that possible cosmologies are necessarily constrained by fundamental laws of physics. A Machian origin of inertia, for example, would allow a universe to evolve by creating inertial mass at the same time as its space-time structure, with the creation process also generating the force which drives its evolution.

It is significant, of course, that

$$\frac{Gm_H}{c^2 r} \frac{dv}{dt} = \frac{Gm_H}{r_H^2}$$

is simply another version of

$$a = \frac{dv}{dt} = \frac{c^2 r}{r_H^2}$$

and does not *require* the development we suggested for its derivation. These serve to provide a *possible* context, but none is needed to generate the equation, and the form of the equation is in itself significant. However, prior prediction leading to experimental confirmation remains one of the strongest arguments available for any theoretical construction, and, in the present case, there is also a prior prediction.

A version of the calculation was done in reverse as part of a larger study when values of Ω_Λ looked less favourable to its conclusions [10], and this was preceded by a series of calculations deriving the red-shift acceleration as $H_0^2 r$ on the basis of a flat universe, some of which predated the experimental discovery of the dark energy. The most accessible, though not the earliest version of $a = H_0^2 r$, from a series of publications beginning over thirty years ago, was incorporated into the book "A Revolution Too Far" [12].

In this predictive context, the dark energy would seem to have a possible explanation in both physical and cosmological contexts. The calculations seem to imply that there may be a critical value for Ω_Λ just as there is for Ω. This potentially critical value now falls within the limits of the data provided by the Planck probe, and future experimental findings may converge on this value, just as they have converged on the physically significant value of unity for Ω. Even a value which came very close would require explanation in the same way as values of Ω close to 1 were thought to be too close for coincidence even before observations were able to establish an exact value. It would be interesting to see how the constraints on other cosmological parameters would be affected by applying an exact value of two thirds for the dark energy density to the Planck data, and how any possible deviations in the assumed universal isotropy and uniformity might be manifested.

The possibility that a particular value for dark energy might contribute to an explanation of inertia is intriguing, but it also leads to a profound question about the nature of inertia itself: is it intrinsically real or fictitious? Sciama thought of his inductive inertial force as a real force, and Newton had earlier described mass as *vis inertia* because he thought of it as a kind of 'force' opposing real ones according to the third law of motion. 'Real' or 'active' forces such as gravity led to the 'passive' response which we call inertia. However, in an unpublished Proposition 43, intended for Book III of the *Principia*, Newton argued that the Earth could be at the centre of a Tychonic Solar System with the Sun and its planets in orbit round it if gravity was countered by an equal and opposite *inertial* force proportional to mass originating from outside the Solar System [12]. A *fictitious* force of this kind occurs if the Earth is taken as an inertial frame with respect to the rest of the universe, but a real force would only exist if the uniformity of nature was not obeyed, and the Earth was singular with respect to the other planets. So, if we take inertia as representing the 'magnetic' aspect of the gravitational force, we may be obliged to take it as fictitious if the Earth's situation in space is not to be regarded as special.

References

[1] Friedmann A 1922 Z. *Phys.* **10** 377
[2] Bernardis P *et al* 2000 *Nature* **405** 955
[3] Schmidt B P *et al* 1998 *ApJ* **507** 46
[4] Perlmutter S *et al* 1999 *ApJ* **517** 565
[5] *Planck Collaboration I* 2013 *Preprint* arXiv: 1303.506
[6] *Planck Collaboration XVI* 2013 *Preprint* arxiv: 1303.5076
[7] Mach E 1911 'History and Root of the Principle of the Conservation of Energy' *The Science of Mechanics; a Critical and Historical Account of its Development, by Ernst Mach: supplement to the 3rd English edition containing the author's additions to the 7th German edition* (Chicago: The Open Court Publishing Co.)
[8] Sciama D W 1953 *MNRAS* **113** 34
[9] Sciama D W 1972 *The Physical Foundations of General Relativity* (New York: Doubleday)
[10] Rowlands P 2007 *Zero to Infinity The Foundations of Physics* (Singapore, London and Hackensack, NJ: World Scientific)
[11] Rowlands P 1994 *A Revolution Too Far* (Liverpool: PD Publications)
[12] Weinberg S 1998 'Physics & history' *Daedalus* **234** 4

IOP Conf. Series: Journal of Physics: Conf. Series **1051** (2018) 012003 doi:10.1088/1742-6596/1051/1/012003

Einstein/Newton duality: An ontological-phase topological field theory

R L Amoroso

Noetic Advanced Studies Institute, Los Angeles, USA
E-mail: amoroso@noeticadvancedstudies.us

Abstract. Newton claimed instantaneous G-influence; Einstein insisted no influence propagated faster than c. Quantum Mechanics (QM) the so-called *basement of reality*, posits a Quantum Gravity, for which no *a priori* science exists. We propose a *paradigm shift* with duality between a semi-quantum Standard Model (SM) limit and Large-Scale Additional Dimensionality (LSXD) in a modified M-Theoretic Unified Field (UF) brane arena as the regime of integration described by an Ontological-Phase Topological Field Theory (OPTFT) requiring fundamental changes in the concept of dimensionality and matter. OPTFT is developed to formally describe 3rd regime Unified Field Mechanics (UFM) (classical-Quantum-UFM) to relate Newton-Einstein duality by added degrees of freedom in a semi-quantum limit enabling topological Dirac-Majorana doublet fusion supervening the uncertainty principle.

1. Introduction – Imminent 3rd Regime Paradigm Shift

An Ontological-Phase Topological Field Theory (OPTFT) requires fundamental changes in the concept of dimensionality and matter. From suggestions by Rowlands, two processes emerge to create XDs (Additional Dimensionality) [1]: 1) duality, with Ds fundamentally different, and 2) anticommutativity, with Ds fundamentally the same [2]. Yang-Mills (YM) Kaluza-Klein (KK) correspondence drives Physics beyond the SM. Horizontal and vertical subspaces in the tangent bundle of M, ($M = M \times G$) defined by YM connections are orthogonal with respect to a KK metric suggesting orthogonal extension to XD beyond the 4D limit of the SM. CERN LHC research seeks KK XD beyond the SM. Current thinking posits XD as $\hbar$-scale since they are unobserved; however, this is not the only interpretation. A LSXD alternative hidden by *subtractive interferometry* is proposed [3-5]. Albeit, our OPTFT iteration of M-Theory is based on radical extensions of the original hadronic string theory because of inherent key elements: virtual tachyon/tardyon interactions and a variable concept of string tension, $T_S = T_0 + \Delta\hbar$ [3,6]. A and B-type topological string theories, and a related Topological M-Theory with mirror symmetry, while interesting, since they allow sufficient XD by Calabi-Yau mirror symmetry, essential for developing UFM; a distinction between these theories causes our model to diverge, as compactification must be continuous. A key parameter is *topological charge* in brane dynamics which by definition makes correspondence to a de-Broglie-Bohm super-quantum potential synonymous with an ontological *Force of Coherence*, an inherent aspect of UF dynamics [3-5]. Thus, UFM predicts no phenomenal graviton (perceived artifact of incompleteness of Gauge Theory, i.e. Gauge Theory is approximate suggesting new physics).

The difference between 4D quantum field phenomenology and LSXD topological field ontology is the *energyless* exchange process. Information (Shannon related) is transferred ontologically by the

dynamics of *topological switching* in M-Theoretic branes carrying topological charge [3-5]. Completing Geometrodynamics inherently includes Newton/Einstein duality [5]; evidenced by interpreting quasar luminosity as *G*-shock waves [3] countering Big Bang interpretations of large redshift, Z based on Doppler recession. Instead, redshift results from periodic photon mass-anisotropy by coupling to a covariant polarized Dirac vacuum [3]. A further conundrum exists by defining a Manifold of Uncertainty (MOU) of finite dimensional radius, allowing a wave-particle-like duality with a quantal-like virtual graviton in the *semi-quantum limit* – an intermediary between *field phenomenology* and *topological ontology*. This has increasing importance for the new field of Relativistic Information Processing (RIP) which introduces G-effects in *parallel transport* of brane topological switching [3-5]. From the broad context above our central theme is the introduction of a topological formalism for a new set of UF transformations beyond the Galilean, Lorentz-Poincairé. An empirical protocol falsifying the model is developed [3-5]; which if successful has far reaching consequences for experimentally validating XD, M-Theory and leads to obsolescence of usual TeV $\rightarrow$ PeV supercollider particle sprays by providing a new form of *table-top low-energy* UFM XD brane collision (LSXD topological fusion) *cross section* alternatives for *viewing* putative SUSY partners in a trans-D *slice* rather than standard TeV/PeV cross section collision physics [7].
OPTFT, is a paradigm shift beyond limits of the Standard Model (SM). Required tenets include:

- Redefinition of Matter: The current 0D fermionic singularity becomes cyclic M-Theoretic mirror-symmetric LSXD dual quaternion/octonion dynamic brane topology.
- Complex (LCU) Least Cosmological Unit Tessellating Space/Spacetime: LCU simplistic similitude of *Unit Cell* building Crystal Structure.
- Semi-Quantum Limit: Uncertainty a Manifold of Finite Radius: Where Duality of SM & LSXD – Newton - Einstein Enters.
- Described by New Topological Transformation: Beyond the Galilean, Lorentz-Poincairé named Noetic Transform because ultimately it must include physics of the Observer...

2. Quasar Luminosity as Gravitational Shock Waves – Newton Einstein G Duality

Conflicts within the SM call into question the fundamental interpretation of the Doppler component of the putative Hubble Expansion Law and the nature of events in associated with conventional coordinates of the line element as attached to the physical basis of the observer. Also of paramount importance is that Einstein's Classical Geometrodynamics is not a complete theory of gravity, as stated by Einstein himself. We postulate nonlinear effects associated with the propagation of light in an intense G-field produce shock waves creating *light-booms* along boundary conditions at cosmological distances approaching the limit of observation that if correct would explain anomalous Quasi-Stellar Object (QSO) luminosity. These G-shock waves are considered observationally manifest in the spectrum of QSOs and Supernova as a continuous front of *light booms* produced by superluminal boosts associated with continuous coordinate transformations relative to a distant observer, suggesting that QSOs are a form of Seifert spiral galaxy with Active Galactic Nuclei (AGN) in the vicinity of the putative observational limit of the Hubble radius, H_R, creating an issue of fundamental basis of Geometrodynamics. Newton's formulation of the G-force law requires each particle to respond instantaneously to every other massive particle regardless of the distance between them which he *proved*; but the proof is only valid in Euclidian space. Today this would be described by the Poisson equation,

$$\left(\frac{\partial^2}{\partial x^2}+\frac{\partial^2}{\partial y^2}+\frac{\partial^2}{\partial z^2}\right)\phi(x,y,z,)=f(x,y,z)$$

according to which, when the mass distribution of a system changes, its G-field instantaneously adjusts. Therefore, theory requires the speed of G to be infinite (instantaneous). Einstein's Geometrodynamics, $G_{\mu\nu}+\Lambda g_{\mu\nu}=\frac{8\pi G}{c^4}T_{\mu\nu}$ a classical extension of Newtonian-G is thus incomplete.

IOP Conf. Series: Journal of Physics: Conf. Series **1051** (2018) 012003 doi:10.1088/1742-6596/1051/1/012003

Physical theory incorporates an upper limit on the propagation speed of an interaction, maintaining that *instantaneous* action-at-a-distance is impossible. However, quantum entanglement between separated quanta enables instantaneous EPR correlations which led to the puzzle as to whether causality or locality is abandoned in transit to 3rd regime natural science.

2.1. Note on Cosmological Principle and G-Duality

In summarizing the *Cosmological Principle* (universe homogeneous and isotropic) [8] events are idealized spacetime instants defined by arbitrary time and position coordinates $t, x, y, z,$ written collectively as x^i with i, 0 to 3. The standard line element is $ds^2 = \sum_{ij} g_{ij} dx^i dx^j = g_{ij} dx^i dx^j$, where metric tensor $g_{ij}(x) = g_{ji}(x)$ is symmetric [8]. In local Minkowski form all first derivatives of g_{ij} vanish at the event taking the form $ds^2 = dt^2 - dx^2 - dy^2 - dz^2$. The Cosmological Principle generally suggests that clocks of all observers are synchronized throughout all space because of inherent homogeneity and isotropy. Because of this synchronization of clocks for the same world time t, for commoving observers the line element becomes, $ds^2 = dt^2 + g_{\alpha\beta} dx^\alpha dx^\beta = dt^2 - dl^2$, where dl^2 represents spatial separation of events at the same world time, t. This spatial component of event dl^2 can be represented as an Einstein 3-sphere (compatible with dual 6D Calabi-Yau 3-tori) $dl^2 = dx^2 + dy^2 + dz^2 + dw^2$ which is represented by the set of points (x, y, z, w) at a fixed distance R from origin: $R^2 = x^2 + y^2 + z^2 + w^2$ where $w^2 = R^2 - r^2$ and $r^2 = x^2 + y^2 + z^2$, so finally we may write the line element of the Einstein 3-sphere as $dl^2 = dx^2 + dy^2 + dz^2 + r^2 dr^2 / R^2 - r^2$ [8]. By imbedding an Einstein 3-sphere in a flat HD space, specifically as a subspace of a new complex 12D superspace [3,4,9], new theoretical interpretations of standard cosmological principles are feasible. This is the line element most compatible with the oscillatory spacetime boundary parameters required by our model of nonlinear dual G-shock waves in QSO luminosity [3].

According to MTW [10] junction conditions may act as generators of G-shocks; the dynamics of spacetime geometry for a 3-surface, Σ which includes *intrinsic* Riemann scalar curvature invariants, R, also includes an *extrinsic* curvature tensor, K_{ij}. When imbedded in an enveloping 4-geometry hypersurface it can change (shrinkage and deformation) in vector, n parallel transported as junction conditions applicable to the G-field (spacetime curvature) and the stress-energy generating it. A discontinuity in K_{ij} across a null surface without stress-energy producing it is a geometric manifestation of a G-shock-wave generated by a different embedding in spacetime *above* Σ than *below* Σ [3,10].

Dray and 't Hooft [11] found conditions for introducing G-shock waves in a class of vacuum solutions to Einstein's equations by coordinate shift. Their model generalizes G-shock waves for a massless particle moving in flat Minkowski space formulated as two Schwarzschild black holes of *equal* masses glued together at the horizon. For a spherical shell of *unequal* masses moving along $u = u_0 \neq 0$; their solution [12] represents two Schwarzschild black holes glued together at $u = u_0$. By infinitely boosting the Dray-'t Hooft solutions various forms of G-shock waves have been found [13-14]. Sfetsos [15] extends these results to the case with matter fields and a non-vanishing cosmological constant. Using the d-D spacetime metric:

$$ds^2 = 2A(u,v)\,du\,dv + g(u,v)\,h_{ij}(x)\,dx^i dx^j, \text{ with } (i, j = 1, 2, \ldots, d-2) \tag{1}$$

he uses a string based dilatonic black hole G-solution [16] from the perspective of a conformal background field theory of coset $SL(2,\mathbf{R}) / \mathbf{R} \otimes \mathbf{R}^2$ to achieve a differential shift factor

$$\left(\frac{d^2}{d\rho^2} + \frac{1}{\rho}\frac{d}{d\rho} - \varepsilon \right) f(\rho) = -16\varepsilon \frac{1}{\rho} \delta(\rho), \tag{2}$$

where $\rho^2 = x^2 + y^2$ and for a black hole singularity case with $\varepsilon = 1$, (2) is a modified Bessel equation [15]. Spitkovsky [17] simulates a relativistic Fermi emission shock process that could provide an alternative to, or component process for our G-shock work. His simulations on relativistic collisionless shocks propagating in initially unmagnetized electron-positron pair plasmas showed natural production of accelerated particles as part of shock evolution. He studied the mechanism that populates the suprathermal tail for particles gaining the most energy. The simulation showed the main acceleration occurs near the shock, where for each reflection these particles gain energy, $\Delta E \sim E$ as is expected in relativistic shocks [18].

Newton's G required instantaneous action at a distance or the conservation of angular momentum would be violated; but for Einstein's GR an instantaneous influence would violate causality and SR and so must be mediated by a field. This is the dual nature of gravity that we have put as the basis for our model. We tried to show that it is possible with further study to relate shock phenomena to G-waves especially for narrow axis massive cosmological objects such as AGN QSOs that readily lend themselves to *light-boom* effects that could therefore be used to explain QSO luminosity as further evidence of the insurmountable shortcomings of Big Bang cosmology. Our model works best contrasting both modes of the intrinsic dual nature of G because nonlinear jumps in flow occur with discontinuity. From the 2nd Law of Thermo-dynamics entropy is only increases when particles cross a shock. The duality of the propagation of the G-influence is evident in Birkhoff's theorem [3,9] where a spherically symmetric G-field is produced by a massive object such as a QSO at the origin; if there were another concentration of mass-energy elsewhere, this would disturb spherical symmetry. This effect could occur if interference occurs between the usual modes of the G-influence by shock parameters.

3. The Phasor (Phase Vector) Complex Probability Amplitude

Ontological-Phase Topological Field Theory (OPTFT) introduces fundamental 3rd regime postulates: 1) A semi-quantum mirror symmetric Calabi-Yau finite radius manifold of uncertainty, 2) with a 4D Minkowski-Riemann subspace, and 3) cyclical duality of *phenomenological* (quantal) field mediation and an *ontological charge* (energyless) topological switching unified field. As initial simplistic modeling of Ontological-Phase we adapt the *phasor* or *phase vector concept* as a precursor to ontological topological phase. In general, a phasor is a complex number for a sinusoidal (π rotation) function with amplitude A, angular frequency ω and initial phase θ, with all time invariant. The complex constant is the phasor [4]. Euler's formula can represent sinusoids as the sum of two complex-valued functions: $A \cdot \cos(\omega t + \theta) = A \cdot \left(e^{i(\omega t+\theta)} + e^{-i(\omega t+\theta)} / 2\right)$, or as the real part of function: $A \cdot \cos(\omega t + \theta) = \mathrm{Re}\left\{A \cdot e^{i(\omega t+\theta)}\right\} = \mathrm{Re}\left\{Ae^{i\theta} \cdot e^{i\omega t}\right\}$. The function, $A \cdot e^{i(\omega t+\theta)}$ is the analytic representation of $A \cdot \cos(\omega t + \theta)$. Multiplication of the phasor, $Ae^{i\theta}e^{i\omega t}$ by a complex constant, $Be^{i\phi}$, produces another phasor changing the amplitude and phase of the underlying sinusoid: $\mathrm{Re}\left\{\left(Ae^{i\theta} \cdot Be^{i\phi}\right) \cdot e^{i\omega t}\right\} = \mathrm{Re}\left\{\left(ABe^{i(\theta+\phi)}\right) \cdot e^{i\omega t}\right\} = AB\cos\left(\omega t + (\theta + \phi)\right)$. If function, $A \cdot e^{i(\omega t+\theta)}$ is depicted in a complex plane, the vector formed by imaginary and real parts rotates around the origin. A is magnitude, i is the imaginary unit, $i^2 = -1$; one cycle is completed every $\frac{2\pi}{\omega}$ seconds, and θ is the angle formed with a real axis at $t = n\frac{2\pi}{\omega}$, for integer values of n [4].

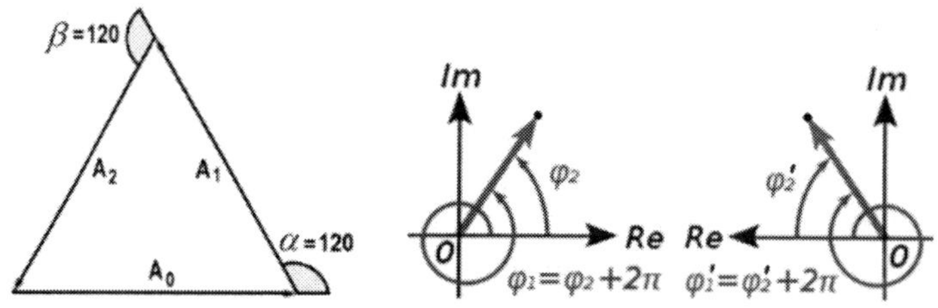

Figure 1. a) Phasor diagram of three waves in *perfect* destructive interference. b) Left-Right phase argument, preparation for phase transition channels in Dirac-Majorana duality modes.

This type of addition (Figure 1a) occurs when sinusoids interfere constructively or destructively. Three identical sinusoids with a specific phase difference may perfectly cancel. To illustrate, we place three equal length vectors matching up head to tail to form an equilateral triangle with a 120° ($2\pi/3$ radians) angle between each phasor of $1/3$ wavelength, $\lambda/3$, so the phase difference between each wave is 120°, $\cos(\omega t)+\cos\left(\omega t+\frac{2\pi}{3}\right)+\cos\left(\omega t-\frac{2\pi}{3}\right)=0$. In the three waves example, the phase difference between 1st and last waves is $240°$, In the many waves limit, phasors must form a circle for destructive interference, so that the 1st phasor is nearly parallel with the last. Thus, for many sources, destructive interference happens when the 1st and last wave differ by 360°, a full wavelength, λ [4]. For any complex number in polar form, such as $re^{i\theta}$, the phase factor is the complex exponential factor, $e^{i\theta}$. As such, *phase factor* relates more generally to term phasor, which may have any magnitude (i.e., need not be part of circle group). A phase factor is a unit complex number of absolute value 1 commonly used in quantum mechanics (QM). The variable θ is referred to as the phase. Multiplying the equation for a plane wave $A^{ei(k\cdot r-\omega t)}$ by a phase factor shifts the phase of the wave by θ: $e^{i\theta}Ae^{i(k\cdot r-\omega t)}=Ae^{i(k\cdot r-\omega t+\theta)}$. In QM, a phase factor is a complex coefficient $e^{i\theta}$ that multiplies a ket $|\psi\rangle$ or bra $\langle\phi|$, not, in itself, having any physical meaning in *standard* QM, since introducing a phase factor does not change the expectation values of a Hermitian operator. That is, the values of $\langle\phi|A|\phi\rangle$ and $\langle\phi|e^{-i\theta}Ae^{i\theta}|\phi\rangle$ are the *same* [4].

3.1. Berry Phase – Precursor to Ontological Phase

However, *differences* in phase factors between two interacting quantum states *can be measurable* under certain conditions such as in Berry phase, which has important consequences [4]. The argument for a complex number $z=x+iy$, denoted $\arg z$, is defined as:

- Geometrically, in the complex plane, as the angle ϕ from the positive real axis to the vector representing z. The numeric value given by the angle in radians is positive if measured counterclockwise (Figure 1b).
- Algebraically, the argument is defined as any real quantity, ϕ such that $z=r(\cos\phi+i\sin\phi)=re^{i\phi}$ for some positive real r (Euler's formula). The quantity r is the *modulus* of z, as $|z|$: $r=\sqrt{x^2+y^2}$.

Use of the terms *amplitude* for the modulus and *phase* for the argument are often used equivalently; by both definitions, the argument of any (non-zero) complex number has many possible values: firstly, as a geometrical angle, whole circle rotations do not change the point, so angles differing by an integer multiple of 2π radians are the same. Similarly, from the periodicity of sin and cos, the 2nd definition also has this property. An N-particle system can be represented in non-relativistic QM by a wave function, $\psi(x_1,x_2,\ldots x_n)$, where each x_i is a point in 3D space. A classical phase-space contains a real-valued function in 6N Ds (each particle contributes 3-spatial coordinates and 3-momenta. Quantum phase-space involves a complex-valued function on a 3N dimensional space. Position and momenta are represented by non-commuting operators, and ψ lives in the maths structure of a Hilbert space. Aside from these differences, the analogy holds. In physics, this addition occurs with constructively or destructively interfering sinusoids. The static vector concept provides useful insight into questions like: *What phase difference is required for three identical sinusoids to perfectly cancel* (again Figure 1a)? Waves are characterized by amplitude and phase, and both may vary as a function of those parameters. According to Berry [19], if parameter of the Hamiltonian of quantum system undergoes adiabatic changes, cyclically returning to original values, the wave function can acquire geometrical and dynamical phase. This additional *Berry phase* is $\neq 0$ when the trajectory in parameter space is near a point of degenerate states. Berry assumed the Hamiltonian is Hermitian (linear) in deviations of parameters from a point. He considered such points to be *monopole-like* when calculating geometrical phase. Thus, such points *generate* a field coinciding in monopole-like form, and the flux of Berry's field through a contour gives the geometrical phase of the system. Berry phase occurs in Aharonov–Bohm effects, where the adiabatic parameter is the magnetic field enclosed by two cyclical interference paths forming a loop and conical intersections (adiabatic parameters are molecular coordinates) of two potential energy surfaces, a set of geometrical points where the two potential energy surfaces are degenerate (intersect) and the non-adiabatic couplings between these two states are non-vanishing. Generally, geometric phase occurs whenever at least two wave parameters in the vicinity of a singularity/hole in the topology; two are required because either the set of nonsingular states will not be simply connected (shrink closed curve to point), or there will be nonzero holonomy. A Berry phase difference is acquired over the course of a cycle, when a system is subjected to cyclic adiabatic processes resulting from the geometrical properties of the parameter space of the Hamiltonian [4,19]. In addition to QM it can occur whenever there are at least two parameters describing a wave in the vicinity of a singularity or topological hole.

In a quantum system at the nth eigenstate, if adiabatic (adapts to gradually changing external conditions; but for rapidly varying conditions there is insufficient time, so the spatial probability density remains unchanged) evolution of the Hamiltonian evolves the system such that it remains in the nth eigenstate, while also obtaining a phase factor. The phase obtained has a contribution from the state's time evolution and another from the variation of the eigenstate with the changing Hamiltonian. The 2nd term is Berry phase which for non-cyclical variations of the Hamiltonian can be made to vanish by a different choice of the phase associated with the eigenstates of the Hamiltonian at each point in the evolution. But if variation is *cyclical*, Berry phase cannot be cancelled, as it is invariant and becomes an observable property of the system. From the Schrödinger equation, the Berry phase γ is: $\gamma[C]=i\oint_C \langle n,t|(\nabla_R|n,t\rangle)dR$, where R parameterizes the cyclic adiabatic process. It follows a closed path C in the appropriate parameter space. Geometric phase along the closed path C can also be calculated by integrating the Berry curvature over surface enclosed by C [4]. The Foucault pendulum is a simple example of geometric phase. The pendulum precess when it is taken around a general path C. For transport along the equator, the pendulum does not precess. But if C is made up of geodesic segments, precession arises from the angles where the segments of the geodesics meet; the total precession is equal to the net deficit angle, which equals the solid angle enclosed by C modulo 2π. We can approximate any loop by a sequence of geodesic segments, from which the most general result is that the net precession is equal to the enclosed solid angle. Since there are no inertial forces on the pendulum precess, precession, relative to the direction of motion along the path, is entirely due to the turning of the path. Thus, the orientation of the pendulum undergoes parallel transport [4].

Topological quantum field theories (TQFT) were created to avoid infinities in quantum field theory. In topological field theory, the concern is topological invariants, objects computed from a topological space (smooth manifold) without any metric. Topological invariance is invariance under the diffeomorphism group of the manifold. TQFT flourished through the work of Witten and Atiyah [4].

To experimentally move from SM Hilbert space to UFM ontological-phase space we must define *topological switching* [3-5]. We begin looking at the ambiguous Necker cube [4] where the central vertices *switch* ontologically (energyless) by *topological charge.*

4. Tight Bound States Below the Lowest Bohr Orbit

Recently, Tight Bound States (TBS) due to em-interactions at small distances below the lowest Bohr orbit have been postulated for the Hydrogen atom [20-21]. Summarizing this seminal work: in the usual atomic physics spin-orbit and spin-spin coupling perturbations give rise to only small corrections in classic Bohr energy levels. However, with distances in the $1/r^3$ and $1/r^4$ range these interaction terms, until now overlooked, can be much higher than the Coulomb term at distances << than the Bohr radius - predicting new HD physics [20]. In a further development, Corben noticed motion of a point charge in a magnetic dipole field at rest is highly relativistic with orbits of nuclear dimensions. Investigation by [20-21] for extending the Pauli equation to a two-body system defined by the Hamiltonian, $H=\frac{1}{2m_1}\left(P_1-e_1A(r_1)\right)^2+\frac{1}{2m_2}\left(P_2-e_2A(r_2)\right)^2+\frac{1}{4\pi\varepsilon_0}\frac{e_1e_2}{|r_1-r_2|}+V_{dd}$ with, m_1 mass, P_1 momentum, e_1 charge, r_1 position of the particles ($i=1,2$), A is electromagnetic vector potential and V_{dd}, the dipole-dipole interaction term:

$$V_{dd}=-\left(\frac{\mu_0}{4\pi}\right)\mu_1\mu_2\delta(r_1-r_2)+\left(\frac{\mu_0}{4\pi}\right)\left[\frac{\mu_1\mu_2}{|r_1-r_2|^3}-\frac{3\left[\mu_1(r_1-r_2)\right]\cdot\left[\mu_2(r_1-r_2)\right]}{|r_1-r_2|^5}\right].$$

In a center-of-mass frame with normal magnetic moment, $\mu=\frac{e}{m}s$ Hamiltonian, H above is:

$$H=\frac{1}{2m_1}p^2-\left(\frac{\mu_0}{4\pi}\right)\frac{e_1e_2}{m_1m_2}\frac{SL}{r^3}+\left(\frac{\mu_0}{4\pi}\right)^2\frac{e_1^2e_2^2\hbar^2}{4m_1m_2m}\frac{1}{r^4}+\frac{1}{4\pi\varepsilon_0}\frac{e_1e_2}{r}-$$
$$\left(\frac{\mu_0}{4\pi}\right)\frac{e_1e_2}{m_1m_2}s_1s_2\delta(r)+\left(\frac{\mu_0}{4\pi}\right)\frac{e_1e_2}{m_1m_2}\left[\frac{s_1s_2}{r^3}-\frac{3(s_1r)\cdot(s_2r)}{r^5}\right]. \tag{3}$$

The possibility of TBS physics as derived from Hamiltonian (3) is shown in simplified form when limited to spherically symmetric terms by the radial Schrödinger equation [20]: $\frac{d^2X}{dr^2}+\frac{2m}{\hbar^2}\left[E-V(r)\right]X=0$ and contains a form for the effective potential in the inverse power law: $V(r)=\frac{A}{r^4}+\frac{B}{r^3}+\frac{C}{r^2}+\frac{D}{r}$. At large distances this potential is an attractive Coulomb tail with a repulsive core at small distances due to the $\frac{A}{r^4}$ term [20]. For proper values of potential V its coefficients could have another potential well in addition to the one at distances of the order of the Bohr radius (location of new physics). Additional theoretical details on the seminal development of TBS by Vigier can be found in [4-5,21]. Idealization of SM elementary particles as 0D points/charge in coordinate context with no known composite subparticles, arose because *size* is considered irrelevant. Paraphrasing Rowlands: *fundamental physics reduces to explaining the structures and interactions of fermions. Fermions appear as singularities not extended objects, with no obvious way of creating such structures within 3D observation space. But, the Dirac equation suggests fermions require a double, rather than single, vector space, confirmed by the double rotation of spin ½ objects, and associated zitterbewegung and Berry phase shift. The 2nd 'space' reveals that it is an 'antispace', with the same information as real space but in less accessible form. The two spaces cancel forming a norm 0 (nilpotent) object with the exact mathematical structure required for a fermionic singularity* [5]. He further notes that fermions as singularities exist in a multiply-connected space requiring double rotations to return to starting position. Fermions also undergo *zitterbewegung* continually switching

IOP Conf. Series: Journal of Physics: Conf. Series **1051** (2018) 012003 doi:10.1088/1742-6596/1051/1/012003

between real space and complex vacuum space. The double circuit in real space is required because a fermion only exists in this space for half its existence. It is not coincidental that fermion algebra (gamma matrices) requires a commutative combination of two vector spaces for full representation; thus, obviously constructing a *singularity* requires a dual space [5,22]. The nilpotent space-antispace model extends understanding of a singularity in terms of the SM, but quaternionic algebra is not a penultimate description of nature; Rowlands' model, *avant garde* to the SM is not sufficiently radical to satisfy the needs of UFM [4-7]; but inspires a basis for correspondence to LSXD UFM OPTFT.

4.1. TBS Access Requires Violation of the Quantum Uncertainty Principle

Demonstrating new TBS spectral lines requires experimentally surmounting the quantum uncertainty principle. A new Noetic Transform will tell us whether one or two additional doublings of a Roland's type space anti-space model are required for up to five additional spectral lines which complete the radius of the MOU. Yang-Mills Kaluza-Klein equivalence provides an empirical path extending standard model particle physics. Einstein realized (1905) that Maxwell's equations obey a special relativity principle – Physical law is the same for all observers in uniform relative motion, $x^\mu = \left(x^0, x^1, x^2, x^3\right) = \left(t, x, y, z\right)$. Then, his General Relativity required two indices: $g_{\mu\nu}(x)$ with line element $ds^2 = g_{\mu\nu}(x)dx^\mu dx^\nu$. In 1919, Kaluza made his attempt to combine electromagnetism and general relativity by postulating a 5th dimension with the additional HD coordinate, $x^M = \left(x^0, x^1, x^2, x^3, x^4\right) = \left(t, x, y, z, \theta\right)$, with line element: $d\hat{s}^2 = \hat{g}_{MN}(x)dx^M dx^N$ where he then made a 4D+1D split. Klein assumed the 5th dimension had circular topology so that the coordinate, θ is periodic: $0 \le \theta \le 2\pi$. The KK 5th dimension was assumed to be curled up at the Planck scale because it was unobserved – As mentioned above this is not the only interpretation for *hidden* XD!

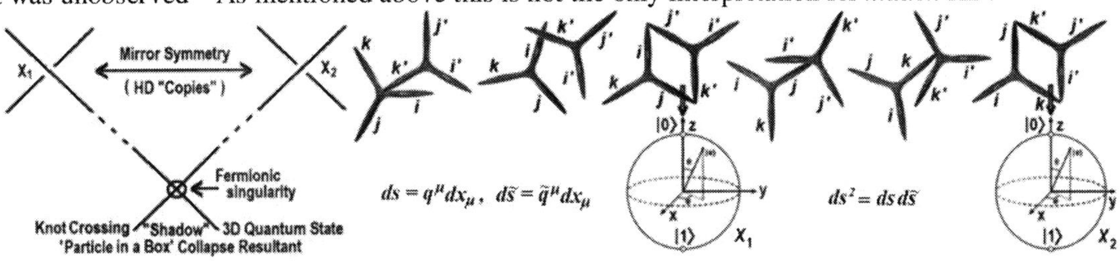

Figure 2. a) Bottom, uncertainty principle arises by a 3-space *knot shadow* of XD topological M-brane degrees of freedom. b) SM line element, locus of semi-classical Riemann Bloch 2-spheres, X_1, X_2, as *basement of reality*; Top, 1st space-antispace mirror symmetric UFM step of quaternionic vertices cycling from 5D QM chaos to topological order as faces of 3-cube.

Figure 2 shows how a localized knot shadow hides XD behind the uncertainty principle. In Figure 2b we see the beginning of separation of the standard 3-space line element into an extended KK cyclicality from order to chaos at the semi-quantum limit. The structure has an inherent *beat frequency* revealed by rf-modulation of the Dirac polarized vacuum.

IOP Conf. Series: Journal of Physics: Conf. Series **1051** (2018) 012003 doi:10.1088/1742-6596/1051/1/012003

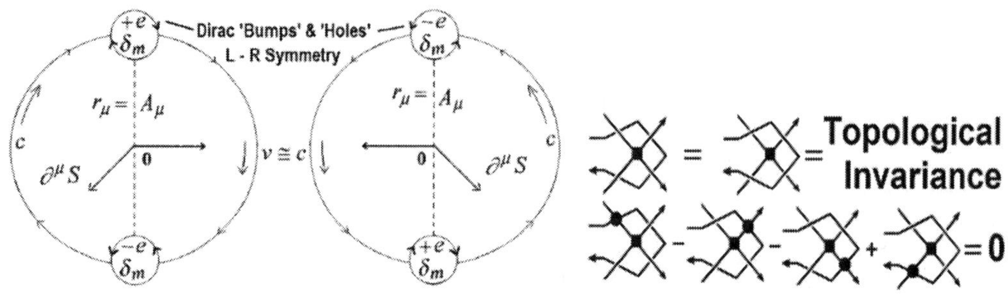

Figure 3. a) Oppositely charged sub-elements rotating at $v \cong c$ around center 0 behaving as dipole bumps and holes on the surface of a covariant polarized Dirac vacuum, allowing rf-modulation. b) Topological Invariance must be included in any phase labeling therein.

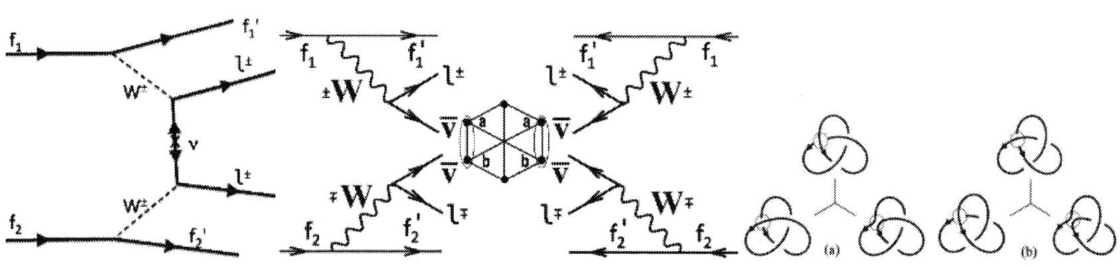

Figure 4. a) Fundamental diagram changing lepton number transitions by two units, generalized for Majorana modes (MJM). b) Dual MJM cyclic modes for fusion of a-b Berry phase cycles (center) in graphene. c) Reduction schemes for L&R-handed trefoil knots.

The Fano snowflake configuration (Figure 6) involutes to form a 2D hexagon (graphene) or vertices of a Euclidean ambiguous Necker 3-cube used to explore possible topological moves for *fusion* of ontological-phase transitions. In the context of graphene, Berry phase is the phase an eigenstate acquires after p is forced to evolve a full circle at constant energy around a Dirac vortex point. When parallel transport creates a deficit angle in brane raising and lowering dynamics, in addition to Reidemeister moves, rotations, reflections or any other topological moves, other types of phase transition with lattice charge in anyon braid fusion channels apply. Half of the leptons are neutrinos, but unknown if they are Dirac or Majorana; finding neutrinoless double β -decay would demonstrate existence of the Majorana nature of neutrinos. Neutrinoless double β -decay occurs when two neutrons in a nucleus decay simultaneously, a fundamental diagram changing lepton number by two units (Figure 4a). We begin to explore a plethora of crossover links and moves cataloging various transformations applicable to anyon fusion channels studied to supervene the inaccessibility of topological braiding, $a \times b = \sum_c N_{a,b}^{c} C$, where $a \,\&\, b \rightarrow c$ [23-24]. We wish to illustrate fusion-duality as a *Principle*, by taking the more simplistic case of de Broglie *fusion*, coordinates x_1, y_1, z_1 and x_2, y_2, z_2 become $X = x_1 + x_2 / 2, Y = y_1 + y_2 / 2, Z = z_1 + z_2 / 2$. Then for identical particles of mass, m without distinguishing coordinates, the Schrödinger equation (center of mass) is

$$-i\hbar \frac{\partial \psi}{\partial t} = \frac{1}{2M} \Delta \psi, \quad M = 2m. \tag{4}$$

Eq. (4) corresponds to the present, Eq. (5a) the advanced wave and (5b) the retarded wave [3].

$$-i\hbar \frac{\partial \varphi}{\partial t} = \frac{1}{2M} \Delta \varphi, \quad -i\hbar \frac{\partial \phi}{\partial t} = \frac{1}{2M} \Delta \phi. \tag{5}$$

Extending Rauscher's concept for a complex 8-space differential line element $dS^2 = \eta_{\mu\upsilon} dZ^{\mu} dZ^{*\upsilon}$,

IOP Conf. Series: Journal of Physics: Conf. Series **1051** (2018) 012003 doi:10.1088/1742-6596/1051/1/012003

where indices run 1 to $4, \eta_{\mu\nu}$ is the complex 8-space metric, Z^{μ} the complex 8-space variable and where $Z^{\mu} = X^{\mu}_{\mathrm{Re}} + iX^{\mu}_{\mathrm{Im}}$ and $Z^{*\nu}$ is the complex conjugate, to 12D continuous-state UFM multiverse space-time; we write just the dimensions for simplicity for space constraints $x_{\mathrm{Re}}, y_{\mathrm{Re}}, z_{\mathrm{Re}}, t_{\mathrm{Re}}, \pm x_{\mathrm{Im}}, \pm y_{\mathrm{Im}}, \pm z_{\mathrm{Im}}, \pm t_{\mathrm{Im}}$, where $\pm$ signifies Wheeler-Feynman/Cramer type future-past/retarded-advanced dimensions. This framework allows the application of hierarchical harmonic oscillator parameters for surmounting uncertainty [3,9].

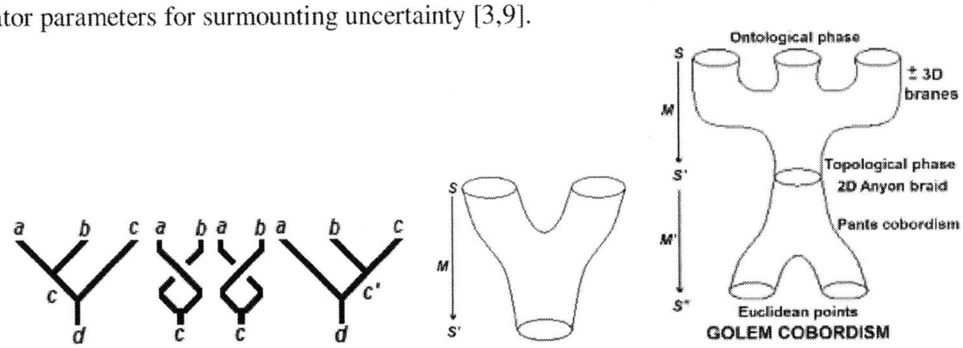

Figure 5. a) Anyon topologically protected fusion channels, b) Basic pants cobordism, c) Golem, composition of cobordisms designed for ontological-phase fusion transformations.

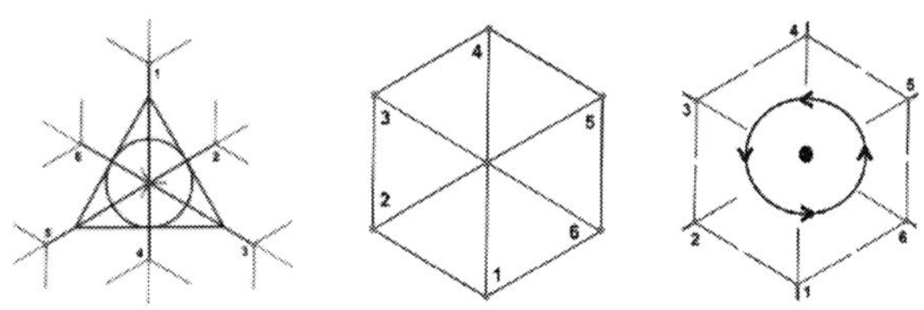

Figure 6. a) *Antennas* (snowflakes) on a Fano plane represent b) vertices on the circumference of a hexagon/cube where c) center rotates unconnected so position 1 or 2 can create the front/rear vertices of an ambiguous Necker cube. b) Antennas 1-6 combine to form the outer vertices of a cube/hexagon depending on what dimensional phase the state is in.

An important feature of TQFTs is they do not presume fixed topology for space/spacetime; in dealing with an n-D TQFT, one is free to choose any $(n-1)$-D manifold to represent space at a given time. Given two such manifolds, S and S', one is free to choose any nD-manifold M to represent the spacetime between S and S'. Mathematicians call M a *cobordism* from S to S' (Figure 5b,c). We write $M: S \to S'$, because M can be the process of time from moment S to moment S'. For example, Figure 5b depicts a 2D manifold M going from a 1D manifold S (pair of circles) to a 1D manifold S' (single circle). Crudely, M represents two separate spaces colliding to form a single one. Seemingly *outré*, but physicists are willing to speculate about processes in which the topology of space changes with time [25]. Various operations can be performed on cobordisms; we describe two:
1) *Compose* two cobordisms $M: S \to S'$ and $M': S' \to S''$, obtaining cobordism $M'M: S \to S''$, Figure 5c. The idea is that the passage of time corresponding to M followed by the time corresponding to M' equals the time corresponding to $M'M$. This is analogous to the idea that waiting t seconds followed by waiting t' seconds is the same as waiting $t+t'$ seconds. The difference in TQFT is we cannot measure time in seconds, because no background metric exists to let us count the passage of time. We track topological changes. Just as ordinary addition is associative, so is the composition of cobordisms: $(M''M')M = M''(M'M)$. But, cobordism composition is not commutative - the famous noncommutativity of observables in QT [25].

IOP Conf. Series: Journal of Physics: Conf. Series **1051** (2018) 012003 doi:10.1088/1742-6596/1051/1/012003

2) Any $(n-1)$D manifold S representing space, there is a cobordism $1_S : S \to S$ called the *identity* cobordism, representing passage of time without topological change. For example, if S is a circle, the identity cobordism 1_S is a cylinder. In general, the identity cobordism 1_S has the property that for any cobordism $M : S' \to S$ we have $1_S M = M$, while for any cobordism $M : S \to S'$ we have $M 1_S = M$ [25]. These properties say that an identity cobordism is analogous to waiting 0 seconds: if you wait 0 seconds and then wait t more seconds, or wait t seconds and then wait 0 more seconds, this is the same as waiting t seconds. These operations just formalize of the notion of *the passage of time* in a context where the topology of spacetime is arbitrary and there is no background metric.
Atiyah's axioms relate this notion to QT as follows:

- A TQFT must assign a Hilbert space $Z(S)$ to each $(n-1)$D manifold S. Vectors in this Hilbert space represent possible states of the universe given that space is the manifold S.
- The TQFT must assign a linear operator $Z(M): Z(S) \to Z(S')$ to each nD cobordism $M : S \to S'$. This operator describes state change given that the portion of space-time between S and S' is the manifold M : If space is initially manifold S and the state of the universe is ψ, with time passage corresponding to M, the state of the universe is $Z(M)\psi$ [25].

TQFTs must satisfy a list of properties. Two are: 1) A TQFT preserves composition: given cobordisms $M : S \to S'$ and $M' : S' \to S''$, we must have $Z(M'M) = Z(M')Z(M)$, where the right-hand side denotes the composite of the operators $Z(M)$ and $Z(M')$. 2) It must preserve identities; given any manifold S representing space, we must have $Z(1_S) = 1_{Z(S)}$, where the right-hand side denotes the identity operator on the Hilbert space $Z(S)$ [25]. These axioms are not unreasonable if one ponders them a bit. The *first* says that the passage of time corresponding to the cobordism M followed by the passage of time corresponding to M' has the same effect on a state as a combined passage of time corresponding to $M'M$. The *second* says that a passage of time in which no topology change occurs has no effect at all on the state of the universe. This seems paradoxical at first, since it seems we regularly observe things happening even in the absence of topology change. However, this paradox is easily resolved: a TQFT describes a world quite unlike ours, one without local degrees of freedom. In such a world, nothing local happens, so the state of the universe can only change when the topology of space itself changes.
Loosely speaking, they all say that a TQFT maps structures in differential topology (study of manifolds) to corresponding structures in quantum theory. Atiyah took advantage of power between differential topology and quantum theory [25]. This analogy between differential topology and QT is the clue we should pursue for a deeper understanding of quantum gravity. At first glance, GR and QT look very different mathematically: one deals with space and spacetime, the other with Hilbert spaces and operators, not easy to combine; but TQFT suggests they are not so different. Quantum topology is technical, but it is obvious that differential topology and QT must merge to understand a background-free QFT. Physics ignoring GR, treats space as a background for displaying world states. Similarly, spacetime is treated as a background for the process of change; these idealizations must be overcome in a background-free theory. In fact, concepts of *space* and *state* are two aspects of a unified whole; likewise, the concepts of *spacetime* and *process* [25]. In an alternative derivation of string tension, T_S we met this challenge finding a unique background independent M-Theory [3], that after another decade led to OPTFT as the putative 3rd regime integrating *unquantized* GR and UFM [4].
A photon, 2-component, 2D traveling plane *wave* projecting at right angles to the direction of propagation has a *particulate radius* not able to pass a slit $> \lambda$. We propose that behind the inherent backcloth of cyclic bumps and holes in the polarized Dirac vacuum (Figure 2a) [4], the uncertainty principle is hiding the XD topology of the MOU, which is not singular as in the SM because cyclic boost-compactification occurs continuously from asymptotic virtual $\hbar$ (shadow of uncertainty, Figure 2), to the Larmor radius of the hydrogen atom, making correspondence to dynamical Type-II M-

theoretic Calabi-Yau florets (multiply-connected Kahler manifold) undergoing translation, rotation, reflection as part of the process. Spectral lines characterize atoms by, $E = \hbar\nu = \hbar c / \lambda$ or wave number, $\sigma \equiv 1/\lambda = E/\hbar c$ by discrete wavelengths confirmed by monochromatic x-ray bombardment. Excited states, E_2 decay to lower states, E_1 by emission of photon energy, $E_1 - E_2$ frequency, ν, wavelength, λ and wave number, $E_2 - E_1 = \hbar\nu = \hbar c/\lambda = \hbar c\sigma$. By conditions hinted at in Figure 2 we propose new hyperspherical spectral lines *below* the lowest (ground state) Bohr orbit. Kowalski's interpretation from laser experiments [26] shows that emission and absorption between Bohr states takes place within a time interval equal to one period of the emitted-absorbed photon wave, the corresponding transition time is the time needed for the orbiting electron to travel one full orbit around the nucleus. Note that the same Lorentz conditions denoted in our tachyon measurement experiment apply directly to the TBS experiment with slight phase control alterations in the Cramer-like standing-wave oscillation of the HD Calabi-Yau mirror symmetries [6]. Standard Hypervolume values for increasing n-dimensionality and radius, r of a unit sphere or n-ball equal to 1 can be used to initially predict two TBS spectral lines hidden within the 6D Calabi-Yau dual 3-torus, the putative wavelengths can be calculated from the general hyperspherical n-volume equation, $\frac{1}{2\pi^2}$, 4,9346 units for 4D, and $\frac{8}{15\pi^2}$, 5,2638 units for 5D. If Randall-Sundrum are correct, the 6D cavity will be degenerate, and the signal escape to infinity. We postulate a Manifold of Uncertainty (MOU) with finite XD-radius corresponding to string theories T-Duality [3-5]. For preliminarily predictions we could calculate hyperspherical volume or surface area of 2D-5D MOU. The general n-volume equation is $V(n,r) = \pi^{\frac{n}{2}} r^n / \Gamma\left(\frac{n}{2}+1\right)$, where $V_{n,r}$ is volume per number of dimensions, n of radius r and Γ a factorial constant. We relate these n-volume equations to volumetric properties of the MOU for calculating an HD C-QED *volume hierarchy* for predicting new Tight-Bound State (TBS) spectral lines in hydrogen [4,21]. If LSXD exist, degeneracy would occur at the limit of r discovered in the same manner the outermost energy level of an atom is detected when an outer electron acquires sufficient energy to escape to infinity.

5. Experimental Design – Rf-Resonance Parameters on the Dirac Polarized Vacuum

Modulated harmonic Sagnac Effect rf-pulses oscillate electrons to couple with nucleons. Additional spin-spin pulse modulation couples to a putative de Broglie-Bohm-Cramer Kaluza-Klein cyclical beat-frequency, an inherent aspect of UFM semi-quantum limit space-time. With Dubois incursive oscillator parameters added, a QED cavity opens into XD/LSXD topology. These conditions in conjunction with a Bessel function *discover* additional TBS spectral lines.

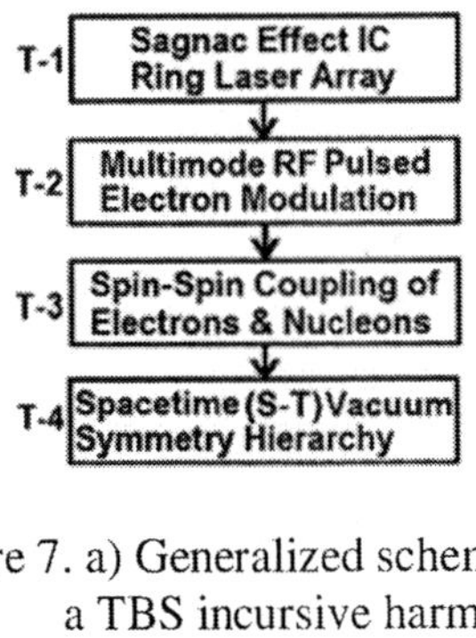

Hierarchical Harmonic Oscillator Parameters	
classical	$X = A\cos(\omega t)$
quantum	$\frac{\hbar^2}{2m}\frac{d^2\psi}{dx^2} + \left(E - \frac{kx^2}{2}\right)\psi = 0$
annihilation creation	$x(t) = x_0\left[a\exp(-i\omega t) + a^{\dagger}\exp(i\omega t)\right]$
future-past retarded-advansed	$F_1 = F_0 e^{-ikx}e^{-2\pi i f}$, $F_2 = F_0 e^{ikx}e^{-2\pi i f}$, $F_3 = F_1 e^{-ikx}e^{2\pi i f}$, $F_4 = F_2 e^{ikx}e^{2\pi i f}$
Incursive	$\frac{dx(t+\Delta t)}{dt} - v(t) = 0, \frac{dv(t+\Delta t)}{dt} + \omega^2 = 0$

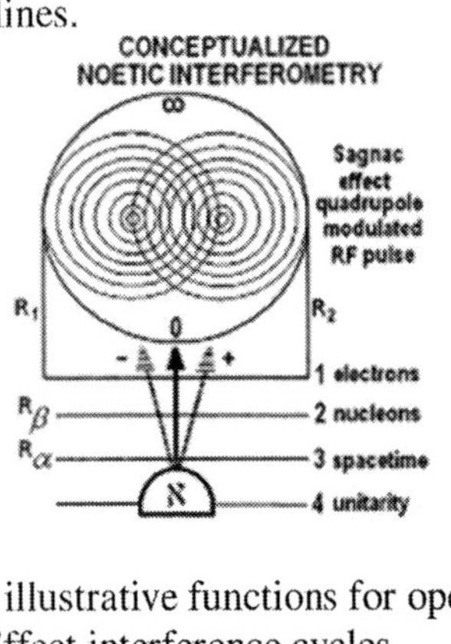

Figure 7. a) Generalized schema for a TBS expriment. b) Simplistic illustrative functions for operting a TBS incursive harmonic oscillator. c) Rf-pulsed Sagnac Effect interference cycles.

From a common general example of a Bessel function, with α an arbitrary complex number:

$x^2 \frac{d^2y}{dx^2} + x\frac{dy}{dx} + \left(x^2 - \alpha^2\right)y = 0$, we solve the Helmholtz equation in spherical coordinates by variable separation such that the radial equation takes the form

$$x^2 \frac{d^2y}{dx^2} + 2x\frac{dy}{dx} + \left[x^2 - n(n+1)\right]y = 0.$$

The spherical Bessel functions, j_n, y_n are the two linearly independent solutions that relate to the ordinary Bessel functions, J_n, Y_n as:

$$j_n(x) = \sqrt{\frac{\pi}{2x}} J_{n+\frac{1}{2}}(x),\ \ y_n(x) = \sqrt{\frac{\pi}{2x}} Y_{n+\frac{1}{2}}(x) = \left(-1\right)^{n+1} \sqrt{\frac{\pi}{2x}} J_{-n-\frac{1}{2}}(x). \quad (6)$$

When written as Rayleigh's formulas

$$j_n(x) = (-x)^n \left(\frac{1}{x}\frac{d}{dx}\right)^n \frac{\sin(x)}{x},\ \ y_n(x) = (-x)^n \left(\frac{1}{x}\frac{d}{dx}\right)^n \frac{\cos(x)}{x}, \quad (7)$$

the first spherical Bessel functions are

$$j_0(x) = \frac{\sin(x)}{x},\ \ j_1(x) = \frac{\sin(x)}{x^2} - \frac{\cos(x)}{x},\ \ j_2(x) = \left(\frac{3}{x^2} - 1\right)\frac{\sin(x)}{x} - \frac{3\cos(x)}{x^2} \quad [27]\ (8)$$

and

$$y_0(x) = -j_{-1}(x) = -\frac{\cos(x)}{x},\ \ y_1(x) = -j_{-2}(x) = -\frac{\cos(x)}{x^2} - \frac{\sin(x)}{x},$$
$$y_2(x) = -j_{-3}(x) = \left(-\frac{3}{x^2} + 1\right)\frac{\cos(x)}{x} - \frac{3\sin(x)}{x^2}. \quad [27].\ (9)$$

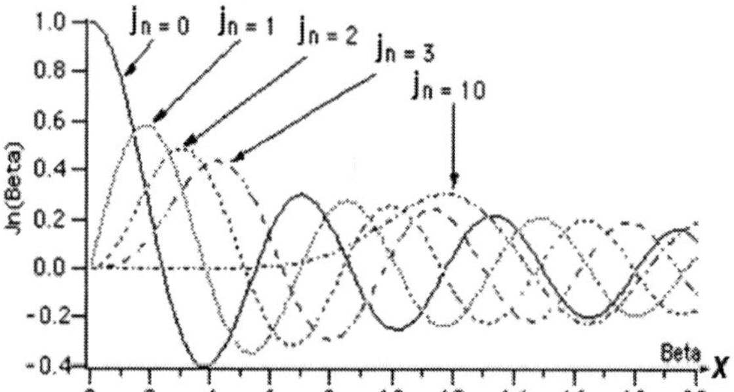

Figure 8. Bessel Function example necessary to synchronize coupling with the Dubois incursive oscillator in order access additional TBS beyond the first 4D QED cavity to 5 and 6D MOU cavities. We think we know how, but surmounting uncertainty will not be trivial.

We show the 1st three Bessel function solutions of the 1st and 2nd kind to illustrate our incursive oscillator process to locate the first TBS spectral line in hydrogen, expected to be relatively easy to find in comparison with finding the 2nd and 3rd. The additional lines will be more challenging as there may be some unexpected complexity in the Bessel harmonic oscillator that must be overcome to completely surmount uncertainty. Restrictions related to this refinement have not revealed themselves to us as yet. They will require additional adjustment to the spin-spin coupling parameters of the algebra describing the HD hyperspherical volume. Parallel transport of the gravitational curvature deficit angle will kick in for the XD mirror symmetric brane topology, *with noeon topological charge corrections* (not quantized gravity) [3-5,7].

Completing the Noetic Transform (beyond Lorentz-Poincaré), clarifies the choice of linear combinations of Bessel solutions (6), which depends on their asymptotic behavior at ∞,

$$J_n(x) \cong \sqrt{\frac{\pi}{2x}} \cos\left(x - \frac{\pi}{2}n - \frac{\pi}{4}\right);\ Y_n(x) \cong \sqrt{\frac{\pi}{2x}} \sin\left(x - \frac{\pi}{2}n - \frac{\pi}{4}\right), \tag{10}$$

thus

$$H_n^{\pm}(x) \cong \sqrt{\frac{\pi}{2x}} \exp\left[\pm i\left(x - \frac{\pi}{2}n - \frac{\pi}{4}\right)\right]. \qquad [28]\ (11)$$

With the proper harmonic oscillator Bessel function solutions, the next step in the experimental design is to apply the Dubois incursive oscillator parameters as the final step in designing the synchrony of Sagnac effect rf-pulses. The incursive algorithms are numerically stable and the numerical simulation of the oscillator will show the conservation of the energy. Let us consider the harmonic pendulum as an illustrative example, with m the oscillating mass and k the spring constant, represented by the ordinary differential equations:

$$dx(t)/dt = v(t),\ \ dv(t)/dt = -\omega^2 x(t)\mathrm{dv}(\mathrm{t})/\mathrm{dt}. \tag{12}$$

where $x(t)$ is the position and $v(t)$ the velocity as functions of the time t, and where the pulsation ω is related to k and m by $\omega^2 = k/m$ [28]. The solution is given by

$$x(t) = x(0)\cos(\omega t) + [v(0)/\omega]\sin(\omega t),\ \ v(t) = -\omega x(0)\sin(\omega t) + v(0)\cos(\omega t) \tag{13a,b}$$

with the initial conditions $x(0)$ and $v(0)$. In the phase space, given by $(x(t), v(t))$, the solutions are given by closed curves (orbital stability). The period of oscillations is given by $T = 2\pi/\omega$. The energy $e(t)$ of the harmonic oscillator is constant and is given by

$$e(t) = kx^2(t)/2 + mv^2(t)/2 = kx^2(0)/2 + mv^2(0)/2 = e(0) = e_0. \tag{14}$$

The simulation of differential equations is impossible. This is only the discrete transformation which is computable with a recursive function. In differential equations, there is only the current time. In discrete systems, there is the current time t and the interval of time $\Delta t = h$. The discrete time is defined as: $t_k = t_0 + kh$ with $k = 0,1,2,\ldots$ where t_0 is the initial value of the time and k is the counter of the number for the interval of time h [29]. The discrete variables are defined as $x_k = x(t_k)$ and $y_k = y(t_k)$. The discrete equations used in the harmonic oscillator case for computing the position and the velocity at consecutive moments have the general form

$$x_{k+1} = Ax_k + Bv_k,\ \ v_{k+1} = Cv_k - D\omega^2 x_k, \tag{15a,b}$$

where A, B, C and D are coefficients with values specific to the numerical integration methods applied. In eliminating v_k of Eq. (15a) in Eq. (15b), a second order discrete equation in x_k is given by $x_{k+2} - (A+C)x_{k+1} + (AC + BD\omega^2)x_k = 0$. The stability analysis for this discrete system can be performed by using the Z-transform, $z^2 - (A+C)z + (AC + BD\omega^2) = 0$ which presents two poles:

$$z_{1,2} = ((A+C) \pm i\sqrt{-(A+C)^2 + 4(AC + BD\omega^2)})/2 \tag{16}$$

that are complex when

$$(A+C)^2 < 4(AC + BD\omega^2). \qquad [29]\ (17a)$$

The position of the poles relative to the unit circle defines the system stability: A system is stable if the poles lie *inside* the unit circle, is unstable if the poles lie *outside* the unit circle and shows an orbital stability if the poles lie *on* the unit circle. The condition for stability is:

$$((A+C)^2 - (A+C)^2 + 4(AC + BD\omega^2))/4 \le 1 \tag{17b}$$

or $AC + BD\omega^2 \le 1$ and the orbital stability must satisfy the strict equality

$$AC + BD\omega^2 = 1, \tag{17c}$$

so, for the harmonic oscillator, the conditions for obtaining an orbital stability are given by relations (17a) and (17c), rewritten as

$$(A+C)^2 < 4 \text{ and } AC + BD\omega^2 = 1 \tag{18a,b}$$

in using the equality from the relation (17c), in the relation (18a) [29].

Considering 1st the well-known Euler and Runge-Kutta integration methods, e.g. Scheid [30]; and then incursive methods will be analysed. In terms of incursive discrete algorithms, Dubois defines a generalized forward-backward discrete derivative, $D(w) = wDf + (1-w)Db$, where w is a weight taking the values between 0 and 1, and where the discrete forward and backward derivatives on a function f are defined by $D_f(f) = \Delta^+ f / \Delta t = [f_{k+1} - f_k]/h$ and $D_b(f) = \Delta^- f / \Delta t = [f_k - f_{k-1}]/h$. Then, a generalized incursive discrete harmonic oscillator is:

$$(1-w)x_{k+1} + (2w-1)x_k - wx_{k-1} = hv_k,\ wv_{k+1} + (1-2w)v_k + (w-1)v_{k-1} = -h\omega^2 x_k. \tag{19a,b}$$

When $w = 0$, $D(0) = D_b$, this gives the first incursive equations:

$$x_{k+1} - x_k = hv_k,\quad v_k - v_{k-1} = -h\omega^2 x_k. \tag{20a,b}$$

When $w = 1$, $D(1) = D_f$, this gives the second incursive equations:

$$x_k - x_{k-1} = hv_k,\quad v_{k+1} - v_k = -h\omega^2 x_k. \tag{21a,b}$$

When $w = \frac{1}{2}$, $D\left(\frac{1}{2}\right) = \frac{[D_f + D_b]}{2}$, this gives the averaged (hyperincursive) equations:

$$x_{k+1} - x_{k-1} = 2hv_k,\quad v_{k+1} - v_{k-1} = -2h\omega^2 x_k. \tag{22a,b}$$

These Eqs. (22a,b) integrate the two incursive equations. This deals with a deduction of this forward-backwards discrete derivative, with the deduction of this time-symmetric discretization of the harmonic oscillator [30].

Next, simulation of the incursive and hyperincursive algorithms of the classical harmonic oscillator: First, for the simulation of the classical harmonic oscillator, the dimensionless variables X, V and H, will be used [30], for the variables, x, v and $h: X(k) = \sqrt{[k/2]}x_k$, $V(k) = \sqrt{\frac{m}{2}}v_k$ $\tau = \omega t$ with $\omega = \sqrt{\frac{k}{m}}$ and $\Delta\tau = \omega\Delta t = \omega h = H$. So, the two incursive dimensionless harmonic oscillators are given by

$$X_1(k+1) = X_1(k) + HV_1(k),\quad V_1(k+1) = V_1(k) - HX_1(k+1) \tag{23a,b}$$

$$V_2(k+1) = V_2(k) - HX_2(k),\quad X_2(k+1) = X_2(k) + HV_2(k+1) \tag{24a,b}$$

and the hyperincursive dimensionless harmonic oscillator is given by

$$X(k+1) = X(k-1) + 2HV(k),\quad V(k+1) = V(k-1) - 2HX(k). \tag{25a,b}$$

So, this confirms that the incursive and hyperincursive algorithms are totally numerically stable with the conservation of energy [31]. When the parameters for the experiment are coordinated and the rf-pulse sent into the MOU HD QED TBS hydrogen cavity, a positive result will retrieve a spectroscopic signal like the one represented in Figure 9. A negative result would send back 0 amplitude (no spectral line) [4,21].

5.1. Experimental UFM OPTFT Protocols

Tight-Bound States (TBS): New spectral lines in hydrogen *below* lowest Bohr orbit. Until now spectral lines are measured in 3-Space. UFM predicts 2 to 5 new TBS Spectral Lines depending on the hyperspherical duality of the Manifold of Uncertainty (MOU) in 4D, 5D & possibly 6D.

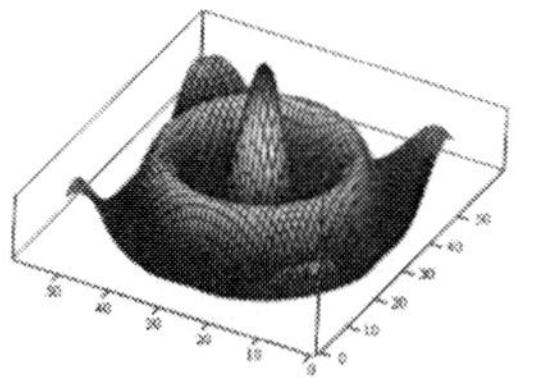

n	V(n,1)	
0	0	0
1	2	2
2	π	$\approx$ 3.1416
3	$4/3\pi$	4.1888
4	$1/2\pi^2$	4.9348
5	$8/15\pi^2$	5.2638
6	Degenerate ?	∞

Figure 9. a) Illustrating First 4D TBS spectral line in hydrogen emerging (Emitted) from the 4D spherical potential well in the Manifold of Uncertainty for $\alpha = \pm 1$ (adapted from [29]).
b) Standard Hypervolume values for increasing n-dimensionality and radius, r of a unit sphere or n-ball equal to 1. The 4D and 5D volumes can predict new TBS spectral lines.

The 1st Bohr orbit in hydrogen is at .5Å & the 2nd at 2Å. We predict two new spectral lines between .5Å & 2Å; with a 3rd line likely to be degenerate, with the test signal escaping to infinity, or so-called LSXD *Bulk*. Experimental success is the gateway to 3rd regime natural science would demonstrate M-Theory and cause CERN type accelerators to become obsolete. Note: The protocol is low *energy and tabletop* by *completely bypassing* uncertainty [4,21].

References

[1] Randall L Sundrum R 1999 Alternative to compactification *Phys Rev L.* **83** 23 p 4690
[2] Rowlands P 2016 How many dimensions are there? eds R L Amoroso *et al Unified Field Mechanics, Natural Science Beyond Veil of Spacetime* (London: World Scientific)
[3] Amoroso R L and Rauscher E A 2009 *The Holographic Anthropic Multiverse: Formalizing the Complex Geometry of Reality* (London: World Scientific)
[4] Amoroso R L 2017 *Universal Quantum Computing: Supervening Decoherence-Surmounting Uncertainty* (London: World Scientific)
[5] Amoroso R L, Rowlands P and Kauffman L H 2014 *Physics of Reality: Space, Time, Matter, Cosmos* (London: World Scientific)
[6] Amoroso R L and Rauscher E A 2010 Empirical protocol for measuring virtual tachyon/tardon interactions in a Dirac vacuum *AIP Conference Proceed*ings **1316** p 199
[7] Amoroso R L 2015 Yang-Mills Kaluza-Klein equivalence: an empirical path extending the standard model of particle physics *Preprint* http://vixraorg/pdf/1511.0257v1.pdf
[8] Peebles PJE 1993 *Principles of Cosmology* (Princeton: Princeton Univ. Press)
[9] Rauscher E A and Amoroso R L 2012 *Orbiting the Moons of Pluto: Complex Solutions to the Einstein, Maxwell, Schrödinger and Dirac Equations* (London: World Scientific)
[10] Misner C W, Thorne K S and Wheeler J A 1973 *Gravitation* (San Fran: WH Freeman)
[11] Dray T and 't Hooft G 1985 *Nucl. Phys. B* **253** 173
[12] Dray T and 't Hooft G 1985 *Communications Mathematical Physics* **99** 613
[13] Lousto C O and Sanchez N 1992 *Nuclear Physics* B 383 377
[14] Hayashi K and Samura T 1994 *Physical Review* D **50** 3666
[15] Sfetsos K 1994 *Preprint* arXiv: 9408169v3
[16] Gibbons G and Maeda K 1988 *Nucl. Phys.* B **298** 741
[17] Spitkovsky A *Preprint* arXiv: 0802.3216v2
[18] Kaloper N 2005 Brane-induced gravity shocks *Preprint* arXiv: 0501028v4
[19] Berry M V 1984 *Proc. Roy. Soc. London* A **392** 45
[20] Dragic´ A, Maric´ Z and Vigier J-P 2000 *Physics Letters* A **265** pp 163-167
[21] Amoroso R L and Vigier J-P 2013 Evidencing 'tight bound states' in the hydrogen atom: empirical manipulation of large-scale XD in violation of QED *Physics of Reality: Space, Time, Matter, Cosmos* eds R L Amoroso *et al* (London: World Scientific)
[22] Rowlands P 2007 *Zero to Infinity: The Foundations of Physics* (Singapore: World Sci)

[23] Wooten J R 2010 *Dissecting Topological Quantum Computation* (University of Leeds, School of Physics and Astronomy)

[24] Aaij R *et al* 2012 Searches for Majorana neutrinos in B$^-$ decays *Phys Rev* D **85** 11

[25] Baez J C and Dolan J 1995 *J. Math. Phys.* **36** 6073 (*Preprint* arXiv: 9503002)

[26] Kowalski M 1999 Photon emission from atomic hydrogen *Phys Essays* **12** pp 312-331

[27] Abramowitz M and Stegun I A 1983 *Handbook of Mathematical Functions with Formulas, Graphs, and Mathematical Tables* (Applied Mathematics Series 55) (National Bureau of Standards; New York: Dover)

[28] Gurarie D 2013 Bessel zoo and harmonic oscillator *Preprint* http://www.cwru.edu/artsci/math/gurarie/classes/445/lecture/bessel.pdf

[29] Scheid F 1986 *Theory and Problems of Numerical Analysis* (McGraw-Hill)

[30] Gurarie D 2013 Bessel zoo and harmonic oscillator *Preprint* http://www.cwru.edu/artsci/math/gurarie/classes/445/lecture/bessel.pdf.

[31] Dubois D M 2016 Hyperincursive algorithms of classical harmonic oscillator applied to quantum harmonic oscillator separable into incursive oscillators *Unified Field Mechanics: Natural Science Beyond the Veil of Spacetime* eds R L Amoroso, L H Kauffman and P Rowlands (London: World Scientific)

IOP Conf. Series: Journal of Physics: Conf. Series **1051** (2018) 012004 doi:10.1088/1742-6596/1051/1/012004

The relativistic inertial coordinate reference frames, synchronized the observed radio emission of pulsar

A Avramenko

Senior Researcher, P.N. Lebedev Physical Institute of Russian Academy of Sciences, Moscow, Russia
E-mail: avr@itaec.ru

Abstract. We obtain the es – as series of simultaneous and joint rotational motions of the planets. Copernicus (1473-1543 analytical coordinate-independent pulsar time scales, which are determined by the observed rotation parameters of pulsar. The scales extent into barycentric and any topocentric coordinate systems, providing simultaneity of the observed physical phenomena, including the periodic radiation of a pulsar as well, in any point of the three-dimensional space.

1. Introduction

For many centuries astronomers have studied the movement of the planets and the Sun, enjoyed the coordinate system associated with the Earth, which seems to be immovable and is considered then as the center of the world. Thus, the visible movements of the planets, Ptolemy (II century a new era) in this system is described in the form of epicycl) has given its significantly different from the description of the Solar System, considered the same motion of the planets in the coordinate system associated with the Sun. Since that time, began the development of a common planetary theory, the contents of which were both analytical and numerical solutions of the tasks of the movement of celestial bodies with mathematical methods of celestial mechanics, based on the postulates of absolute time, absolute space, the laws of Newtonian mechanics and Newton's law of gravitation.

However, the difficulties that began in middle of the 19th century, led to the crisis of Newtonian physics in the early 20th century, when trying to explain the experimental data in the field of optics and electrodynamics of moving bodies. The crisis has stimulated the emergence of the theory of relativity, the special (SRT) by 1905 and general (GRT), formulated by Einstein in 1916 publication. Later relativistic celestial mechanics developed mainly on the principles of general relativity, expressed by the equations of the gravitational field, which in any reference frame are written the same way. Unlike the metric of SRT, in which the mathematical coordinates in a flat (no curvature) Euclidean space inertial systems are a Galilean, and have a physical meaning of time and the three spatial coordinates, in the Riemannian metric of GRT the properties of space and time in the form of event space metrics, are determined by the motion and distribution of mass, the movement and distribution of mass, in turn, are determined by the metric of field. Galilean coordinates in the GRT does not exist, and space of events under the influence of gravitating masses in different coordinate systems look as different. The mathematical relationship which are carried out in any space of events in the SRT, in space of events in the SRT are satisfied only within infinitesimal vicinities of the selected point defined by its coordinates. This means that the principle of equivalence flowing of all the physical processes that apply to any inertial system in the space of events in the SRT, in the space

IOP Conf. Series: Journal of Physics: Conf. Series **1051** (2018) 012004 doi:10.1088/1742-6596/1051/1/012004

of events in the GRT is valid only locally. Therefore, the solutions of the equations in the space of events in the GRT inevitably differ from each other in different coordinate systems, and, therefore, no GRT coordinates for finite (not infinitesimal) regions of space-time have no physical meaning and, in such a way, can not be compared with observational data [1]. In the special theory of relativity, as in Newtonian mechanics, this problem of comparison the observational and theoretical data does not exist, since the use of inertial coordinate systems from the beginning as a baseline, or go to him in the final stage of processing observational data obtained in GR coordinates, automatically gives a solution to the equations of time-space directly in terms of the measured quantities.

The paper focuses on the interconnection of the metric of space-time relations in different coordinate reference systems, both dynamic and kinematic ones. A comparison of mathematical ephemeris time scales obtained in the form of a numerical solution of the movements of celestial bodies, and physical pulsar time scales expressed analytically as the intervals of the observed events of the periodic radio emission from galactic sources – pulsars, is done. Synchronize on the periodic emission of pulsar gives the inertial properties of the coordinate reference frames and, as a consequence, leads to metric compatibility and physical equivalence of the measuring and data processing in the relativistic positional astronomy, regardless of the initial choice of coordinate systems.

2. The hierarchy of the relativistic coordinate reference systems

In positional astronomy of space-time, it is important and takes into account the whole set of observational data obtained by different observers at different times. For further processing of this set essential theoretical results related both to the movement of celestial bodies (dynamic task), and the propagation of light (kinematic task). It is important that both of these tasks are solved at the same coordinates. Coordinate method in the relativistic celestial mechanics is implemented using four-dimensional coordinate system (CS) - three spatial coordinates and coordinate time scale. Although the framework of general relativity all the CS equivalent, to solve a specific problem, there are astronomical more preferred or less preferred CS. For example, if from the barycentric CS with the beginning of the spatial axes in the center of mass of the Solar System go to the relativistic geocentric CS with the beginning of the spatial axes in the center of mass of the Earth, where the main attracting body is the Earth, while the effect of all other heavenly bodies (Sun, Moon, planets) appears only in the form of tidal forces, so the contribution of the relativistic components is two orders of less [1]. Earth parameters, which are determined by the ratio V^2 / c^2 (non-spherical shape of the Earth, Earth's angular velocity, etc.), in the geocentric CS are closer to the measured values than in the barycentric CS. The fourth coordinate such relativistic systems is the scale of the corresponding coordinates time (barycentric, geocentric, topocentric time), which is an argument in relevant theories of motion or rotation of celestial bodies.

Practical implementation of the coordinate systems in astronomy attained by setting the coordinate values of the reference celestial objects. Thus, the coordinate system as a mathematical construct is transformed into meaningful astronomical reference frames (RF). The modern positional astronomy holds a permanent work to maintain the International Celestial Reference Frame (ICRF) and the International Terrestrial Reference Frame (ITRF), The ICRF is defined by the provisions of the quasars in the International Celestial Reference System (ICRS), which is a particular barycentric RF. The ITRF is defined by the provisions of ground reference stations in the International Terrestrial Reference System (ITRS), which is a rotating with the Earth geocentric RF. The relationship of these systems, on the one hand, is given theoretically by solutions relativistic equations of motion the celestial bodies and the Earth's rotation, and on the other hand, is determined from observations.

A special place in the hierarchy takes topocentric coordinate system, the beginning of which, unlike the geocentric system, can be selected at any observation point on Earth's surface, approximated as the geoid, which is associated with the system of astronomical coordinates, or ellipsoid, connects with the system of geodetic coordinates. On average, both approximations are almost identical, but in some areas of the surface differences can be substantial. Therefore geodetic methods on the surface built a network of local reference ellipsoids, which are tied to the topocentric coordinate reference system

according to the average terrestrial ellipsoid. With the help of these coordinates is given by the mutual position of the observation points.
Thus, the topocentric system, the beginning of which, coinciding with the location of the observer in the phase center of the radio telescope, has the absolute physical significance as a for the primary observational data based on direct measurements, as well as to check the consistency of reference coordinate systems on physical criteria in the framework of existing theories and models (ephemeris calculations and monitoring, conservation laws, etc.).
Reference coordinates systems are built so that they adhered to the principle of equivalence, mathematically it means that in such systems, all physical processes and phenomena occur equally and physical equations have the same form, that is, all reference systems should be equal, and observed them in the same conditions of physical phenomena and processes are indistinguishable. Such reference systems are inertial.
In contrast to the inertial coordinate Newtonian mechanics and SRT, the GRT coordinates for the finite (not infinitesimal) space of events have no physical meaning and cannot be directly compared with observational data. Therefore, the problem of comparing of the measured and calculated values becomes methodologically important and physically essential task of the modern relativistic astronomy.
The basis of the choice of coordinate methods for constructing theories of the motion of celestial bodies in GRT puts the mathematical approach, taking into account the convenience of the various coordinates for purely mathematical solving of dynamic task, adopted, for example, in the calculation of the coordinates in the equations of planetary ephemeris of the Solar System [2]. The subsequent transition from ephemeris coordinates to the coordinate-independent physically measured values is achieved by combination of solutions of the dynamic task (the motion of bodies) and the kinematic task (propagation of light) in the same coordinates [1]. To comply with the principle of equivalence is only important that both tasks have to be solved in the same coordinates.

3. Relativistic time scales

The fourth coordinate of a relative reference frame is a coordinate time scale, which is the mathematical argument in the appropriate theories of the motion or rotation of celestial bodies in the Solar System. In the article [3], the basic principles of constructing a self-consistent relativistic theory of time scales in the Solar System on the basis of the hierarchy of dynamical nonrotating reference systems of the General Theory of Relativity (GRT), is described. In accordance with this approach, the barycentric dynamic time (TDB) at a fixed point – the center of mass of the Solar System, and the terrestrial Earth's time (TT) are considered as the coordinate time of the barycentric (BRS) and geocentric (GRS) reference system, respectively.
The problem of physical realization of the fourth coordinate – time is caused by the fact that the variable of time in the equations of motion of General relativity theory is just a mathematical argument and is not an observable and directly measurable quantity that can be directly compared with the calculated values. The physical realization of TT is tied to the International Atomic Time (TAI), which is the result of averaging the readings of atomic clocks distributed over the Earth's surface and synchronized with TT. TAI transfer outside the Earth's surface can be performed within any three-dimensional hypersurface under the condition $TT = const$. The unit of measurement TT, defined by the TAI scale, is the same as the unit of measurement TDB and is equal to the second SI. The ephemeris Earth's time TT, which is the argument of the time of geocentric ephemerides, is reduced to the ephemeris scale of the TDB, which is the argument of heliocentric ephemeris, whose unit is also the SI second [2] of the TAI atomic time.
The relationship between TAI and TT in the form $TT = TAI + 32{,}184$ s, refers to the geocentre, the point at which the averaged result of the readings of atomic clocks distributed over the real Earth's surface is localized. For a given unit of time, the scale TDB differs from the TT scale (TAI) only by relativistic nonlinear and periodic components resulting from the theories of the motion of the Earth, planets and the Moon. The proper time τ_0 of the observer in any point on the Earth's surface coincides

with the coordinate time τ of the corresponding topocentric reference system (TRS), estimated at its beginning. The time τ_0 is associated with the TT (TAI) relativistic transformation, including the GRS observer speed, its height above the geoid and the quadrupole tidal gravitational potential of the outer masses. This involves visualizing ideal clocks on the Earth's surface, the TAI of which extends within the hypersurface *TT* = *const* and is synchronized with the TT.

It should be noted that this assumption does not explicitly take into account the coordinate topocentric time τ, that is, the observer's proper time, depending on its location on the Earth's surface. For each observer, it is different, whereas ideal atomic clocks that reproduce the standard SI second-on-time cannot by themselves react to the relativism of topocentric coordinate time and, therefore, cannot perform the synchronization function of topocentric time scales in accordance with the principle of simultaneity of events observed in different coordinate systems.

Thus, the development of astronomical time scales is one of the most urgent problems of applied astronomy. This requires improving the quality of theoretical models, including mathematically accurate definitions and physically meaningful interpretations of all quantities involved in time measurements. Given that within the framework of GRT, no coordinates for any finite region of 4-dimensional time-space have no physical meaning and, therefore, cannot be directly compared with observational data, it is necessary to explicitly indicate the corresponding metric form that meets the requirements of direct comparison of measured and calculated values in the coordinate reference frames and formulate the conditions that must be met to construct this form.

The implementation of the principle of simultaneity of observed events, confirmed by the coordinate time scales, corresponds to the synchronise in the inertial systems and, consequently, to the physical equivalence of the observations in any of them.

4. The principle of simultaneity in the coordinate systems

In a non-inertial coordinate system, the interval $d\sigma$ is expressed in terms of local physical quantities dl and $d\tau$ as follows: $d\sigma^2 = c^2 d\tau^2 - dl^2$. The physical time determines the course of the physical processes, but the quantity $d\tau$ in the non-inertial coordinate system has a local character, since it is not a complete differential. In this case, there is no single physical time, the lines of which would be orthogonal to the three-dimensional space of the variable τ [4-5]. Therefore, the concept of simultaneity of events in different points of space loses meaning in non-inertial systems, because synchronization of clocks in different spatial points with the help of physical quantities (light signal) cannot be carried out, since it depends on the signal propagation path. There does not exist a variable l, because dl also has a local character. In the non-inertial systems, with necessarily there are the coordinate quantities, which in the space of events of the General relativity theory describe the phenomena with allowance for constraints due to the local nature of mathematical quantities, whether in barycentric or topocentric frames.

In the inertial system in Galilean coordinates, the value $d\tau$ coincides with the differential dt, therefore one can introduce a single time t in Minkowski space. It will be physical time. The introduction of simultaneity for all points of a three-dimensional space is a consequence of the pseudo-Euclidean geometry of the four-dimensional space of events. The field at each point of the Minkowski space, according to the local characteristics of the dl and $d\tau$, has a velocity equal to the electrodynamic constant: $dl / d\tau = c$. From this follows the axiom formulated by G.Minkovsky thus: "A substance located at any world point, always, with a proper definition of space and time can be considered as being at rest. The axiom expresses the idea that at every world point the algebraic sum of squares $c^2dt^2 - dx^2 - dy^2 - dz^2$ is always positive or, in other words, that any velocity V is less than c" (that is, with an appropriate choice of the reference frame in Minkowski space [4]).

It is necessary to distinguish coordinate values of physical quantities. The change in coordinate values leads only to a change in the relationship of the physical time with the coordinate values. The conditional agreement on simultaneity always corresponds to a definite choice of the coordinate

system in the inertial reference frame of the Minkowski space. In the inertial frame of reference, the physical values of time and distance do not depend on the choice of the agreement on simultaneity.
Thus, the conditional agreement on simultaneity is nothing more than a definite choice of a coordinate system in the inertial reference frame of Minkowski space. In this space, in the inertial system in Galilean coordinates there exists a single physical time t, the lines of which are orthogonal to the three-dimensional space; we can speak of a constant speed of light that is the same in all directions and coincides in magnitude with the electrodynamic constant c [4].

5. Determination of the observer's proper time by the radio emission of a pulsar

The time in the coordinate reference system is determined by the local time scale. As a result of transition from one coordinate system to another, the time of the same event changes. Lorentz defined the coordinate transformations for inertial systems in such a way that the physical equations of the observable processes with respect to these transformations remained invariant. He found such relationship of coordinates and changed local time for which the condition of simultaneity in all systems under such coordinate transformations is satisfied on all points of space, and this modified local time is the observer's proper physical time. This time, being relativistic, reflects the position, the speed of the observer's movement and the time of light propagation into different points of space. For each observer it is different.
The ephemeris coordinates time TCB and TCG, expressed by the arguments in the mathematical equations of the motion of the Sun and the rotating Earth, refers to the unobserved computed points of the barycenter and geocenter, respectively, which are inaccessible to direct measurements and comparison. Meanwhile, direct observation with the help of a radio telescope, taking into account the geographical location of the observation point, can be measured the observer's proper time τ, which coincides with the coordinate time τ of the corresponding topocentric reference system (TRS) estimated at its beginning. This involves visualizing ideal clocks within the Earth's surface, the TAI readings of which are synchronized with the observer's proper time.
Using the equation of motion of celestial bodies of the planetary theory VSOP87, including the transformation of the BRS–GRS dynamic coordinate systems, including the coordinate time transformation TCB–TCG–τ, one can determine the observer's own time, designated as the τ –TCG difference in dynamic coordinate systems. These measurements allow estimating the relativistic effects related to the reference systems and time scales. This approach, based on the difference measurements of the coordinate time, was applied in VLBI measurements of pulsar coordinates (Fomalont, 1984) [8], and later S. Kopeikin (1990) [9] within the framework of this problem, the difference between pulsar time at two different points observations of the pulsar with terrestrial radio telescopes. Reduction of the VLBI measurements of the pulsar coordinates to the form of the difference of the topocentric time $\tau_2 - \tau_1$ was performed by solving the dynamic task of General relativity in the pulsar reference system (PRS) for two points on the Earth's surface. To calculate the difference, the initial data necessary for such a task were taken, including the first and the second antenna coordinates in the GRS, the barycentric velocity of the Earth, velocity of the second radio telescope, Geopotential and Solar System's gravitational potential excluding the Earth, a relativistic correction for different time rates between observer clocks caused by distinct geographical positions of the radio telescopes.
It should be noted that in this task the time of propagation of an electromagnetic signal from a galactic source to the different points on Earth's surface is defined in a form of the solution of a dynamic equation of GRT, without involving of the Lorentz transformations.
Thus, the coordinate time measured from the observed pulsar pulse at the radio telescope at one point can be transformed into any other point with given local geodetic coordinates. By the difference $\tau_2 - \tau_1$ of the same pulse of pulsar radiation observed in two different topocentric points, the coordinated values of the coordinate time of these ground stations are found, having, as will be shown below, the physical meaning of the observer's proper time in these points.

6. Synchronization of coordinate reference systems on the observed periodic pulsar radiation

By observations of a pulsar on the radio telescope, the topocentric time t_{TRS} of arrival of the radiation pulse in a point of the Earth's surface with the geodetic coordinates of the radio telescope, is measured. By solution of the dynamic task of motion of celestial bodies, the observed topocentric time t_{TRS} is reduced to the ephemeris barycentric time t_{BRS}, taking into account the relativism of the observer's proper time on the radio telescope [9]. Figure 1a shows the difference $t_{TRS} - t_{BRS}$ from observations of the pulsar J1509 + 5531 on the phased array radio telescope BSA LPI (Pushchino). Here, the coordinate topocentric and barycentric pulsar time, which in both these non-inertial reference frames are defined as the solution of the dynamic task, refers only to two local points of space. The principle of simultaneity for them is not valid initially, in contrast to the inertial reference frames.

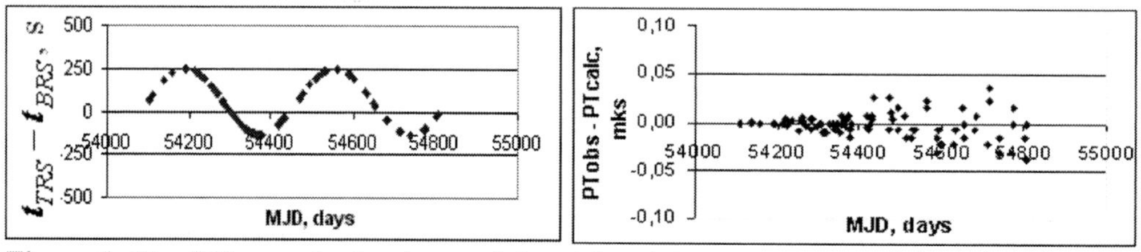

Figure 1. The difference in the observed time (a) and the convergence of the pulsar time (b) of the pulsar J1509 + 5531.
a) the difference of the topocentric and the barycentric time (non-inertial frame), b) the convergence of the pulsar time (inertial frame).

Meanwhile, the observed pulsed radio emission of a neutron star detects a strict periodicity of the pulses, which is generated due to the losses of the rotational energy of the star with a monotonous deceleration. The slowdown is shown in the gradual increase of the observed rotation period P, which is completely determined by its derivatives $\dot{P}$, $\ddot{P}$, and is detected in the measured intervals PT_{obs} of the observed pulses, counted from some pre-selected pulse that fixes the initial observation epoch. It is shown [10] that for any initial epoch there exists the combination of numerical values of the rotation parameters P_0^*, $\dot{P}$, $\ddot{P}$, unique for each pulsar, such that the observed intervals PT_{obs} represent the convergent power series which are expressed by the analytical function of the observed rotation parameters of pulsar.

The equation of the observed on the radio telescope intervals PT_i, written in the form of the convergent power series, is followed:

$$PT_i = (1+\alpha_i)(P_0^* N + \frac{1}{2}P_0^*\dot{P}N^2 + \frac{1}{6}(P_0^{*2}\ddot{P} - 2P_0^*\dot{P}^2)N^3)_i. \qquad (1)$$

Here are: PT_i are the numerical values of the observed intervals coordinated with planetary ephemeris; P_0^*, $\dot{P}$, $\ddot{P}$ are the pulsar rotation parameters obtained by solving of the equation (1); α_i is divergence of the series of the PT_i. By the parametric approximation of the observed intervals PT_i, the consistency numerical values of the period P_0^* for the initial epoch and its derivatives $\dot{P}$, $\ddot{P}$, are determined. The rotation period $P(t)$ within any duration of observations is calculated as follows:

$$P(t) = P_0^* + \dot{P}_0 \cdot t + \frac{1}{2}\ddot{P}\cdot t^2. \qquad (2)$$

Equation (1) expresses the relationship between the observed intervals PT_{obs} and the calculated intervals of periodic pulsar radiation in any coordinate reference system:

$$PT_{obs} = PT_i;\ PT_{calc} = (P_0^* N + \frac{1}{2} P_0^* \dot{P} N^2 + \frac{1}{6}(P_0^{*2} \ddot{P} - 2P_0^* \dot{P}^2) N^3)_i. \quad (3)$$

Figure 1b shows the differences between the observed intervals PT_{obs} and the intervals PT_{calc} calculated from the observed rotation parameters of the PSR J1509 + 5531.
The equation of the intervals of coherent radiation of pulsar (1), which is form invariant in the both barycentric and topocentric frames, determines the physical time in these coordinate systems in the form of an analytic function of the observed period and its derivatives. As numerical values of the observed period for coinciding local time epochs in any coordinate system are identical, therefore, in general, on the observed divergence of local coordinate time $t_{TRS} - t_{BRS}$ (Figure 1a) such value of the rotation period is defined which corresponds to the rotation parameters, measured at the epochs of observation in these coordinates. So in each inertial reference frame, there is its proper physical time – the changed local time of Lorentz. The physical time obtained in this way in inertial systems corresponds to the implementation of the principle of simultaneity of the observed events and, as a consequence, automatically - the synchronization conditions of the coordinate systems from the observed periodic radio emission of the pulsar. The amendment in the coordinate values leads only to a change in the relationship of the physical time with the coordinate values. In each inertial system in Galilean coordinates there exists a single physical time t, whose lines are orthogonal to the three-dimensional space.

7. Extension of the synchronous pulsar time scales into Geo-cosmic coordinate systems

In order to obtain a physical pulsar time at any point with given geographic coordinates, it is sufficient to extend the topocentric PT scale obtained analytically from observations of a pulsar on a radio telescope from the observation point to this ground point. By analogy with the VLBI measurement of the difference $\tau_2 - \tau_1$ of the coordinate time in the pulsar reference system (PRS) for two spaced points on the Earth's surface, there were calculated the differences for three spaced points, one of which (τ_1) coincides with the phase center of the BSA FIAN radio telescope (Pushchino), and the other two ones were determined by the flat coordinates of arbitrarily chosen points on the surface of the Earth, one in the north, and the other in the east direction of the radio telescope. For these two points, the coordinate time was calculated from the equations of motion of celestial bodies and the rotation of the Earth in the framework of the solution of the dynamic task of the GRT with subsequent approximation in the observed parameters of the rotation of the pulsar by solving the kinematic task within of the SRT.
This numerical experiment on the propagation of pulsar time scales into two points distant from the radio telescope, taken as a test case. It was carried out in 2015 with the active encouragement and kind assistance of the Laboratory of Ephemeris Astronomy of the IPA RAS (Prof. E.Pitjeva, Dr. D.Pavlov). For coordinate calculations, a specialized ERA software package, developed by the IPA RAS [11], was used.
The graphs in Figure 2 illustrates the results of a comparison of the physical coordinate time at the selected geodetic points calculated in inertial reference frames from the observed parameters of the rotation of the pulsar J1509+5531 in approximately three years of observations at the BSA FIAN (Pushchino). Barycentric TD (TDB) and topocentric TT coordinate pulsar time intervals are calculated from the observed parameters of rotation of the pulsar, consistent within any chosen extention, the same in any coordinate system for coinciding epochs of local coordinate time. The intervals are measured from the epoch of the initial observable event MJD 55960.1561 (TB) and MJD 55960.1542 (TT) they are obtained by the formula (2), which follows from equation (1). Solutions of equation (1) are fixed rotation parameters of pulsar, their values are shown under the graphs of the Figure 2. Both analytical scales TT and TB are expressed in the absolute calculus of intervals from the initial event with subnanosecond resolution. Graphs (b, c) show the differences of topocentric pulsar scales for three geodetic points, including two points distant from Pushchino, for which the TT scales were calculated with respect to the topocentric Pushchino pulsar scale for the same rotation parameters.

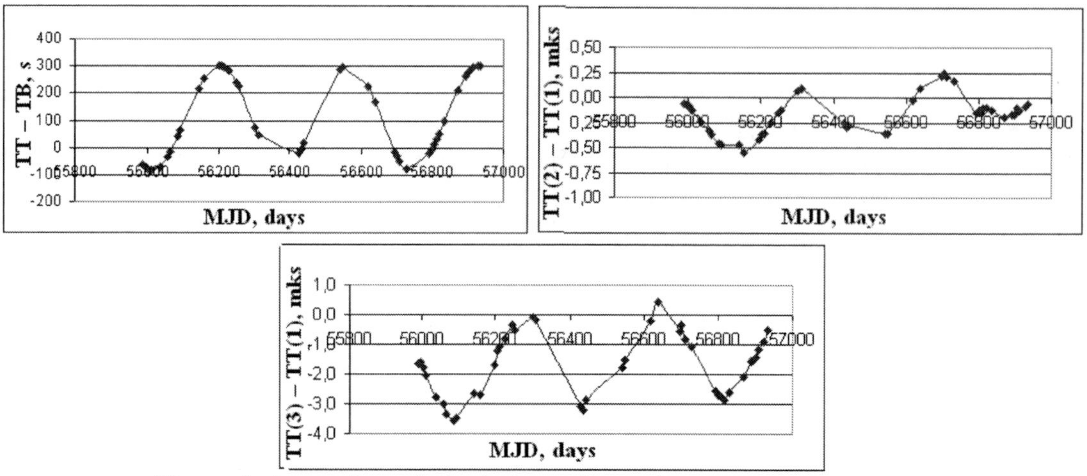

Figure 2. Comparison of physical coordinate time on the geodetic points.
a) Pushchino **(1)**: **54°50'00"N; 37°37'00"E** (BSA LPI); $P_0(TRS) = 0,739684538225754$ s (MJD 55960.1542); $P_0(BRS) = 0,739684538226603$ s (MJD 55960.1561); $\dot{P} = 4,99821 \cdot 10^{-15}$; $\ddot{P} = 9,821177 \cdot 10^{-29}$ s^{-1};
b) North point **(2): 62°57'36"N; 40°41'00"E**; $P_0(TT) = 0,739684538225754$ s (MJD 55960.1542);
c) East point **(3): 51°52'37"N; 128°20'47"E**; $P_0(TT) = 0,739684538225754$ s (MJD 55960.1542).

Comparison of the coordinate time based on the pulsar scales, shows their parametric and numerical consistency. The condition for the convergence of analytic pulsar scales, expressed by a polynomial power series (1), is satisfied by a single combination of rotation parameters P_0^*, $\dot{P}$, $\ddot{P}$. It can be seen that TT and TB scales, which differ by hundreds of seconds due to the Earth's orbital motion, show significant differences between $P_0(TT)$ and $P_0(BT)$ in smaller decimal signs (in the Figure 2 it is allocated with a boldface type). However, at ground stations where the current coordinate differences of topocentric scales is only about 1 μs or less, the values of the $P_0(TT)$ have no apparent differences at all geodetic points within the required subnanosecond accuracy of the approximation of intervals set by the quantity of decimal signs in a numerical format of the period. It indicates the parametrical and the numerical consistency of topocentric pulsar scales in any geodetic points, each of which corresponds to a single barycentric scale.

8. Conclusion

The relationship of spatial coordinates and time scales in different coordinate frames, both dynamic and kinematic, is considered. Metric compliance of mathematical ephemeris time scales in the chosen points of space and physical time scales obtained by the approximation of observed intervals of a periodic radio emission of pulsars in inertial systems with the coordinate beginning in the same points of space, is shown. Analytic coordinate–independent pulsar time scales extent unchanged to any topocentric coordinates system with a center given by three locally geodetic coordinates on the geoid.
This proves that the periodic radiation from the galactic source of the pulsar determines the "true" physical time of any inertial system within the Solar system, which is determined by the form-invariant equations and localized in the form of time scales in any kinematic inertial system satisfying to coordinate transformations of Lorentz. In this case, all coordinate systems are equal. In each such system, there is a single physical time t, the lines of which are orthogonal to the three–dimensional space, and the time scale corresponds to the same conditions for observing the motion of celestial

bodies, in particular, the rotation of pulsars, and, consequently, the simultaneity for all points of the three-dimensional space.

The extension of the physical pulsar time scales into inertial coordinate systems confirms, firstly, the homogeneity of time, which means the same flowing of physical phenomena (under the same conditions) at different times of their observation and, as a consequence, the compliance of the law of conservation of energy. Secondly, it means the isotropy of space, that is the same properties of space at each point in all directions: the physical processes do not change when the coordinate axes rotate through the space to any angle and, as a consequence, the compliance of the law of conservation of the angular momentum (the pulsar as well).

Analytic pulsar scales TB and TT in barycentric and topocentric coordinate systems, synthesized on the radio telescope BSA FIAN in Pushchino from the observed rotation parameters of pulsar, reach of machine accuracy and subnanosecond resolution of the intervals, which are the same in any coordinate system for coinciding epochs of local coordinate time. The relative instability of pulsar scales, which are defined by the coherence periodic radio emission of pulsar, does not exceed $10^{-20} - 10^{-21}$ within the near 50-year long extension of the observations of pulsars.

References

[1] Brumberg V A 2013 *Solar System Research* **47** 5 p 347
[2] Pitjeva E 2012 *Proc. IAA RAS* **26** 54
[3] Brumberg V A and Kopejkin S M 1990 *Celestial Mechanics and Dynamical Astronomy* **48** 1 p 23
[4] Logunov A A 2004 Henri Poincare and the Relativity Theory *Nauka,* (Moscow)
[5] Logunov A A 1990 *Lectures in Relativity and Gravitation* (Pergamon Press)
[6] Brumberg V A *et al* 1993 *Astronomy and Astrophysics* **275** 651
[7] Fomalont E R *at al* 1984 *Monthly Notices Royal Astronomical Society* 210 113
[8] Kopejkin S M *at al* 1990 *Inertial Coordinate System on the Sky* 213
[9] Doroshenko O V and Kopejkin S M 1990 *Astron. Zh.* **67** 5 986
[10] Avramenko A E 2015 *Proc. of the Journées 2014 "Systèmes de référence spatio-temporels"* 81
[11] Pavlov D A and Scripnichenko V I 2015 *Proc. of the Journées 2014 "Systèmes de référence spatio-temporels"* 243

Vaidya Collapse with Nonzero Radial Pressure

A Beesham[1] and S G Ghosh[2]

[1] University of Zululand, Kwa-Dlangezwa, South Africa
[2] Centre for Theoretical Physics, Jamia Millia Islamia, Delhi, India
E-mail: beeshama@unizulu.ac.za; sghosh2@jmi.ac.in

Abstract. We discuss the collapse of a fluid with nonzero pressure within the context of the Vaidya spacetime considering a decaying cosmological parameter. It is found that both black holes and naked singularities can form, depending upon the initial conditions.

1. Introduction

The cosmic censorship hypothesis (CCH) [1] is regarded as one of the most important unsolved problems in classical general relativity theory. If we look at a typical star and ask what happens after it has exhausted its nuclear fuel, there are three possibilities, depending upon the mass of the star. The first two possibilities are, in order of increasing starting mass, a white dwarf or a neutron star. However, if the star is massive enough (typically of the order of about 10 solar masses or so, depending upon whether it rotates or not) after it has exhausted its nuclear fuel, there is nothing to halt the collapse process. According to the Hawking-Penrose singularity theorems [2], it will end up in a singularity. However these theorems do not give information about the nature of this singularity itself.

According to the cosmic censorship hypothesis, the singularity will be a black hole, covered by an event horizon. Despite much investigation, as yet, there is no proof of the hypothesis. However, there is another possibility, and that is that naked singularities can form, which can in principle be observed by a distant observer. Nowadays, there are numerous examples where the hypothesis is violated, although there is still no agreement if these are physically realistic or not [3-4] (and references therein). One of the first examples that were studied to exhibit a naked singularity is the Vaidya solution [5], which represents a spherically symmetric body consisting of null dust that is radiating (or collapsing). The metric is given by

$$ds^2 = -\left[1 - \frac{2m(v)}{r}\right]dv^2 + 2dvdr + r^2 d\Omega^2, \tag{1}$$

where v is the advanced ingoing time coordinate, r is the usual radial coordinate and $d\Omega^2 = d\theta^2 + \sin^2\theta d\varphi^2$ represents the metric of the unit two-sphere. Dwivedi and Joshi [6] have given a good treatment of the occurrence of the naked singularity for this spacetime.

One of the criticisms that has been levelled against the Vaidya metric is that it consists of null dust which has no pressure. In a realistic gravitational collapse scenario, we expect there to be pressure. The question that arises is whether the introduction of pressure can help to avoid the naked singularity and hence restore the validity of the cosmic censorship hypothesis. Recently, an interesting generalization of the Vaidya spacetime has been given [7] which incorporates nonzero radial pressure:

$$ds^2 = -\left[1 - \frac{2m(v)}{r} - \frac{\Lambda(v)r^2}{3}\right]dv^2 + 2dvdr + r^2 d\Omega^2, \tag{2}$$

where $\Lambda(v)$ is the cosmological parameter, considered variable.

There are nowadays many reasons for considering a nonzero cosmological constant. With the recent discovery [8-9] of the accelerated expansion of the universe, the cosmological constant has made a comeback in the currently accepted ΛCDM concordance model of cosmology. Wagh and Maharaj [10] had studied gravitational collapse in the Vaidya spacetime with a cosmological constant, and they found that the introduction of the cosmological constant did not make any difference to the situation as obtained without the constant. Apart from this, there are now strong reasons to consider a decaying cosmological parameter since this can solve the cosmological constant problem, and it is claimed that this can give a better fit to cosmological observations than the usual concordance model [14].

2. Generalised Vaidya Solution

In this work, the gravitational collapse of the Vaidya model is studied with a variable cosmological parameter, to see if the hypothesis is still violated. By assuming some reasonable forms for $m(v)$ and $\Lambda(v)$, and studying analytically the geodesics of the spacetime following the methods as espoused in [11], we find that both black holes and naked singularities can form, depending upon fairly general initial conditions. We consider the mass function to be a linear function of the time given by

$$2m(v) = \varsigma v \quad (\varsigma > 0, const). \tag{3}$$

It can easily be verified that there is a singularity at $r = 0$, e.g., by examining the Kretschmann scalar $\left(K = R_{abcd}R^{abcd}\right)$ which works out to be

$$K = \frac{48m^2(v)}{r^6} + \frac{8\Lambda^2(v)}{3}. \tag{4}$$

This scalar diverges as $r \to 0$.

Let us first look at the energy momentum tensor for the problem under discussion. For the metric (2), the complex null vectors may be taken as follows:

$$l_a = -\delta_a^1,\ n_a = -\frac{\Delta\delta_a^1}{2r^2} + \delta_a^2,\ m_a = -\frac{r}{\sqrt{2}}\left(\delta_a^3 + i\sin\theta\delta_a^0\right), \tag{5}$$

where $\Delta = r^2 + 2m(v) - \frac{\Lambda(v)r^4}{3}$. Here l_a and n_a are real null vectors, and m_a is complex. They satisfy the usual normalisation conditions $l_a n^a = -1 = m_a \overline{m}^a$, with the other products of the null vectors being zero. From Einstein's field equations $R_{ab} - \frac{1}{2}Rg_{ab} = T_{ab} - \Lambda(v)g_{ab}$, we find that the energy momentum tensor describing the matter content for the spacetime (2) is given by

$$T_{ab} = \mu l_a l_b + 2\rho l_{(a}n_{b)} - 2pm_{(a}\overline{m}_{b)}, \tag{6}$$

where the coefficients ρ, p and μ are the density, pressure and null density, respectively. These quantities are given by:

$$\rho = -p = -\Lambda(v),\ \mu = -\frac{\dot{m}(v)}{r^2} + \frac{r\dot{\Lambda}(v)}{3}. \tag{7}$$

Here, and over dot represents a derivative with respect to v.

3. Existence of naked singularities

Let $K^a = dx^a/dk = \left(\dot{v}, \dot{r}, \dot{\theta}, \dot{\varphi}\right)$ be the tangent vector to the null geodesics, where k is an affine parameter. The following geodesic equations may then be derived:

$$\frac{dK^a}{dk}+\frac{1}{r}\left[\frac{m(v)}{r}-\frac{\Lambda(v)r^2}{3}\right]=0, \tag{8}$$

$$\frac{dK^a}{dk}+\frac{\dot{m}(v)}{r}(K^v)^2=0. \tag{9}$$

We introduce [11]

$$K^v=\frac{P}{r}. \tag{10}$$

From the null condition $K^aK_a=0$, and the geodesic equations we obtain

$$K^r=\left[1-\frac{2m(v)}{r}-\frac{\Lambda(v)r^2}{3}\right]\frac{P}{2r}, \tag{11}$$

$$\frac{dP}{dk}-\left[1-4\frac{m(v)}{r}+\frac{\Lambda(v)r^2}{3}\right]\frac{P}{2r^2}=0. \tag{12}$$

We shall integrate the above geodesic equations for a linear mass function and a variable cosmological parameter.
Most Λ decay laws come from phenomenological considerations and a decaying Λ term is proving to provide a better fit to observations than the conventional ΛCDM model [14]. The actual form of the decay law for Λ is likely to be complicated, but for the purposes of this work, we consider a fairly simple form that can allow meaningful conclusions to be drawn from it. We consider a simple generalisation of the commonly assumed form $\left(\Lambda\alpha\frac{1}{t^2}\right)$:

$$\Lambda(v)=\frac{3\eta}{v^2}, \tag{13}$$

where η is a constant. With this choice, the spacetime is self-similar, i.e., it admits homothetic Killing vector field ξ:

$$\xi^a=r\frac{\partial}{\partial r}+v\frac{\partial}{\partial v}, \tag{14}$$

which satisfies the condition $L\xi g_{ab}=\xi_{a;b}+\xi_{b;a}=2g_{ab}$, where L denotes the Lie derivative. It follows that ξ^aK_a is constant along the radial null geodesics and hence a constant of motion:

$$\xi^aK_a=rK_r+vK_v=C. \tag{15}$$

Using equations (3), (10), (11) and (13), equation (15) becomes

$$P=\frac{2yC}{\varsigma y^3-y^2+2y+\eta} \tag{16}$$

and thus we can determine the geodesics completely. In this equation y is a similarity variable. It may be seen that P satisfies the differential equation (12).
For the singularity to be naked there must be a light ray that emerges from the singularity. We will investigate the behaviour of the radial null geodesics near the singularity to determine if outgoing geodesics meet the singularity in the past. By equations (10) and (11), radial null geodesics satisfy the equation

$$\frac{dr}{dv}=\frac{1}{2}\left[1-\frac{2m(v)}{r}-\frac{\Lambda(v)r^2}{3}\right]. \tag{17}$$

On using equations (3) and (13), equation (17) becomes

$$\frac{dr}{dv}=\frac{1}{2}\left[1-\varsigma y^2-\frac{\eta}{y^2}\right]. \tag{18}$$

In order to determine the nature of the limiting value of y at $r=0,\ v=0$ on a singular geodesic, we let $\lim_{r\to 0, v\to 0} y = \lim_{r\to 0, v\to 0} \frac{v}{r}$. From equation (18), we find that

$$\varsigma y_0^3 - y_0^2 + 2y_0 + \eta = 0. \tag{19}$$

For the singularity to be naked, equation (19) must exhibit at least one positive root. This will then indicate that there is at least one outgoing geodesic which terminates at the singularity in the past. The absence of positive roots indicates that the singularity is a black hole. This occurrence of positive roots implies that the strong CCH is violated, though not necessarily the weak CCH.

We will discuss the two cases $\eta > 0$ and $\eta < 0$ separately. For $\eta > 0$, in the limit as $\eta \to 0$, equation (19) has positive roots in the range $\varsigma \in (0, 1/8]$. For $\varsigma = 0$, naked singularities form in this range, and black holes outside. Thus, $\varsigma_c = 1/8$ is called a critical value because at this value, a transition occurs and the end state changes from naked singularities to black holes. This is in agreement with earlier work [11]. The weak energy condition is satisfied for $\varsigma \geq \frac{2\eta}{y^3}$. We wish to determine the effect of a nonzero Λ on the collapse, at least qualitatively. It can be verified that $\varsigma_c \to 0$, as $\eta \to \infty$ i.e., the naked singularity spectrum of the collapsing Vaidya region continuously shrinks with increasing η. Hence, the initial data space $(0, \varsigma_c]$ for a naked singularity contracts once Λ is taken into account. Thus, at a qualitative level, the presence of Λ favours black holes over naked singularities.

For the other case, $\eta < 0$, it can easily be verified that equation (19), being a cubic equation, must have at least one real root. From the theory of such equations, it follows that for $\varsigma > 0$ and $\eta < 0$, any real root of equation (19) must be positive. Thus gravitational collapse for $\eta < 0$ always leads to a naked singularity.

4. Conclusion

Thus it appears that the introduction of pressure is not sufficient to halt the occurrence of naked singularities in the Vaidya spacetime with a variable cosmological parameter. It is not too difficult to introduce charge into this model, and this investigation is ongoing. It is also worthwhile to mention that another fertile area for research is that of finding regular black holes, viz., black holes that resemble known ones, such as Schwarzschild, but which are regularly behaved at the origin [12-13]. Of course, all of these regular black holes violate one of the energy conditions, and require a modification of general relativity, e.g., nonlinear electrodynamics or a scalar field.

References

[1] Penrose R 1965 Gravitational collapse: The role of general relativity. *Riv. Nuovo Cim.* **1** pp 252-276
[2] Hawking S W and Ellis G F R 1973 *The Large Scale Structure of Spacetime* (Cambridge University Press)
[3] Hamid A I M, Goswami R and Maharaj S D 2014 Cosmic censorship conjecture revisited: Covariantly *Class Quantum Grav.* **31** 135010
[4] Hamid A I M, Goswami R and Maharaj S D 2016 Notes on cosmic censorship conjecture revisited: Covariantly *Preprint* arXiv: 1604.01157
[5] Vaidya P C 1943 The external field of a radiating star in general relativity *Current Sci.* **12** 183
[6] Dwivedi I H and Joshi P S 1989 On the nature of naked singularities in Vaidya space- times

Class. Quantum Grav. **6** pp 1599-1606

[7] Ishwarchandra N and Singh K Y 2013 Vaidya Solution in Non-Stationary de Sitter Background: Hawking's Temperature *Int. J. Astron. Astrophys.* **3** pp 494-499

[8] Reiss A G *et al* 1998 Observational Evidence from Supernovae for an Accelerating Universe and a Cosmological Constant *Astron. J.* **116** pp 1009-1038

[9] Perlmutter S *et al* 1998 Discovery of a supernova explosion at half the age of the Universe Nature **391** pp 51-54

[10] Wagh S M and Maharaj S D 1999 Naked Singularity of the Vaidya-deSitter Spacetime and Cosmic Censorship Conjecture *Gen. Relativ. Grav.* **31** pp 975-979

[11] Joshi P S 1994 *Global aspects in Gravitation and Cosmology* (Clarendon Press, Oxford)

[12] Fan Z and Wang X 2016 Construction of Regular Black Holes in General Relativity. *Preprint* arXiv: 1610.02636v3

[13] Ansoldi S 2008 Spherical Black Holes with regular centre *Preprint* arXiv: 0802.0330

[14] Sola J 2017 Towards the first compelling signs of vacuum dynamics in modern cosmological observations *Preprint* arXiv: 1703.08218v3

IOP Conf. Series: Journal of Physics: Conf. Series **1051** (2018) 012006 doi:10.1088/1742-6596/1051/1/012006

Phenomenology of cosmological particle creation, Dirac sea and all that

V A Berezin[1], V I Dokuchaev[1,2] and Yu N Eroshenko[1]

[1] Institute for Nuclear Research of Russian Academy of Sciences, Moscow, Russia
[2]National Research Nuclear University (Moscow Engineering Physics Institute), Moscow, Russia
E-mail: berezin@inr.ac.ru; dokuchaev@inr.ac.ru; eroshenko@inr.ac.ru

Abstract. We constructed the conformally invariant model for scalar particle creation induced by strong gravitational fields. Starting from the "usual" hydrodynamical description of the particle motion written in the Eulerian coordinates we substituted the particle number conservation law (which enters the formalism) by "the particle creation law", proportional to the square of the Weyl tensor (following the famous result by Ya.B. Zel`dovich and A.A. Starobinsky). Then, demanding the conformal invariance of the whole dynamical system, we have got both the (Weyl)-conformal gravity and the Einstein-Hilbert gravity action integral with dilaton field. Thus, we obtained something like the induced gravity suggested first by A.D. Sakharov. It is shown that the resulting system is self-consistent. We considered also the vacuum equations. It is shown that, beside the "empty vacuum", there may exist the "dynamical vacuum", which is nothing more but the Dirac sea. The latter is described by the unexpectedly elegant equation which includes both the Bach and Einstein tensors and the cosmological terms. All the vacuum solutions violate the local conformal invariance.

1. Introduction

The discovery of the relic (CMB) radiation proved that the early universe was hot and filled with the soup of elementary particles. Tracing this situation back in time (and replacing the elementary particles by more and more elementary ones) we come inevitably to the Big Bang scenario.

It seems more plausible that the universe was creating quantum mechanically from "nothing" [1] or reemerged after some previous circle [2-3] in a pure quantum state. And only then it was filled with some matter constituents. This reminds the Starobinsky inflationary model [4], in which the present Friedman stage of the universe is originated from decay of the initial de Sitter space-time (which is unstable due to the presence of terms in the Lagrangian quadratic in the curvature).

The cosmological particle creation in the framework of General Relativity were studied extensively in 70-s of the last century by many authors [5-16]. Due to results of their works we know much about the structure of the counter-terms, the importance of the trace anomaly in the particle creation processes, the rate of particle production and so on.

All the above-mentioned investigations were confined to considering the quantum scalar field on the given background metrics, namely, cosmological homogeneous, but slightly anisotropic, space-times. What about the back reaction? The main obstacle in accounting for the back reaction is that the rigorous solution of the quantum problem requires the knowledge of the boundary conditions, while the latter can be imposed only after solving the (classical) Einstein equations. Thus, we have got the "vicious circle".

The full self-consistent treatment of the pure quantum phenomenon of particle creation is impossible when the space-time is considered as the classical entity. This is because we need to impose the boundary conditions in order to define the very notion of the quantum particle, but to do so we have to know the global space-time geometry, which, in turn, can be constructed only after solving the gravitational field equations with these very particles (among the other constituents) as the source. To avoid this difficulty we use phenomenological description of the particle creation process, which should be quite applicable when the fields responsible for it are strong enough, as well as the gravitational field serving as a trigger. All these requirements are met in the very early universe.
We describe the already created particles by some hydrodynamics, the "creation law" being incorporated into the hydrodynamical action integral (written using Eulerian variables). For this "creation law" we adopted the well known result for cosmological particle creation [8] that the particle production rate is proportional to the square of the Weyl tensor. As the quantum field responsible for the particle creation we considered the specific scalar field and wrote the corresponding Lagrangian, as simplest as possible, i.e., it consists of the kinetic and "mass" terms only with no ad hoc self-interaction. But the scalar field in the Lagrangian is not the fundamental one, since one part of the fundamental field is already exists in the form of the created particles, while another one, namely, the square of the Weyl tensor is some part of the quantum trace anomaly.
We proceeded by demanding the local conformal invariance of the total action. This was inspired by the appearance of the square of the Weyl tensor in the total action and some speculations by R. Penrose and G.'tHooft [2,17] on this subject. The immediate result is the emergence of the scalar curvature in the scalar field Lagrangian. Together with the square of the Weyl tensor this is in spirit of the A.D. Sakharov's idea about the induced gravity [18]. The other result appeared completely unexpected. Namely, the "mass" term becomes converted into the self-interaction term which exponent (=4) is due to the space-time dimensionality. There are other consequences of the local conformal invariance in our model, which will be discusses in due course. Throughout the paper we use the units $\hbar = c = 1$, the signature of the metric tensor $g_{\mu\nu}$ is $(+,-,-,-)$, the Riemann curvature tensor is defined as

$$R^{\mu}_{\lambda\nu\sigma} = \frac{\partial \tilde{A}^{\mu}_{\nu\sigma}}{\partial x^{\lambda}} - \frac{\partial \tilde{A}^{\mu}_{\nu\lambda}}{\partial x^{\sigma}} + \tilde{A}^{\mu}_{\kappa\lambda}\tilde{A}^{\kappa}_{\nu\sigma} - \tilde{A}^{\mu}_{\kappa\sigma}\tilde{A}^{\kappa}_{\nu\lambda}, \tag{1}$$

while the Ricci tensor is the convolution $R_{\nu\sigma} = R^{\mu}_{\nu\mu\sigma}$. The scalar curvature $R = g^{\nu\sigma}R_{\nu\sigma}$, and $\tilde{A}^{\mu}_{\nu\sigma}$ are the metric connections, i.e., the covariant derivatives of the metric tensor are zero.

2. Phenomenology of particle creation

We start with construction of the hydrodynamical part of our model. In the "classical" hydrodynamics there exist two different sets of dynamical variables, the so called Lagrangian and Eulerian coordinates. The first of them are comoving, i.e., the observer is sitting on some world-line. So, using the least action principle, one has to vary the trajectory of the (quasi)-particles. Since in such a case we cannot take into account the very processes of both creation and annihilation of particles (i.e., trajectories), it is not appropriate for our purposes. Therefore, we need to use the Eulerian description, when the dynamical variables are fields, namely, the particle number density $n(x)$ and the four-velocities u^{μ}. The action integral in this case is [19] (for details see also [20]):

$$S_{hydro} = -\int \varepsilon(X,n)\sqrt{-g}dx + \int \lambda_0\left(u^{\mu}u_{\mu} - 1\right)\sqrt{-g}dx + \int \lambda_1\left(nu^{\mu}\right)_{;\mu}\sqrt{-g}dx + \int \lambda_2 X_{,\mu}u^{\mu}\sqrt{-g}dx, \tag{2}$$

where $\varepsilon(X,n)$ is the invariant energy density, $n(x)$ - invariant particle number density, $u^{\mu}(x)$ - four-velocity of the particle flow, $X(x)$ is the auxiliary dynamical variable introduced in order to avoid the identically zero vorticity of particle flow. It enters the action integral with the Lagrange multiplier λ_2, indicating the constraint $X_{,\mu}u^{\mu} = 0$, i.e., $X(x) = const$ on the trajectories, thus enumerating them. The other two Lagrange multipliers, $\lambda_0(x)$ and $\lambda_1(x)$ are responsible, respectively, for the constraints

IOP Conf. Series: Journal of Physics: Conf. Series **1051** (2018) 012006 doi:10.1088/1742-6596/1051/1/012006

$u^\mu u_\mu = 1$ (natural normalization of the four-velocities) and $(nu^\mu)_{;\mu} = 0$ - particle number conservation law. The semicolon ";" denotes a covariant derivative with respect to the metric $g_{\mu\nu}$.

Our aim is to incorporate into the formalism the particle "creation law" $(nu^\mu)_{;\mu} = \hat{O}(inv) \neq 0$ Evidently, the function $\hat{O}$ should depend on some invariants of the fields causing this particle creation. Here we would like to explore the fundamental result by Ya.B. Zel`dovich and A.A. Starobinsky [15] obtained for the cosmological particle production

$$(nu^\mu)_{;\mu} = \beta \tilde{N}^2,$$

where $\tilde{N}^2$ is the square of the Weyl tensor $C^\mu_{\nu\lambda\sigma}$ (its definition as well as some most important properties see e.g., in [21-23]) and the coefficient β depends on the type of particles under consideration. We will consider this "creation law" as our first postulate. So, the hydrodynamical part of the action integral now becomes

$$S_{hydro} = -\int \varepsilon(X,n)\sqrt{-g}dx + \int \lambda_0 (u^\mu u_\mu - 1)\sqrt{-g}dx + \\ + \int \lambda_1 \left((nu^\mu)_{;\mu} - \beta C^2\right)\sqrt{-g}dx + \int \lambda_2 X_{,\mu} u^\mu \sqrt{-g}dx. \tag{3}$$

Very important note. The Lagrange multiplier λ_1 is, actually, defined up to the additive constant. Indeed, let us replace $\lambda_1 \to \lambda_1 + \gamma_0, \gamma_0 = const$, then

$$\gamma_0 \int \left((nu^\mu)_{;\mu} - \beta C^2\right)\sqrt{-g}dx = \gamma_0 \int \left((n\sqrt{-g}u^\mu)_{,\mu} - \beta C^2 \sqrt{-g}\right)dx. \tag{4}$$

Due to the identity $(nu^\mu)_{;\mu}\sqrt{-g} = (n\sqrt{-g}u^\mu)_{;\mu}$, the corresponding volume integral transforms into the surface integral with no effect on the dynamical equations. In result, we are left with the same "creation law" as before plus the Weyl gravitational action

$$S^{Weyl}_{grav} = -\gamma_0 \beta \int C^2 \sqrt{-g}dx. \tag{5}$$

This means that we can start with the action that includes the Weyl gravity and then incorporate it into the hydrodynamical action simply by redefinition of the Lagrange multiplier λ_1. Such a procedure, of course, will not reduce the number of dynamical degrees of freedom.

3. Scalar field and conformal invariance

By the conformal transformation we will understand the space-time dependent scaling of the metric tensor $g_{\mu\nu}$,

$$ds^2 = g_{\mu\nu}dx^\mu dx^\nu = \Omega^2 \hat{g}_{\mu\nu}dx^\mu dx^\nu = \Omega^2(x)d\hat{s}^2. \tag{6}$$

The conformal invariance means

$$\frac{\delta S_{tot}}{\delta \Omega} = 0. \tag{7}$$

Therefore, we can (and will) consider the conformal factor Ω as a dynamical variable and make variations independently in Ω and in $\hat{g}_{\mu\nu}$ [17]. Let us $S_{tot} = S_{grav} + S_{matter}$. By definition

$$\delta S_{matter} = \frac{1}{2}\int T_{\mu\nu}\sqrt{-g}\delta g^{\mu\nu}dx \text{ or } \delta S_{matter} = \frac{1}{2}\int \hat{T}_{\mu\nu}\sqrt{-\hat{g}}\delta \hat{g}^{\mu\nu}dx, \tag{8}$$

where $T_{\mu\nu}(\hat{T}_{\mu\nu})$ is the matter energy-momentum tensor. Consider, first, the following transformation of the metric tensor

$$\delta \hat{g}^{\mu\nu} = -\frac{1}{2\Omega^3}\hat{g}^{\mu\nu}\delta\Omega = -\frac{1}{2\Omega}g^{\mu\nu}\delta\Omega. \tag{9}$$

Suppose $\frac{\delta S_{tot}}{\delta\Omega}=0$, then

$$0=\delta S_{matter}=\int T_{\mu\nu}g^{\mu\nu}\frac{\delta\Omega}{\Omega}\sqrt{-g}dx\,, \qquad (10)$$

that is, the trace of the energy-momentum tensor should be zero: $Tr\left[T_{\mu\nu}\right]=Tr\left[\hat{T}_{\mu\nu}\right]=0$. If one considers the metric tensor transformation of the kind $\delta g^{\mu\nu}=\Omega^2\delta\hat{g}^{\mu\nu}$, then, as can be easily seen, $\hat{T}_{\mu\nu}=\Omega^2 T_{\mu\nu}, \hat{T}^{\mu}_{\nu}=\Omega^4 T^{\mu}_{\nu}, \hat{T}^{\mu\nu}=\Omega^6 g^{\mu\nu}$.

Let us go further on. The question arises: quanta of what kind a field is creating? The simplest choice is the scalar field. And the simplest action integral is

$$\delta S_{scalar}=\int\left(\frac{1}{2}\chi^{\mu}\chi_{\mu}+-\frac{R}{12}\chi^2-\frac{1}{2}m^2\chi^2\right)\sqrt{-g}dx\,. \qquad (11)$$

Here $\chi_{\mu}=\chi_{,\mu}$ (comma denotes the partial derivative), $\chi^{\mu}=g^{\mu\nu}\chi_{\mu}\chi_{\nu}$, m is some constant with the dimension of mass. The term $\frac{R}{12}\chi^2$, where R is the scalar curvature, constructing from the metric $g_{\mu\nu}$, is added to make this action conformally covariant. After the "standard" conformal transformations, namely, $g^{\mu\nu}=\Omega^2\hat{g}^{\mu\nu}$ and $\chi=\frac{\hat{\chi}}{\Omega}$, one gets

$$\delta S_{scalar}=\int\left(\frac{1}{2}\hat{\chi}^{\mu}\hat{\chi}_{\mu}+\frac{\hat{R}}{12}\hat{\chi}^2-\frac{1}{2}m^2\Omega^2\hat{\chi}^2\right)\sqrt{-\hat{g}}dx-\frac{1}{2}\int\left(\hat{\chi}^2\frac{\Omega^{\lambda}}{2}\right)_{|\lambda}\sqrt{-\hat{g}}dx\,. \qquad (12)$$

Now indices are raising and lowering with the metric $\hat{g}_{\mu\nu}$ and vertical line "|" denotes the covariant derivative also with respect to the metric $\hat{g}_{\mu\nu}$. The last term can be transformed to the surface integral, it does not effect the dynamics. Remarkably enough, that started with no gravitational action at all, we have got now both the conformal gravity (as a part of the "creation law") and the Einstein-Hilbert-dilaton gravity (as a part of the conformally covariant scalar field Lagrangian).

It seems that if one puts $m=0$, everything else will be all right. But, it is not so easy; there exists a problem [17]. This problem concerns the signs. With the "correct" sign for the kinetic term $\frac{1}{2}\chi^{\mu}\chi_{\mu}$ we have the "wrong" sign for the Einstein-Hilbert-dilaton part, $+\frac{\hat{R}}{12}\chi^2$) (with our sign convention there should be "$-$" instead of "$+$"), and vice-versa. Our choice is the "correct" sign for $\hat{R}$, i.e., $-\frac{\hat{R}}{12}\chi^2$, and the "wrong" sign for the kinetic term, i.e., $-\frac{1}{2}\chi^{\mu}\chi_{\mu}$. This requires some explanation. First of all, we do not care about the "correct" sign for the kinetic term, because our scalar χ is not the "genuine" (i.e., fundamental) one. Some part of it we have already "used" as the created particles. The residual part can be viewed as the vacuum fluctuations that consist of virtual particles, including the conformal anomaly, which is responsible for the creation process. Moreover, the "wrong" sign in the kinetic term means the absence of the lower bound for the energy and allows even infinite number of the created particles (let us remember the C-field in the "steady state" cosmological model by [24]). Besides, we are not going to consider our field χ as an independent dynamical variable. One more thing. If the scalar field χ is an independent dynamical variable, then, why it "knows" about the conformal transformation $g^{\mu\nu}=\Omega^2\hat{g}^{\mu\nu}$ and adjusts itself properly, i.e., $\chi=\frac{\hat{\chi}}{\Omega}$? Only, when this field is a part of it! Fortunately, in our case it is not so, and one can always choose the conformal factor,

$\Omega = \phi$, in such a way that $\chi = \frac{\phi}{l}$, where l is some factor having dimension of length (it is introduced in order to keep the action integral dimensionless). Then, the action integral for the scalar field takes the form

$$\delta S_{scalar} = -\frac{1}{2}\int\left(\frac{1}{2}\phi^{\mu}\phi_{\mu} + \frac{\hat{R}}{12}\phi^2 + \frac{1}{2}m^2\phi^4\right)\sqrt{-\hat{g}}dx\,. \tag{13}$$

There appears the self-interaction term, ϕ^4. It must be noted that the power 4 in this term is only in the case of the four-dimensional space-time (it depends on the space-time dimensions). Here, two comments are in order. First, the above action is covariant under the conformal transformation

$$\varphi(new) = \hat{\Omega}\varphi(old),\quad g^{\mu\nu}(old) = \hat{\Omega}^2 g^{\mu\nu}(new),\quad \sqrt{-g(old)} = \hat{\Omega}\sqrt{-g(new)}\,,$$

what can be easily checked. Second, it is now evident, that $3m^2 = \Lambda$ plays the role of the (bare) cosmological term.

To finish this Section we write down the energy-momentum tensor $T_{\mu\nu}$ for our (new) scalar field φ obtained by varying S_{scalar} in $\hat{g}_{\mu\nu}$:

$$\begin{aligned}\hat{T}^{scalar}_{\mu\nu} &= -\frac{1}{l^2}\phi_{\mu}\phi_{\mu} + \frac{1}{2l^2}\phi^{\sigma}\varphi_{\sigma}\hat{g}_{\mu\nu} + \frac{1}{2l^2}m^2\phi^4\hat{g}_{\mu\nu} - \\ &-\frac{1}{6l^2}\left(\phi^2\left(\hat{R}_{\mu\nu} - \frac{1}{2}\hat{R}\hat{g}_{\mu\nu}\right) - 2\left(\left(\phi\phi_{\mu}\right)_{|\mu} - \left(\phi\phi^{\sigma}\right)_{|\sigma}\hat{g}_{\mu\nu}\right)\right).\end{aligned} \tag{14}$$

Note the appearance of the second derivatives. The trace of this tensor equals

$$Tr\left[\hat{T}^{scalar}_{\mu\nu}\right] = -\frac{1}{l^2}\left(\phi^{\sigma}{}_{|\sigma} - \frac{\hat{R}}{6}\phi^2 - 2m^2\phi^4\right). \tag{15}$$

4. Hydrodynamics and conformal covariance

Since we consider now the conformal factor ϕ and transformed metric tensor $\hat{g}_{\mu\nu}$ as the independent dynamical variables, the above-written hydrodynamical action integral should be properly "updated". Let us start with analyzing the "creation law",

$$0 = \left(\left(nu^{\mu}\right)_{;\mu} - \beta\tilde{N}^2\right)\sqrt{-g} = \left(nu^{\mu}\right)_{,\mu} - \beta\tilde{N}^2\sqrt{-g}\,. \tag{16}$$

It is well known that in the four-dimensional space-time the combination $C^2\sqrt{-g}$ is invariant under conformal transformation, i.e., $C^2\sqrt{-g} = \hat{C}^2\sqrt{-\hat{g}}$. So should be the full derivative $\left(nu^{\mu}\right)_{,\mu}$. The square of the interval ds^2 transforms as

$$ds^2 = g_{\mu\nu}dx^{\mu}dx^{\nu} = \phi^2\hat{g}_{\mu\nu}dx^{\mu}dx^{\nu} = \phi^2 d\hat{s}^2\,. \tag{17}$$

Therefore, the four-velocity behaves as $u^{\mu} = \frac{dx^{\mu}}{ds} = \frac{u^{\mu}}{\phi}$, and, respectively, $u_{\mu} = g_{\mu\nu}u^{\nu} = \phi\hat{u}_{\mu}$. Thus $n\sqrt{-g}u^{\mu} = n\phi^3\sqrt{-\hat{g}}\hat{u}^{\mu} = \hat{n}\hat{u}^{\mu}$, where we introduced the new notation $\hat{n} = n\phi^3\sqrt{-\hat{g}}$.

It is clear that in the comoving coordinate system $\hat{n}$ is nothing but the particle number per unit spatial coordinate volume, and, thus, the conformally invariant quantity. So, the "creation law" does not contain the conformal factor ϕ explicitly. Therefore, the hydrodynamical part of the total action integral becomes now

IOP Conf. Series: Journal of Physics: Conf. Series **1051** (2018) 012006 doi:10.1088/1742-6596/1051/1/012006

$$S_{hydro} = -\int \varepsilon\left(X, \frac{\hat{n}}{\phi^3\sqrt{-\hat{g}}}\right)\phi^4\sqrt{-\hat{g}}dx + \int \lambda_0\left(\hat{u}^\mu\hat{u}_\mu - 1\right)\phi^4\sqrt{-\hat{g}}dx + \\ +\int \lambda_1\left(\left(\hat{n}\hat{u}^\mu\right)_{,\mu} - \beta\hat{C}^2\sqrt{-\hat{g}}\right)dx + \int \lambda_2 X_{,\mu}\hat{u}^\mu\phi^3\sqrt{-\hat{g}}dx. \tag{18}$$

and now the hydrodynamical variables are $\hat{n}$, $\hat{u}^\mu$ and. Let us write down the corresponding equations of motion

$$\frac{\delta S_{hydro}}{\delta\hat{n}} = -\frac{\partial\varepsilon}{\partial n}\frac{1}{\phi^3\sqrt{-\hat{g}}} - \lambda_{1,\sigma}\hat{u}^\sigma = 0, \tag{19}$$

$$\frac{\delta S_{hydro}}{\delta\hat{u}^\sigma} = 2\lambda_0\hat{u}_\mu\phi^4 + \lambda_2\phi^3X_{,\mu} - \lambda_{1\mu}\frac{\hat{n}}{\phi^3\sqrt{-\hat{g}}} = 0, \tag{20}$$

$$\frac{\delta S_{hydro}}{\delta X} = -\frac{\partial\varepsilon}{\partial X}\phi^4 - \frac{\left(\lambda_2\phi^3\sqrt{-\hat{g}}\hat{u}^\sigma\right)_{,\sigma}}{\sqrt{-\hat{g}}} = 0. \tag{21}$$

To these we should add, of course, the constraints that follow from variation of the action integral in Lagrange multipliers λ_0, λ_1 and λ_2: $\hat{u}^\sigma\hat{u}_\sigma = u^\sigma u_\sigma = 1$, $X_{,\sigma}\hat{u}^\sigma = X_{,\sigma}u^\sigma = 0$, $\left(\hat{n}\hat{u}^\mu\right)_{,\mu} = \beta\hat{C}^2\sqrt{-\hat{g}}$. The last of them being equivalent to $\left(nu^\mu\right)_{,\mu} = \beta C^2\sqrt{-g}$. The above equations of motion can be written in terms of the quantities without "hats". Namely,

$$-\frac{\partial\varepsilon}{\partial n} - \lambda_{1,\sigma}u^\sigma,\ 2\lambda_0u_\mu + \lambda_2X_{,\mu} - \lambda_{1,\mu} = 0,\ \frac{\partial\varepsilon}{\partial X} + \left(\lambda_2u^\sigma\right)_{,\sigma} = 0. \tag{22}$$

It is not difficult to extract the Lagrange multiplier λ_0 from these equations. Indeed, by making the convolution of the second of the equations with the four-velocity vector u^μ and using the constraints, we get, after comparing the results with the first of the equations, that $2\lambda_0 = -n\frac{\partial\varepsilon}{\partial n}$. Then, introducing the pressure p in the usual way, $p = -\varepsilon + n\frac{\partial\varepsilon}{\partial n}$, one obtains $2\lambda_0 = -(\varepsilon + p)$. The next step is to compute the hydrodynamical part of the total energy-momentum tensor. Omitting the details, we present here the result:

$$\hat{T}^{hydro}_{\mu\nu} = (\varepsilon + p)\phi^4\hat{u}^\mu\hat{u}^\nu - p\phi^4\hat{g}_{\mu\nu} - 4\beta\left(\lambda_1\hat{C}_{\mu\sigma\nu\lambda}\right)^{|\lambda|\sigma} + \frac{1}{2}\lambda_1\hat{C}_{\mu\sigma\nu\lambda}\hat{R}^{\lambda\sigma}, \tag{23}$$

with the trace, equal to

$$Tr\left[\hat{T}^{hydro}_{\mu\nu}\right] = (\varepsilon - 3p)\phi^4. \tag{24}$$

In result, the total trace equals

$$Tr\left[\hat{T}^{hydro}_{\mu\nu}\right] = -\frac{1}{l^2}\left(\phi^\sigma{}_{|\sigma} - \frac{\hat{R}}{6}\phi^2 - 2m^2\phi^4\right) + (\varepsilon - 3)\phi^4. \tag{25}$$

Finally, let us write the result of the variation of the total action integral in ϕ, which can be considered as one of the equations of motion as well as the consequence of the postulated conformal invariance. One gets:

$$\frac{1}{l^2}\left(\phi^\sigma{}_{|\sigma} - \frac{1}{6}\hat{R}\phi - 2m^2\phi^3\right) + (\varepsilon - 3p)\phi^3 = 0, \tag{26}$$

as it should be: $Tr\left[\hat{T}^{tot}_{\mu\nu}\right] = 0$! The relation $\hat{T}_{\mu\nu} = \phi^2T_{\mu\nu}$ can be also easily verified. This proves the self-consistency of our model. Note that in no way ϕ can be zero value, since this would lead to to the degeneracy of the whole space-time, i.e., actually, to its disappearance.

Introducing new notations, $6l^2 = 8\pi G$ and $3m^2 = \Lambda$, we are able to write our equations in a more familiar form (without "hats", corresponding to $\phi = 1$).

$$R_{\mu\nu} - \frac{1}{2} g_{\mu\nu} R - \Lambda g_{\mu\nu} = 8\pi G T_{\mu\nu}^{hydro}. \tag{27}$$

These equations look like the ordinary Einstein equations with a cosmological constant, but now the hydrodynamical energy-momentum tensor is modified by the presence of terms originated from the "creation law", namely

$$T_{\mu\nu}^{hydro} = (\varepsilon + p) u_\mu u_\nu - p g_{\mu\nu} - 4\beta B_{\mu\nu}[\lambda_1], \tag{28}$$

where

$$B_{\mu\nu}[\lambda_1] = \left(C_{\mu\sigma\nu\lambda}\right)^{;\lambda;\sigma} + \frac{1}{2}\lambda_1 C_{\mu\lambda\nu\sigma} R^{\lambda\sigma}, \tag{29}$$

for $\lambda_1 = 1$ it is just the Bach tensor $B_{\mu\nu}$. Note, that $\hat{B}_{\mu\nu} = \phi^2 B_{\mu\nu}$ and $\hat{T}_{\mu\nu} = \phi^2 T_{\mu\nu}$. It can be checked that derived equations are conformally covariant, i.e., if one makes the conformal transformation $\hat{g}_{\mu\nu} = \Omega^2 \hat{\hat{g}}_{\mu\nu} = g_{\mu\nu} = \left(\phi^2 \Omega^2\right) \hat{\hat{g}}_{\mu\nu}$, then the equations written in terms of $= \left\{\phi\Omega, \hat{\hat{O}}\right\}$ the same as for $= \left\{\varphi\Omega, \hat{O}\right\}$.

5. Vacuum equations and empty vacuum

By the empty vacuum we mean the space-time both without particles (i.e., $\varepsilon = p = 0$) and without particle creation (i.e., $C^2 = 0$). Then, in the case of spherical symmetry the Weyl tensor itself is also zero, $C_{\mu\nu\lambda\sigma} = 0$ [22]. The vacuum equations become simply the Einstein equations with the cosmological term. The only solution is the (anti-) de Sitter ones. It can be shown that there are no other spherically symmetric vacuum metrics (i.e., conformal to (anti-) de Sitter). The detailed calculations will be published elsewhere. The gravitational equations now look as follows

$$32\pi G\beta B_{\mu\nu}[\lambda_1] = R_{\mu\nu} - \frac{1}{2} g_{\mu\nu} R - \Lambda g_{\mu\nu}. \tag{30}$$

The equation for scalar field (=conformal factor) becomes

$$\psi^{\sigma}{}_{|\sigma} = \frac{1}{6}\varphi\left(\hat{R} + 4\Lambda\varphi^2\right) = -\frac{\partial V_{eff}}{\partial\varphi}, \tag{31}$$

where V_{eff} – the effective potential and $\hat{R}$ is the scalar curvature constructed of the conformally transformed metric $\left(ds^2 = \phi^2 d\hat{s}^2, d\hat{s}^2 = \hat{g}_{\mu\nu} dx^\mu dx^\nu\right)$, we will call it the "reference metric". For this reference metric it is more convenient to choose the solution of the vacuum equation, corresponding to $\phi = 1$ (i.e., without "hats"). The equation (24) for the scale factor becomes now

$$\phi^{\sigma}{}_{|\sigma} + \frac{2}{3}\phi\left(1 - \phi^2\right). \tag{32}$$

Then, due to the tracelessness of the "Bach tensor" $B_{\mu\nu}[\lambda]$ we have immediately $\hat{R} = -4\Lambda$, and

$$V_{eff} = V_0 - \frac{2}{3}\Lambda\left(\frac{\varphi^2}{2} - \frac{\varphi^4}{4}\right), \tag{33}$$

where V_0 is an arbitrary constant. We see that for the negative cosmological constant $(\Lambda < 0)$ it is just the famous Ginzburg-Landau potential (Higgs potential), while for $\Lambda > 0$ it is turned upside down.

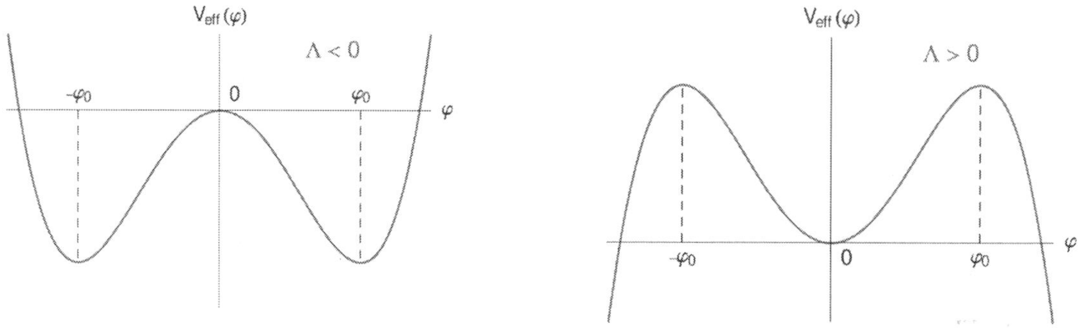

Figure 1. Effective potential V_{eff}.

It is evident that for $(\Lambda < 0)$ we have the (stable) minima at $\phi = \pm 1$ and the (unstable) maximum at $\phi = 0$ (forbidden in our model). Respectively, we for $(\Lambda > 0)$ we have the (unstable) maxima at $\phi = \pm 1$ and the (stable) minimum at $\phi = 0$. The instability in the case of the cosmological term means that such a vacuum can exist only one single moment, the moment of the creation "from nothing". And then the universe starts filling with some matter (particles). In the case of the negative cosmological term, the vacuum is stable, but the infinite volume prevents it from the creation "from nothing".

6. Vacuum equations and Dirac sea (dynamical vacuum)

We found that there may to exist much more general vacuum solutions when the requirement $C^2 = 0$ is removed. Namely, the requirement $C^2 = 0$ may appear too restrictive (e.g., in the case of the baby universe creation in the dilatonic inflation [25-27]). Instead, one can imagine the particle creation of both positive and negative energies.

This is something like the Dirac sea. For the vacuum solutions they must compensate each other. One should not be afraid of fluctuations having negative energies above the vacuum state. Due to the self-antigravitation they will be gone away, while those with positive energies will undergo the usual gravitational instability and form the structures. Thus we need two (instead of one) parts of hydrodynamical action with, correspondingly, two sets of dynamical variables (labeled by "$\pm$". Let us write down the corresponding equations of motion

$$\left(\varepsilon_{(\pm)} + p_{(\pm)}\right) u^{(\pm)}_{\mu} + \lambda_2^{(\pm)} X^{(\pm)}_{,\mu} - n_{(\pm)} \lambda^{(\pm)}_{1,\mu} = 0, \qquad (34)$$

$$\frac{\partial \varepsilon_{(\pm)}}{\partial X^{(\pm)}} - \left(\lambda_2^{(\pm)} u^{\sigma(\pm)}\right)_{;\sigma} = 0. \qquad (35)$$

In the vacuum, exactly as in the Dirac sea, from $\varepsilon_+ = -\varepsilon_-$ and $n_+ = n_-$, it follows that $p_+ = -p_-$. Since there must be no energy or particle number flows in the vacuum, we get

$$u^{\mu(+)} = u^{\mu(-)}, \qquad (36)$$

i.e., the trajectories of these two types of "matter" are the same. For this reason, the auxiliary variables $X^{(\pm)}$ are also the same. Therefore, the second (scalar) equation of motion (35) gives us

$$X^{(+)} = X^{(-)} \Rightarrow \lambda_2^{(+)} = -\lambda_2^{(-)}, \qquad (37)$$

and from the first (vectorial) equation of motion (34) it follows that

$$\lambda^{(+)}_{1,\mu} = -\lambda^{(-)}_{1,\mu}. \qquad (38)$$

The Lagrange multiplier $\lambda_1^{(+)}$ will enter as a sum in our vacuum equation, so

$$\lambda_1^{(+)} - \lambda_1^{(-)} = const. \qquad (39)$$

Finally, we obtain the following equation for what can be called "the dynamical vacuum"

$$4\alpha_0 B_{\mu\nu} + \frac{1}{16\pi G} G_{\mu\nu} - \frac{\Lambda}{16\pi G} g_{\mu\nu} = 0 , \tag{40}$$

where

$$G_{\mu\nu} = R_{\mu\nu} - \frac{1}{2} g_{\mu\nu} R , \tag{41}$$

is the Einstein tensor and

$$B_{\mu\nu} = C_{\mu\sigma\nu\lambda}{}^{;\lambda;\sigma} + \frac{1}{2} C_{\mu\lambda\nu\sigma} R^{\lambda\sigma} \tag{42}$$

is the Bach tensor. The equation (40) appeared unexpectedly elegant.

References

[1] Vilenkin A 1982 Creation of universes from nothing *Phys. Lett.* B **117** pp 25-28
[2] Penrose R 2010 *Cycles of time: An extraordinary new view of the Universe* (The Random House, London)
[3] Agullo I, Ashtekar A and Nelson W 2013 The pre-inflationary dynamics of loop quantum cosmology: Confronting quantum gravity with observations *Class. Quant. Grav.* **30** 085014
[4] Starobinskij A A 1981 Evolution of small perturbations of isotropic cosmological models with one-loop quantum-gravity corrections *Pis'ma ZhETF* **34** pp 460-463
[5] Parker L 1969 Quantized Fields and Particle Creation in Expanding Universes. I. *Phys. Rev.* **183** pp 1057-1068
[6] Grib A A and Mamaev S G 1970 On field theory in the Friedman space *Sov. J. Nucl. Phys.* **10** pp 722-728
[7] Zel'dovich Ya B 1970 Particle production in cosmology *JETP Lett.* **307**
[8] Zel'dovich Ya B and Starobinskii A A 1972 Particle production and vacuum polarization in an anisotropic gravitational field *Sov. Phys. JETP* pp 1159-1166
[9] Zel'dovich Ya B and Pitaevsky L P 1971 On the possibility of the creation of particles by a classical gravitational field *Comm. Math. Phys.* **23** pp 185-188
[10] Parker L and Fulling S A 1973 Quantized Matter Fields and the Avoidance of Singularities in General Relativity *Phys. Rev.* D **7** pp 2357-2374
[11] Hu B L, Fulling S A and Parker L 1973 Quantized Scalar Fields in a Closed Anisotropic Universe *Phys. Rev.* D **8** pp 2377-2385
[12] Fulling S A, Parker L and Hu B L 1974 Conformal energy-momentum tensor in curved spacetime: Adiabatic regularization and renormalization *Phys. Rev.* D **10** pp 3905-3924
[13] Fulling S A and Parker L 1974 Renormalization in the theory of a quantized scalar field interacting with a Robertson-Walker spacetime *Ann. Phys.* D **87** pp 176-204
[14] Lukash V N and Starobinskii A A 1974 The isotropization of the cosmological expansion owing to particle production *Sov. Phys. JETP* **39** pp 742-747
[15] Zel'dovich Ya B and Starobinskii A A 1977 Rate of particle production in gravitational fields *JETP Lett.* **26** pp 252-255
[16] Berezin V A, Kuzmin V A and Tkachev I I 1983 Could the metastable vacuum burn? *Phys. Lett.* B **124** 6 pp 479-483
[17] 'tHooft G 2015 Singularities, horizons, firewalls, and local conformal symmetry *Preprint* arXiv: 1511.04427
[18] Sakharov A D 1968 Vacuum quantum fluctuations in curved space and the theory of gravitation *Sov. Phys. Doklady* **12** pp 1040-1041
[19] Ray J R 1972 Lagrangian density for perfect fluids in general relativity *J. Math. Phys.* **13** pp 1451-1453
[20] Berezin V A 1987 Unusual hydrodynamics *Int. J. Mod. Phys.* A **2** pp 1591-1615
[21] Berezin V A, Dokuchaev V I and Eroshenko Yu N 2016 Spherically symmetric conformal gravity and "gravitational bubbles" *JCAP* **01** 019

[22] Berezin V A, Dokuchaev V I and Eroshenko Yu N 2016 Conformal gravity and "gravitational bubbles" *Int. J. Mod. Phys.* A **31** 1641004
[23] Berezin V A, Dokuchaev V I and Eroshenko Yu N 2017 Particle creation phenomenology, Dirac sea and the induced Weyl and Einstein–dilaton gravity *JCAP* **01** 018
[24] Hoyle F and Narlikar J V 1963 Mach's principle and the creation of matter *Proc. Roy. Soc. London. Ser* A **273** pp 1-11
[25] Linde A D 1984 Quantum creation of an inflationary universe *ZhETF* **87** pp 369-374
[26] Kofman L A, Einasto J and Linde, A D 1987 Cosmic bubbles as remnants from inflation *Nature* **326** pp 48-49
[27] Berezin V A, Kozimirov N G, Kuzmin V A and Tkachev I I 1988 On the quantum mechanics of bubbles *Phys. Lett.* B **212** pp 415-417

Acknowledgments
The reported study was partially supported by RFBR, research project No. 15-02-05038a. The authors would like to thank Alexey Smirnov for numerous discussions and comments. One of us (VB) is also grateful to Artyom Starodubtsev, Michail Smolyakov and Igor Gulamov for helpful discussions.

IOP Conf. Series: Journal of Physics: Conf. Series **1051** (2018) 012007 doi:10.1088/1742-6596/1051/1/012007

The rest momentum as an additional property of a massive particle in Finsler space-time

G Bogoslovsky

Lomonosov Moscow State University, Skobeltsyn Institute of Nuclear Physics, Moscow, Russia
E-mail: bogoslov@theory.sinp.msu.ru

Abstract. It is shown that a flat relativistically-invariant Finslerian space-time with partially broken 3D isotropy, i.e. an axially-symmetric space of events is a natural generalization of the Minkowski space of special relativity. The respective Finsler metric depends on two constant parameters, one of which determines the magnitude of spatial anisotropy or, in other words, the degree of deviation of the Finsler metric from the Minkowski metric, while the second parameter represents the physically preferred direction in 3D space. The replacement of the Minkowski line element in the action integral for a free massive particle by the Finsler line element leads to the Lagrange function which describes the relativistic particle in partially anisotropic 3D space. The respective function allows us to conclude that, like the situation in Minkowski space, if particle's velocity tends to zero, energy of the particle tends to its absolute minimum, i.e. to ordinary rest energy. As for momentum of the particle, unlike the situation in Minkowski space, it does not tend to zero, if particle's velocity tends to zero. Thus, in the anisotropic space, in addition to the rest energy, any massive particle obtains one more observable parameter - the rest momentum directed along the locally preferred direction in 3D space. Due to the directly proportional dependence of the rest momentum on the magnitude of local space-time anisotropy and also on mass of a particle and speed of light, the fact that in our time the magnitude of the local space-time anisotropy is extremely small does not at all mean the impossibility of detecting anisotropy in astrophysical processes with a giant energy release and, accordingly, with a giant mass defect. Obviously, in the processes with giant mass defects, one can also expect noticeable momentum defects. With allowance for the law of conservation of momentum, this implies a possible self-acceleration of matter and local isotropization of the Universe.

1. Introduction

The impressive successes of the contemporary cosmology are based to a large extent on the use of the Riemannian space-time model and the Einstein equations with Λ term

$$R_{ik} - \frac{1}{2} R g_{ik} = \Lambda g_{ik} + T_{ik}. \tag{1}$$

In connection with the Λ term let us direct our attention to the Weyl theory [1-2]. Trying to construct a unified theory of gravitation and electromagnetism, Weyl proceeded from the requirement of invariance of the theory with respect to the local conformal transformations of the Riemannian metric tensor $g_{ik} \to \exp\left[2\sigma(x)\right] g_{ik}$. The appearing Abelian gauge field, denoted below by A_i, was identified by Weyl with an electromagnetic field. The basis for such identification was the gradient character of the corresponding gauge transformations $A_i \to A_i + a\sigma_{;i}$, where a is a constant with the dimension of length. It is easy, however, to show [3] that in reality in the Weyl theory the A_i is not an electromagnetic field but acquires mass due to the violation of local conformal invariance. This result is essentially contained in the Weyl variational principle

$$\delta\left[\frac{1}{16\Lambda}\int\left(R+\frac{6}{a}A^i_{;i}-\frac{6}{a^2}A_iA^i\right)^2\sqrt{-g}d^4x-\frac{1}{4}\int F_{ik}F^{ik}\sqrt{-g}d^4x\right]=0, \tag{2}$$

where $F_{ik}=A_{k;i}-A_{i;k}$ and Λ is a constant. In fact, using the Weyl gauge condition

$$R+\frac{6}{a}A^i_{;i}-\frac{6}{a^2}A_iA^i=-4\Lambda \tag{3}$$

and his variational principle (2), we get the following system of equations

$$F^{ik}{}_{;k}-\frac{6}{a^2}A^i=0, \tag{4}$$

$$R_{ik}-\frac{1}{2}Rg_{ik}-\Lambda g_{ik}=-F_{il}F^l{}_k+\frac{1}{4}F_{lm}F^{lm}g_{ik}+\frac{6}{a^2}\left(A_iA_k-\frac{1}{2}A_lA^lg_{ik}\right). \tag{5}$$

Since, by virtue of (4), the equality $A^i_{;i}=0$ holds and, by virtue of (5), the equality $R-\frac{6}{a^2}A_iA^i=-4\Lambda$ holds, it is clear that the Weyl gauge condition (3) is fulfilled for the solutions of the system of equations (4) and (5). Thus, a system of fourth-order field equations, following from the Weyl gauge invariant variational principle (2), together with the condition (3) fixing the gauge, is equivalent to the system of equations (4) and (5). In other words, the fixing of the gauge, which implies in the Weyl theory the incorporation of length and mass standards, converts the Abelian vector field A_i into a massive vector field and reduces the equations for gravitational field to the Einstein equations with Λ term.

Let us return to the Einstein equations (1). The vacuum Lorentz-invariant energy- momentum tensor Λg_{ik} can be written in the form $\Lambda g_{ik}=diag(\rho_\upsilon,-\rho_\upsilon,-\rho_\upsilon,-\rho_\upsilon)$, where the so-called dark energy density ρ_υ is responsible for the accelerated expansion of the Universe. Thus, the corresponding vacuum space turns out to be filled with specific medium (scalar condensate, or quintessence), the pressure in which is negative and determined by the equation of state $p=-\rho_\upsilon$.

Apart from scalar condensate, the dynamic rearrangement of vacuum of quantized fields under spontaneous violation of initial gauge symmetry may give rise to anisotropic condensates. The flat space-time keeps being the Minkowski space in the only case of scalar condensate. But, if an anisotropic condensate arises, the corresponding anisotropy also appears in space-time [4]. Speaking of local anisotropy of space-time we have in mind its Finslerian geometric structure. In contrast to Riemann metrics, Finsler ones do not reduce to a quadratic form in the differentials of the coordinates. Being homogeneous functions of the differentials of the coordinates of the second degree of homogeneity, such metrics serve as a natural generalization of Riemann metrics.

In order to arrive naturally at the flat relativistically invariant Finslerian space-time we first confine ourselves to a two-dimensional space and show that it is possible to generalize the Lorentz transformations

$$\begin{cases} x'_0=x_0\cosh\alpha-x\sinh\alpha; \\ x'=-x_0\sinh\alpha+x\cosh\alpha; \end{cases} \qquad \tanh\alpha=\frac{\upsilon}{c}, \tag{6}$$

so that the new linear transformations will also form a group with a single parameter α and will keep invariance of the wave equation. Guided by the conformal invariance of the electrodynamic equations, we insert the additional scale transformations $e^{-b\alpha}$ into the standard Lorentz ones (6). As a result we obtain the generalized Lorentz transformations in the form

$$\begin{cases} x'_0 = e^{-b\alpha}\left(x_0\cosh\alpha - x\sinh\alpha\right); \\ x' = e^{-b\alpha}\left(-x_0\sinh\alpha + x\cosh\alpha\right). \end{cases} \tag{7}$$

where b is a dimensionless parameter of the scale transformations. Since the relation of the group parameter α to the velocity υ of the primed frame remains unchanged, i.e. $\tanh\alpha = \frac{\upsilon}{c}$, the generalized Lorentz transformations can be rewritten as follows

$$\begin{cases} x'_0 = \left(\dfrac{1-\upsilon/c}{1+\upsilon/c}\right)^{b/2} \dfrac{x_0 - (\upsilon/c)x}{\sqrt{1-\upsilon^2/c^2}}; \\ x' = \left(\dfrac{1-\upsilon/c}{1+\upsilon/c}\right)^{b/2} \dfrac{x - (\upsilon/c)x_0}{\sqrt{1-\upsilon^2/c^2}}. \end{cases}$$

Obviously, in contrast to the standard Lorentz transformations (6), the generalized ones (7) do not leave invariant the Minkowski metric $ds^2 = dx_0^2 - dx^2$ but conformally modify it. Therefore, the question arises as to what the metric of an event space invariant under such generalized Lorentz transformations is. A rigorous solution of this problem leads to the following metric

$$ds^2 = \left[\frac{(dx_0 - dx)^2}{dx_0^2 - dx^2}\right]^b \left(dx_0^2 - dx^2\right). \tag{8}$$

Not being a quadratic form but a homogeneous function of the coordinate differentials of degree two, this metric falls into the category of Finslerian metrics. It describes a flat but anisotropic event space.

2. Metric and group of relativistic symmetry of the flat Finslerian space-time with partially broken 3D isotropy

Proceeding from the Finslerian metric of the two-dimensional space of events (8) and using the substitutions $\left(dx_0^2 - dx^2\right) \to \left(dx_0^2 - d\boldsymbol{x}^2\right)$; $\left(dx_0 - dx\right) \to \left(dx_0 - \boldsymbol{n}\, d\boldsymbol{x}\right)$, we arrive at the corresponding Finslerian metric [5] of the four-dimensional space of events

$$ds^2 = \left[\frac{(dx_0 - \boldsymbol{n}\, d\boldsymbol{x})^2}{dx_0^2 - d\boldsymbol{x}^2}\right]^b \left(dx_0^2 - d\boldsymbol{x}^2\right). \tag{9}$$

Here the unit vector $\boldsymbol{n}$ indicates a preferred direction in 3D space while the parameter b determines the magnitude of space anisotropy and characterizes the degree of deviation of the metric (9) from the Minkowski metric. Thus, the flat Finslerian space-time (9) proves to be a generalization of the isotropic Minkowski space of conventional special relativity. Instead of the 3-parameter rotation group, the flat anisotropic event space (9) admits only 1-parameter group of rotations around the unit vector $\boldsymbol{n}$. This is why one can speak of a partially broken 3D isotropy. As to relativistic symmetry, it is realized by means of 3-parameter group of the generalized Lorentz transformations (of the generalized Lorentz boosts) which link the physically equivalent inertial reference frames in the flat anisotropic space-time and leave its metric (9) invariant. The generalized Lorentz boosts have the form

$$x'^i = D(\upsilon,\boldsymbol{n})R^i_j(\upsilon,\boldsymbol{n})L^j_k(\upsilon)x^k\,, \tag{10}$$

where υ denotes the velocities of moving (primed) inertial reference frames, the matrices $L^j_k(\upsilon)$ represent the ordinary Lorentz boosts, the matrices $R^i_j(\upsilon,\boldsymbol{n})$ represent additional rotations of the spatial axes of the moving frames around the vectors $[\upsilon\boldsymbol{n}]$ through the angles

$$\phi = \arccos\left\{1-\frac{\left(1-\sqrt{1-\upsilon^2/c^2}\right)[\upsilon\boldsymbol{n}]^2}{(1-\upsilon\boldsymbol{n}/c)\upsilon^2}\right\}$$

of relativistic aberration of $\boldsymbol{n}$, and the diagonal matrices

$$D(\upsilon,\boldsymbol{n}) = \left(\frac{1-\upsilon\boldsymbol{n}/c}{\sqrt{1-\upsilon^2/c^2}}\right)^b I$$

stand for the additional dilatational transformations of the event coordinates.

It should be noted that at $b=0$ the Finslerian metric (9) reduces to the Minkowski metric, but the generalized Lorentz boosts, i.e. transformations (10) do not reduce to ordinary Lorentz boosts and take the form

$$x'^i = R^i_j(\upsilon,\boldsymbol{n})L^j_k(\upsilon)x^k\,. \tag{11}$$

At $b=0$ i.e. in the case of Minkowski space where all directions in 3D space are equivalent, $\boldsymbol{n}$ has no physical meaning. In this case, each of the transformations (11) is differed from the respective Lorentz boost $x'^i = L^j_k(\upsilon)x^k$ by the additional rotation $x'^i = R^i_j(\upsilon,\boldsymbol{n})x^k$ the spatial axes. These additional rotations are adjusted in such a way that if a ray of light has the direction $\boldsymbol{n}$ in one frame, then it will have the same direction in all frames under consideration. Thus, at $b=0$, i.e. within the framework of conventional special relativity, the transformations (11) represent an alternative to the Lorentz boosts, however, in contrast to the boosts, they constitute a 3-parameter noncompact group which turns out to be a subgroup of the 6-parameter homogeneous Lorentz group. Physically such noncompact transformations are realized as follows. First choose as $\boldsymbol{n}$ a direction towards a preselected star and then perform an arbitrary Lorentz boost by complementing it with such a turn of the spatial axes that in a new reference frame the direction towards the star remains unchanged. The set of the transformations described has just been given by the above-displayed relation (11).

Now let us consider an inhomogeneous group of relativistic symmetry or, in other words, inhomogeneous group of isometries of the flat Finslerian space-time (9). Since the respective homogeneous noncompact group (10) is 3-parametric, with inclusion of 1-parameter group of rotations around the preferred direction $\boldsymbol{n}$ and 4-parameter translation group, the inhomogeneous group of relativistic symmetry of the space-time (9) appears to have eight parameters. To obtain the simplest representation for its generators, it is enough to send the third spatial axis along $\boldsymbol{n}$ and rewrite the homogeneous transformations (10) in the infinitesimal form. As a result, we come to the following eight generators

$$X_1 = -(x^1p_0 + x^0p_1) - (x^1p_3 - x^3p_1),$$
$$X_2 = -(x^2p_0 + x^0p_2) + (x^3p_2 - x^2p_3),$$
$$X_3 = -bx^ip_i - (x^3p_0 + x^0p_3), \qquad p_i = \partial/\partial x^i\,.$$
$$R_3 = x^2p_1 - x^1p_2;$$

These generators satisfy the commutation relations

$$[X_1, X_2] = 0, \qquad [R_3, X_3] = 0,$$

$[X_3, X_1] = X_1$, $[R_3, X_1] = X_2$,
$[X_3, X_2] = X_2$, $[R_3, X_2] = -X_1$;
$[p_i, p_j] = 0$;
$[X_1, p_0] = p_1$, $[X_2, p_0] = p_2$, $[X_3, p_0] = bp_0 + p_3$, $[R_3, p_0] = 0$,
$[X_1, p_1] = p_0 + p_3$, $[X_2, p_1] = 0$, $[X_3, p_1] = bp_1$, $[R_3, p_1] = p_2$,
$[X_1, p_2] = 0$, $[X_2, p_2] = p_0 + p_3$, $[X_3, p_2] = bp_2$, $[R_3, p_2] = -p_1$,
$[X_1, p_3] = -p_1$, $[X_2, p_3] = -p_2$, $[X_3, p_3] = bp_3 + p_0$, $[R_3, p_3] = 0$.

It is clear that the homogeneous isometry group of the flat Finslerian space-time (9) contains four parameters (generators X_1, X_2, X_3 and R_3). This group is a subgroup of the 11-parametric Similitude (Weyl) group [6], and it is isomorphic to the corresponding 4-parametric subgroup (with generators X_1, X_2, $X_3|_{b=0}$ and R_3) of the homogeneous Lorentz group. Since the 6-parametric homogeneous Lorentz group does not have any 5-parametric subgroup [7], while its 4-parametric subgroup is unique up to isomorphisms, the passage from the Minkowski space-time to the Finslerian space-time (9) implies a minimum possible violation of Lorentz symmetry. With this, the relativistic symmetry represented now by generalized Lorentz boosts (10) remains valid. Besides, in spite of a new geometry of event space the Lobachevsky geometry of 3-velocity space and, respectively, the relativistic law of addition of 3-velocities remain unchanged. Finally, it should be noted that the above-described 8-parameter inhomogeneous group of relativistic symmetry (the group of motions) of the anisotropic event space (9) was first scrutinized in [8-10]. At the present time, thanks to [11], this group is known as *DISIM$_b$*(2), i.e. Deformed Inhomogeneous SIMilitude group that includes 2-parameter Abelian homogeneous noncompact subgroup.

3. Relativistic point mechanics and the rest momentum as a new observable in the flat Finslerian space-time with partially broken 3D isotropy

A remarkable property of the partially anisotropic space of events (9) is the fact that it keeps the conformal structure (light cones) of the Minkowski space, i.e. light propagates according to the equation $dx_0^2 - d\boldsymbol{x}^2 = 0$. Therefore, the speed of light is independent of the direction of its propagation and is equal to c. It thus appears that the square of the distance dl between adjacent points of 3D space, determined by means of exchange of light signals, is expressed by the formula $dl^2 = d\boldsymbol{x}^2$. Thus, although in the 3D space there is a preferred direction $\boldsymbol{n}$ its geometry remains Euclidean. But, what does the anisotropy of space physically manifest itself in? First of all, the space anisotropy affects the dependence of proper time of a moving clock on the direction of clock's velocity: according to (9), the interval $d\tau$ of proper time read by the clock moving with a velocity υ, is related to the time interval dt read by a clock at rest by the relation $d\tau = \left(\dfrac{d\tau}{dt}\right) dt$, where

$$\frac{d\tau}{dt} = \left(\frac{1 - \upsilon\boldsymbol{n}/c}{\sqrt{1 - \upsilon^2/c^2}}\right)^b \sqrt{1 - \upsilon^2/c^2}\,.$$

Thus, in contrast to Minkowski space (for which: $b = 0$, $\left.\left(\dfrac{d\tau}{dt}\right)\right|_{b=0} = \sqrt{1 - \upsilon^2/c^2} \le 1$ and, hence, the moving clock is always slow in comparison with the clock at rest), in the anisotropic space the time dilatation factor $\left.\left(\dfrac{d\tau}{dt}\right)\right|_{b>0}$ can take on values greater than unity.

IOP Conf. Series: Journal of Physics: Conf. Series **1051** (2018) 012007 doi:10.1088/1742-6596/1051/1/012007

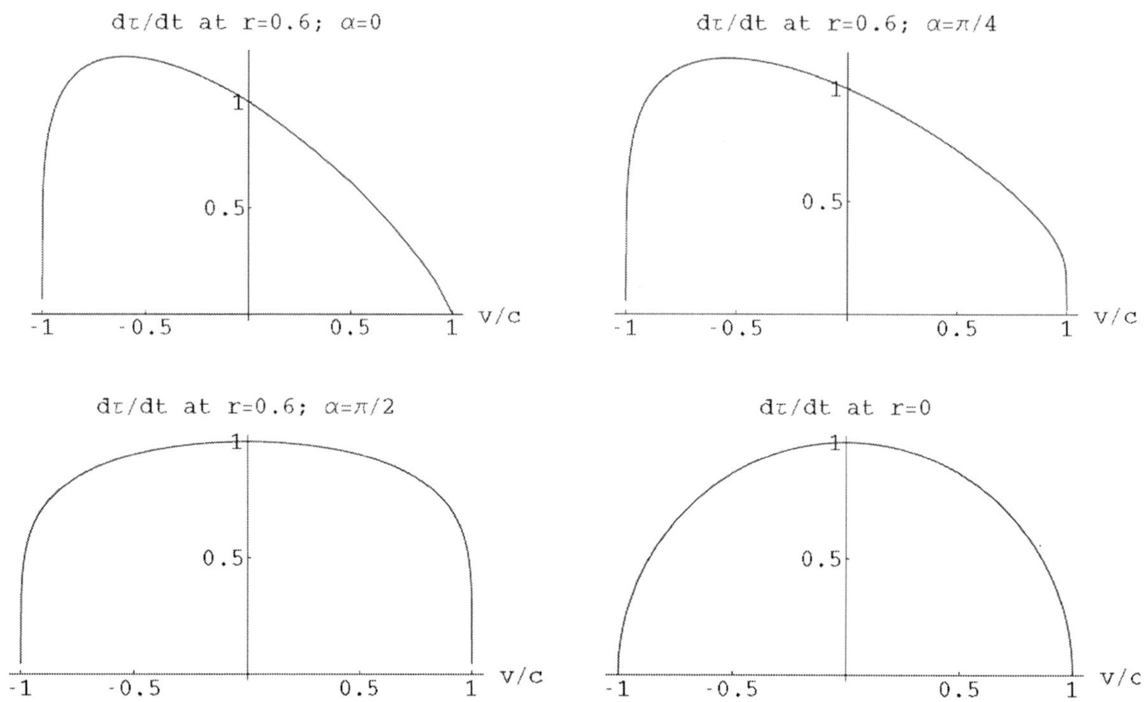

Figure 1. Plots illustrating the dependence of time dilatation factor $\left.\left(\dfrac{d\tau}{dt}\right)\right|_{b\equiv r=0.6}$ on the magnitude of clock's velocity at angles α between $\boldsymbol{v}$ and $\boldsymbol{n}$ equal to 0, $\dfrac{\pi}{4}$ and $\dfrac{\pi}{2}$. The semi-circumference corresponds to the isotropic Minkowski space for which $b \equiv r = 0$.

In order to generalize conventional relativistic point mechanics for the case of the flat partially anisotropic space-time, we replace the Minkowski line element $ds = \sqrt{dx_0^2 - d\boldsymbol{x}^2}$ with the Finslerian element in the usual action integral. As a result, the action integral for a free relativistic particle in the flat partially anisotropic space-time takes, according to (9), the form

$$S = -mc\int_i^f ds\,, \tag{12}$$

where $ds = \left[\dfrac{dx_0 - \boldsymbol{n}\, d\boldsymbol{x}}{\sqrt{dx_0^2 - d\boldsymbol{x}^2}}\right]^b \sqrt{dx_0^2 - d\boldsymbol{x}^2}$. Now we have to make sure that formula (12) correctly determines the action for a free particle. To this end, consider, along with the clock at rest, another clock which starts to move from the clock at rest and, having executed nonuniform motion along a closed path, returns to the initial point. Since at certain values of the velocity $\boldsymbol{v}$ the time dilatation factor in the anisotropic space is greater than unity (see Figure 1) then in the corresponding sections of its path the moving clock will lead that at rest. Yet, on returning to the initial point, it will necessarily be slow. The validity of this statement was rigorously proved in [12]. Thus in the flat anisotropic space-time (9) the integral $\int_i^f ds$ is maximum while the action (12) is minimum on a straight world line connecting the points i and f. Therefore, (12) really correctly determines the action for a free particle in the flat anisotropic space-time (9).

The action (12) leads to the following relativistic Lagrange function

$$L=-mc^2\left(\frac{1-\boldsymbol{\upsilon n}/c}{\sqrt{1-\upsilon^2/c^2}}\right)^b\sqrt{1-\upsilon^2/c^2}\,. \tag{13}$$

It follows from (13) that

$$E=\frac{mc^2}{\sqrt{1-\upsilon^2/c^2}}\left(\frac{1-\boldsymbol{\upsilon n}/c}{\sqrt{1-\upsilon^2/c^2}}\right)^b\left[1-b+b\frac{1-\upsilon^2/c^2}{1-\boldsymbol{\upsilon n}/c}\right]. \tag{14}$$

Thus, the energy E of a free particle in the anisotropic space depends both on the magnitude and on the direction of its velocity $\boldsymbol{\upsilon}$. At $\boldsymbol{\upsilon}=0$ the energy reaches its absolute minimum, i.e the rest energy $E\big|_{\upsilon=0}=mc^2$. As regards the momentum

$$\boldsymbol{p}=\frac{m}{\sqrt{1-\upsilon^2/c^2}}\left(\frac{1-\boldsymbol{\upsilon n}/c}{\sqrt{1-\upsilon^2/c^2}}\right)^b\left[(1-b)\boldsymbol{\upsilon}+bc\boldsymbol{n}\frac{1-\upsilon^2/c^2}{1-\boldsymbol{\upsilon n}/c}\right], \tag{15}$$

its direction does not coincide with the direction of the velocity $\boldsymbol{\upsilon}$ of a massive particle. Even in the case $\boldsymbol{\upsilon}=0$ the momentum of the particle does not vanish; there remains its rest momentum $\boldsymbol{p}\big|_{\upsilon=0}=bmc\boldsymbol{n}$. Massless particles have no such property; for them, as in conventional special relativity, $\upsilon=c$ and $\frac{E^2}{c^2}-\boldsymbol{p}^2=0$.

The energy (14) and the momentum (15) are related by the following modified dispersion relation (modified mass-shell equation)

$$\left[\frac{(p_0-\boldsymbol{pn})^2}{p_0^2-\boldsymbol{p}^2}\right]^{-b}\left(p_0^2-\boldsymbol{p}^2\right)=(mc)^2(1-b)^{(1-b)}(1+b)^{(1+b)}, \tag{16}$$

where $p_0=p^0=\frac{E}{c}$. This modified relation describes a deformed two-sheeted hyperboloid. Such a hyperboloid is relativistically invariant owing to the fact that the generalized Lorentz transformations (10) of the event coordinates x^i, i.e. the transformations $x'^i=DR^i_jL^j_kx^k$ are accompanied by the corresponding transformations $p'^i=D^{-1}R^i_jL^j_kp^k$ of 4-momenta $p^i=\left(p^0,\boldsymbol{p}\right)$. It is obvious that the deformation of a two-sheeted hyperboloid depends on the magnitude b of space anisotropy. The corresponding two-sheeted hyperboloids at $b\equiv r=0, 0.3, 0.6$ and 0.8 are shown in Figure 2.

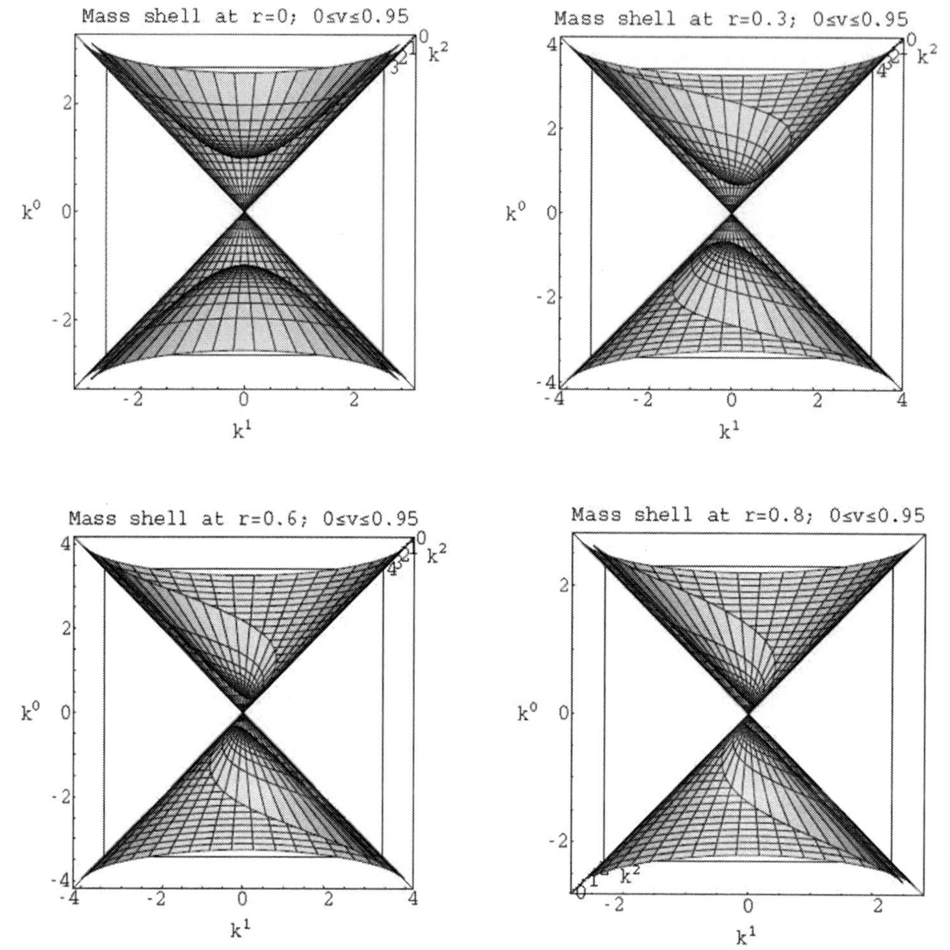

Figure 2. Parametric 3D plots of the mass shells in a space of kinematic 4-momenta. Any of the deformed two-sheeted hyperboloids remains inscribed into a light cone and like a light cone it is an invariant of the generalized Lorentz transformations.

The Lagrange function (13) takes on the following form in the non relativistic limit

$$L=-mc^2+bmc(\boldsymbol{vn})+(1-b)\frac{m\boldsymbol{v}^2}{2}+b(1-b)\frac{m(\boldsymbol{vn})^2}{2}. \tag{17}$$

Since in (17) the expression $-mc^2+bmc(\boldsymbol{vn})$ is a total derivative with respect to time, it can be omitted. As a result, we see that in the case of a non relativistic particle in the anisotropic space its kinetic energy $T=\frac{m_{\alpha\beta}v^\alpha v^\beta}{2}$ and momentum $p_\alpha=m_{\alpha\beta}v^\beta$ are determined by the inertial mass tensor

$$m_{\alpha\beta}=m(1-b)(\delta_{\alpha\beta}+bn_\alpha n_\beta). \tag{18}$$

The rest momentum $\boldsymbol{p}|_{v=0}=bmc\boldsymbol{n}$ of a particle means that the particle is in an anisotropic physical vacuum. Such a vacuum is actually filled with an anisotropic condensate or, in other words, an anisotropic quintessence, which in turn generates the Finsler geometry (9) of a flat event space.

4. Conclusion

Alongside with the phenomena, for explanation of which the concept of dark matter or dark energy is commonly used, there are the data of astrophysical observations and particle physics, indirectly indicating the existence of a local space-time anisotropy in the Universe. In this respect, we should mention first of all the anisotropy of the accelerated expansion of the Universe and the possible anisotropy of the fine-structure constant, a unified Finslerian description of which was proposed in [13]. Note also the Greisen-Zatsepin-Kuzmin effect in the physics of ultra-high energy cosmic rays and other effects considered in [14]. In particular, the absence of the GZK cutoff has yet not been explained convincingly and still remains the main empirical fact which indirectly speaks in favor of violation of Lorentz symmetry and, therefore, in favor of local space-time anisotropy.

Actually, the first direct evidence of the existence of local space-time anisotropy was obtained by the CMS collaboration at LHC within the framework of the so-called Ridge/CMS effect [15]. Let us note in passing that the Ridge/CMS effect is characteristic only to the high multiplicity events. Such events take place in case of the central collision of the initial protons where the energy density at the moment of the collision is comparable to the energy density shortly after the Big Bang, when instead of hadrons there was quark-gluon plasma. Since the initial total momentum within the CMS experiment is equal to zero, the appearance of a preferred direction, which coincides with the direction of an elongated ridge of measured correlation function and with the protons collision axis, speaks of vacuum rearrangement with the appearance of anisotropic axially-symmetric condensate. On the one hand, quantum-field vacuum filled with the anisotropic condensate is the physical carrier of the local anisotropy of space-time and it can be regarded as an anisotropic quintessence, on the other - it imparts all the particles the properties of quasi-particles in the crystalline environment. In particular, in addition to the rest energies, all massive particles acquire the respective rest momenta $\boldsymbol{p}|_{v=0} = bmc\,\boldsymbol{n}$ and the anisotropy (18) of their inertial properties.

Generally speaking, in contrast to the Minkowski space with its constant quintessence, the flat Finslerian space (9) and its anisotropic quintessence are able to change their anisotropy if b is changeable. It easy to see that, at $b \to 1$ the Finslerian metric $ds = \left[\dfrac{dx_0 - \boldsymbol{n}\,d\boldsymbol{x}}{\sqrt{dx_0^2 - d\boldsymbol{x}^2}}\right]^b \sqrt{dx_0^2 - d\boldsymbol{x}^2}$ degenerates into the total differential $ds = dx_0 - \boldsymbol{n}\,d\boldsymbol{x}$. This means that the notion of spatial extent disappears and in the space-time there remains the single physical characteristic, namely, time duration and it should be regarded as an interval of absolute time. Besides at $b \to 1$ in accordance with (18) there also disappears the inertial mass of any particle. All this suggests that absolute time, where the very notions of spatial extent and inertial mass together with unobservable primordial quintessence become meaningless, is not a stable degenerate state of space-time. As a result of geometric phase transition, which accompanies a spontaneous breaking of the original gauge symmetry, such a primordial space-time may turn into anisotropic space-time, the anisotropy of which and, accordingly, the anisotropy of the quintessence decrease in the course of the accelerated expansion of the Universe. If this scenario proves to be correct, then the question posed by the authors of work [11], namely, why are the magnitude of the local space anisotropy b and the Λ-term simultaneously so small, may be answered.

References

[1] Weyl H 1918 Gravitation und Elektrizität *Sitzber. preuss Akad. Wiss., Phys. Math. Kl.* 465

[2] Weyl H 1919 Eine neue Erweiterung der Relativitätstheorie *Ann. Phys., Lpz.* **59** 101

[3] Bogoslovsky G Yu 1992 From the Weyl theory to a theory of locally anisotropic space-time *Class. Quantum Grav.* **9** 569

[4] Bogoslovsky G Yu 2012 Dynamic rearrangement of vacuum and the phase transitions in the geometric structure of space-time *Int. J. Geom. Methods Mod. Phys.* **09** 1250007

[5] Bogoslovsky G Yu 1973 On a special relativistic theory of anisotropic space-time *Dokl. Akad. Nauk SSSR* **213** 1055
[6] Patera J, Winternitz P and Zassenhaus H 1975 Continuous subgroups of the fundamental groups of physics *J. Math. Phys.* **16** 1597
[7] Winternitz P and Fris I 1965 Invariant expansions of relativistic amplitudes and subgroups of the proper Lorentz group *Yadern. Fiz.* **1** 889
[8] Bogoslovsky G Yu 1977 A special relativistic theory of the locally anisotropic space-time. Part 1. The metric and group of motions of the anisotropic space of events *Nuovo Cimento* B **40** 99
[9] Bogoslovsky G Yu 1977 A special relativistic theory of the locally anisotropic space-time. Part 2. Mechanics and electrodynamics in the anisotropic space *Nuovo Cimento* B **40** 116
[10] Bogoslovsky G Yu 2005 Subgroups of the group of generalized Lorentz transformations and their geometric invariants *Symmetry, Integrability and Geometry: Methods and Applications* **1** 017
[11] Gibbons G W, Gomis J and Pope C N 2007 General very special relativity is Finsler geometry *Phys. Rev.* D **76** 081701
[12] Bogoslovsky G Yu 1983 On the local anisotropy of space-time, inertia and force fields *Nuovo Cimento* B **77** 181
[13] Li X, Lin H-N, Wang S and Chang Z 2015 A unified description for dipoles of the fine-structure constant and SnIa Hubble diagram in Finslerian universe *Eur. Phys. J.* C **75** 181
[14] Balan V *et al* 2012 Geometrical models of the locally anisotropic space-time *J. Mod. Phys.* **3** 1314
[15] The CMS collaboration, Khachatryan V, Sirunyan A M *et al* 2010 Observation of long-range near-side angular correlations in proton-proton collisions at the LHC *Journal of High Energy Physics* **9** 1

Induced actions for higher spin fields

P Dominis Prester[3], L Bonora[1], M Cvitan[2], S Giaccari[2] and T Štemberga[2]

[1]International School for Advanced Studies (SISSA), Trieste, Italy, and INFN, Sezione di Trieste
[2]Theoretical Physics Division of Particles and Fields, Faculty of Science, University of Zagreb, Zagreb, Croatia
[3] Department of Physics, University of Rijeka, Rijeka, Croatia
E-mail: bonora@sissa.it, mcvitan@phy.hr, pprester@phy.uniri.hr, sgiaccari@phy.hr, tstember@phy.hr

Abstract. We have calculated two-point correlators of higher spin currents, with spins up to six, in free massive QFT's with scalar and Dirac fields in $D = 3,4,5,6$ spacetime dimensions. The infrared and ultraviolet limits are analysed, with the motivation for improving our understanding of the interacting higher spin theories.

1. Introduction

The construction of interacting quantum field theories with massless higher spin ($s > 2$) fields still poses an interesting theoretical problem. On one hand, there are different no-go theorems putting serious constraints on such theories, especially in flat spacetime. On the other hand, free fields can be constructed in the same manner as in lower spin cases, and in AdS background there is a fully consistent covariant HS theory [1]. In addition, free lower spin quantum field theories possess conserved higher spin currents which simply beg to be coupled to higher spin fields. To gain a deeper insight into the structure of QFT, it is important to understand what happens when higher spin fields are included, either by constructing and developing such theories or finding obstructions which forbid such couplings.

Differences between lower spin and higher spin theories emerge already at the level of classical free field theories. The simplest way to construct a free theory of a HS field is provided by the Fronsdal equation [2]:

$$F \equiv \Box\phi - \partial\partial\cdot\phi + \partial^2\phi' = 0, \tag{1}$$

where spin-s field is described by the completely symmetric rank-s tensor field $\varphi \equiv \varphi_{\mu_1\cdots\mu_s}$. In this expression standard conventions from [3-4] are assumed. The Fronsdal equation (1) is invariant under local transformations parameterized by a traceless completely symmetric rank-($s-1$) tensor fields $\Lambda \equiv \Lambda_{\mu_1\cdots\mu_{s-1}}$

$$\delta_\Lambda \phi = \partial\Lambda, \tag{2}$$

when $\Lambda' = 0$. Moreover, the Fronsdal equation can be written in Lagrangian formulation only if one assumes $\phi'' = 0$ off-shell.

As constraining local symmetries in interacting theories usually have dramatic and unwanted consequences, one would like to have a formulation of HS theories which has: (i) unrestricted gauge symmetry, (ii) Lagrangian formulation with interaction with matter included. This implies that the linearised equations of motion (LEoM) must be of the form

$$E_{\mu_1\ldots\mu_s}[\phi] = J_{\mu_1\ldots\mu_s}\;;\; \partial\cdot E = 0\,,\; \delta_\Lambda E = 0, \tag{3}$$

where symmetric LEoM-tensor E_{μ^s} is off-shell conserved (i.e., divergence-free) and linear in the HS field φ, and the current J_{μ^s} is the matter current which is on-shell conserved (i.e., when EoM for matter fields are assumed). It is important to understand that even if two LEoM tensors E_{μ^s} are equivalent in the free case (for $J=0$), this does not imply that they lead to equivalent EoM when coupled to the generic external current.

For $s>2$ a LEoM-tensor E_{μ^s} which gives an EoM equivalent to the Fronsdal equation (1) in the case of a free HS field must be either higher-derivative or nonlocal. This can be traced to the property that generalisation of field strength tensor to HS (g-Riemann tensor) is of s-derivative type, which means that it is a higher derivative object for $s>2$. The least non-local LEoM tensor, which is second-order in derivatives, is given by [3]

$$E_{\mu_1\ldots\mu_s} = \Box^{-n} G_{\mu_1\ldots\mu_s}\,, \qquad s = \begin{cases} 2(n+1), & s \text{ even} \\ 2n+1, & s \text{ odd} \end{cases}, \tag{4}$$

where G_{μ^s} is g-Einstein tensor, the lowest derivative fully gauge invariant divergence free symmetric tensor of rank-s.[1] The non-local operator $\Box^{-n}$ makes LEoM (3) 2nd order in derivatives, and at the same time makes the free LEoM equivalent to Fronsdal equation (1).

In [4] a surprising result was obtained - for $s>2$ LEoM (3) with (4) is not consistent with propagation of only spin-s excitations for generic conserved currents. In the same paper it was shown that this problem can be circumvented in two ways. One is by realising that for every $s>2$ there is a $[s/2]$ parameter family of possible HS kinetic terms, all giving LEoM equivalent to the Fronsdal equation for a free HS field. In our notation, the family can be written as

$$E_{\mu^s} = \Box^{-n}\left(G_{\mu^s} + \sum_{j=1}^{[s/2]} \xi_j \left(\pi_{\mu\mu}\right)^j G^{[j]}_{\mu^{s-2j}} \right), \tag{5}$$

where $\pi_{\mu\nu}$ is a symmetric transverse projector defined in (9), ξ_{sj} are arbitrary real numbers, and n is defined as in (4). The choice (4), which is obtained from (5) by taking $\xi_j=0$, is special in having a non-locality of the lowest degree. The important result of [4] was that for every s there is a choice of parameters ξ_j for which equation (3) is consistent in the above mentioned sense. However, contrary to (4), this consistent choice for the kinetic term always has the highest degree of non-locality. The second way to cure the problem is to assume that the current in (3) is not generic. In spin-3 case it was shown in [4] that if the current can be written as

$$J_{\mu\nu\rho} = \bar{J}_{\mu\nu\rho} - \frac{3d \pm \sqrt{3d^2-6d}}{d(d+1)} \pi_{(\mu\nu} \bar{J}'_{\rho)}\,,\; \partial\cdot\bar{J} = 0\,, \tag{6}$$

then LEoM tensor (4) is viable as a kinetic term in (9). Note that this redefinition is nonlocal.

[1] In $d=3$ for odd spins there is a parity-odd tensor which has one derivative less then g-Einstein tensor. In the main part of the paper we assume $d>3$ and we treat the $d=3$ case separately.

In [5] we proposed to study HS actions by using the induced action method [6] as an inspiration and developed this idea further in [7-8]. In these papers we calculated two-point correlators of higher spin currents, with spins up to six, in free massive QFT's with scalar and Dirac fields in $D=3,4,5,6$ spacetime dimensions. In the following sections we present some of the main results of these calculations, and its implications for the linearised HS actions. In Appendix A we briey review the HS covariant formalism we use in the main part.

2. Higher spin currents

All free relativistic theories possess conserved currents of all spins, represented by totally symmetric tensor fields $J_{\mu_1 ... \mu_s}$. We shall concentrate here on currents with integer spin s, and consider currents in free scalar field and free Dirac field (Majorana field in $d=3$ spacetime dimensions) QFT's, which we denote as bosonic and fermionic matter, respectively. It can be shown that conserved currents in these theories, in the simplest form, are given by

$$J^{(s)b}_{\mu_1 ... \mu_s} = i^s \varphi^+ \overleftrightarrow{\partial}_{\mu_1} ... \overleftrightarrow{\partial}_{\mu_s} \varphi, \quad J^{(s)f}_{\mu_1 ... \mu_s} = i^{s-1} \overline{\psi} \gamma_{(\mu_1} \overleftrightarrow{\partial}_{\mu_2} ... \overleftrightarrow{\partial}_{\mu_s} \psi. \tag{7}$$

Note that simple currents of the bosonic matter are s-th order in derivatives, while currents of the fermionic matter are of $(s-1)$-th order. We shall call these form simple currents.

As is well known from the spin-2 case, there is a degeneracy present in the definition of conserved currents with $s>1$. Once we have a full set of currents, and all relativistic free QFT's have such full set, one can construct a $[s/2]$ parameter family in the following way

$$J^{(s)}_{\mu^s} \to J^{(s)}_{\mu^s} + \sum_{j=1}^{[s/2]} \zeta_j (\Box \pi_{\mu\mu})^j J^{(s-2j)}_{\mu^{s-2j}}, \tag{8}$$

where ζ_j are arbitrary real dimensionless parameters, and $\pi_{\mu\nu}$ is the symmetric transverse projector defined by

$$\pi_{\mu\nu} \equiv \eta_{\mu\nu} - \Box^{-1} \partial_\mu \partial_\nu \quad \Rightarrow \quad \partial \cdot \pi = 0, \quad \pi^2 = \pi, \quad \pi^\mu_\mu = (D-1). \tag{9}$$

In this paper we shall consider also the traceless currents, defined by $J'_{\mu^{s-2}} = 0$ in the massless limit, for which the coefficients in (8) are given by

$$\zeta^{f,t}_j = \frac{(s-1)!\Gamma\left(s + \frac{d-3}{2} - j\right)}{4^j j!(s-2j-1)!\Gamma\left(s + \frac{d-3}{2}\right)}. \tag{10}$$

In view of (5) and (6) one should also consider the following nonlocal redefinitions of currents

$$J^{(s)}_{\mu^s} \to J^{(s)}_{\mu^s} + \sum_{j=1}^{n} \xi_j (\pi_{\mu\mu})^j J^{(s)[j]}_{\mu^{s-2j}}, \tag{11}$$

where ξ_j are arbitrary real numbers, and n is defined as in (22).

One of our goals is to understand the differences among forms of currents, in context of building proper couplings of matter to HS fields.

3. Two-point correlators

3.1. Definitions

We are interested in two-point time ordered correlators of HS currents discussed in previous section. We have calculated also mixed correlators, i.e., correlators of two currents with different spins s_1 and s_2. We use interchangeably coordinate and momentum space representations

$$\tilde{J}_{\mu_1\ldots\mu_{s_1}\nu_1\ldots\nu_{s_2}}(k) \equiv \int d^d x e^{ik\cdot x} \langle 0|T\left\{J_{\mu_1\ldots\mu_{s_1}}(x) J_{\nu_1\ldots\nu_{s_2}}(0)\right\}|0\rangle. \tag{12}$$

We divide correlators into conserved (i.e., divergence-free) part and non-conserved parts. Conserved part is of the form

$$\tilde{J}^{(c)}_{\mu_1\ldots\mu_{s_1}\nu_1\ldots\nu_{s_2}}(k) = m^{d+s_1+s_2-4} \sum_r \tau_r^{s_1 s_2}\left(\frac{k}{m}\right)^2 \underset{r}{\Pi}(k)_{\mu_1\ldots\mu_{s_1}\nu_1\ldots\nu_{s_2}}, \tag{13}$$

where $\tau_r^{s_1 s_2}(z)$ are the form factors and $\underset{r}{\Pi}$ are the complete sets of divergence-free polarization tensors. Form factors are analytic functions for $|z|<1$, with branch-cuts extending from $z=\pm 4$ to infinity. In $d \neq 3$ there is only a parity even sector, and conserved part is spanned by properly symmetrized products of transverse projectors (9). For diagonal correlators, $s_1 \equiv s$, there are $[s/2]+1$ independent polarisation tensors.

In $d=3$ one has, in addition, a parity-odd contribution to the 2-point correlators. In the case of diagonal correlators, in this sector there are $\left[\frac{s+1}{2}\right]$ independent polarisation tensors.

The full expressions for spins $s \le 5$ in $d=3,4,5,6$ together with a detailed analysis of the behaviours in different limits can be found in [8]. Here we present some generic properties and relations we found particularly interesting.

An application of these correlators can be seen as follows. One couples the matter quantum fields to background higher spin fields, represented by completely symmetric rank-s tensor fields $\phi_{\mu^s}(x)$, by linear coupling to the corresponding conserved currents

$$L_1 = \sum_s J^{(s)}_{\mu_1\ldots\mu_s} \phi^{\mu_1\ldots\mu_s}. \tag{14}$$

Then from 2-point correlators one calculates the quadratic part of an effective action (or induced action, in Sakharov's language) for higher spin field

$$S_2^{eff}[\phi] = \sum_{s_1}\sum_{s_2} \int d^d k \tilde{\phi}(-k)^{\mu^{s_1}} \tilde{J}(k)_{\mu^{s_1}\nu^{s_2}} \tilde{\phi}(k)^{\nu^{s_2}} \tag{15}$$

or, in other words, the linear part of the corresponding induced equations of motion (LEoM)

$$\left\langle\left\langle \tilde{J}^{(S)}_{\mu^{S_1}} \right\rangle\right\rangle_{\phi,1} = \sum_{s_2} \tilde{J}_{\mu^{S_1}\nu^{S_2}} \tilde{\phi}^{\nu^{S_2}}. \tag{16}$$

Let us mention that in the even spin case 1-point correlators are generically non-vanishing. They contribute to the constant part of the effective EoM

$$\left\langle\left\langle \tilde{J}^{(S)}_{\mu^S} \right\rangle\right\rangle_{\phi,0} = \left\langle J^{(S)}_{\mu^S} \right\rangle = m^{d+s-2} c(s,d) (\eta_{\mu\mu})^{s/2}, \tag{17}$$

where $c(s,d)$ are dimensionless constants. We see that this can be interpreted as (the 0th-order part of) a generalised cosmological constant term. For odd spins there is no such term.

3.2. Non-conserved part of the 2-point correlators

In this paper we mainly ignore non-transverse parts of correlators. We can always choose them to be local, so they could in principle be subtracted out from the effective action by local counter-terms. Also, we extract the non-conserved part in the "minimal way". We expect that non-conserved terms play a major role in the analysis of the seagull (quadratic coupling of HS fields to matter fields) and tadpole (generalised cosmological constant) terms, which we plan to do in the near future. For now, we ignore these terms, and leave the detailed analysis of the non-covariant part to our future work in which we plan to investigate non-linear HS structures.

3.3. Non-diagonal terms in LEoM

For the currents we considered in this paper, 2-point correlators of currents with different spins are generally non-vanishing. The exceptions are only correlators of traceless currents of massless matter. This introduces non-diagonal terms in LEoM's, which significantly complicates analysis of the propagating degrees of freedom.

3.4. IR analysis - general d

The infrared (IR) expansion of the conserved part of the correlators, valid for $\frac{|k|}{m}<1$, can generally be written as[2]

$$\tilde{J}^{(c,IR)}_{\mu^s\nu^s}(k)=m^{d-2+2[s/2]}k^{2\left[\frac{s-1}{2}\right]}\sum_{j=1}^{\infty}a_j(s,d)\left(\frac{k^2}{m^2}\right)^j\sum_r b_{jr}(s,d)\tilde{\prod_r}(k)_{\mu^s\nu^s}, \tag{18}$$

where $a_j(s,d)$ and $b_{jr}(s,d)$ are dimensionless coefficients. When plugged into (15) the conserved part contributes weakly non-local contribution to the effective action.[3]

It can be shown that for the simple conserved currents, and all currents whose definition do not include d explicitly (e.g., lowest derivative ones), the form of the polarisation tensors at each order n is independent of d when expressed in terms of π-projectors, which means that the coefficients $b_{jr}(s,d)$ do not depend on d.

Let us take the spin-3 case as an example. One has

$$\tilde{J}^{c,\mathrm{IR}}_{\mu\mu\mu\nu\nu\nu}=\sum_{j=1}^{\infty}a_j(3,d)m^{d-2j}k^{2+2j}\pi_{\mu\nu}\left(\pi^2_{\mu\nu}+b_j(3,d)\pi_{\mu\mu}\pi_{\nu\nu}\right), \tag{19}$$

where for the simple form of the currents we obtain

$$b^f_j(3,d)=\frac{j-3}{2j};\; b^b_1(3,d)=-1;\; b^b_j(3,d)=\frac{3}{2},\; j>1. \tag{20}$$

The leading order term of the IR expansion of the conserved part is universal for all spins. The simplest way to present it is in the language of LEoM. We obtain

$$\left\langle\left\langle J_{\mu^s}\right\rangle\right\rangle_{\phi,1}=a_1(s,d)m^{d-4+2[s/2]}G_{\mu^s}+(\text{higher}-\text{derivative terms}), \tag{21}$$

2 For the sake of economy, we take diagonal case $s_1=s_2\equiv s$.
3 Weakly non-local means that the action is an infinite sum of local terms.

where $G_{\mu^s}(x)$ is the g-Einstein tensor (see Appendix A for definition). Going to next-to-leading term in IR expansions, we find no universal behaviour.
Considering that the effective actions are weakly nonlocal in the IR region, and that g-Einstein is the lowest-derivative LEoM-tensor, this is a natural result. Note that in the lowest order in IR, Eq. (21) is almost identical to the LEoM (3)-(4), proposed in [3]. The difference is in the presence of nonlocal operators $\Box^{-n}$ in (4). This difference may be avoided if, instead of (14), we define the coupling as

$$L_1 = \sum_s J^{(s)}_{\mu_1 \ldots \mu_s} \Box^n \varphi^{\mu_1 \ldots \mu_s}. \tag{22}$$

Redefinitions of current (11) are not expected to change the form of the right hand side of (21), because it is what one generically expects for the conserved part. This means that instead of (21) one expects to obtain in the leading order in IR

$$\Box^n \langle\langle J_{\mu^s} \rangle\rangle_{\varphi,1} \propto m^{d+4(n-1)+2[s/2]} G_{\mu^s}, \tag{23}$$

which is the same as (3)-(4). As noted in Sec. 1, for generic currents and $s > 2$ solutions of this LEoM do not yield only physical spin-s excitations expected from the analysis of the Poincare group. In [4] it was shown that, at least in the spin-3 case (see Eq. (6)), the resolution can be found through a proper choice of the conserved current, by using redefinition of the type (11). We plan to investigate this point in the future.
Let us summarise our findings from the IR regime. Based on our analysis, the following procedure has the potential to reconcile our results, based on the induced action, with the analyses from [4]:

(i) The linear coupling of matter to HS fields should be of the higher-derivative form (22).
(ii) One should use HS current which satisfies proper current exchange equation, as described in [4]. Based on the spin-3 example, we assume that this can be achieved by a proper choice of coefficients ξ_j, for all spins. We emphasize that the coupling is local.
(iii) The remaining freedom in the choice of the currents should be used to achieve vanishing of the non-diagonal ($s_1 \neq s_2$) correlators.

We plan to investigate this program in the future.

3.5. IR analysis for $d = 3$

The specialty of $d = 3$ dimensions is that beside parity-even sector there is also a parity-odd sector of 2-point correlators, present for Majorana fermion matter QFT. The parity-even sector has been already described above, so we turn our attention here to the parity-odd sector. There is an important difference in the IR behaviour between odd spin and even spin cases.
For odd spins the LEoM term produced in the IR limit is

$$\langle\langle J_{\mu^s} \rangle\rangle_{P-odd} \sim \operatorname{sign}(m) m^{s-1} B_{\mu^s}, \tag{24}$$

where the tensor B_{μ^s} is defined in (A.10). The B_{μ^s} tensor is by definition the lowest derivative local LEoM-tensor in $d = 3$.
In the case of even spins we obtain

$$\langle\langle J_{\mu^s} \rangle\rangle_{P-odd} \sim \operatorname{sign}(m) m^{s-2} \varepsilon_{\mu}{}^{\nu\sigma} \partial_\sigma G_{\nu\mu^{s-1}}. \tag{25}$$

We see that in the spin-odd case the leading IR term is s-th order in derivatives, while in spin even case it is of ($s+1$)-th order. We can compare this with the parity even part (18). We see that the parity-odd sector is dominant for odd spins and subdominant for even spins, which generalises the known behaviour for spins 1 and 2. Note that for the Majorana field in $d = 3$, the sign of the mass term is relevant, and the action of parity transformation in the free Lagrangian amounts to $m \to -m$.

Note that while in spin 1 and 2 cases the tensor B_{μ^s} is equal to the generalised Cotton tensor, this is not so for $s>2$. The reason is that the number of independent parity-odd LEOM tensors is $\left[\frac{s+1}{2}\right]$, which for $s>1$ is larger then one. By definition the g-Cotton tensor is of the highest order type in derivatives, so it is not natural to expect that it appears as the lowest order parity-odd term in the IR expansion for $s>2$.

3.6. UV limit and massless matter

For massless matter it is natural to consider traceless currents. The contribution to the induced LEoM for a HS field is

$$\left\langle\left\langle J^{(t)}_{\mu^s}\right\rangle\right\rangle \propto \square^{\frac{d}{2}-2} E^{(W)}_{\mu^s}, \tag{26}$$

where $E^{(W)}$ is the lowest-derivative local traceless spin-s LEoM tensor (A.9). Note that the right hand side in (26) is non-local for odd d and $d=2$. As it is of $(2s)$-derivative type, it is easy to see that the corresponding effective action for even $d\geq 4$ should be

$$S^{(m=0)}[\phi] \propto \int d^d x W\cdot \square^{\frac{d}{2}-2} W, \tag{27}$$

where $W_{\mu_1\nu_1\dots\mu_s\nu_s}$ is the spin-s g-Weyl tensor, the generalisation of the linearised Weyl tensor to spin-s (see Appendix A.1 for definition and properties). This result suggests that in $d=4$ dimensions there should be an action term of the form

$$S_W = \int W^2, \tag{28}$$

which is g-Weyl invariant not only in linearised theory, but also in some hypothetical full nonlinear HS theory. Indeed, in the spin-2 case we know that this is true.

If one considers the conserved part of 2-point correlators of the traceless currents for massive matter in the UV limit $\left(\frac{|k^2|}{m^2}>>1\right)$, then the leading term is the same as in the massless case. However, in $d=3$ for fermionic matter there is an interesting sub-leading contribution in the parity-odd sector

$$\tilde{J}^{(t,UV)}_{\mu^{s_1}\nu^{s_2}}(k) = (-1)^{\frac{s_1+s_2}{2}} \frac{imk^{s_1+s_2-3}}{2^{s_2+1}} \varepsilon_{\sigma\mu\nu} k^\sigma \times \tilde{\pi}_{\nu\nu}^{\frac{s_2-s_1}{2}} \tilde{\pi}_{\mu\nu}^{s_1-1} {}_2F_1\left(\frac{1-s_1}{2}, -\frac{s_1}{2}, 1-s_1; \frac{\tilde{\pi}_{\mu\mu}\tilde{\pi}_{\nu\nu}}{\tilde{\pi}^2_{\mu\nu}}\right), \tag{29}$$

where spins s_1 and s_2 have the same parity, and $s_2>s_1$. The hypergeometric function is a polynomial in its last argument. The polarisation tensor is a unique traceless and conserved one in the parity-odd sector. From (29) we obtain the following leading order contribution to LEoM in the UV limit contribution for the diagonal case $s_1=s_2\equiv s$:

$$\left\langle\left\langle J^{(t)}_{\mu^s}\right\rangle\right\rangle_{\text{P-odd,UV}} \propto m\,\square^{-1/2} C_{\mu^s}, \tag{30}$$

where C_{μ^s} is the g-Cotton tensor (see Appendix A.3 for definition and properties). Observe that the above expression is non-local. However, if we assume that there are N free Majorana fermion fields with the same mass m, and take particular large-N UV limit defined by $N \to \infty$ and $\left|\frac{m}{k}\right| \to 0$ with $\lambda \equiv N\left|\frac{m}{k}\right| = \text{fixed}$, then from (30) follows

$$\left\langle\left\langle J^{(t)}_{\mu^S}\right\rangle\right\rangle_{P-oss,UV} \propto \lambda C_{\mu^S} + (\text{higher}-\text{order terms}), \tag{31}$$

which is a *local* LEoM term. As for $s = 2$ this term follows from the (gravitational) Chern-Simons Lagrangian term, we expect that the same is true for all spins. Note that if the same limit is applied to the non-diagonal case, one obtains Chern-Simons-like terms of the mixed kind.

Appendix A. Linearised covariant formalism for higher-spin fields

In this section we first introduce well-known definitions and properties about higher spin tensors, their linearized EoM's and their possible geometrical formulations. We complete and review the covariant formalism for higher spin (HS) gauge fields in the linear approximation in the flat Minkowski background, originally introduced in [9] and further developed in spin-3 case in [10].

Appendix A.1. Generalized curvature tensors

A spin-s field is described by a totally symmetric rank-s tensor field $\varphi_{\mu_1\ldots\mu_S}$. We assume there is the gauge symmetry, which we call g-diff, with transformations defined in (2).[4] The idea is to generalise the covariant formalism of Maxwell theory (spin-1) and linearised General Relativity (spin-2) to arbitrary spin.

While one usually starts with generalised connections, as we shall not use them here we skip directly to the generalisation of Riemann (g-Riemann) tensor. The g-Riemann tensor is

$$R_{\alpha_1\mu_1\ldots\alpha_S\mu_S} \equiv \partial_{\alpha_1}\ldots\partial_{\alpha_S}\varphi_{\mu_1\ldots\mu_S}\Big|_{|\alpha_j,\mu_j|}, \tag{A.1}$$

where anti-symmetrization (without weight factors) in all pairs of indices $\{\alpha_j, \mu_j\}$, $j = 1,\ldots,s$ is understood. Symmetries of such defined g-Riemann tensor straightforwardly generalise those of the ordinary Riemann tensor - it is antisymmetric under the one exchange $\mu_j \leftrightarrow \nu_j$, symmetric under the exchange of any two pairs $(\alpha_j, \mu_j) \leftrightarrow (\alpha_k, \mu_k)$, and satisfies cyclic and Bianchi identities.

Note that the g-Riemann tensor is of s-th order in derivatives. It is easy to see that in the spin-1 case (A.1) gives the field strength tensor $F_{\alpha\mu} \equiv \partial_\alpha\varphi_\mu - \partial_\mu\varphi_\alpha$, while in the spin-2 case (A.1) it gives the linearised Riemann tensor. The important property of the g-Riemann tensor is encoded in the theorem stating that if it vanishes in some region then the field φ is a pure gauge in that region.

For even spin one can construct a symmetric rank-s tensor by tracing all α indices

$$R_{\mu_1\ldots\mu_S} = \eta^{\alpha_1\alpha_2}\ldots\eta^{\alpha_{s-1}\alpha_s} R_{\alpha_1\mu_1\ldots\alpha_s\mu_S}, \tag{A.2}$$

which we call the g-Ricci tensor. By repeatedly tracing this tensor one obtains the tower of symmetric g-Ricci tensors, $R_{\nu_1\ldots\nu_{2k}}$, with even ranks ranging from s to 0.

For odd spin $s = 2n+1$, the g-Ricci tensor is obtained by tracing g-Riemann tensor n times

[4] In $d > 4$ one also need to take into account fields with mixed symmetries. Also, keep in mind that the formalism is linearised both in the HS field φ and g-diff parameter field Λ. Symmetrizations (...) and antisymmetrisations [...] are in the Appendix defined with the weight 1, if not mentioned otherwise.

$$R_{\alpha\mu\nu_1\ldots\nu_{s-1}} = \eta^{\beta_1\beta_2}\ldots\eta^{\beta_{s-1}\beta_{s-1}} R_{\alpha\mu\beta_1\nu_1\ldots\beta_{s-1}\nu_{s-1}}. \tag{A.3}$$

Here g-Ricci tensor is antisymmetric in the first pair of indices (α,μ), symmetric in the remaining indices (ν) and satisfies $R_{[\alpha\mu\nu]\nu_1\ldots\nu_{2k}}$. By successive tracings over ν indices one can define a tower of g-Ricci tensors, $R_{\alpha\mu\nu\nu_1\ldots\nu_{2k}}$, with even ranks ranging from $(s+1)$ to 2.[5]
For $s>2$ one can introduce a generalisation of the Weyl tensor, as the completely traceless part of the Riemann tensor. The g-Weyl tensor is invariant on the g-Weyl transformations

$$\delta_\omega \varphi_{\mu_1\ldots\mu_S} = \frac{s(s-1)}{2}\eta_{(\mu_1\mu_2}\omega_{\mu_3\ldots\mu_S)}, \tag{A.4}$$

where the gauge parameter $\omega_{\mu_1\ldots\mu_{S-1}}$ is an arbitrary symmetric rank-$(s-2)$ tensor field. It can be shown that vanishing of the g-Weyl tensor implies that the HS field is a pure g-Weyl gauge.

Appendix A.2. LEoM-tensors
For the purpose of constructing linearised equations of motion which are gauge-invariant under unrestricted g-diffs, one needs to classify divergence-free symmetric tensors of rank s which are linear in φ, which we call LEoM-tensors.[6] Using transverse projector (9) one can classify LEoMtensors, as they must be expressible as linear combinations of the terms which can be written in the form $\Box^r \pi \ldots \pi\varphi$, where for spin s there are s projectors in the product, and there are s contractions which include all indices belonging to HS field φ. In [7] we showed that there are $[s/2]+1$ independent LEoM-tensors (i.e., independent polarisation structures).
The $d=3$ spacetime dimensions presents a special case in which there are in addition parity-odd LEoM-tensors. This is developed in Appendix A.3.
The lowest derivative local parity-even LEoM-tensor we call g-Einstein tensor. Using the formalism of π-projectors it is easy to show that this tensor is unique, and it is given by

$$G_{\mu_S} = \Box^{\left[\frac{s+1}{2}\right]}\pi_{\mu\nu}^{s \bmod 2}\left(\pi_{\mu\nu}^2 - \pi_{\mu\mu}\pi_{\nu\nu}\right)^{\left[\frac{s}{2}\right]}\varphi^{\nu_S}, \tag{A.5}$$

where [ ] denotes integer part and a symmetrisation over $\mu_S \equiv \{\mu_1\ldots\mu_S\}$ indices is assumed. Obviously, the g-Einstein tensor is of order $s+(s \bmod 2)$ in derivatives. We want to express the g-Einstein tensor using the covariant formalism. As is visible from the number of derivatives in (A.5), there is a difference between even and odd spin cases.
For even spin the g-Einstein tensor has the form

$$G_{\mu_S} = R_{\mu_S} + a_1^{(s)}\eta_{\mu\mu}R_{\mu_S-2} + \ldots + a_n^{(s)}\left(\eta_{\mu\mu}\right)^n R, \qquad s=2n. \tag{A.6}$$

The coefficients a_j are uniquely fixed by the condition $\partial\cdot G=0$.
For odd spin one needs an extra derivative aside those present in the Riemann tensor

$$G_{\mu_S} = \partial^\alpha\left(R_{\alpha\mu_S} + a_1^{(s)}\eta_{\mu\mu}R_{\alpha\mu_S-2} + \ldots + a_n^{(s)}\left(\eta_{\mu\mu}\right)^n R_{\alpha\mu}\right), \qquad s=2n+1, \tag{A.7}$$

where for $s\geq 3$ symmetrisation over μ indices is understood. Again, the coefficients a_j are uniquely fixed by the condition $\partial\cdot G=0$.

[5] As $R_{\alpha\mu}$ is antisymmetric, there is no g-Ricci scalar in spin-odd cases.
[6] This follows if the EoM is generated from an action invariant on gauge transformations (2).

Once the Einstein tensor is constructed, a full basis for LEoM-tensors can be defined by the following set of tensors

$$E^{(r)}_{\mu^s} = \Box^r \pi^r_{\mu\mu} G^{|r|}_{\mu^{s-2r}}, \quad 0 \le r \le \left[\frac{s}{2}\right]. \tag{A.8}$$

Every LEoM-tensor can be written as a linear combination of tensors $\Box^n E^{(r)}_{\mu^s}$.

If the linearised equations of motion are obtained from an action invariant on g-Weyl transformations (A.4), then LEoM-tensors are in addition traceless. Again, the π-projectors formalism offers a practical way to study and classify such Weyl-LEoM tensors. In fact, there is only one such tensor for given s in d dimensions, which is given by

$$E^{(W)}_{\mu^s} = \Box^s \pi^s_{\mu\nu}\, {}_2F_1\left(\frac{1-s}{2}, -\frac{s}{2}, \frac{5-d}{2} - s; \frac{\pi_{\mu\mu}\pi_{\nu\nu}}{\pi^2_{\mu\nu}}\right)\varphi^{\nu^s}, \tag{A.9}$$

where ${}_2F_1$ is the hypergeometric function which is a polynomial of the order $[s/2]$ in its last argument. Observe that Weyl-LEoM tensors are of $(2s)$-th order in derivatives, i.e., they are at the top level in the tower (A.8).

Appendix A.3. Specifics in $d = 3$

It is well-known that for spins 1 and 2 there are some specific properties present in three dimensional spacetime. In particular: (1) the Riemann tensor is reducible, i.e., it can be expressed in algebraic fashion using the Ricci tensor $R^{(1)}$ and the metric tensor, (2) the Weyl tensor vanishes, and the Cotton tensor assumes its role, (3) there are parity-odd Chern-Simons terms which may contribute to LEoM. In [10] it was shown that these properties straightforwardly generalise to spin-3, and later in [11] to all integer spins.

For a start, let us analyse the parity-odd sector of LEoM-tensors in 3D using of π-projectors. It is generated by the tensors of the form $\Box^r(\varepsilon\cdot\partial)\pi...\pi\varphi$, where $\varepsilon_{\mu\nu\rho}$ is the Levi-Civita tensor. For spin s there are $(s-1)$ projectors π in the product, and there are s contractions which include all indices belonging to HS field φ. In [7] we showed that there are $\left[\frac{s+1}{2}\right]$ independent tensors of this type, so the total number of independent LEoM-tensors, of both parities, is $(s+1)$.

A particularity of $D=3$ is that one can easily construct the full set of LEoM-tensors in covariant formalism, by using Levi-Civita tensor. One starts by observing that

$$B_{\rho_1\cdots\rho_s} \equiv \varepsilon_{\rho_1\mu_1\nu_1}\cdots\varepsilon_{\rho_s\mu_s\nu_s} R^{\mu_1\nu_1\cdots\mu_s\nu_s} \tag{A.10}$$

is a LEoM-tensor.[7] This tensor is s-derivative, and it is the lowest derivative local LEoM-tensor.

For even spins the B LEoM-tensor must be proportional to the g-Einstein tensor, due to uniqueness of the latter. However for odd spins it has one derivative less than the g-Einstein tensor, and, having the odd number of Levi-Civita's, it is parity-odd. Its interpretation becomes clearer after observing that in spin-1 case it is the standard gauge Chern-Simons LEoM-term $\varepsilon_{\mu\nu\rho}F^{\nu\rho}$.

It is easy to understand the existence of B tensor for odd spins using transverse projectors. For odd s a local parity-odd rank-s tensor of s-th order in derivatives can be constructed as

[7] That $B_{\rho_1\cdots\rho_s}$ is divergence-free follows by applying Bianchi identity.

$$B_{\mu^s} = \varepsilon_{\mu}{}^{\nu\sigma} \partial_{\sigma} \Box^{\frac{s-1}{2}} \left(\pi_{\mu}^{\nu} \pi_{\mu}^{\nu} - \pi_{\mu\mu} \pi^{\nu\nu} \right)^{\frac{s-1}{2}} \varphi_{\nu^s} . \qquad \text{(A.11)}$$

By using the B LEoM-tensor one can construct a tower of LEoM-tensors in the following way

$$B^{(n)}_{\rho_1 \ldots \rho_s} \equiv \partial_{\mu_1} \ldots \partial_{\mu_n} B_{\nu_1 \ldots \nu_n (\rho_{n+1} \ldots \rho_s} \varepsilon_{\rho_1}{}^{\mu_1 \nu_1} \ldots \varepsilon_{\rho_n)}{}^{\mu_n \nu_n} , \; 1 \le n \le s. \qquad \text{(A.12)}$$

The tensor $B(n)$ is $(s+n)$-th order in derivatives. Together with B, we have constructed $(s+1)$ independent LEoM-tensors, which means it is a complete set (both parities).
In spin-2 case there is one parity-odd LEoM-tensor, the (linearised) Cotton tensor generated by gravitational CS Lagrangian term, which has some important properties including Weyl covariance. It can be generalised to all spins, and it sits at the top of the tower (A.12). As shown in [11], g-Cotton tensor is the *unique* tensor satisfying the following properties: (i) rank-s completely symmetric traceless tensor, (ii) divergence-free, (iii) invariant under g-Weyl transformations, (iv) replaces g-Weyl tensor in controlling g-Weyl invariance (vanishing of g-Cotton tensor is necessary and sufficient condition for gauge field to be pure g-Weyl gauge).
The properties (i) and (ii) indicate that g-Cotton tensor is an LEoM-term. g-Cotton tensor can be expressed in different ways, the most useful for our purposes is

$$C_{\mu^s} = \varepsilon_{\mu\nu\sigma} \partial^{\sigma} \Box^{s-1} {}_2F_1\left(\frac{1-s}{2}, -\frac{s}{2}, 1-s; \frac{\pi_{\mu\mu}\pi_{\nu\nu}}{\pi_{\mu\nu}^2} \right) \varphi^{\nu^s} . \qquad \text{(A.13)}$$

References

[1] Vasiliev M A 1990 *Phys. Lett.* B **243** 378
[2] Fronsdal C 1978 *Phys. Rev.* D 18 3624
[3] Francia D and Sagnotti A 2003 *Class. Quant. Grav.* **20** S473
[4] Francia D, Mourad D J and Sagnotti A 2007 *Nucl. Phys.* B **773** 203
[5] Bonora L, Cvitan M, Dominis Prester P, Lima de Souza B and Smolic I 2016 *J. High Energy Phys. JHEP* **05** 072
[6] Sakharov A D 1967 *Dokl. Akad. Nauk Ser. Fiz.* **177** 70
[7] Bonora L, Cvitan M, Dominis Prester P, Giaccari S, Lima de Souza B and Stemberga T 2016 *J. High Energy Phys. JHEP* **12** 084
[8] Bonora L, Cvitan M, Dominis Prester P, Giaccari S and Stemberga T 2017 *Preprint* arXiv:1709.01738
[9] de Wit B and Freedman D Z 1980 *Phys. Rev.* D **21** 358
[10] Damour T and Deser S 1987 *Annales Poincare Phys. Theor.* **47** 277
[11] Henneaux M, Hörtner S and Leonard A 2015 *J. High Energy Phys. JHEP* **01** 073
[12] Fang J and Fronsdal C 1978 *Phys. Rev.* D **18** 3630
[13] Francia D and Sagnotti A 2002 *Phys. Lett.* B **543** 303

Acknowledgments

The research has been supported by the Croatian Science Foundation under the project No. 8946 and by University of Rijeka under the research support No. 13.12.1.4.05.

IOP Conf. Series: Journal of Physics: Conf. Series **1051** (2018) 012009 doi:10.1088/1742-6596/1051/1/012009

Super-Poynting vector and comoving observers in the Einstein-Rosen spacetime

P J Domínguez and H Vargas
University of Guadalajara, Lagos de Moreno, México
E-mail: pjdominguezc@gmail.com

Abstract. In this work we used the concept of super-energy flux in General Relativity and the monad formalism of Ehlers-Zelmanov to find the comoving observers with the Einstein-Rosen cylindrical gravitational waves and obtained their corresponding 3-velocity of propagation with respect to a family of observers locally at rest. Therefore, we found that the aforementioned waves propagate at subluminal speed in vacuum.

1. Introduction

A number of alternative theories to General Relativity have the feature of subluminal propagation of gravitational waves (e.g. [2,6-7]). Nevertheless in General Relativity it is assumed that gravitational radiation propagates at the speed of light in vacuum [6] for instance this is the case of exact solutions like PP-waves. A special case of these solutions are gravitational plane waves, they are the gravitational analogue of plane electromagnetic waves. It is clear that gravitational waves propagate at the speed of light also in the linearized version of General Relativity. Previous works have shown subluminal propagation of gravitational waves in General Relativity in special cases. Bretón, et al [1] demonstrated that two colliding gravitational plane waves slow down their propagation speed as they approach to the singularity that appear after their collision. Now, if one take into consideration cylindrical gravitational waves (Einstein-Rosen waves) [3,8] the situation is somewhat different. In this case one cans find commoving observers with the waves which of course means subluminal propagation in vacuum. In electromagnetic theory, the Poynting vector is used to measure the flux of energy-momentum density. The corresponding gravitational analog is the super-Poynting vector [4] although the latter is a measure of the flux of a rather different physical quantity, the super-energy. If the super-Poynting vector vanishes in any reference frame, then this frame is at rest or comoving with respect to the gravitational field. Using the tools of the monad method [5,9] we determined the propagation speed of the comoving observers with respect to a family of observers at rest with respect to the symmetry axis in the Einstein-Rosen spacetime.

2. Einstein-Rosen spacetime

The Einstein-Rosen solution found in 1937, describes the propagation in vacuum of cylindrically symmetric monochromatic gravitational waves or pulses, they approach to the axis of symmetry, and then move away of it [3,8]. This solution in cylindrical coordinates is

$$ds^2 = e^{2(K-U)}(dt^2 - d\rho^2) - e^{2U}dz^2 - e^{-2U}\rho^2 d\varphi^2 , \tag{1}$$

where K and U are functions of ρ and t alone, describes the propagation of cylindrical gravitational waves in vacuum. This metric satisfies the vacuum field equations

$$U'' + \rho^{-1}U' - \ddot{U} = 0 , \tag{2}$$

$$K' = \rho(U'^2 + \dot{U}^2), \quad \dot{K} = 2\rho U'\dot{U}. \tag{3}$$

The function U can be in principle any solution to the cylindrical wave equation (2) and K can be obtained by simple integration using equations (3). Two special cases are of physical importance; the first one is that of a monochromatic wave propagating in vacuum, the metric functions are

$$U_m = AJ_0(\omega\rho)\cos(\omega t) + AY_0(\omega\rho)\sin(\omega t) \tag{4}$$

$$\begin{aligned} K_m = \frac{1}{2}A^2\omega\rho\{ & J_0(\omega\rho)J_0'(\omega\rho) + Y_0(\omega\rho)Y_0'(\omega\rho) + \\ & + \omega\rho[(J_0(\omega\rho))^2 + (J_0'(\omega\rho))^2 + (Y_0(\omega\rho))^2 + (Y_0'(\omega\rho))^2] + \\ & + [J_0(\omega\rho)J_0'(\omega\rho) - Y_0(\omega\rho)Y_0'(\omega\rho)]\cos(2\omega t) + \\ & + [J_0(\omega\rho)Y_0'(\omega\rho) + Y_0(\omega\rho)J_0'(\omega\rho)]\sin(2\omega t)\} - \frac{2}{\pi}A^2\omega t, \end{aligned} \tag{5}$$

where $J_0(\omega\rho)$ and $Y_0(\omega\rho)$ are Bessel functions of first and second kinds respectively, A and ω are the wave amplitude and frequency. The second case is known as the Weber-Wheeler-Bonnor pulse, the metric functions are

$$\begin{aligned} U_p &= 2C\int_0^\infty e^{-a\omega} J_0(\omega\rho)\cos(\omega t)d\omega, \\ &= \frac{C}{[(a-it)^2+\rho^2]^{1/2}} + \frac{C}{[(a+it)^2+\rho^2]^{1/2}}, \end{aligned} \tag{6}$$

$$\begin{aligned} K_p = \frac{1}{2}C^2\{ & a^{-2} - \rho^2[(a-it)^2+\rho^2]^{-2} - \rho^2[(a+it)^2+\rho^2]^{-2} \\ & - a^{-2}(t^2+a^2-\rho^2)[t^4+2t^2(a^2-\rho^2)+(a^2+\rho^2)]^{-\frac{1}{2}}\}. \end{aligned} \tag{7}$$

3. Gravito-electromagnetism and invariants of the Weyl tensor

Now we introduce the well-known electric and magnetic Weyl tensors

$$X_{\alpha\beta} = C_{\mu\alpha\nu\beta}\tau^\mu\tau^\nu, \qquad Y_{\alpha\beta} = -\overset{*}{C}_{\mu\alpha\nu\beta}\tau^\mu\tau^\nu, \tag{8}$$

where C is the Weyl conformal curvature tensor, the star over indices denotes dual conjugation and τ is a unit time-like 4-covector. There are only four independent algebraic invariants of the Weyl tensor and they can be expressed through the fields X and Y

$$\begin{aligned} I_1 &= C_{\alpha\beta\gamma\delta}C^{\alpha\beta\gamma\delta} = 8(X_{\beta\delta}X^{\beta\delta} - Y_{\beta\delta}Y^{\beta\delta}), \\ I_2 &= \overset{*}{C}_{\alpha\beta\gamma\delta}C^{\alpha\beta\gamma\delta} = -16X_{\beta\delta}Y^{\beta\delta}, \\ I_3 &= C_{\alpha\beta\gamma\delta}C^{\alpha\beta\delta\eta}C_{\delta\eta}{}^{\gamma\delta} = 16X^\alpha_\beta(X^\beta_\gamma X^\gamma_\alpha - 3Y^\beta_\gamma Y^\gamma_\alpha), \\ I_4 &= \overset{*}{C}_{\alpha\beta\gamma\delta}C^{\alpha\beta\delta\eta}C_{\delta\eta}{}^{\gamma\delta} = 16Y^\alpha_\beta(Y^\beta_\gamma Y^\gamma_\alpha - 3X^\beta_\gamma X^\gamma_\alpha). \end{aligned} \tag{9}$$

Since the quadratic constructions exclusively formed by X o Y are positive definite, it is clear that the sign of I_1 determines the electric type (plus) or magnetic (minus) of the field, whereas $I_1 = 0$ corresponds to the null type. The field is purely electric if also $I_2 = 0, I_4 = 0$, on the other hand the field is purely magnetic if also $I_2 = 0$, $I_3 = 0$. In this work we have softened the purity condition described by Mitskievich in [5], requiring only the vanishing of two invariants rather than three. We propose to call ultrapure fields those for whom a third invariant vanishes ($I_2 = 0, I_3 = 0, I_4 = 0$). When we conclude that the field is purely electric or purely magnetic, means that the alternative field can be transformed away by a choice of the reference frame.

The Einstein-Rosen solution has only two non-vanishing invariants:

$$I_1 = \frac{16}{\rho^2} e^{4(U-K)} (U'^2 - 6\rho^3 U'U''\dot{U}^2 + 3\rho U'\dot{U}^2 + \rho U'U'' - 2\rho^3 U''U'^3 + 3\rho^2 U''\dot{U}^2$$
$$+ 3\rho^2 U''U'^2 - 9\rho^2 \dot{U}^2 U'^2 + 3\rho^4 U'^2 \dot{U}^4 - 3\rho^4 \dot{U}^2 U'^4 - 3\rho^3 U'\dot{U}^4 + 6\rho^3 \dot{U}^2 U'^3 \qquad ((10)$$
$$+ 2\rho^3 \dot{U}'\dot{U}^3 + \rho^2 U''^2 + 2\rho^2 U'^4 + \rho^4 U'^6 - 3\rho^3 U'^5 + 3\rho^2 \dot{U}^4 - \rho^2 \dot{U}'^2 - \rho^4 \dot{U}^6$$
$$- 6\rho^2 \dot{U}U'\dot{U}' + 6\rho^3 \dot{U}\dot{U}'U'^2).$$

$$I_3 = \frac{48}{\rho^2} e^{6(U-K)} [(\rho U'^2 - U' - \rho \dot{U}^2)(3\rho^3 U'^2 \dot{U}^4 - \rho^3 \dot{U}^6 - 3\rho^2 U'\dot{U}^4$$
$$+ 2\rho \dot{U}^4 + 2\rho^2 \dot{U}'\dot{U}^3 - 3\rho^3 \dot{U}^2 U'^4 + 6\rho^2 \dot{U}^2 U'^3 - 6\rho^2 U'U''\dot{U}^2 - 7\rho \dot{U}^2 U'^2 \qquad (11)$$
$$+ 3\rho U''\dot{U}^2 + U'\dot{U}^2 + 6\rho^2 \dot{U}\dot{U}'U'^2 - 6\rho \dot{U}U'\dot{U}' + \rho^3 U'^6 - 2\rho^2 U''U'^3$$
$$- 3\rho^2 U'^5 + \rho U''^2 - \rho \dot{U}'^2 + \rho U'^4 + 3\rho U''U'^2 + U'U'' + 2U'^3)].$$

The sign in them depends only on the derivatives of $U(t,\rho)$. We analyze the sign for a specific $U(t,\rho)$ more in detail below.

4. Comoving observers

Let us now introduce the monad field

$$\tau = \tau_{(0)}\theta^{(0)} + \tau_{(1)}\theta^{(1)}, \qquad (12)$$

where $\theta^{(0)} = e^{(K-U)}dt$ and $\theta^{(1)} = e^{(K-U)}d\rho$ are elements of the tetrad basis in (1), coordinates will be numbered as $x^0 = t, x^1 = \rho, x^2 = \varphi, x^3 = z$. The co-vector τ can be used as the 4-velocity of the observer comoving with the wave when one of the electric or magnetic fields in this spacetime can be transformed away. The electric Weyl tensor has the components

$$\mathrm{X}_{(0)(0)} = Q(t,\rho)[\tau^{(1)}]^2, \quad \mathrm{X}_{(1)(1)} = Q(t,\rho)[\tau^{(0)}]^2, \quad \mathrm{X}_{(0)(1)} = -Q(t,\rho)\tau^{(0)}\tau^{(1)},$$
$$\mathrm{X}_{(2)(2)} = R(t,\rho)[\tau^{(0)2} + \tau^{(1)2}] + S(t,\rho)\tau^{(0)}\tau^{(1)} + Q(t,\rho)[\tau^{(1)}]^2, \qquad (13)$$
$$\mathrm{X}_{(3)(3)} = -R(t,\rho)[\tau^{(0)2} + \tau^{(1)2}] - S(t,\rho)\tau^{(0)}\tau^{(1)} - Q(t,\rho)[\tau^{(0)}]^2,$$

where

$$Q(t,\rho) := \left(U'^2 - \rho^{-1}U' - \dot{U}^2\right)e^{2(U-K)},$$
$$R(t,\rho) := \left(\rho\, U'^3 - U'' + 3\rho\, \dot{U}^2 U' - 2U'^2 - \dot{U}^2\right)e^{2(U-K)},$$
$$S(t,\rho) := \left(2\rho \dot{U}^3 - 2\dot{U}' - 6\dot{U}U' + 6\rho \dot{U}U'^2\right)e^{2(U-K)}.$$

The only non-vanishing component of the magnetic Weyl tensor

$$Y_{(2)(3)} = M(\rho,t)[(\tau^{(0)})^2 + (\tau^{(1)})^2] + N(\rho,t)\tau^{(0)}\tau^{(1)}, \qquad (14)$$

where

$$M(\rho,t) := \left(3\dot{U}U' - 3\rho \dot{U}U'^2 - \rho \dot{U}^3 + \dot{U}'\right)e^{2(U-K)},$$
$$N(\rho,t) := \left(3\dot{U}^2 + 3U'^2 + 2U'' + \rho^{-1}U' - 6\rho U'\dot{U}^2 - 2\rho U'^3\right)e^{2(U-K)}.$$

It can be found the observer for the vanishing of Y but not for all components of X. Furthermore, in order to define a comoving observer, we have to introduce the super-Poynting vector

$$\mathrm{P}_\alpha = b_{\alpha\gamma} T^{\gamma\beta\lambda\mu} \tau_\beta \tau_\lambda \tau_\mu \equiv 2 b^{\grave{u}}_\alpha E^{\sigma\gamma}_{\grave{u}\lambda} X^{\lambda\nu} Y_{\gamma\nu} \tau_\sigma, \qquad (15)$$

where $T^{\gamma\beta\lambda\mu}$ is the tensor of Bel-Robinson, $b_{\alpha\gamma}$ projects onto the 3-subspace orthogonal to τ_μ and $E^{\sigma\gamma}_{\grave{u}\lambda}$ is the axial tensor of Levi-Civita. This vector represents the flux of super-energy of the gravitational field and we can define a comoving observer as that for whom the super-Poynting vector vanishes. Using the monad field (12), the super Poynting vector can be reduced for the Einstein-Rosen spacetime to

IOP Conf. Series: Journal of Physics: Conf. Series **1051** (2018) 012009 doi:10.1088/1742-6596/1051/1/012009

$$\mathrm{P}_{(\alpha)} = 2b_{(\alpha)(\sigma)}[X_{(2)(2)} - X_{(3)(3)}]Y^{(3)}_{(2)} E^{(\dot{u})(3)(\sigma)(2)} \tau_{\dot{u}} \,. \tag{16}$$

Then two comoving observers arise, the first one can be obtained of the vanishing of $Y^{(3)}_{(2)}$ and the second one can be obtained of the vanishing of the expression $\mathrm{X}_{(2)(2)}$-$\mathrm{X}_{(3)(3)}$. Using $\tau^{(0)} = \cosh(\psi)$ and $\tau^{(1)} = \sinh(\psi)$ as components of the monad field (12) with ψ being a function to be determined, the two comoving observers satisfy the relations

$$\frac{\tanh\psi_{II}}{1+\tanh^2\psi_{II}} = -\frac{M(\rho,t)}{N(\rho,t)}, \quad \frac{\tanh\psi_{I}}{1+\tanh^2\psi_{I}} = -\frac{2R(\rho,t)+Q(\rho,t)}{2S(\rho,t)} \,. \tag{17}$$

5. Propagation speed

Now we are in position to obtain the 3-velocity of propagation of the comoving observers with respect to the frame locally at rest $\theta^{(0)}$

$$\begin{aligned} v_{(\mu)} &= \frac{1}{\tau\cdot\tilde{\tau}} \tilde{b}_{(\mu)(\nu)} \tau^{(\nu)} = \frac{1}{\tau\cdot\tilde{\tau}} \left[g_{(\mu)(\nu)} - \tilde{\tau}_{(\mu)} \tilde{\tau}_{(\nu)} \right] \tau^{(\nu)}, \\ &= \frac{1}{\tau\cdot\tilde{\tau}} \left[\tau_{(\mu)} - (\tau\cdot\tilde{\tau}) \tilde{\tau}_{(\mu)} \right] = -\tanh(\psi) \delta^1_\mu . \end{aligned}$$

And since $v = \tanh(\psi)$, we can write

$$\begin{aligned} v_I &= \tanh\left[\frac{1}{2} \operatorname{arctanh}\left| -\frac{2R(\rho,t)+Q(\rho,t)}{S(\rho,t)} \right| \right], \\ v_{II} &= \tanh\left[\frac{1}{2} \operatorname{arctanh}\left| -2\frac{M(t,\rho)}{N(t,\rho)} \right| \right]. \end{aligned} \tag{18}$$

The invariant I_1 of the Weyl tensor for the monochromatic wave takes positive values and since $I_2 = 0$ this field is of purely electric type. The 3-velocities with respect to the symmetry axis of both comoving observers for whom the super-Poynting vector vanishes are shown in Figure 1. The observer with velocity v_I propagates with the impure field of the wave whereas the observer with velocity v_{II} propagates with the purely electric part.

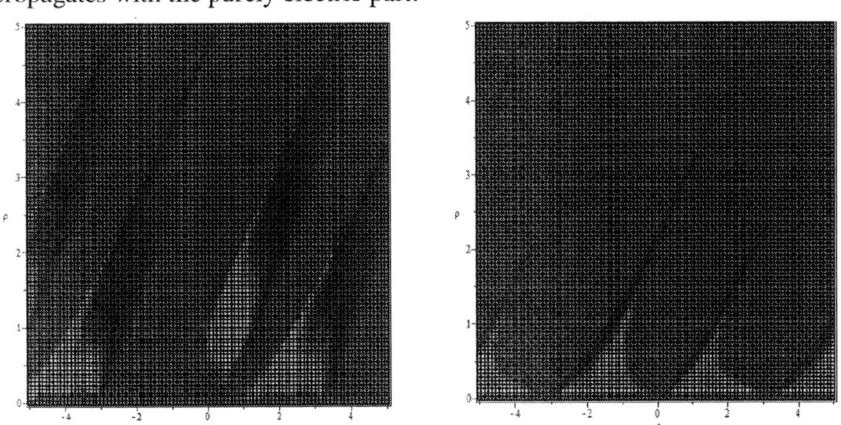

Figure 1. Velocity magnitudes v_I (left) and v_{II} (right) of the Einstein-Rosen monochromatic wave. The wave takes velocities near the speed of light on the magenta region and diminishes its speed near to zero as it approaches the orange region. Constants A and ω were set equal to 1.

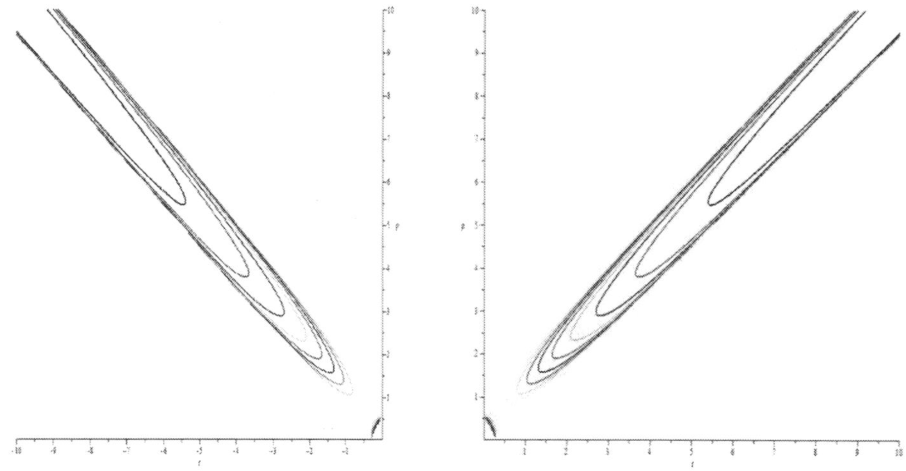

Figure 2. Curves of constant velocities of the Weber-Wheeler-Bonnor pulse for $v_I(\rho,t)$. Negative values are on the left, black represents $v_I = -0.8$, brown $v_I = -0.7$, blue $v_I = -0.6$, green $v_I = -0.5$, orange $v_I = -0.4$, purple $v_I = -0.3$, magenta $v_I = -0.2$, pink $v_I = -0.1$, and yellow $v_I = 0$. Positive values are on the right, black represents $v_I = 0.8$, brown $v_I = 0.7$, blue $v_I = 0.6$, green $v_I = 0.5$, orange $v_I = 0.4$, purple $v_I = 0.3$, magenta $v_I = 0.2$, pink $v_I = 0.1$, and yellow $v_I = 0$. Constants a and C were set equal to 1.

The same analysis is applied to the Weber-Wheeler-Bonnor pulse. The invariant I_1 of the Weyl tensor takes in this case positive, negative or zero values and since $I_2 = 0, I_4 = 0$, this field is of purely electric, purely magnetic or purely null type. The 3-velocities with respect to the symmetry axis of both comoving observers for whom the super-Poynting vector vanishes are shown in Figure 2 and 3. The observer with velocity v_I propagates with the impure field of the pulse whereas the observer with velocity v_{II} propagates with the purely electric part. The pulse slows down as it approaches the symmetry axis and then is reflected. Both observers are mutually complementary and fill all spacetime.

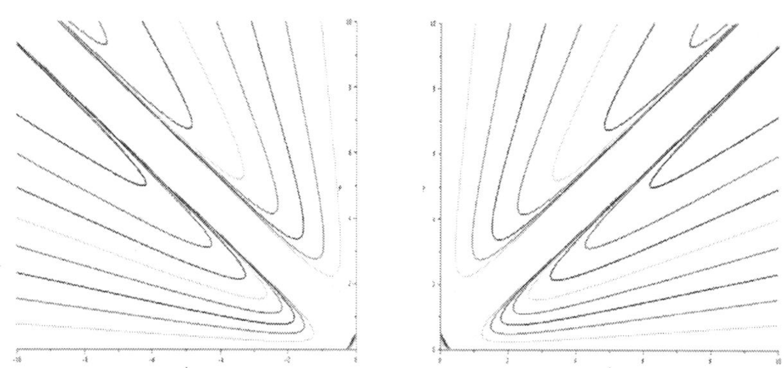

Figure 3. Curves of constant velocities of the Weber-Wheeler-Bonnor pulse for $v_{II}(\rho,t)$. Negative values are on the left, black represents $v_{II} = -0.8$, brown $v_{II} = -0.7$, blue $v_{II} = -0.6$, green $v_{II} = -0.5$, orange $v_{II} = -0.4$, purple $v_{II} = -0.3$, magenta $v_{II} = -0.2$, pink $v_{II} = -0.1$, and yellow $v_{II} = 0$. Positive values are on the right, black represents $v_{II} = 0.8$, brown $v_{II} = 0.7$, blue $v_{II} = 0.6$, green $v_{II} = 0.5$, orange $v_{II} = 0.4$, purple $v_{II} = 0.3$, magenta $v_{II} = 0.2$, pink $v_{II} = 0.1$, and yellow $v_{II} = 0$. Constants a and C were set equal to 1.

6. Concluding remarks

We have shown that it is possible to find the comoving frame for the Einstein-Rosen gravitational waves and therefore they propagate at sub-luminal speed in vacuum. There is a hole family of independent but complementary observers that fill all E-R spacetime for whom the super-Poynting vector vanishes, they propagate with different speeds, a part of them is comoving just with the gravito-electric field since the gravito-magnetic field vanishes completely in their frames of reference.

References

[1] Bretón N *et al* 1993 The Bel-Robinson tensor for the collision of gravitational plane waves *General Relativity and Gravitation* **25** 3
[2] Caves C M 1980 Gravitational Radiation and the Ultimate Speed in Rosens Bimetric Theory of Gravity *Annals of Physics* **125**
[3] Einstein A and Rosen N J 1937 On gravitational waves *J. Franklin Inst.* **223** 43
[4] García-Parrado Gomez-Lobo A 2007 Dynamical laws of superenergy in general relativity *Class. Quantum Grav.* **25** 015006
[5] Mitskievich N V 2006 *Relativistic Physics in Arbitrary Reference Frames* (Nova Science Publishers)
[6] Novello M *et al* 1999 The velocity of gravitational waves *Physics Letters* A **254** pp 245-250
[7] Rosen N 1974 A theory of gravitation *Annals of Physics* **84** pp 455-473
[8] Weber J and Wheeler J 1957 Reality of the Cylindrical Gravitational Waves of Einstein and Rosen *Reviews of Modern Physics* **29**
[9] Zelmanov A L 1976 Orthometric form of monad formalism and its relations to chronometric invariants and kinemetric invariants *Doklady Akad. Nauk SSSR* **227** 78

Acknowledgements

We would like to thank Prof. Vladimir Gladyshev and Natalya Liovina for their hospitality during the PIRT-2017 Conference in Moscow. We also greatly appreciate the financial support granted by the University of Guadalajara and CONACyT in Mexico.

IOP Conf. Series: Journal of Physics: Conf. Series **1051**(2018) 012010 doi:10.1088/1742-6596/1051/1/01201(

Gravitational waves from primordial black holes collisions in binary systems

Yu N Eroshenko

Institute for Nuclear Research of the Russian Academy of Sciences, Moscow, Russia
E-mail: eroshenko@ms2.inr.ac.ru

Abstract. It was shown by (Nakamura et al. 1997), (Ioka et al. 1998), and (Sasaki et al. 2016) that primordial black holes (PBHs) binaries can form effectively at the cosmological stage of radiation dominance, and the merge of the PBHs in pairs can explain the gravitational wave burst GW150914. In this paper, the model is re-examined by considering the effect of inflationary dark matter density perturbations which produce additional tidal forces. As a result, the merge rate of PBHs binaries and the corresponding rate of the gravitational bursts can be suppressed by the factor $\approx 1.5-2$ in comparison with previous calculations. This rate matches the LIGO data if the PBHs constitute the $f \approx 5\times10^{-4} - 5\times10^{-3}$ fraction of dark matter.

1. Introduction

Although gravitational waves were predicted by Albert Einstein in 1916, to date their existence was shown only indirectly through the orbital changes of binary pulsar PSR B1913+16. The first direct detection was done September 14, 2015 by two laser interferometers LIGO [1]. The form of the GW150914 signal corresponds to the general relativity prediction for the merge of two black holes with masses $36M_\odot$ and $29M_\odot$, and the statistical significance of the registration is 5.1σ. This result demonstrates the existence black holes in the binary systems. Such black holes can born at the massive supernova explosions as a result of the standard stellar evolution [2-3]. In particular, the low-metallicity environment boosts the massive double black holes formation [4-6]. It is possible, however, that the black holes were born in the collapse of massive pre-galactic stars, or they were formed by merge of smaller black holes in the dense star clusters.

The merge of primordial black holes (PBH) in a pairs provides the alternative explanation of GW150914 [7-9]. The possibility of PBH formation was predicted in general by [10-11], and there are many particular scenarios of their formation [12-17]. For concreteness we consider in this paper the formation of PBHs from curvature perturbation as in [12].

The PBH pair can form occasionally if two PBH born sufficiently close to each other [7]. The motion of the binary PBH is influenced by neighbouring third PBH, therefore the PBHs do not move exactly towards each other but experience tangential displacement under the influence of the tidal gravitational forces from the third PBH. The [7-8] discussed the MACHOs with masses $\approx 0.1-1M_\odot$, while [9] explained the signal GW150914 as the merge of PBHs with masses $\approx 30M_\odot$. In addition, the possibility of the PBHs clusters

formation in the early universe was proposed by [18], and the PBHs collisions in such clusters can produce the bursts of gravitational waves [19-20].
In this study we consider the additional source of tidal forces – the tidal forces from the usual inflationary density perturbations in dark matter. This effect was not considered previously in the calculations of the PBHs pair evolution. These additional tidal forces alter the distribution of PBHs pairs over their orbital parameters, and finally suppress the rate of gravitational bursts.
The close PBHs pairs form at the cosmological stage of the radiation dominance (RD), but they can also form later at the stage of matter domination. A halo of dark matter is accumulated around the PBHs by the secondary accretion mechanism [21]. The PBHs in the halo experience the dynamic friction, lose angular momentum and move toward the center, there they form the gravitationally bounded binary, even if they were not associated initially [22]. The further fate of this pair is uncertain, because the effectiveness of dynamic friction in this situation was not investigated.
The paper is organised as follows. In the Section 2 we mainly define the notations and repeat some calculations of [7-8] by other method. Section 2 is devoted to the new effect – the tidal forces from adiabatic density perturbations with CMB-normalised spectrum. Then in the Section 4 we calculate the resultant rate of the gravitational bursts. In the Section 5 we consider critically the possibility of the PBHs merge inside dark matter minihalos. And finally we present some conclusions in the Section 6.

2. The evolution of binaries at the stage of radiation domination

Consider the evolution of the binary PBHs at the RD-stage, $t < t_{eq}$, where t_{eq} is the moment of matter-radiation equality. Let the scale factor of the universe $a(t)$ is normalized as $a(t_{eq}) = 1$, then the density of radiation $\rho_r = \rho_{eq} / a^4$. Let us assume that the PBHs constitute the $f \le 1$ fraction of dark matter, $\Omega_{BH} = f\Omega_m$. In the case $f < 1$ the rest of DM can be in some other form, for example in the form of weakly interacting massive particles (WIMPs). Denote by x the comoving space between the components of PBHs pair, while the average distance between PBHs is

$$\bar{x} = n^{-1/3} = \left(\frac{M_{BH}}{f\rho_{eq}}\right)^{1/3}. \qquad (1)$$

Let the binary form at $t < t_{eq}$. The condition for this is [7, 9]

$$\frac{M_{BH}}{x^3 a_m^3} \approx \rho_r, \qquad (2)$$

where

$$a_m = \frac{1}{f}\left(\frac{x}{\bar{x}}\right)^3. \qquad (3)$$

This occurs at the RD-stage if $a_m < 1$ that can be written as $x < x^*$, where

$$x^* = \left(\frac{M_{BH}}{\rho_{eq}}\right)^{1/3}. \qquad (4)$$

The x^* would be the mean distance between the PBH in the case $f = 1$.
According to [7,9] the semi-major axis of orbit is fixed at the time given by (3), and the minor axis is influenced by tidal forces from the third PBH, located at the comoving distance $y < x$. Minor axis is calculated as the product of the tidal force and the time of free fall. As a result, [9] obtained for the major

and minor semi-axes, respectively,

$$A = \alpha \frac{1}{f} \frac{x^4}{\bar{x}^3}, \qquad B = \beta A \left(\frac{x}{y} \right)^3, \tag{5}$$

and the eccentricity of the orbit

$$e = \sqrt{1 - \frac{B^2}{A^2}} = \sqrt{1 - \beta^2 \frac{x^6}{y^6}}. \tag{6}$$

In the paper [9] the expressions (5) and (6) were obtained in the case $\alpha = \beta = 1$. The [8] introduced the additional factors α and β, and obtained them from the numerical solution of the evolution equations. In the rest of this section we solve the similar equations and in the next section we consider additional effects.

The distance of each PBH from the center of masses of the pair is $r/2$. We consider the scales under the cosmological horizon $r = ct$, so we can use the Newtonian dynamics, taking into account the contribution of uniformly distributed radiation by the substitution $\rho \to \rho + 3pc^2$ [23]. The r evolves according to the equation

$$\frac{d^2(r/2)}{dt^2} = -\frac{GM_{BH}}{r^2} - \frac{8\pi G\rho_r (r/2)}{3}, \tag{7}$$

where the last term describes the gravitational effect of uniformly distributed radiation within the sphere with radius $r/2$. Following the approach of [24], we do the following parametrization

$$r = abx. \tag{8}$$

The evolution of b shows the factor of displacement from the comoving distance in the homogeneous universe. If one takes into account that at the radiation-dominated stage $a \approx t^{1/2} / t_{eq}^{1/2}$ the Eq. (7) can be rewritten as

$$a\frac{d^2b}{da^2} + \frac{db}{da} = -\frac{3}{4\pi b^2} \frac{\bar{x}^3}{x^3}. \tag{9}$$

Analogous equation was obtained in [9] by other method. The combination of the quantities in the right-hand side

$$\delta_i = \frac{2M_{BH}}{(4\pi/3)x^3 \rho_{eq}} = \frac{3}{2\pi} \frac{\bar{x}^3}{x^3} \tag{10}$$

is the density perturbation produced by two PBH, so one can conclude that Eq. (9) is equivalent to the equation of [24] (taken in the limit $a = 1$), which describes the evolution of spherically symmetric entropy (isocurvature) perturbations. Initial time t_i is taken when the PBH mass is negligible in comparison with the radiation mass inside the sphere, and the initial conditions at the time t_i have the form $b = 1$, $db/da = 0$. The distance between the PBHs stops growing at the time t_s, when $dr/dt = 0$, which corresponds to the condition $db/da = -b/a$ [24]. It is easy to see that $\alpha = b(a(t_s))$ in the Eq. (5).

Let us consider the evolution of small semi-axis under the influence of tidal forces from the 3rd PBH, located at the comoving distance y from the pair. We consider the transverse displacement δr as a small perturbation to the radial orbit, and we introduce the parametrization

$$\delta r = a\xi. \tag{11}$$

The tidal force has the form

$$F_t = \frac{2GM_{BH}^2}{(ay)^3} r, \tag{12}$$

and the equation $d^2\delta r / dt^2 = F_t$ can be written as

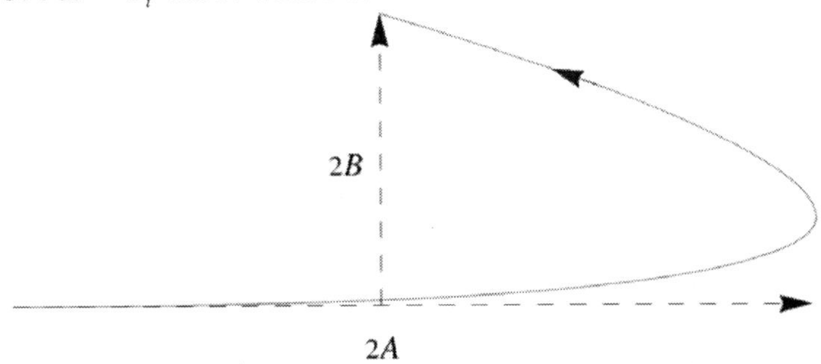

Figure 1. Example of the orbit evolution under the influence of tidal forces from the numerical solution of Eqs. (9) and (13). The vertical and horizontal dashed lines show the doubled major and minor semi-axis of the orbit, respectively.

$$a\frac{d^2\xi}{da^2} + \frac{d\xi}{da} - \frac{\xi}{a} = -\frac{3}{4\pi}\frac{x^{*3}xb}{y^3}. \tag{13}$$

The initial conditions at the time t_i are $\xi = d\xi / da = 0$.

We solve the system of equations (9), (13) numerically until the binary expands up to the maximum distance (this distance is the A major semi-axis) and then approaches twice in radius. The value of $a\xi$ at this moment gives the size of the minor axis B. We assume for simplicity that all PBH have the same masses $M_{BH} = 30M_{\odot}$. An example of the orbit is shown at Figure 1. Solving the equation (9) and (13) for any x (different values of y are taken into account in (13) by the scaling of ξ) and comparing with (5), we obtain the following correction factors in (5)

$$\alpha \approx 0.64, \qquad \beta \approx 2.8. \tag{14}$$

In our calculations these factors are independent of x with high accuracy, however [8] obtained $\alpha \approx 0.4$, $\beta \approx 0.8$ and weak dependence on the fraction x / y. We don't consider here the angle dependence (direction to the 3rd PBH). This dependence was considered in [8].

3. Tidal forces from inflationary perturbations

Neighbourhood PBHs are not the only sources of disturbing. Tidal forces can be produced also by the dark matter density perturbations, and the main contribution comes from the characteristic scale x^*, where dark matter mass is equal to the PBHs mass $M \approx M_{BH}$. Really, the scale of the PBHs binary contains the mass of dark matter

$$M \approx \rho_{eq} x^3 = M_{BH} \frac{x^3}{x^{*3}}. \tag{15}$$

If $M < M_{BH}$ the PBHs pair forms at the RD-stage, and from (15) it follows $x < x^*$. Otherwise, the PBHs do not define the dynamics of the system, but they follow the motion of dark matter. At smaller scales $x < x^*$ the dark matter is highly disturbed by the PBHs and is involved into the motion. The tide from the smaller scales is unclear. In contrast, the larger scales produce regular tidal force, so we consider the effect of the scales $x \geq x^*$ as minimum warranted contribution. Let the density perturbation at the mass scale M

at the moment t_{eq} is δ_{eq}. R.m.s. value of perturbation is (see e.g. [25] and references therein)

$$\sigma_{eq}(M) \approx 8.2\times 10^{3.7(n_s-1)-3}\left(\frac{M}{M_{\odot}}\right)^{\frac{1-n_s}{6}}\left[1-0.06\log\left(\frac{M}{M_{\odot}}\right)\right]^{\frac{3}{2}}, \tag{16}$$

where $n_s = 0.9608 \pm 0.0054$ according to the Planck data. In particular, $\sigma_{eq}(30M_{\odot}) \approx 5.2\times 10^{-3}$. At the RD-stage the adiabatic perturbation δ evolves only logarithmically. We can write the evolution of the dark matter perturbation as follows

$$\delta = \delta_{eq}F(a), \tag{17}$$

where the function [25]

$$F(a) = \frac{\ln\left[g(a)/\sqrt{3}\right]+C-\frac{1}{2}}{\ln\left[g(1)/\sqrt{3}\right]+C-\frac{1}{2}}, \tag{18}$$

$C \approx 0.577$ is the Euler constant, and

$$g(a) = \frac{\pi}{2^{2/3}}\left(\frac{3}{2\pi}\right)^{1/6}\frac{ac}{M^{1/3}G^{1/2}\rho_{eq}^{1/6}}. \tag{19}$$

Then the dark matter excess can be estimated as $\Delta M \approx M\delta$, and the tidal force acting on the pair is

$$F_{\sigma} \approx \frac{GM_{BH}\Delta M(axb)}{(ax^{*})^{3}}, \tag{20}$$

where b is defined by (8). The comparison

$$\frac{F_{\sigma}}{F_t} = \frac{\delta_{eq}}{f}F(a)\frac{y^3}{\bar{x}^3} \tag{21}$$

shows that for the sufficiently isolated PBHs pairs with $y \approx \bar{x}$ the contribution from inflation perturbations can prevail over the contribution from the 3rd PBH. The additional source of the tidal forces changes the value of maximum eccentricity e, calculated in [7,9].
We substitute the (20) into the equation $d^2\delta r/dt^2 = F_{\sigma}$ and solve this equation (instead of (13)) numerically for different x. If we parametrise

$$B = kA\frac{x^3}{x^{*3}}\nu, \tag{22}$$

where $\nu = \delta_{eq}/\sigma_{eq}$, we obtain from the numerical solution that the factor k is approximately constant,

$$k \approx 0.005. \tag{23}$$

In the next Section we use these results for the calculation of the rate of gravitational bursts.

4. Rate of gravitational bursts

Let us first neglect the effect of tidal forces from dark matter inflationary perturbations, but take into account the above correction factors (14). The probability distribution for A and e, obtained in [7] and [9], can be rewritten in this case as

$$dP = \frac{3}{2}\frac{f^{3/2}\beta}{\alpha^{3/2}\bar{x}^{3/2}}A^{1/2}e(1-e^2)^{-3/2}dAde. \tag{24}$$

This distribution is obtained by calculating the Jacobian $\partial(x,y)/\partial(A,e)$ under the assumption of the uniform distribution of x and $y<x$ in the ranges from 0 to $\bar{x}$. Lifetime of PBHs pair due to the radiation of gravitational waves is [9]

$$t_c = \frac{5c^5}{512G^3M_{BH}^3}A^4(1-e^2)^{7/2}. \tag{25}$$

The probability that the time (25) is less than t can be obtained from (24) by integrating over the corresponding curved region in the parameter space of A and e, selected by the condition $t_c<t$:

$$P(<t) = \frac{1}{\beta}\left[\frac{37}{29}\left(\frac{t}{t_{max}}\right)^{3/37} - \frac{8}{29}\left(\frac{t}{t_{max}}\right)^{3/8}\right], \tag{26}$$

where

$$t_{max} = \frac{5c^5}{512G^3M_{BH}^3}\frac{\alpha^4}{\beta^{16/3}}\frac{x^{*4}}{f^{16/3}}. \tag{27}$$

In the case $\alpha=\beta=1$ the distributions (24) and (26) coincide with those obtained in [7].
The rate of gravitational wave bursts near the current moment t_0 is

$$R = \frac{\rho_c\Omega_m f}{M_{BH}}\left.\frac{dP(<t)}{dt}\right|_{t=t_0}, \tag{28}$$

where $\rho_c = 9.3\times10^{-30}$ g cm^{-3} is the critical density, $\Omega_m \approx 0.27$. The result of the calculation is shown at Figure 2 in comparison with the result of [9], which is reproduced with $\alpha=\beta=1$. Our rate is about 40% of [9] value.
Now let us take into account the effect of tidal forces from ordinary inflationary perturbations as it was considered in the Section 3. In this case we have the 3-dimensional probability distribution

$$dP = \frac{18x^2y^2}{\bar{x}^6}\frac{1}{\sqrt{2\pi}}e^{-v^2/2}dxdydv, \tag{29}$$

where we consider not only positive $v>0$ but also negative $v<0$ perturbations. Now we have the sum of (5) and (22)

$$B = A\left(\frac{x}{y}\right)^3 + kA\frac{x^3}{x^{*3}}v. \tag{30}$$

IOP Conf. Series: Journal of Physics: Conf. Series **1051** (2018) 012010 doi:10.1088/1742-6596/1051/1/012010

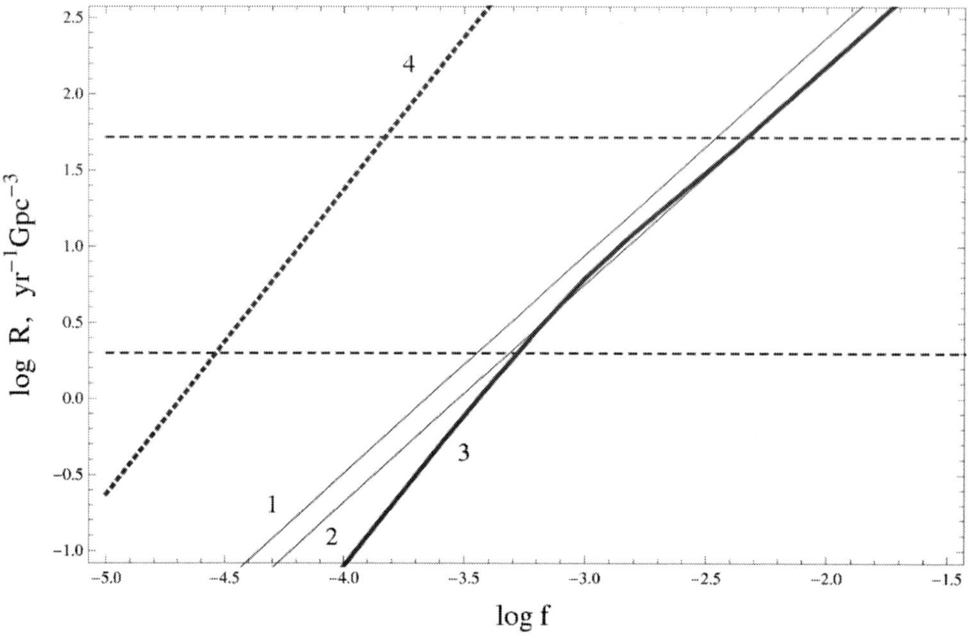

Figure 2. The rate of gravitational waves bursts from the collision of PBHs in pairs, depending on the fraction f of PBHs in dark matter. The upper solid curve 1 shows the result of the work [9]. The middle solid curve 2 was obtained according to (28) with the correction factors (14). The lower bold curve 3 shows the result of the current work. Dotted curve 4 shows the rate of bursts under the assumption of the efficiency of dynamic friction inside minihalos at the last stage of PBHs pair evolution. The horizontal dashed lines limit the area of LIGO observations, obtained in [9].

The tidal force from inflation perturbations prevails if

$$\nu > \nu^* \equiv \frac{\beta x^{*3}}{y^3 k}. \tag{31}$$

The integration of (29) in the 3D space with the condition $t_c < t$ is more complicated in this case. The result of the numerical integration is shown on Figure 2 by the curve 3. One can see that in the region $f \approx 5\times10^{-4} - 5\times10^{-3}$ the tidal forces from dark matter do not change the result significantly, but at smaller f they can suppress the rate of the gravitational bursts by the factor $\approx 1.5 - 2$.

5. Binaries inside dark matter minihalos

Consider the case $f = 1$ and $\delta_i < 1$ in (10). The first of these conditions means that PBHs constitute only small fraction of dark matter and the rest of the dark matter consists of something else, for example, of new elementary particles. The second condition means that we consider sufficiently large region where PBHs create only small perturbation in dark matter. This perturbation at $t > t_{eq}$ evolves into the minihalo of dark matter around the PBHs pair. Similar model was proposed in [22]. Consider the evolution of perturbations at the comoving scale x (the final minihalo contains the mass $M \approx 10^2 M_{BH}$ [25]). After the detachment from the cosmological expansion the perturbed region is compressed twice and virialized at the final size [26]

$$r_v = \frac{x}{2\delta_i}. \tag{32}$$

The mass within this radius is given by (15), so the velocity dispersion

$$v \approx \left(\frac{GM}{r_v}\right)^{1/2} \approx \left(\frac{3GM_{BH}}{\pi x}\right)^{1/2}, \tag{33}$$

and the average density

$$\rho_H \approx \frac{3^4}{4\pi^4}\frac{x^{*9}}{x^9}\rho_{eq}. \tag{34}$$

PBHs experience dynamical friction, lose the orbital angular momenta and approach the minihalos centers. The change of the orbital radius obeys the equation [27]

$$\frac{dr}{dt} = -\frac{4\pi G^2 M_{BH}\rho_H(r)Dr}{v(r)^3}, \tag{35}$$

where numerical constant $D \approx 4.27$. From Eq. (35) we obtain the characteristic time of the fall to the center

$$t_d \approx 5.3 t_{eq}\left(\frac{x}{x^*}\right)^{15/2}. \tag{36}$$

PBHs come to the radius r, within which the mass of dark matter is $M(r) \approx M_{BH}$, and the PBHs sit on the orbit around each other. Dark matter within their orbit and the surrounding material will lead to the evolution of the orbit, but the effectiveness of dynamical friction under these conditions has not been elucidated yet, and it requires further investigation. The lifetime of the pair (25) is many orders of magnitude greater than the age of the universe t_0, therefore, for the further approach and merger the PBHs must continue to lose the orbital angular momenta. Virial velocity in the minihalo is $v \approx 0.36$ km s^{-1}, and the corresponding virial temperature of baryonic gas is $T \approx m_p v^2/(2k_B) \approx 8$ K, where k_B is the Boltzmann constant. So the radiative cooling mechanisms do not work, and the gas does not form stars, which could give additional dynamical friction.

Consider the most optimistic case, when at the radii $< r$ the dynamic friction is as effective as before and is described by (35). Then the significant fraction of the PBHs pairs can merge during the Hubble time and produce gravitational bursts. Distribution of x under the assumption of uniformity [7, 9] has the simple form

$$dP = \frac{3x^2 dx}{\bar{x}^3}. \tag{37}$$

The condition $t_d < t$ selects the region of integration, and

$$P(<t) \approx 64 f\left(\frac{t}{t_0}\right)^{2/5}. \tag{38}$$

Substituting into (28), we obtain the rate of gravitational bursts that is shown at Figure 2 by the point curve. This result was obtained under the assumption of high efficiency of dynamical friction at the last stage of evolution, so it gives the most optimistic scenario, but this model requires further elaboration.

6. Conclusion

In this work we considered the formation and evolution of the PBHs pairs and found the expected rate of LIGO gravitational bursts. Our approach is similar to one of [7-9], but the calculations are made by another method, which takes into account the tidal forces from adiabatic density perturbations of dark matter with CMB-normalised spectrum. These tidal forces alter the probability distribution of PBHs pairs over their orbital parameters. It turned out that the merge rate matches the LIGO signals if the PBHs constitute the fraction $f \approx 5\times10^{-4} - 5\times10^{-3}$ of all dark matter. In the range $f \approx 5\times10^{-4} - 5\times10^{-3}$ the tidal forces from the rest of dark matter do not change the result significantly, but at smaller f the rate of the gravitational bursts can be suppressed by the factor $\approx 1.5 - 2$. The rate of the bursts could be higher due to the formation of dark matter minihalos around PBHs [22], but this scenario is questionable because of the uncertain efficiency of dynamical friction at the last stage of pair evolution.

References

[1] Abbott B P *et al* 2016 Observation of Gravitational Waves from a Binary Black Hole Merger. *Phys. Rev. Lett.* **116** 061102

[2] Grishchuk L P, Lipunov V M, Postnov K A, Prokhorov M E and Sathyaprakash B S 2001 Gravitational Wave Astronomy: in Anticipation of First Sources to be Detected *Phys. Usp.* **44** pp 1-51

[3] Bogomazov A I, Lipunov V M and Tutukov A V 2007 Evolution of close binaries and gamma-ray bursts. *Astronomy Reports* **51** pp 308-317

[4] Belczynski K *et al* 2010 The effect of metallicity on the detection prospects for gravitational waves *The Astrophysical Journal Letters* **715** L138-L141

[5] Belczynski K *et al* 2016 The first gravitational-wave source from the isolated evolution of two 40-100 $M_{\odot}$ stars *Nature* **534** pp 512-515

[6] Abbott B P *et al* 2016 Astrophysical Implications of the Binary Black-Hole Merger GW150914 *The Astrophysical Journal Letters* **818** L22

[7] Nakamura T, Sasaki M, Tanaka T and Thorne K S 1997 Gravitational Waves from Coalescing Black Hole MACHO Binaries *The Astrophysical Journal* **487** L139-L142

[8] Ioka K, Chiba T, Tanaka T and Nakamura T 1998 Black Hole Binary Formation in the Expanding Universe — Three Body Problem Approximation. *Phys. Rev.* D **58** 063003

[9] Sasaki M, Suyama T, Tanaka T and Yokoyama S 2016 Primordial black hole scenario for the gravitational wave event GW150914 *Preprint* arXiv:1603.08338

[10] Zel'dovich Ya B and Novikov I D 1967 The Hypothesis of Cores Retarded during Expansion and the Hot Cosmological Model. *Sov. Astron.* **10** pp 602-603

[11] Hawking S 1971 Gravitationally collapsed objects of very low mass. *MNRAS* **15** pp 75-78

[12] Carr B J 1975 The primordial black hole mass spectrum. *Astrophys. J.* **201** pp 1-19

[13] Garcia-Bellido J, Linde A, and Wands D 1996 Density perturbations and black hole formation in hybrid inflation *Phys. Rev.* D **54** pp 6040-6058

[14] Jedamzik K 1997 Primordial black hole formation during the QCD epoch *Phys. Rev.* D **55** pp 5871-5875

[15] Khlopov M Yu and Polnarev A G 1980 Primordial black holes as a cosmological test of grand unification *Phys. Lett.* B **97** pp 383-387

[16] Berezin V A, Kuzmin V A and Tkachev I I 1983 Thin-wall vacuum domain evolution *Phys. Lett.* B **120** pp 91-96

[17] Dolgov A and Silk J 1993 Baryon isocurvature fluctuations at small scales and baryonic dark matter *Phys. Rev.* D **47** pp 4244-4255

[18] Rubin S G, Sakharov A S and Khlopov M Yu 2001 The Formation of Primary Galactic Nuclei during Phase Transitions in the Early Universe *JETP* **92** pp 921-929

[19] Dokuchaev V I, Eroshenko Yu N and Rubin S G 2009 Gravitational wave bursts from collisions of

primordial black holes in clusters *Astronomy Letters* **35** pp 143-149
[20] Clesse S and Garcia-Bellido J 2016 The clustering of massive Primordial Black Holes as Dark Matter: measuring their mass distribution with Advanced LIGO *Preprint* arXiv:1603.05234
[21] Bertschinger E 1985 Self-similar secondary infall and accretion in an Einstein-de Sitter universe *Astrophys. J. Supp.* **58** pp 39-65
[22] Hayasaki K, Takahashi K, Sendouda Y and Nagataki S 2016 Rapid Merger of Binary Primordial Black Holes *Preprint* arXiv:0909.1738
[23] Mc Crea W H 1951 Relativity Theory and the Creation of Matter. *Proc. Roy. Soc.* A **206** pp 562-575
[24] Kolb E W and Tkachev I I 1994 Large Amplitude Isothermal Fluctuations And High Density Dark Matter Clumps *Phys. Rev.* D **50** pp 769-773
[25] Berezinsky V S, Dokuchaev V I and Eroshenko Yu N 2013 Formation and internal structure of superdense dark matter clumps and ultracompact minihaloes. *JCAP* **11** 059
[26] Peebles P J E 1980 *The Large-Scale Structure of the* Universe (Princeton Univ. Press, Princeton)
[27] Saslaw W C 1987 *Gravitational Physics of Stellar and Galactic Systens* (Cambridge Univ. Press, Cambridge)

Field interpretation of General Relativity

M L Fil'chenkov and Yu P Laptev

Peoples' Friendship University of Russia, Moscow, Russia
E-mail: fmichael@mail.ru

Abstract. A field interpretation of General Relativity, without resorting to Einstein-Hilbert's equations, has been presented for three cases: a static centrally symmetric gravitational field, radiation of gravitational waves and a homogeneous isotropic cosmological model. The possibility of an effective Riemannian space being used in General relativity should not been ruled out.

1. Introduction

According to R. Feynman, gravity has both a field and geometric interpretation [1]. A. Eddington wrote:"... it is likely that some of the phenomena will be determined by comparatively simple equations in which the components of curvature of the world do not appear; such equations will be the same for a curved region as for a flat one..." [2]. Below the field interpretation of General Relativity will be exemplified by three well-known cases: a) a static centrally symmetric gravitational field, b) radiation of gravitational waves, c) a homogeneous isotropic cosmological model, without resorting to Einstein-Hilbert's equations.

2. Static centrally symmetric gravitational field

The action has the form:

$$S = -mc\int ds\,, \tag{1}$$

where

$$ds = c\,d\tau\sqrt{1-\frac{v^2}{c^2}} \tag{2}$$

is the interval,

$$d\tau = dt\sqrt{g_{00}} \tag{3}$$

is the proper time in a static gravitational field defined from the gravitational frequency shift [3]

$$\frac{\Delta\omega}{\omega} = \sqrt{g_{00}} - 1, \tag{4}$$

where

$$g_{00} = 1 + \frac{2\phi}{c^2}. \tag{5}$$

ϕ is the gravitational potential satisfying Poisson's equation

$$\Delta\phi = 4\pi G\rho_{eff}. \tag{6}$$

ρ_{eff} is the effective mass density.

The energy is defined as

$$E = v\frac{\partial L}{\partial v} - L, \tag{7}$$

where

$$L = -mc^2\sqrt{g_{00}}\sqrt{1-\frac{v^2}{c^2}} \tag{8}$$

is the Lagrangian. From formulae (7), (8) we obtain [4]

$$E = \frac{mc^2\sqrt{g_{00}}}{\sqrt{1-\frac{v^2}{c^2}}}. \tag{9}$$

In fact, we are constructing an effective Riemannian space with $g_{00}g_{11} = -1$, using the static gravitational field satisfying Poisson's equation in a flat space. The curvature of such a Riemannian space is simulated by a mutual effect of gravity with the potential ϕ and motion with the velocity v on the proper time of a test particle in Minkowski space.

Taking into account that in the static centrally symmetric field, apart from the energy E, the angular momentum

$$J = \frac{\partial L}{\partial \dot{\varphi}} = \frac{mr^2\dot{\varphi}}{\sqrt{g_{00}}\sqrt{1-\frac{v^2}{c^2}}} \text{ at } \theta = \frac{\pi}{2}, \tag{10}$$

(φ is the polar angle) is conserved, and we obtain the law of motion

$$t = \frac{Jc}{E}\int \frac{dr}{g_{00}\sqrt{1-\left(m^2c^2+\frac{J^2}{r^2}\right)g_{00}\frac{c^2}{E^2}}} \tag{11}$$

and the trajectory

$$\varphi = \frac{Jc}{E}\int \frac{dr}{r^2\sqrt{1-\left(m^2c^2+\frac{J^2}{r^2}\right)g_{00}\frac{c^2}{E^2}}} \tag{12}$$

of a test particle [5].

The gravitational potential $\phi = -\frac{GM}{r}$ for the Newtonian field created by a point source with the mass M is given by Poisson's equation (6). From formula (5) we have the expression

$$g_{00} = 1 - \frac{2GM}{c^2 r}, \tag{13}$$

that completely coincides with the temporal component of Schwarzschild's metric derived from General Relativity. Thus, formula (5) proves to be valid in the general case, but not only in the nonrelativistic case, as often mentioned. Its left side is the temporal component of the metric satisfying Einstein-Hilbert's equations, and the right one contains the gravitational potential satisfying Poisson's equation (6).

Substituting (13) and $E = mc^2$ into (12), we have

$$\frac{mv^2}{2} = \frac{GMm}{r}, \tag{14}$$

wherefrom P. Laplace obtained a formula for the gravitational radius [6]

$$r = \frac{2GM}{c^2} \tag{15}$$

for $v = c$.
The classical gravitational effects [2,5,7] can be found from the formula for the trajectory (12):
a) deflection of light passing near the Sun

$$\Delta\varphi_r = 2\left|\varphi(R_O) - \varphi(\infty)\right| - \pi = \frac{2r_g}{R_O}, \tag{16}$$

where R_O is the Sun radius;
b) Mercury's perihelion shift

$$\Delta\varphi_p = 2\left|\varphi(r_+) - \varphi(r_-)\right| - 2\pi = \frac{3\pi r_g}{p}, \tag{17}$$

where r_+ is the aphelion, r_- the perihelion, p Mercury's orbit parameter, r_g the gravitational radius of the Sun.
Consider one more static field being generated by a charged massive point source. The effective mass of this source is of the form

$$M_{eff} = M - \frac{Q^2}{c^2 r}, \tag{18}$$

where the first term corresponds to an attractive gravitational mass M and the second one to an effective repulsive mass corresponding to the proper electrostatic energy of the charge Q. Poisson's equation (6) for the effective mass density, which is related to the effective mass by the formula

$$M_{eff} = 4\pi \int \rho_{eff} r^2 dr, \tag{19}$$

has the solution as follows:

$$\phi_{RN} = -\frac{GM}{r} + \frac{GQ^2}{2c^2 r^2}. \tag{20}$$

From formula (5) we obtain the expression

$$g_{00} = 1 - \frac{2GM}{c^2 r} + \frac{GQ^2}{c^4 r^2} \tag{21}$$

known as Reissner-Nordström's metric.

3. Gravitational Radiation

Einstein's formula for the intensity of gravitational radiation [8]

$$I = \frac{G}{5c^5} \dddot{D}_{\alpha\beta} \dddot{D}^{\alpha\beta}, \tag{22}$$

where

$$D_{\alpha\beta} = \int \rho_m \left(x_\alpha x_\beta - \frac{1}{3}\delta_{\alpha\beta} \right) dV \tag{23}$$

is the quadrupole moment and ρ_m the mass density, is derivable from the intensity of the electric quadrupole radiation by substituting the photon by the graviton, the charge density by the mass density and introducing the gravitational constant [5]. Based on an analogy between Coulomb's law for charges and Newton's law for masses, we assume that accelerated masses radiate gravitational waves in the same way as accelerated charges radiate electromagnetic ones. The gravitational radiation is a quadrupole one due to the absence of a dipole moment, since the gravitational charge-to-mass ratio is constant. The spin of the graviton $s = 2$ is being equal to the minimum change of the orbital moment $\Delta l = 2$ in the quadrupole transitions. It in no way follows that a gravitational symmetric tensor field is related to a curved space-time. The astronomical data [9-10] validating a reality of gravitational waves support only the quadrupole formula (22) but are, in general, unrelated to a space-time curvature.
From General Relativity we have a zero divergence of the energy momentum tensor:

$$T^{ik}_{;k} = \frac{\partial\left(T^{ik} + t^{ik}\right)}{\partial x^k} = 0, \tag{24}$$

where t^{ik} is the pseudotensor describing a free gravitational field, which is not invariant under coordinate transformations [11].

4. Homogeneous Isotropic Cosmological Model

From Einstein-Hilbert's equations follow Friedmann's equations for a homogeneous isotropic cosmological model [12]

$$\frac{\dot{a}^2}{2} - \frac{4\pi G\varepsilon a^2}{3c^2} = -\frac{kc^2}{2}, \tag{25}$$

$$\ddot{a} = -\frac{4\pi G}{3}\rho_{eff}, \tag{26}$$

where a is the scale factor, $\rho_{eff}c^2 = \varepsilon + 3p$, ε the energy density, p the pressure, $k = 0, \pm 1$ the model parameter. Equation (25) is the first integral of equation (26), providing

$$a\dot{\varepsilon} + 3\left(p + \varepsilon\right)\dot{a} = 0, \tag{27}$$

which is the first law of thermodynamics for an adiabatic expansion of the Universe. From equation (27) it follows that

$$\varepsilon \sim a^{-3(1+w)}, \tag{28}$$

where $p = w\varepsilon$ is the barotropic equation of state for $w = const$.

Friedmann's first equation was obtained by W. McCrea and E. Milne [13] as well in the framework of Newtonian theory of gravity from the energy conservation law of a unit mass on an expanding ball in Euclidean space using the relations:

$$\phi(a) = -\frac{GM}{a},\ M = \frac{4}{3}\pi a^3 \rho, \tag{29}$$

i.e. substituting r by a in the Newtonian gravitational potential.
Consider the case of de Sitter's vacuum:

$$w = -1, p = -\varepsilon, \varepsilon = \varepsilon_0, \rho_{eff} = -\frac{2\varepsilon_0}{c^2}. \tag{30}$$

Substituting ρ_{eff} from (30) into Poisson's equation (6), we have

$$\phi_{dS} = -\frac{c^2 r^2}{2r_0^2}, \tag{31}$$

where

$$\frac{1}{r_0^2} = \frac{8\pi G\varepsilon_0}{3c^2}, \tag{32}$$

r_0 is de Sitter's horizon defining the law of de Sitter's Universe expansion $a(t) \sim \exp\left[\frac{ct}{r_0}\right]$, which follows from (25) for $k = 0$, $\varepsilon = \varepsilon_0$. Substituting $\phi = \phi_{dS}$ into (5), we obtain the expression

$$g_{00} = 1 - \frac{r^2}{r_0^2} \tag{33}$$

known as de Sitter's static metric [14].
Thus, we regain Newton's gravity for Friedmann's cosmological models and Poisson's equation for de Sitter's static model, without using Einstein-Hilbert's equations.

IOP Conf. Series: Journal of Physics: Conf. Series **1051** (2018) 012011 doi:10.1088/1742-6596/1051/1/012011

5. Conclusion

The results of importance for a static field, gravitational waves and cosmology may be obtained without restoring to Einstein-Hilbert's equations in a curved space-time. It is unlikely accidental, there are too many coincidences. More likely, this indicates that General Relativity is overdetermined, i.e. its mathematical formalism is unnecessarily complicated to obtain simple or even trivial results. The field interpretation of gravity, i.e. degeomerization of General Relativity, does not change its results; hence it is quite admissible alongside the geometric one. The possibility of an effective Riemannian space being used in General Relativity should not be ruled out.

References

[1] Feynman R, Moringo F R and Wagner W G 1995 *Feynman Lectures on Gravitation* (Addison--Wesley Pub. Co., Pasadena)
[2] Eddington A S 1929 *The Mathematical Theory of Relativity* (University Press, Cambridge)
[3] Pound R V, Rebka G A 1960 Apparent weighting of photons *Phys. Rev. Lett.* **4** pp 337-341
[4] Landau L D and Lifshitz E M 1971 *The Classical Theory of Fields* (Pergamon Press, Oxford)
[5] Fil'chenkov M L and Laptev Yu P 2016 *Quantum Gravity* (Lenand (URSS), Moscow)
[6] Laplace P S 1796 *Exposition du système du monde* (Cercle-social l'An IV, Paris)
[7] Weinberg S 1972 *Gravitation and Cosmology* (J. Wiley and Sons Inc, New York)
[8] Einstein A 1916 Närungswise Integration der Feldgleichungen der Gravitation *Sitz. Preuss. Akad. Wiss* **1** pp 688-696
[9] Taylor J H 1994 Nobel Lecture: Binary pulsars and related gravity *Re. Mod. Phys.* **66** pp 711-719
[10] Abbott B P *et al* 2016 Observation of Gravitational Waves from a Binary Black Hole Merges *Phys. Rev. Lett.* **116** 061102
[11] Schrödinger E 1918 Die Energiekomponenten des Gravitationsfeldes *Phys. Z.* **19** pp 4-7
[12] Friedmann A 1922 Über die Krümmung des Raumes *Phys. Z.* **10** pp 377-380
[13] McCrea W H and Milne E A 1934 Newtonian Universes and the Curvature of Space *Quart. J. Math.* **5** pp 73-80
[14] de Sitter W 1917 On Einstein's theory of gravitation and its astronomical consequences, III *Mon. Not. Roy. Astron. Soc.* **78** pp 3-28

Acknowledgements

This publication was prepared with the support of the "RUDN University Prorgam 5-100".

The Resonance Photon-Paraphoton Conversion in Media

V S Gorelik[1] and V V Filatov[2]

[1] Lebedev Physical Institute of Russian Academy of Sciences, Moscow, Russia
[2] Bauman Moscow State Technical University, Moscow, Russia
E-mail: gorelik@sci.lebedev.ru

Abstract. We propose the new approach for the direct dark matter particles observation at a laboratory. To do this one should use the violet laser to irradiate the sodium uranyl acetate-filled globular photonic crystal. In this case the laser light quanta will resonantly couple to paraphotons at uranyl unitary polaritons frequencies. Due to high density of states at the uranyl absorption edge this conversion will be extremely efficient. So there is a good opportunity to experimentally observe the paraphotons generation.

1. Introduction

According to the modern astrophysical data the most of the Universe's matter is "dark" [1]. So it's very important to explore this *terra incognita*. We consider paraphotons as candidates for the role of dark matter quanta [2-13]. In this paper we analyze the process of photon-paraphoton conversion to demonstrate the possibilities of dark particles generation at a laboratory.

The paraphoton occurs when the two photons merge into the scalar bound state

$$\gamma + \gamma \rightarrow p\gamma . \tag{1}$$

Because of two photons interaction the Lagrangian must be proportional to the square of the electric field:

$$L \sim g \cdot \psi_s \cdot E^2 . \tag{2}$$

The smallness of the conversion cross section g makes the process (1) practically impossible – one has to imply the superstrong electromagnetic field to observe this in the void.

In the paper we offer the new approach to solve the problem – to involve medium to increase the efficiency of the photon-paraphoton conversion due to the resonant interaction of electromagnetic radiation with matter when the laser radiation's frequency is close to the polar type exciton transition frequencies for single-photon processes of absorption.

2. Theoretical analysis

In the electronic theory of dispersion the motion of an electron in an atom at the external field $E = E_0 \exp(i\omega t)$ is described by the simple model

$$m_0 \ddot{x} = -m_0 \omega_0^2 x + q_0 E_0 e^{i\omega t} . \tag{3}$$

Here x is the electron coordinate; m_0 is its mass; q_0 is the elementary charge; ω_0 is the natural frequency of an electronic oscillator; ω is the external field's photon frequency.

One can see that the solution for (3) is

$$x=\frac{q_0}{m_0\left(\omega_0^2-\omega^2\right)}E_0e^{i\omega t}\ . \tag{4}$$

So, the electric polarization for the medium containing n_0 atoms per cubic meter is

$$P=n_0\cdot q_0x=\frac{n_0q_0^2}{m_0\left(\omega_0^2-\omega^2\right)}E_0e^{i\omega t}\ . \tag{5}$$

On the other hand,

$$P=\varepsilon_0\left(\varepsilon-1\right)E_0e^{i\omega t}\ . \tag{6}$$

Therefore, the material medium dielectric constant depends on the external field frequency the following way:

$$\varepsilon=1+\frac{n_0q_0^2}{\varepsilon_0m_0\left(\omega_0^2-\omega^2\right)}\equiv1+\frac{\omega_p^2}{\omega_0^2-\omega^2}=\frac{\left(\omega_0^2+\omega_p^2\right)-\omega^2}{\omega_0^2-\omega^2}\equiv\frac{\omega_l^2-\omega^2}{\omega_0^2-\omega^2}\ . \tag{7}$$

In this formula

$\omega_p=\sqrt{\frac{n_0q_0^2}{\varepsilon_0m_0}}$ – the plasma frequency; $\omega_l=\sqrt{\omega_0^2+\omega_p^2}$ – the longitudinal mode frequency.

When many oscillators the formula (7) becomes

$$\varepsilon\left(\omega\right)=\varepsilon_\infty\prod_j\frac{\omega_{lj}^2-\omega^2}{\omega_{0j}^2-\omega^2}\ . \tag{8}$$

Here ω_{0j} and ω_{lj} are the characteristic frequencies of the material oscillators; $\varepsilon_\infty=\lim_{\omega\to\infty}\varepsilon\left(\omega\right)$ is the medium high frequency dielectric permittivity

This case the medium electromagnetic field wavenumber is

$$k\left(\omega\right)=\frac{\omega}{c}n=\frac{\omega}{c}\sqrt{\varepsilon\left(\omega\right)\mu}=\frac{\omega}{c}\sqrt{\varepsilon_\infty\prod_j\frac{\omega_{lj}^2-\omega^2}{\omega_{0j}^2-\omega^2}\mu}\ . \tag{9}$$

And the group velocity of the electromagnetic wave is equal to

$$\mathrm{v}\left(\omega\right)=\frac{d\omega}{dk}=\left(\frac{dk\left(\omega\right)}{d\omega}\right)^{-1}\ . \tag{10}$$

Also the effective mass of the relevant quanta (photons and polaritons) is

$$m=\left(\frac{d^2W}{dp^2}\right)^{-1}=\hbar\left(\frac{d^2\omega}{dk^2}\right)^{-1}=\hbar\left(\mathrm{v}\left(\omega\right)\frac{d\mathrm{v}\left(\omega\right)}{d\omega}\right)^{-1}\ . \tag{11}$$

The formulas (8) – (10) show that the group velocity tends to zero when $\omega\to\omega_{lj}$. At known conditions this leads to the formation of polaritons Bose-Einstein condensation with the critical temperature

$$T_C=\left(\frac{n}{\zeta\left(3/2\right)}\right)^{\frac{2}{3}}\frac{h^2}{2\pi mk_B}\ . \tag{12}$$

Here n is the concentration of polaritons; m is its effective mass; h is the Planck constant; k_B is the Boltzmann constant; ζ is the Riemann Zeta-function, $\zeta\left(\frac{3}{2}\right)=2.6124...$

Thus, there is a great opportunity to create the ultra-high density polariton states at material medium oscillators frequencies ω_{lj}. Because of (2) this leads to the resonant amplification for the photon-paraphoton conversion.

Besides, to realize (1) one has to meet the energy and the momentum conservation. This case the medium refractive index n must be equal to one. It is possible at the medium unitary polaritons [4] frequencies: the equation (8) shows that in media, there are the frequencies ω_u at which the absolute value of dielectric constant is equal to one:

$$\left|\varepsilon(\omega_u)\right| = \left|\varepsilon_\infty \prod_j \frac{\omega_{lj}^2 - \omega_u^2}{\omega_{0j}^2 - \omega_u^2}\right| = 1. \tag{13}$$

Thus, at these frequencies the medium is absolute transparent and the process (1) is welcomed. Therefore, the photon to paraphoton conversion (1) in medium is best observable for the photons of $\omega = \omega_u$ when ω_u is close to one of ω_{lj}. As seen from (13), the more factors involved in the product, the higher probability that there is such j that $\omega_{lj} \approx \omega_u$. So, it is necessary to use a material medium with the greatest possible number of electronic oscillators. These elements are located at the bottom of the Periodic table and known as actinides or *f*-elements. Unfortunately these elements are unstable, and with great difficulty we managed to find an affordable non-radioactive uranium-containing sample–sodium uranyl acetate $NaU^{238}O_2(CH_3COO)_3$.

It is well-known [14-16] that the uranyl acetate absorption edge is located near 480 nm. It means that at the violet spectral region, there are a lot of oscillators, so there is the one with $\omega_{lj} \approx \omega_u$. In the paper [17] the uranyl resonant absorption wavelengths λ_{0j} was found (see table 1). So, to use the data we managed to rewrite (8) following way:

$$\varepsilon(\lambda) = 1 + \sum_j \frac{f_j}{\lambda^2 - \lambda_{0j}^2}. \tag{14}$$

Since the oscillator strengths f_j are unknown, we assume that all of f_j are equal. The value of f_j was found by equating the calculated refractive index at sodium *D*-line ($\lambda_D = 589.3$ nm) to the known value of $n_D = 1.504$ [18]:

$$f_j = \frac{\varepsilon(\lambda_D) - 1}{\sum_j \frac{1}{\lambda_D^2 - \lambda_{0j}^2}} = \frac{n_D^2 - 1}{\sum_j \frac{1}{\lambda_D^2 - \lambda_{0j}^2}}. \tag{15}$$

This way $f_j = 1.3 \cdot 10^{-14}$ m^2.

Table 1. The input data for the calculations

j	λ_{0j}, nm	f_j,, 10^{-14} m^2
1	480.5	
2	462.2	
3	446.7	
4	432.4	1.3
5	419.0	
6	414.5	
7	409.5	

j	λ_{0j}, nm	f_j, 10^{-14} m^2
8	402.5	1.3
9	398.0	
10	391.2	
11	388.3	
12	380.5	
13	364.5	
14	361.8	
15	351.5	
16	346.0	
17	343.0	

We use the table 1 data to calculate the dependence of the uranyl acetate dielectric permittivity on the external electromagnetic field wavelength (8) – Figure 1. One can see the poles (λ_{0j}) and the zeros (λ_{lj}). There are at least dozen resonances located at the absorption edge. Note the wavelengths at which $\varepsilon = 1$ – it's the unitary polaritons ones.

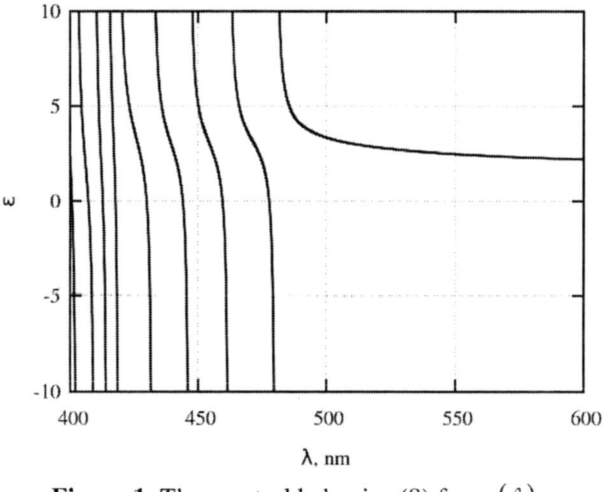

Figure 1. The spectral behavior (8) for $\varepsilon(\lambda)$.

Then we calculated the dispersion law (9) – Figure 2. At the picture the thick lines represent the medium polaritons dispersion (9) and the thin one illustrates the dispersion law in the vacuum: $\omega = ck$. The intersections mark the unitary polaritons positions (ω_u) and the zeros – the LO-modes locations (ω_{lj}).

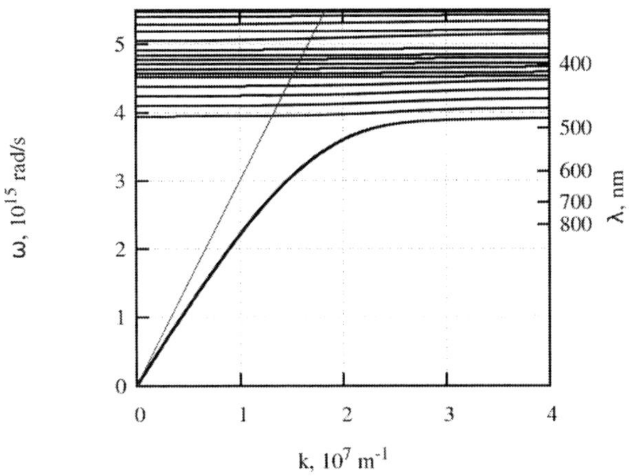

Figure 2. The dispersion of polaritons in the sodium uranyl acetate (thick lines). The thin line illustrates the dispersion law in the vacuum: $\omega = c \times k$. The intersections correspond to the unitary polaritons.

Next we found the electromagnetic waves group velocity (10) in the crystal – Figure 3 (Figure 3a is plotted at the linear scale, Figure 3b at the logarithmic one). One can see the anomalous speed of light slowing near the absorption edge. Due to the conservation of the energy flux density $\langle S \rangle = \varepsilon_0 E^2 = wv$; so, for the slow light $E^2 = E_0^2 \frac{c}{v}$ and the Lagrangian (2) is proportional to the deceleration: $L \sim (c / v)$. It means that the light deceleration to 1 m/s increases the conversion (1) probability by 8 orders of magnitude.

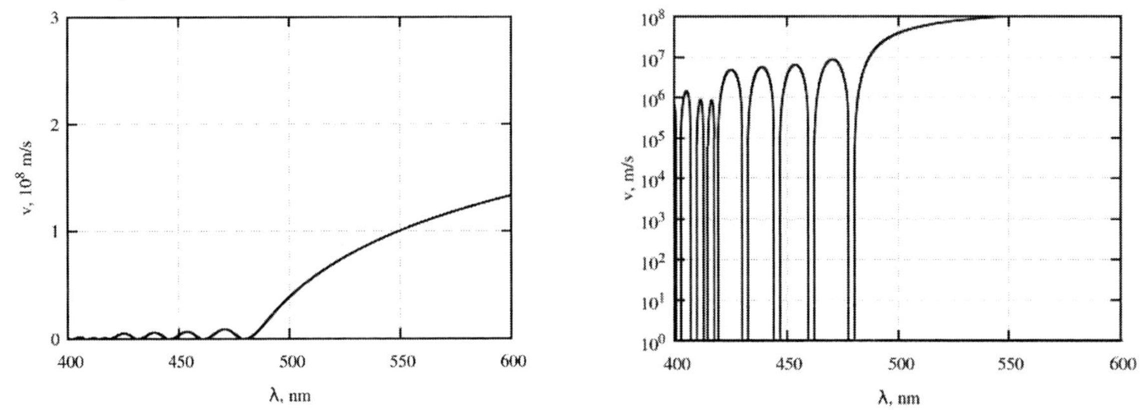

(a) (b)

Figure 3. The light group velocity dispersion in the sodium uranyl acetate calculated by (10)

At last we established the effective masses of light quanta in the sample – Figure 4. At unitary polaritons frequencies the light quanta have mass the order of 10^{-34} kg.

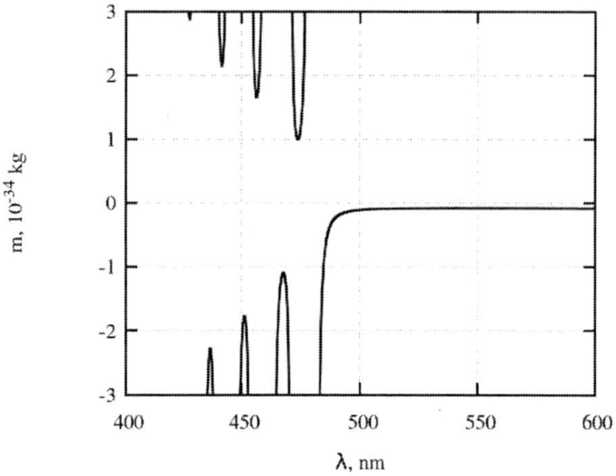

Figure 4. The polaritons effective mass dispersion (11) in the sodium uranyl acetate.

In the paper [19] we shown that it is possible to adjust medium natural resonant frequencies by using the photonic crystal – according the Bragg's law one should to rotate crystal to shift the lines:

$$2an\sin\vartheta = \lambda_m \,. \tag{16}$$

So it is very intriguing to analyze the conversion process (1) in the photonic crystal filled by the sodium uranyl acetate.

The simplest 1D-photonic crystal is the layered medium consists of two alternating layers. Let's call them layer №1 and layer №2. Each layer has its own thickness (a_i) and refractive index (n_i). And for the 1D-photonic crystal the dispersion law is [20]

$$\cos(ka) = \cos(k_1 a_1)\cdot\cos(k_2 a_2) - \frac{1}{2}\left(\frac{n_1}{n_2}+\frac{n_2}{n_1}\right)\cdot\sin(k_1 a_1)\cdot\sin(k_2 a_2). \tag{17}$$

In case of the globular photonic crystal based on the artificial opal [21] at the practically mostly used direction at the normal to natural growth lattice plane (111):

$$k(\omega) = \frac{1}{D\sqrt{\frac{2}{3}}}\arccos\left\{\cos\left(\frac{\omega}{c}n_1(\omega)\cdot(1-\eta)D\sqrt{\frac{2}{3}}\right)\cdot\cos\left(\frac{\omega}{c}n_2(\omega)\cdot\eta D\sqrt{\frac{2}{3}}\right)\right\} - \\ -\frac{1}{2}\left(\frac{n_1(\omega)}{n_2(\omega)}+\frac{n_2(\omega)}{n_1(\omega)}\right)\cdot\left(\sin\left(\frac{\omega}{c}n_1(\omega)\cdot(1-\eta)\cdot\sqrt{\frac{2}{3}}\right)\cdot\sin\left(\frac{\omega}{c}n_2(\omega)\cdot\eta D\sqrt{\frac{2}{3}}\right)\right). \tag{18}$$

Here D is the quartz globules diameter (typically $D \approx 250$ nm); $\eta = 0.26$ is the porosity of the crystal gcc-lattice; n_1 and n_2 are layers refractive indices (the index 1 corresponds to the crystal matrix, 2 – for the substance in pores); ω is the wave frequency and $c = 3\cdot10^8$ m/s is the speed of light in the vacuum.

For the n_1 there is the Sellmeier equation [22] (λ is in μm):

$$n_1^2(\lambda) = 1+\frac{0.6961663\lambda^2}{\lambda^2-0.0684043^2}+\frac{0.4079426\lambda^2}{\lambda^2-0.1162414^2}+\frac{0.8974794\lambda^2}{\lambda^2-9.896161^2}. \tag{19}$$

And n_2 is calculated via (14):

$$n_2^2(\lambda) = \varepsilon(\lambda) = 1+\sum_j \frac{f_j}{\lambda^2-\lambda_{0j}^2}. \tag{20}$$

The simulation (18) - (20) results are shown by Figure 5. One can see the increased density of modes at the uranyl absorption edge due to the photonic crystal lattice.

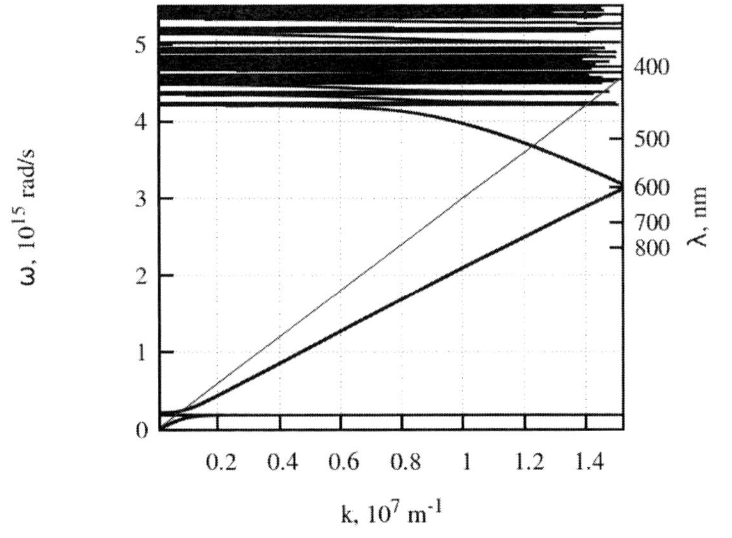

Figure 5. The dispersion law $\omega = \omega(k)$ for the globular photonic crystal on the artificial opal with $D = 250$ nm filled by the sodium uranyl acetate.

Let's analyze the group speed of light (10) in the photonic crystal with uranyl. The spectral dependences for the electromagnetic waves quanta propagation speed are presented at Figure 6. Note that in photonic crystal light deceleration is greater than without it. Moreover there are additional zero light quanta speed regions due to the photonic crystal band structure.

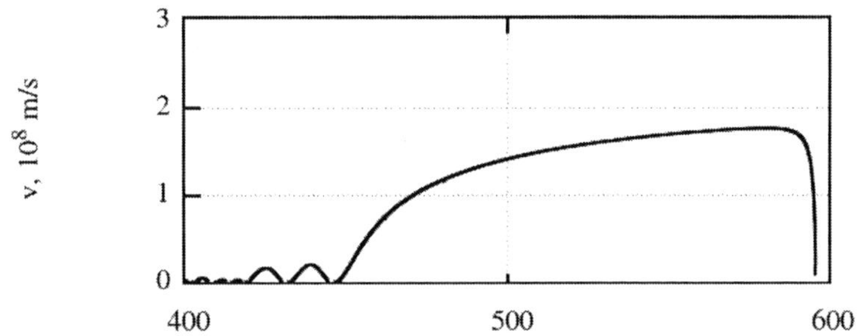

Figure 6. The light quanta speed in the globular photonic crystal filled by the sodium uranil acetate. The crystal's globules diameter is $D = 250$ nm.

3. Discussion

The simulation shows that in medium it is possible to obtain deceleration of light velocity over to 8 orders of magnitude. Moreover, in the photonic crystal there are the zero light quanta speed spectral bands. Because of the Lagrangian proportionality to (c / v), it means that this case the conversion (1) becomes resonant and the process probability tends to 1. The theory predicts the best external field frequencies for this: it is the unitary polaritons ones at the uranyl absorption edge at the violet edge of the visible spectral region.

Moreover, in discussed case we wait to observe not only spontaneous but the stimulated conversion which probability increases resonant way. We propose the sodium uranyl acetate-filled globular photonic crystal with globules of 250 nm as an active media for the potential dark matter generator under powerful laser emission excitation.

4. Conclusion

The results show that there is a opportunity to observe photon-paraphoton conversion at a laboratory by the using of specific media: *f*-element compounds. The best of them is the non-radioactive commercially available uranyl compounds. We predict the best conditions for the photon to paraphoton transformation (1) at the uranyl absorption edge.

The advantage of process (1) over photon to axion conversion is the unnecessarily of the magnetic field. So it is the opportunity to make the proper conditions for dark matter quanta generation at a laboratory.

References

[1] Young B L 2017 *Front. Phys.* **12** 121201
[2] Gorelik V S and Izmailov G N 2011 *Bull. Lebedev Phys. Inst.* **38** 177
[3] Gorelik V S, Dresvyannikov M A and Zherikhina L N *et al* 2014 *Bull. Lebedev Phys. Inst.* **41** 18
[4] Gorelik V S and Pyatyshev A Y 2014 *Bull. Lebedev Phys. Inst.* **41** 316
[5] Gorelik V S 2015 *Bull. Lebedev Phys. Inst.* **42** 55
[6] Lagarkov A N, Ryzhikov I A, Vaskin A V *et al* 2015 *Moscow Univ. Chem. Bull.* **70** 93
[7] Gorelik V S and Sverbil' P P 2015 *Inorg Mater* **51** 1104
[8] Gorelik V S, Pyatyshev A Y, Belyanchikov M A *et al* 2016 *Phys. Wave Phen.* **24** 87
[9] Gorelik V S, Pyatyshev A Y and Krylov A S 2016 *Bull. Lebedev Phys. Inst.* **43** 167
[10] Gorelik V S, Pyatyshev A Y and Krylov A S 2016 *Bull. Russ. Acad. Sci. Phys.* **80** 1127
[11] Belyanchikov M A, Gorelik V S, Gorshunov B P *et al* 2017 *J. Exp. Theor. Phys.* **124** 77
[12] Gorelik V S, Sidorov N V and Vodchits A I 2017 *Phys. Wave Phen.* **25** 10
[13] Belyanchikov M A, Gorelik V S, Gorshunov B P *et al* 2017 *Crystallogr. Rep.* **62** 290
[14] Volod'ko L V, Komyak A I and Sleptsov L E 1965 *J Appl Spectrosc* **3** 97
[15] Burkov V I, Kizel V A, Krasilov Y I *et al* 1967 *J Appl Spectrosc* **7** 527
[16] Alikhanova, Z M, Burkov, V I, Ivanova-Korfini, I N *et al* 1971 *J Appl Spectrosc* **15** 1019
[17] Sanwal D N and Pant D D 1969 *Proc. Indian Acad. Sci.* **69** 324
[18] Locquin M and Langeron M 1983 *Handbook of Microscopy* (Butterworth-Heinemann; London, Boston, Durban, Singapore, Sydney, Toronto, Wellington)
[19] Gorelik V S, Pudovkin A V and Filatov V V 2016 *J Russ Laser Res.* **37** 604
[20] Yariv A and Yeh P 2003 *Optical Waves in Crystals: Propagation and Control of Laser Radiation* (Wiley; Hoboken, New Jersey)
[21] Gorelik V S 2008 *Laser Phys.* **18** 1479
[22] Malitson I H 1965 *J Opt Soc Am* **55** 1205

Acknowledgements

The work is supported by the Russian Foundation for Basic Researches (grants nos. 15-02-02882, 16-08-00618, 16-02-00488, 16-52-00026).

IOP Conf. Series: Journal of Physics: Conf. Series **1051** (2018) 012013 doi:10.1088/1742-6596/1051/1/012013

Geometrothermodynamics of a gravitating system with axially symmetric metric

H V Grushevskaya and N G Krylova

Belarusian State University, Minsk, Belarus
E-mail: grushevskaja@bsu.by, krylovang@bsu.by

Abstract. Theory of the first-order phase transition in contact statistical manifold is proposed to describe interface systems. The theory has not limitations of known Van der Waals-like phase transition theories. Based on this approach, geometrothermodynamics of gravitating systems with axially symmetric metrics is investigated. It has been shown that such geometrothermodynamics occurs in space-time with Newman–Unti–Tamburino-like metric.

1. Introduction

To date, there are known important experimental facts in cosmology: a nonlinear dependence of galaxies recession on magnitude of the red shift, a nonlinear dependence of supernova luminosity on the red shift, gravitational lensing of galaxies which does not associated with fluctuations, the existence of voids in galaxies distribution [1]. These facts can be explained through the existence of space-time universe regions which is expanding at an accelerating rate. Gravitating matter in such regions does not interact (or interacts weakly) with ordinary substance and is known as dark matter. Beside the fact that our universe is in the phase of accelerated expansion, the universe is a flat one, and therefore vacuum energy cosmological constant Λ has to be nonzero: $\Lambda \neq 0$ [2-5]. Λ-term within Einstein equations describes the dark energy (see, for example, [6] and references therein] which can be interpreted as a fluid with negative pressure p, $p<0$ [7]. It is known that the pressure takes on a negative value at the first-order (1st-order) phase transition while the matter is in a metastable state. ΛCDM (Lambda cold dark matter) cosmological models [8] and electrically charged AdS (Anti-de Sitter) black holes [9-10] exhibit a liquid-gas like 1st-order phase transition culminating in critical points that resemble the phase diagram of a Van der Waals fluid [11]. The thermodynamics has been proposed to model gravitating systems [12-13]. Based on the assumption that there exists a correspondence between AdS models and Conformal Field Theory Models (CFT), cosmological models are analyzed as thermodynamic objects with a Bekenstein-Hawking entropy and a Hawking temperature [14-16].

However, the Van der Waals–Maxwell gas model is not invariant with respect to the physically meaningful Legendre transformations [17-18]. Moreover, Einstein equations are not conformally invariant because of Weyl tensor does not vanish. The thermodynamics of cosmological models, such as Friedmann–Lemaître–Robertson–Walker ones, includes effects of the non-null Weyl tensor (anisotropic pressure) by introducing a viscosity parameter and relaxation time (inverse expansion coefficient) [19-20]. But, relaxation processes are absent in the Van der Waals–Maxwell gas model. Thus, a more realistic theory of the 1st-order phase transition has to be used to develop geometrothermodynamic approaches to cosmology.

IOP Conf. Series: Journal of Physics: Conf. Series **1051** (2018) 012013 doi:10.1088/1742-6596/1051/1/012013

In [21] we have proposed a theory of the 1st-order phase transition in a contact statistical manifold that describes interface systems with an electrocapillary mechanism of the energy dissipation and with relaxation time distributions; and this theory has not limitations which are in the Van der Waals-like phase transition theories. The triad $\left\{\vec{r},\dot{\vec{r}},\vec{\xi}\right\}$ for space-time (r, t) defines our contact manifold which is called the thermodynamic phase space $\Gamma=\left\{\vec{r},\dot{\vec{r}},\vec{\xi}\right\}$, $\dot{\vec{r}}=\vec{r}_{\xi}\left(\frac{ds}{d\xi}\right)^{-1}$ [22-24]. The tangent space $T_{\Gamma}=\left\{\dot{\vec{r}},\vec{\xi}\right\}$ is a subspace of the bundle Γ. The tangent space

T_{Γ} has the vertical (hidden) v_{ξ} and horizontal (H) direct sum components: $T_{\Gamma}=v_{\xi}\oplus H$. Thus, v_{ξ} is the vertical subspace generated by $\vec{\xi}$, and H is the horizontal distribution. The bundle Γ of our problem can be redefined by the dynamical-system phase-space $\left\{\vec{r},\vec{r}_{\xi}\right\}$, which is augmented by the subspace spanned by the tangent vector $\frac{ds}{d\xi}$: $\Gamma=\left\{\vec{r},\vec{r}_{\xi},\frac{ds}{d\xi}\right\}$. The bundle Γ includes the four-dimensional (4D) space of the two pairs $\left\{\vec{r},\vec{r}_{\xi}\right\}$ and admits a non-trivial connection determined by the vector $\frac{\partial}{\partial s}$. Here $\frac{\partial}{\partial s}$ is a Reeb vector field of the contact manifold (see [25-27] and references therein). Therefore, Γ is a fifth-dimension (5D) contact manifold. Because the 1st-order phase transition proceeds on the interface, the metric is an axially symmetric one.

In the paper, utilizing approach considered above we study the geometrothermodynamics of gravitating systems with an axially symmetric metric which is similar to Newman–Unti–Tamburino (NUT) one.

2. A geometro-thermodynamical background of problem

An entropy production ΔS in the first-order phase transition proceeding on the interface [21] reads

$$\Delta S=\sum_{a}\Delta S_{a}=-\sum_{a}F_{a}\Delta t_{a}+\sum_{a}E_{a}\Delta t_{a}, \tag{1}$$

where ΔS_{a} is the entropy production for the a-th relaxation process,

$$-\sum_{a=1}^{N}F_{a}\Delta t_{a}=\Delta s\sum_{a=1}^{N}\tau_{a}F_{a}=\Delta sF=\ln Z, \tag{2}$$

$$\sum_{a=1}^{N}E_{a}\Delta t_{a}=-\Delta s\sum_{a=1}^{N}\tau_{a}E_{a}=-\sum_{a=1}^{N}\frac{\partial\ln Z}{\partial\Delta t_{a}}\Delta t_{a}=-N\Delta s\frac{\partial\ln Z}{\partial\Delta s}. \tag{3}$$

Z is the statistical sum

$$Z=\int\prod_{a=1}^{N}DX_{a}p(\vec{r}_{1},...,\vec{r}_{N},t)\exp\left\{\Delta s\sum_{a}\tau_{a}\left[H(X_{a})+i\Gamma(X_{a})\right]\right\},$$

$\int\prod_{a=1}^{N}DX_{a}$ represents the path integral $\int DX=\int\prod_{a=1}^{N}DX_{a}$ over configurations X_{a}, $X_{a}=\left\{\vec{r}_{1}(t_{1});...;\vec{r}_{a-1}(t_{a-1});\vec{r}_{a}(t_{a}+\Delta t_{a});\vec{r}_{a+1}(t_{a+1});...;\vec{r}_{N}(t_{N})\right\}$. The Hamiltonian $H(X_{a})$ and the decay rate $\Gamma(X_{a})$ of each individual relaxation process are renormalized by its relaxation time τ_{a}, the role of inverse temperature β in the transition state would have been played by the parameter $(-\Delta s)$. The formulas (2) and (3) are determined the Helmholtz free energy F and the internal energy E for the first-order phase transition as

$$F=\sum_{a=1}^{N}\tau_a F_a=\frac{1}{\Delta s}\ln Z,\ E=\sum_{a=1}^{N}\tau_a E_a/N=\frac{\partial \ln Z}{\partial \Delta s}. \tag{4}$$

Expanding the expression (1) in series over small parameter Δs as

$$-\sum_a F_a\Delta t_a=\ln Z\big|_{\Delta t_a=0}+\sum_a \Delta t_a\left[\frac{\int p\{X_a\}(H(X_a)+i\Gamma(X_a))DX_a}{Z}\right]_{\Delta t_a=0}=\sum_a \Delta t_a\, C, \tag{5}$$

$$\sum_a E_a\Delta t_a=-\sum_a\frac{\partial \ln Z}{\partial \Delta t_a}\Delta t_a=-\frac{1}{Z}\sum_a \Delta s\int p\{X_a\}[\tau_a(H(X_a)+i\Gamma(X_a))]DX_a; \tag{6}$$

and passing to the limit $\dot{\xi}=\frac{\partial t}{\partial s}=\lim\limits_{\Delta s\to 0}\frac{\Delta t}{\Delta s}$ one can find an action on the statistical manifold of the 1st-order phase transition:

$$\Delta S=\int\left[\dot{\xi}C-L(\dot{\xi})\right]ds.$$

In a case of electrocapillary loss, the Lagrangian $L(\dot{\xi})\equiv L(\vec{r},\dot{\vec{r}},\dot{\xi})$ reads in polar coordinates (r,φ)

$$L(\vec{r},\dot{\vec{r}},\dot{\xi})=-p|V|r^5 e^{\frac{2|V|\xi}{r}}\frac{\dot{\xi}^2}{\dot{r}}+U(\xi,r)\dot{\xi}+m\frac{(\dot{r}^2+r^2\dot{\varphi}^2)}{2\dot{\xi}} \tag{7}$$

and, respectively, the action dl on the statistical manifold is written in the following form:

$$dl=L(\vec{r},\dot{\vec{r}},\dot{\xi})ds=\left(-p|V|r^5 e^{\frac{2|V|\xi}{r}}\frac{\dot{\xi}^2}{\dot{r}}+U(\xi,r)\dot{\xi}+m\frac{(\dot{r}^2+r^2\dot{\varphi}^2)}{2\dot{\xi}}\right)ds, \tag{8}$$

where $\dot{r}=\frac{dr}{ds}$ and $\dot{\varphi}=\frac{d\varphi}{ds}$,

$$U(\xi,r)=p\left\{\left[-\frac{4}{3}r^5+\frac{16}{15}(|V|\xi)r^4+\frac{1}{30}(|V|\xi)^2r^3+\frac{1}{45}(|V|\xi)^3r^2+\right.\right.$$

$$\left.\left.+\frac{1}{45}(|V|\xi)^4r+\frac{2}{45}(|V|\xi)^5\right]e^{\frac{2|V|\xi}{r}}-\frac{4}{45}\frac{(|V|\xi)^6}{r}\mathrm{Ei}\left(\frac{2|V|\xi}{r}\right)\right\},$$

$|V|$ is a modulus of compression (or expansion) rate, p and m are model parameters.

3. Generalized NUT-metric

The NUT metric reads

$$dl^2=\left(1-\frac{r_g r+2n^2}{r^2+n^2}\right)\left(dx^0+4n\sin\frac{\theta}{2}d\varphi\right)^2-\frac{dr^2}{1-\frac{r_g r+2n^2}{r^2+n^2}}-(r^2+n^2)(d\theta^2+\sin^2\theta d\varphi^2), \tag{9}$$

where n is a NUT parameter, r_g is a gravitational radius, x_0 is a time coordinate; r, θ, φ are spatial spherical coordinates. In the case of $r^2, r_g r>>n^2$ this metric behaves like Schwarzschild metric or Reissner– Nordström one if a term $\frac{q^2+g^2}{r^2+n^2}$ to $\Phi(r)=1-\frac{r_g r+2n^2}{r^2+n^2}$ is added. Here q and g are electric and magnetic charges, respectively. Since a following approximation $\Phi^{-1}(r)r^2\approx(r^2+n^2)$, holds, one can transform the expression (9) as

$$dl^2=\Phi(r)\left(dx^0+4n\sin^2(\theta/2)d\varphi\right)^2-\frac{dr^2}{\Phi(r)}-\frac{r^2}{\Phi(r)}(d\theta^2+\sin^2\theta\, d\varphi^2). \tag{10}$$

At $r^2, r_g r < n^2$ one can add a correction $(\Lambda/3)\left(r^2+5n^2\right)-\dfrac{8(\Lambda/3)n^4-q^2-g^2}{r^2+n^2}$ to $\Phi(r)$ in metric (10). Then the NUT-metric transforms to the form:

$$dl^2=\tilde{\Phi}(r)\left(\mathrm{dx}^0+4n\sin^2\frac{\theta}{2}d\phi\right)^2-\frac{dr^2}{\tilde{\Phi}(r)}-\frac{dr^2}{\tilde{\Phi}(r)}\left(d\theta^2+\sin^2\theta d\phi^2\right), \tag{11}$$

where $\tilde{\Phi}(r)=1-\dfrac{r_g r+2n^2}{r^2+n^2}+\dfrac{\Lambda}{3}\left(r^2+5n^2\right)-\dfrac{8\left(\Lambda/3\right)n^4-q^2-g^2}{r^2+n^2}$, Λ is a cosmological constant. While $r^2 \ll n^2$, $r \sim r_g$, $r_g^2 \sim -2n^2-\Lambda n^4+q^2+g^2$ (or $r^2 \ll \left|n^2\right|$, $r \sim r_g$, $r_g^2 \sim -2\left|n^2\right|-\Lambda n^4+q^2+g^2$ for $n^2<0$), the substitution $\dfrac{r^2}{\tilde{\Phi}(r)} \approx \dfrac{r^2}{1+\dfrac{\Lambda}{3}r^2} \to r^2$ leads to the metric (11) obtains the features of anti de Sitter metric ($\tilde{\Phi}(r) \to 1+\dfrac{\Lambda}{3}r^2$).

4. An axially symmetric metric of thermodynamic phase space entering a 5-dimensional contact manifold

Based on the metric function (8) one can construct the following metric:

$$dl^2=L(\vec{r},\dot{\vec{r}},\dot{\xi})ds^2=-p\left|V\right|r^5 e^{\frac{2|V|\xi}{r}}\frac{\dot{\xi}}{\dot{r}}d\xi\, ds+U(\xi,r)\dot{\xi}ds^2+m\frac{(\dot{r}^2+r^2\dot{\varphi}^2)}{2\dot{\xi}}ds^2. \tag{12}$$

A space-time metric can be obtained from the metric (12) by the Wick rotation $\xi=-i\tau$ and $s=-ix^0$:

$$d\tilde{l}^2=p\left|V\right|r^5\exp\left(\tfrac{-i2|V|\tau}{r}\right)\frac{\dot{\xi}}{\dot{r}}d\tau\, dx^0-U(-i\tau,r)\dot{\xi}\left(dx^0\right)^2-\frac{m}{2\dot{\xi}}(dr^2+r^2d\varphi^2), \tag{13}$$

where $dr=\dot{r}dx^0$, $d\varphi=\dot{\varphi}dx^0$. Let $\dot{\xi}$ be a known function $F(r)$: $\dot{\xi}=F(r)m/2$. Then, the metric can be considered as a metric of 4D-surface $\theta=\pi/2$ in the contact 5D-manifold:

$$d\tilde{l}^2=\frac{m}{2}F(r)\left[-U(-i\tau,r)\left(dx^0\right)^2+\exp\left(\tfrac{-i2|V|\tau}{r}\right)\frac{p\left|V\right|r^5}{\dot{r}}d\tau\, dx^0\right]-\frac{\left[dr^2+r^2\left(d\theta^2+\sin^2\theta d\varphi^2\right)\right]}{F(r)},$$

$$\theta=\pi/2. \tag{14}$$

Let us choose some another 4D-surface in the contact 5D-manifold in a such way to the angle θ takes arbitrary values, while ranges of variables r, φ are dynamically limited by the variable τ. This can be realized in the following way. Let us impose a gauge condition on the scalar field $\phi \equiv e^{\frac{-i|V|\tau}{r}}$:

$$\frac{4p}{45}\frac{\left(|V|\tau\right)^6}{r}=1. \tag{15}$$

We consider the case of small values of rate $\left|V\right| \to 0$ when the function $\mathrm{Ei}\left(-i\dfrac{2|V|x^0}{r}\right)$ turns to a constant value $\operatorname{Re}\mathrm{Ei}\left(i0\right)=-\infty$, $\operatorname{Im}\mathrm{Ei}\left(i0\right)=\pi/2$. Then, while the condition (15) is imposed, the scalar field ϕ oscillates strongly,

$$U(-i\tau,r)=-\mathrm{Ei}\left(i0\right) \text{ at } \left|V\right| \to 0, \text{ and } r=\frac{4p}{45}\left(|V|\tau\right)^6. \tag{16}$$

Due to the constraint (15), one has $\dot{r}=\frac{dr}{d\tau}\frac{d\tau}{-i\,dx^0}=\frac{24p}{45}|V|^6\tau^5\frac{d\tau}{-i\,dx^0}$. Then changing $\left(dx^0\right)^2\to(m/2)\left(dx^0\right)^2$ the metric (14) takes the form

$$dl^2_{renorm}=F(r)\left[\operatorname{Im}\operatorname{Ei}(i0)-i\exp\left(\tfrac{-i2|V|\tau}{r}\right)\frac{45}{24}\frac{r^5}{|V|^5\tau^5}\right]\left(dx^0\right)^2-\frac{\left[dr^2+r^2\left(d\theta^2+\sin^2\theta d\varphi^2\right)\right]}{F(r)}, \qquad (17)$$

where $dl^2_{renorm}=d\tilde{l}^2-\operatorname{Re}\operatorname{Ei}(i0)F(r)\left(dx^0\right)^2$.

Now, one has to exclude the 5-th coordinate τ. A phase $\frac{2|V|\tau}{r}$ in exponential term $\exp\left(\frac{-i2|V|\tau}{r}\right)$ of 5-th dimension varies in the range from 0 up to π. Therefore, the phase $\frac{|V|\tau}{r}$ is defined by the angle $\theta/2$:

$$dl^2_{renorm}=F(r)\left\{\operatorname{Im}Ei(i0)-i\left(\cos\frac{\theta}{2}-i\sin\frac{\theta}{2}\right)^2\frac{45}{24}\frac{r^5}{|V|^5\tau^5}\right\}\left(dx^0\right)^2-\frac{dr^2+r^2\left(d\theta^2+\sin^2\theta d\varphi^2\right)}{F(r)}. \qquad (18)$$

At $\tau\to\infty$ the expression (18) is an expansion over a small parameter $1/\tau$:

$$dl^2_{renorm}=F(r)\left[d\tilde{x}^0-i\frac{\left(\cos\frac{\theta}{2}-i\sin\frac{\theta}{2}\right)^2}{48\operatorname{Im}\operatorname{Ei}(i0)}\frac{45r^5d\tilde{x}^0}{|V|^5\tau^5}\right]^2-\frac{\left[dr^2+r^2\left(d\theta^2+\sin^2\theta d\varphi^2\right)\right]}{F(r)}, \qquad (19)$$

where $d\tilde{x}^0=(\operatorname{Im}\operatorname{Ei})^{1/2}(i0)dx^0$. Then, taking into account that $0\le\frac{r}{|V|\tau}\le1$ to impose the constraint of the form $2\pi\left(\frac{r}{|V|\tau}\right)^5d\tilde{x}^0=d\varphi$ on the angle φ, one gets

$$dl^2_{renorm}=F(r)\left[d\tilde{x}^0-\frac{45i}{24\pi}\left(\cot\frac{\theta}{2}-i\right)^2\sin^2\frac{\theta}{2}d\varphi/(2\pi)\right]^2-\frac{\left[dr^2+r^2\left(d\theta^2+\sin^2\theta d\varphi^2\right)\right]}{F(r)}. \qquad (20)$$

Comparison between (11) and (20) gives the expression for the NUT-parameter, which becomes an imaginary constant at $\theta\to\pi,0$:

$$n\sin^2\frac{\theta}{2}\bigg|_{\theta\to\pi,0}\equiv-\frac{45i}{192\pi^2}\sin^2\frac{\theta}{2}\left(\cot\frac{\theta}{2}-i\right)^2\bigg|_{\theta\to\pi,0}=\pm\frac{45i}{192\pi^2}.$$

Geodesic in the contact manifold with the metric (12) and an estimation of dependence of $\dot{\xi}$ at $|V|\to0$ are shown in Figure 1. The simulation results and the analytic calculations performed above prove that such Schwarzschild-like geodesics behave similarly to that of the NUT metric.

IOP Conf. Series: Journal of Physics: Conf. Series **1051** (2018) 012013 doi:10.1088/1742-6596/1051/1/012013

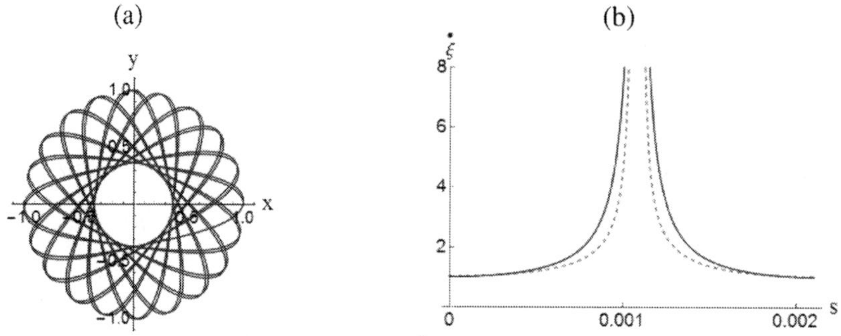

Figure 1. (a) Geodesic; (b) the dependence of $\dot{\xi}(s)$ (blue solid curve) and its approximation by

$$\dot{\xi}(s) = -0.8 + \frac{1}{r(s) - 0.45} \text{ (red dashed curve).}$$

5. Discussion and conclusion

So, using of the contact statistical 5-dimensional manifold, which describes phase transitions proceeding in the interface systems with an electrocapillary mechanism of energy dissipation, allows us to construct the theory of cosmological models with axially symmetric metrics. These metrics are that of 4-dimensional surfaces in the 5D manifold. We have shown that the NUT-theory parameter n is the gauge parameter of the scalar field which plays a role of fifth dimension. The metrics in 5D world are similar to the (anti) de Sitter metrics at large values of n.

References

[1] Heymans C 2017 *The Dark Universe* (Bristol: IOP Publishing)
[2] Weinberg S 1989 *Rev. Mod. Phys.* **61** 1
[3] Sahni V and Starobinski A 2000 *Int. J. Mod. Phys.* D **9** 373
[4] Tegmark M *et al* (the SDSS collaboration) 2004 *Phys. Rev.* D **69** 103501
[5] Vladimirov Y S 2010 *Geometrophysics* (Moscow: BINOM, Laboratory of knowledge)
[6] Rao V U M, Suryanarayana G and Aditya Y 2016 *Advances in Astrophysics* **1** 62
[7] Dutta S, Jain A and Soni R 2013 *JHEP* 060
[8] Padmanabhan T 2003 *Phys. Rept.* **380** 235
[9] Banerjee R, Modak S K and Roychowdhury D 2012 *JHEP* **1210** 125
[10] Chaturvedi P, Das A and Sengupta G 2017 *Eur. Phys. J.* C **77** 110
[11] Sahay A, Sarkar T and Sengupta G 2010 *JHEP* **10** 118
[12] Hawking S W 1975 *Commun. Math. Phys.* **43** 199
[13] Bekenstein J D 1973 *Phys. Rev.* D **7** 2333
[14] Hawking S W and Page Don N 1983 *Commun. Math. Phys.* **87** 577
[15] Cvetič M and Gubser S S 1999 *JHEP* **07** 010
[16] Chaturvedi P and Sengupta G 2015 *Preprint* arXiv:1501.07748v1
[17] Bravetti A, Lopez-Monsalvo C S, Nettel F and Quevedo H 2014 *J. Geom. Phys.* **81** 1
[18] Bravetti A, Lopez-Monsalvo C S and Nettel F 2015 *Entropy* **17** 6150
[19] Hawking S 1966 *Astrophys. J.* **145** 544
[20] Bittencourt E, Salim J and dos Santos G B 2014 *Int. J. Nonlin. Phenom. in Complex Sys.* **17** 352
[21] Balan V, Grushevskaya H V, Krylova N G and Neagu M 2016 *Int. J. Nonlin. Phenom. in Complex Sys.* **19** 223
[22] Quevedo H 2007 *J. Math. Phys.* **48** 13506
[23] Quevedo H and Quevedo M N 2011 *EJTP - Zacatecas Proc. II* 1
[24] Quevedo H, Sánchez A, Taj S and Vázquez A 2011 *Gen. Rel. Grav.* **43** 1153
[25] Boyer C P 2011 *SIGMA* **7** 058

[26] Arnold V I and Novikov S P 2001 *Dynamical Systems IV: Symplectic Geometry and Its Applications* (Berlin: Springer)
[27] Mrugała R 2000 *Rep. Math. Phys.* **46** 461

IOP Conf. Series: Journal of Physics: Conf. Series **1051** (2018) 012014 doi:10.1088/1742-6596/1051/1/012014

Probabilities in the topos approach to branching space-time

L Il'ichov

IA&E SB RAS, Novosibirsk, Russia
Novosibirsk State University, Novosibirsk, Russia
E-mail: leonid@iae.nsk.su

Abstract. There was previously proposed a topos unified view-point both to relativistic and quantum physics [4-5] on the basis of Belnapian branching space-time [3]. In the context of this approach, to any event of the branching space-time there can be associated a local orthocomplemented lattice (logic) of 'physically reasonable' propositions made by a localized observer. Elements of these local orthologics are special collections of the so-called Belnapian worlds (maximal upward directed subsets of events) which the local event belongs to. In the ordinary quantum setting, the counterpart of any local orthologic is the lattice of closed subspaces of Hilbert space of the observer's environment. The present work suggests a natural scheme of inserting truth probabilities of the 'physically reasonable' propositions into the topos framework. To this end, the concept of Chu spaces is used. The role of hypothesis for the state of environment is played by an element of local 'retro-causal' orthologic. Sufficient condition of functorial character of the law by which Chu space is related to any event is pointed and proved, i.e. a functor is constructed from the branching space-time to the category of all Chu spaces. There presented a theorem on category nature of Bayesian reassessment of hypothesis likelihood made by the local observer.

1. Introduction

Fundamental phenomena of micro- and macro-worlds are presently described by different but equally successful physical theories -- quantum mechanics (QM) and general relativity (GR). Creation of a unified theory will definitely demand new frame of notions and new branch of mathematics as a basis of the theory. The machinery of linear operators in Hilbert spaces, the mathematics of quantum physics, as well as geometry of smooth manifolds, the basis of GR, should find natural places within this new still unknown construction.

Category theory is the most universal branch of mathematics. We may suppose it to be the basis of the desired physical theory in a broad sense. The notion of sets and accompanying Boolean logic lie in the foundation of almost all special branches of mathematics and physics. There is a notion of topos in category theory as a deep generalization of category of sets *Set* [1]. Theory of topoi is considered for a long time as a powerful tool for unification of mathematics. Recent works [2] have demonstrated the topoi theory being a promising instrument in physics -- in construction of a so-called neo-realistic approach to QM with consistent quantum gravity as a final goal.

Generalizing the notion of set and map between sets, topoi provide their own non-classical intuitionistic logic. This logic must evidently be inherited by physical theories base on topos mathematics. In some a sense, one may expect the formation of logical paradigm in physics in parallel with geometric paradigm which was formulated a century ago.

The conceptual frame of the expected topos theory must naturally retain the main features both of quantum world with its immanent randomness and macroscopic theory of space-time with its nontrivial structure of causal relations between events. Not long ago in the framework of purely logical (non-topos) approach to unification of the aforementioned basic features of quantum and relativistic physics Nuel Belnap proposed the model of branching space-time [3]. Partly ordered (by means of causal relations) set C of *all possible* events lies in its basis. Not a single outcome which has been actualized but all possible ones of any quantum measurement belong to C. Event is a primitive of the Belnap's model; the main derived notion is a `history' -- maximal causally directed set of events. In our approach we call it `Belnapian world'.

Being translated into the topos context, the model of branching space-time reveals new interesting features [4-5]. All the play takes place in the topos $Set^{\mathrm{C}^{op}}$. It is the category of presheaves – contravariant functors from the poset category C into the category of sets. The basic object of the topos is $\mathbf{Loc}$. The presheaf $\mathbf{Loc}$ associates to any event e the set $\mathbf{Loc}_e$ of all Belnapian worlds which e belongs to, and to any causal arrow (relation) e_1 Î e_2 the inclusion map $\mathbf{Loc}_{e_2} \to \mathbf{Loc}_{e_1}$ is associated.

2. Local orthologics

The set $\mathbf{Loc}_e$ is a local site used by an observer in e when constructing the picture of his (her) surrounding world. Certain subsets in $\mathbf{Loc}_e$ were shown in [5] to naturally form orthocomplemented lattice OL_e as a model of some orthologic. This is a non-classical logical scheme known also as minimal quantum logic. In quantum physics to the proposition of the type `The value of observable $O$ lies in Borel set $\Delta$' (`$O \in \Delta$') there associated some subspace $\mathrm{H}_{O\in\Delta}$ of a total Hilbert space H. This subspace is twofold closed. First, $\mathrm{H}_{O\in\Delta}$ is orthoclosed, i.e. $\mathrm{H}_{O\in\Delta} = (\mathrm{H}_{O\in\Delta}^{\perp})^{\perp}$. Secondly, it is closed in the norm topology of H. Elements of OL_e were shown to be twofold closed as well. First, they were constructed in such a way that any $A \in \mathrm{OL}_e$ is orthoclosed by definition: $A = (A^{\perp_e})^{\perp_e}$, where $\perp_e$ is local ortocomplementation. Secondly, there is a natural topology in OL_e and elements of OL_e are closed with respect to this topology [5]. The conspicuous similarity between subspaces of H and elements of OL_e let one identify the last ones with physical propositions made by the localized observer. It is worth to stress that the lattice OL_e with its internal connectives acts as a prototype of the lattice of closed subspaces of some *local* Hilbert space H. The vectors of H are states of the external world seen by the observer from e.

3. Probabilities in local orthologics

Developing the aforementioned analogy, we face the problem of ascribing truth values to propositions from OL_e. In QM the proposition `$O \in \Delta$' is true with probability

$$Prob(O \in \Delta) = Tr\hat{\rho}_0 \hat{P}_{O\in\Delta}, \tag{1}$$

where $\hat{P}_{O\in\Delta}$ is the projector onto the subspace $\mathrm{H}_{O\in\Delta}$ and $\hat{\rho}_0$ is the statistical operator (state) of the considered system. It encapsulates the entire observer's knowledge on the system. Generally, the statistical operator may be presented as

$$\hat{\rho}_0 = \sum_{i\in I} w_i \hat{P}^{(i)}. \tag{2}$$

Here I is some set of indices, $\hat{P}^{(i)}$ are projectors onto subspaces of H and w_i are weights (probabilities) ascribed to various $\hat{P}^{(i)}$. Naturally, one has $w_i > 0$ and $\sum_{i\in I} w_i = 1$. Hence, both multipliers in the rhs of (1) have the same nature -- they are constructed of projectors. Yet there is a significant difference between the forms of their time behavior. In the simplest case a unitary operator $\hat{U}$ gives the evolution of the system from zero time up to t. Truth probability of the proposition `$O \in \Delta$' at the moment t

$$Prob_t(O \in \Delta) = Tr\hat{\text{ñ}}_t \hat{P}_{O\in\Delta}, \tag{3}$$

where the transformed state is the result of evolution:

$$\hat{\text{ñ}}_0 \mapsto \hat{\text{ñ}}_t = \hat{U}\hat{\text{ñ}}_0\hat{U}^\dagger. \tag{4}$$

Note that the rhs of (3) may also be presented as

$$Prob_t(O \in \Delta) = Tr\hat{\text{ñ}}_0(\hat{U}^\dagger \hat{P}_{O\in\Delta}\hat{U}). \tag{5}$$

If (4) is interpreted as forward *t*-transformation (Schrodinger arrow) than the transformation

$$\hat{P}_{O\in\Delta} \mapsto \hat{U}^\dagger \hat{P}_{O\in\Delta}\hat{U} \tag{6}$$

is *in this sense* backward *t*-transformation (Heisenberg arrow). In the general case the transformations (4) and (6) are replaced by some weighted mixtures

$$\hat{\text{ñ}} \mapsto \sum_\alpha q^{(\alpha)}\hat{U}_\alpha \hat{\text{ñ}}_0 \hat{U}_\alpha^\dagger,\ \hat{P} \mapsto \sum_\alpha q^{(\alpha)}\hat{U}_\alpha^\dagger \hat{P}\hat{U}_\alpha,$$

where $q^{(\alpha)} > 0, \sum_\alpha q^{(\alpha)} = 1$ and $\hat{U}_\alpha$ are unitary operators.

The oppositely directed transformations will serve us as a hint when introducing probabilities in the topos approach where physical propositions made in e are elements of OL_e. To any causal arrow e'Î e a natural map $\mathrm{OL}_e \to \mathrm{OL}_{e'}$ can be associated. We clarify its structure by transforming some $A \in \mathrm{OL}_e$. By virtue of the presheaf **Loc** we have $A \subset \mathbf{Loc}_{e'}$. A might be not an element of $\mathrm{OL}_{e'}$ but its local algebraic orthoclosure

$$A \mapsto Cl_{e'}(A) =_{df} (A^{\perp_{e'}})^{\perp_{e'}} \tag{7}$$

pertains to $\mathrm{OL}_{e'}$. It is this operation which will be used as a map associated to e'Î e:

$$\mathrm{OL}_{e'}^{e} : \mathrm{OL}_e \to \mathrm{OL}_{e'}, \tag{8}$$

where $\mathrm{OL}_{e'}^{e} =_{df} Cl_{e'}$.

The behavior of local observer in the branching space-time is conditioned by his (or her) conjectures about the external world. These conjectures along with weights-probabilities which the observer ascribes to them make up a prototype of quantum state. So, looking at the resemblance between propositions and states in QM, it is reasonable to expect some similarity of their nature with that of propositions from OL_e (the local character first) but with the mentioned peculiarity in the directions of *t*-transforms. That means that for any causal arrow e'Î e we should expect to have a map for the sets of local hypothesis (conjectures) in reverse direction with respect to (8). Keeping this fact in the mind, it is natural to construct the system of local hypothesis by means of opposed category C^{op} with reversed directions of all causal arrows. The main topos is $Set^{(\mathrm{C}^{op})^{op}} = Set^{\mathrm{C}}$ -- the category of all covariant functors from C to Set. Objects of Set^{C} and derived structures will get the same notations as their counterparts in $Set^{\mathrm{C}^{op}}$ but with over-lines. For instance, $\overline{\mathbf{Loc}}$ is the presheaf of Belnapianretroworlds -- maximal directed subsets of C^{op}. The sets of local conjectures are $\overline{\mathrm{OL}}_e$, i.e. orthocomplemented lattices which are akin to OL_e in their construction. In $\mathrm{P}(\overline{\mathbf{Loc}}_e)$ the closure $\overline{Cl}_e$ operates. It gives the map

$$\overline{\mathrm{OL}}^{e}_{e'} : \overline{\mathrm{OL}}_{e'} \to \overline{\mathrm{OL}}_{e}, \tag{9}$$

which resembles (8) in all respects but the direction (along the arrow e' Î e).
Assuming a hypothesis $H_e \in \overline{\mathrm{OL}}_e$, the observer in *e* gets the conditional probability $p_e(A_e \mid H_e)$ of a chosen physical proposition A_e being true. This probability is every bit as objective as the probability (1) is. Taking into account the mutually opposite directions of the maps (8) and (9), we subordinate the conditional probability to a restriction (just in parallel with (3) and (5)). Namely, for any e' Î e, $H_{e'} \in \overline{\mathrm{OL}}_{e'}$ and $A_e \in \mathrm{OL}_e$ we assume

$$p_e(A_e \mid \overline{\mathrm{OL}}^{e}_{e'}(H_{e'})) = p_{e'}(\mathrm{OL}^{e}_{e'}(A_e) \mid H_e). \tag{10}$$

In accordance with the definition of Chu spaces [6], the triples $(p_e, \mathrm{OL}_e, \overline{\mathrm{OL}}_e)$ and $(p_{e'}, \mathrm{OL}_{e'}, \overline{\mathrm{OL}}_{e'})$, the ordered pair of maps $(\overline{\mathrm{OL}}^{e}_{e'}, \mathrm{OL}^{e}_{e'})$ provide the example of Chu space and a morphism between them:

$$(\overline{\mathrm{OL}}^{e}_{e'}, \mathrm{OL}^{e}_{e'}) : (p_{e'}, \mathrm{OL}_{e'}, \overline{\mathrm{OL}}_{e'}) \to (p_e, \mathrm{OL}_e, \overline{\mathrm{OL}}_e). \tag{11}$$

The maps (8) and (9) are associated to the arrow e' Î e. It is worth to find conditions which endow this association with functorial property, i.e. which make the composition of causal arrows to be transformed into the composition of respective associated maps. Hence, for the chain of three events e_1 Î e_2 Î e_3 one has to get

$$\mathrm{OL}^{e_3}_{e_1} = \mathrm{OL}^{e_2}_{e_1} \cdot \mathrm{OL}^{e_3}_{e_2}, \overline{\mathrm{OL}}^{e_3}_{e_1} = \overline{\mathrm{OL}}^{e_3}_{e_2} \cdot \overline{\mathrm{OL}}^{e_2}_{e_1}. \tag{12}$$

We shall consider the first relation. It is equivalent to

$$Cl_{e_1}\left[Cl_{e_2}(A)\right] = Cl_{e_2}(A) \tag{13}$$

for any $A \in \mathrm{OL}_{e_3}$. The inclusion $Cl_{e_2}(A) \subseteq Cl_{e_1}\left[Cl_{e_2}(A)\right]$ is trivially fulfilled, since $A \subseteq Cl_{e_2}(A)$. We shall explore the conditions of the inverse inclusion

$$Cl_{e_1}\left[Cl_{e_2}(A)\right] \subseteq Cl_{e_2}(A). \tag{14}$$

To this end, we are to take into account the structure of closure operation (see [5]). There is used a special presheaf **Acc** from $Set^{C^{op}}$. For any event e $\mathbf{Acc}_e \subseteq \mathbf{Loc}_e \times \mathbf{Loc}_e$ is the so-called accessibility relation for Belnapian worlds: for w_1 and w_2 from $\mathbf{Loc}_e$ one has $\langle w_1, w_2 \rangle \in \mathbf{Acc}_e$ iff *e* is not the last common event in these worlds. The following expression for Cl_e was evaluated in [5]:

$$Cl_e(X) = \bigcup\{Y \subseteq \mathbf{Loc}_e : Y \bullet \mathbf{Acc}_e \subseteq X \bullet \mathbf{Acc}_e\}, \tag{15}$$

where $\bullet$ stands for relational product:

$$\mathrm{X}' \bullet \mathbf{Acc}_e =_{df} \{w \in \mathbf{Loc}_e : \exists w' \left(w' \in \mathrm{X}', \langle w, w' \rangle \in \mathbf{Acc}_e\right)\}. \tag{16}$$

The inclusion (14) takes place if $w \in Cl_{e_1}\left[Cl_{e_2}(A)\right]$ (i.e. $\{w\} \bullet \mathbf{Acc}_{e_1} \subseteq Cl_{e_2}(A) \bullet \mathbf{Acc}_{e_1}$) implies $w \in Cl_{e_1}(A)$ (i.e. $\{w\} \bullet \mathbf{Acc}_{e_1} \subseteq A \bullet \mathbf{Acc}_{e_1}$).
Let us consider the conditions which guarantee

$$Cl_{e_2}(A)\bullet \mathbf{Acc}_{e_1} \subseteq A\bullet \mathbf{Acc}_{e_1}. \tag{17}$$

By virtue of (15) we have the equivalence

$$\left(w' \in Cl_{e_2}\right) \Leftrightarrow \left(\{w'\}\bullet \mathbf{Acc}_{e_2} \subseteq A\bullet \mathbf{Acc}_{e_2}\right). \tag{18}$$

Hence, the inclusion (17) may be given the form of implication

$$\left(\{w'\}\bullet \mathbf{Acc}_{e_2} \subseteq A\bullet \mathbf{Acc}_{e_2}\right) \Rightarrow \left(\{w'\}\bullet \mathbf{Acc}_{e_1} \subseteq A\bullet \mathbf{Acc}_{e_1}\right). \tag{19}$$

Now let us assume $\mathbf{Acc}$ to be the presheaf of equivalences. Then the lhs of (19) asserts that the equivalence class of the world w' belongs to a set of classes. Since $\mathbf{Acc}_{e_2} \subseteq \mathbf{Acc}_{e_1}$, substitution of $\mathbf{Acc}_{e_2}$ by $\mathbf{Acc}_{e_1}$ expands the basis set ($\mathbf{Loc}_{e_2} \subseteq \mathbf{Loc}_{e_1}$) and generally expands any equivalence class. The affiliation of some class to a set of classes stays invariant. This is asserted by the rhs of (19). So, if $\mathbf{Acc}$ is the presheaf of equivalences, the law of association $e \to \mathrm{OL}_e$ (forgetting the orthocomplemented lattice structure of OL_e) can be written as a contravariant functor

$$\mathrm{OL} : \mathrm{C}^{op} \to Set, \tag{20}$$

i.e. OL is an object of the main topos $Set^{\mathrm{C}^{op}}$. In a similar manner, if $\overline{\mathbf{Acc}}$ is the presheaf of equivalences of retroworlds, we get the covariant functor

$$\overline{\mathrm{OL}} : \mathrm{C} \to Set, \tag{21}$$

i.e. $\overline{\mathrm{OL}}$ is an object of Set^{C}. Since both functor appear in the Chu spaces $(p_e, \mathrm{OL}_e, \overline{\mathrm{OL}}_e)$, we have proven the following

3.1. Proposition

If $\mathbf{Acc}$ and $\overline{\mathbf{Acc}}$ are presheaves of equivalences and if (10) is fulfilled for local Chu spaces $(p_e, \mathrm{OL}_e, \overline{\mathrm{OL}}_e)$, we have contravariant functor from C into the category of all Chu spaces.

4. Bayesianism in the branching space-time

Now it is worth to discuss the model of local observer in the branching space-time. The picture of external world is formulated in the observer's mind as a set of hypothesis -- conjectures to which the observer ascribes weights -- probabilities. These subjective probabilities are taken into account by the observer when he faces the problem of making choice between various activities. The result of the observer's acts provides him with a new information and make him re-evaluate the subjective probabilities. So, the proposed model of the observer in a great extent follows the well-known Bayesian concept [7]. Local nature of propositions and conjectures is the main peculiarity caused by the branching. This peculiarity will be proven to get adequate and natural representation in the topos approach. Yet the new generalized form of Bayesian rule is of independent interest beyond the the context of the topos approach. Hence, it is reasonable to give the most general form to this modified Bayesian rule. To this end, we consider two Chu spaces and some morphism between them:

$$(f^*, f_*) : (p_1, \mathbf{A}_1, \mathbf{H}_1) \to (p_2, \mathbf{A}_2, \mathbf{H}_2), \tag{22}$$

where $f_* : \mathbf{H}_1 \to \mathbf{H}_2$, $f^* : \mathbf{A}_2 \to \mathbf{A}_1$ and

$$p_2(A_2 \mid f_*(H_1)) = p_1(f^*(A_2) \mid H_1) \tag{23}$$

for all $H_1 \in \mathbf{H}_1$ and $A_2 \in \mathbf{A}_2$.

Probability distributions over the collection of all conjectures $\mathbf{H}$ form a set $\mathrm{D}(\mathbf{H})$, where D is the monad of distributions (special endofunctor $\mathrm{D}: Set \to Set$ in the category of sets) [8]. The original Bayesian rule for re-evaluation of subjective probabilities upon a proposition $A \in \mathbf{A}$ truth revealing will be denoted by the same symbol:

$$A: \mathrm{D}(\mathbf{H}) \to \mathrm{D}(\mathbf{H}). \tag{24}$$

Under this map, any element $q \in \mathrm{D}(\mathbf{H})$ is transformed into $q' \in \mathrm{D}(\mathbf{H})$ by the Bayesian rule

$$q'(H) = p(A \mid H)q(H)\Big[\sum_{H' \in \mathbf{H}} p(A \mid H')q(H')\Big]^{-1}. \tag{25}$$

The map $f_*: \mathbf{H}_1 \to \mathbf{H}_2$ is transformed by D into $\mathrm{D}f_*: \mathrm{D}(\mathbf{H}_1) \to \mathrm{D}(\mathbf{H}_2)$ where

$$[\mathrm{D}f_*(q_1)](H_2) =_{df} \sum_{f_*(H_1)=H_2} q_1(H_1) \tag{26}$$

for any distribution q_1 over $\mathbf{H}_1$.
The introduced notations are used in the following

4.1. Theorem
For any pair of Chu spaces (22) and any morphism between them the following diagram is commutative for any proposition $A_2 \in \mathbf{A}_2$:

$$\begin{array}{ccc} \mathcal{D}(\mathbf{H}_1) & \xrightarrow{\mathcal{D}f_*} & \mathcal{D}(\mathbf{H}_2) \\ {\scriptstyle f^*(A_2)}\downarrow & & \downarrow{\scriptstyle A_2} \\ \mathcal{D}(\mathbf{H}_1) & \xrightarrow{\mathcal{D}f_*} & \mathcal{D}(\mathbf{H}_2) \end{array}$$

Proof is straightforward. Let $q_2 =_{df} \mathrm{D}f_*(q_1)$ and $q_2' =_{df} (A_2 \cdot \mathrm{D}f_*)(q_1)$. By virtue of (25), we have

$$q_2'(H_2) = p_2(A_2 \mid H_2)q_2(H_2)\Big[\sum_{H_2' \in \mathbf{H}_2} p_2(A_2 \mid H_2')q_2(H_2')\Big]^{-1}. \tag{27}$$

From (26) we get

$$q_2'(H_2) = \sum_{f_*(H_1)=H_2} p_2(A_2 \mid H_2)q_1(H_1)\Big[\sum_{H_2' \in \mathbf{H}_2} \sum_{f_*(H_1')=H_2'} p_2(A_2 \mid H_2')q_1(H_1')\Big]^{-1} =$$

$$= \sum_{f_*(H_1)=H_2} p_2(A_2 \mid f_*(H_1))q_1(H_1)\Big[\sum_{H_2' \in \mathbf{H}_2} \sum_{f_*(H_1')=H_2'} p_2(A_2 \mid f_*(H_1'))q_1(H_1')\Big]^{-1}. \tag{28}$$

The relation (23) let us transform the last equation into

$$q_2'(H_2) = \sum_{f_*(H_1)=H_2} p_1(f^*(A_2) \mid H_1)q_1(H_1)\Big[\sum_{H_1' \in \mathbf{H}_1} p_1(f^*(A_2) \mid H_1')q_1(H_1')\Big]^{-1} = \\ = \Big[\big(\mathrm{D}f_* \cdot f^*(A_2)\big)(q_1)\Big](H_2). \tag{29}$$

So, we get

$$A_2 \cdot \mathrm{D}f_* = \mathrm{D}f_* \cdot f^*(A_2). \tag{30}$$

In the context of the elaborated topos approach this theorem states the commutativity of the diagram

$$\begin{array}{ccc} \mathcal{D}(\overline{\mathcal{OL}}_{e'}) & \xrightarrow{\mathcal{D}(\overline{\mathcal{OL}}^{e}_{e'})} & \mathcal{D}(\overline{\mathcal{OL}}_{e}) \\ {\scriptstyle \mathcal{OL}^{e}_{e'}(A)}\downarrow & & \downarrow{\scriptstyle A} \\ \mathcal{D}(\overline{\mathcal{OL}}_{e'}) & \xrightarrow[.]{\mathcal{D}(\overline{\mathcal{OL}}^{e}_{e'})} & \mathcal{D}(\overline{\mathcal{OL}}_{e}) \end{array}$$

for the causal arrow e'Î e and any physical proposition A from OL_e. The commutativity may be interpreted as unambiguously defined generalized Bayesian rule (the main diagonal of the diagram):

$$\mathrm{D}(\overline{\mathrm{OL}}_{e'}) \to \mathrm{D}(\overline{\mathrm{OL}}_{e}) \tag{31}$$

for re-evaluation of subjectively estimated likelihood of the local observer's conjectures about the external world.

5. Conclusion

There are some reasons to consider the notion of probability as a part of logic. Along with logic itself, this notion emerged in evolution. This assertion may in full measure be referred to the Bayesian rule.
In the present work a natural way is suggested of introducing the notion of probability into the topos approach to the Belnar's model of branching space-time. Heuristic considerations were delivered by quantum theory, since local orthologic OL_e may be treated as prototype of lattices of subspaces of quantum states. The notion of Chu spaces appeared to be very helpful as well as constructions based on `retroworlds' in space-time with reversed causal relations.
The first important result is the sufficient condition of the functorial character of the local orthologics OL and $\overline{\mathrm{OL}}$. This provides us with a functor from the branching space-time to the category of Chu spaces, which encompasses the entire probability structure of our topos approach. The second important result deals with generalized Bayesian rule of re-evaluation of subjective probabilities ascribed by local observer to his (her) conjectures about the external world.

References

[1] MacLane S and Moerdijk I 1992 *Sheaves in Geometry and Logic* (Springer-Verlag)
[2] Dring A and Isham C J 2007 A topos foundation for theories of physics *Preprints* arXiv: 0703060, 0703062, 0703064, 0703066
[3] Belnap N 1992 *Synthese* **92** 385
[4] Il'ichov L 2012 *Proc. Int. Conf. PIRT-2011 (Moscow)* p 117
[5] Il'ichov L 2013 *Proc. Int. Conf. PIRT-2013 (Moscow)* p 112
[6] Pratt V 1999 *Chu Spaces* (University of Coimbra)
[7] Howson C and Urbach P 2005 *Scientific Reasoning: the Bayesian Approach* (Open Court Publ. Comp)
[8] MacLane S 1998 *Categories for the Working Mathematician* (Springer)

Acknowledgements

This work is supported by RFBR (grant 16-02-00329) and the Council of the President of the Russian Federation for Support of Young Scientists and Leading Scientific Schools (grant NSh-2979.2012.2).

IOP Conf. Series: Journal of Physics: Conf. Series **1051** (2018) 012015 doi:10.1088/1742-6596/1051/1/012015

Gain of photon-axion conversion in paramagnetics

G N Izmaïlov[1] and V S Gorelik[2,3]

[1] Moscow Aviation Institute (National Research University), Moscow, Russia
[2] Lebedev Physical Institute of the Russian Academy of Sciences, Moscow, Russia
[3] Bauman Moscow State Technical University, Moscow, Russia
E-mail: gorelik@sci.lebedev.ru

Abstract. The possibility of realization of a photon-axion and axion-photon conversion processes in transparent dielectric media within the visible frequency region is analyzed. Materials contain paramagnetic ions with large magnetic moment: Tb^{3+}, Er^{3+}, etc. Therefore an additional strong magnetic field is present in the region of the paramagnetic ion. The reinforced field provides increased probability of the conversion processes. In order to ensure the synchronism condition it is supposed to use laser radiation whose frequency coincides with the frequency of unitary polaritons with a refractive index equals to one. Schemes akin the known scheme "Light Shining through a Wall" can be used to observe the photon-axion conversion process. The estimation of efficiency of conversion processes using real sources of laser radiation and the specific dielectric media with paramagnetic ions is given.

1. Introduction

According to modern concepts [1], pseudoscalar bosons called axions are representative of the Dark Matter (DM) components. This particles are weakly interacting with the visible matter and have very small rest energy ($E_{0a} \sim 1-100\ \mu\text{eV}$). Axions were introduced as a consequence of deviations from the Standard Model. It follows, in particular, from the CP violation, due to the absence of the electric dipole moment of the proton. Axions as pseudo scalar particles, preserving CP invariance, remove the problem existing within the framework of the Standard Model [2-4]. Astrophysical and cosmological problems arising in the analysis of the energy balance of the stars radiation, a halo presence around the galaxies [5-6], the cooling patterns of white dwarfs [7] and to reduce the duration of supernova SN 1987A [8-9] - can also be described on the basis of the existence of axions. According to the estimates of [1], the axions density in an environment is about 10^{21} m^{-3}.

Recently the assumptions [5, 9] that axions included in the DM can be detected not only in the astrophysical processes, but in laboratory have been expressed. Several experimental solutions have been proposed for axion searches in the plausible axion-photon coupling range [10]. For the easy registration of axions, the axion conversion into a photon in an external magnetic field was selected (Primakov's effect, see Figures 1, 2) [11]. As it was shown previously [12], the probability of N_a pseudoscalar bosons emerging, resulting from the conversion of N_γ quanta (photons) of the exciting electromagnetic radiation in the pseudoscalar bosons is given by the expression:

$$P_{\gamma\to a} = \frac{N_a}{N_\gamma} = \frac{1}{4}\cdot\left(\frac{Q}{\pi}\right)\cdot\left(\frac{\omega}{k_a}\right)\cdot\left(dB_0 l\right)^2\left(\frac{2}{ql}\sin\frac{ql}{2}\right)^2. \quad (1)$$

Here $d = 2.7 \cdot 10^{-11}$ GeV^{-1} is an axion-photon coupling constant, $q = k_a - k_\gamma$ is the change of axion momentum k_a during the emission of a photon, Q is a resonator quality factor. It is assumed that $\hbar = c = 1$.

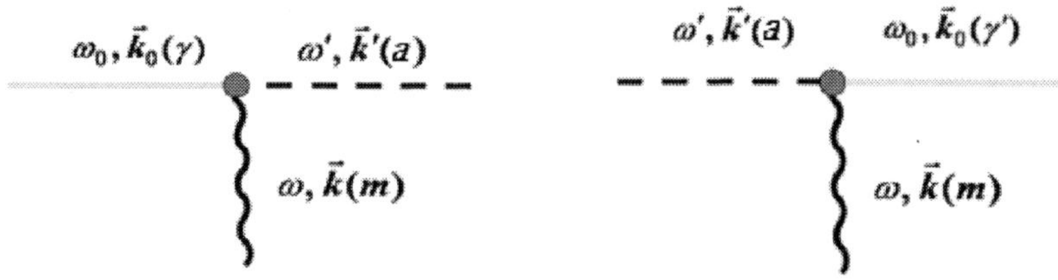

Figure 1. Feynman diagrams illustrating the effect of a magnetic field in Primakov effect.

It is possible also to reverse the process of turning photons into axions in the presence of the magnetic field. In the effect of "Light shining through walls" (LSTW) the intense laser light as the photon source should be used [12]. (See Figure 2.).

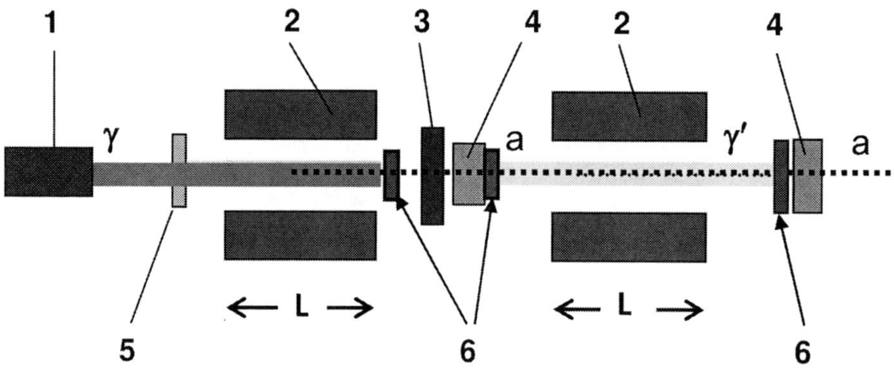

Figure 2. A schematic diagram of the setup for the observation of the effect of conversion of laser radiation photons in axions (γ → a) and inverse process - reconversion - ($a \to \gamma'$) using the resonators; 1 - a laser light source; 2 - coils; 3 - an opaque wall; 4 - receivers of secondary radiation; 5- an input translucent mirror of the resonator; 6 - resonators as Fabry-Perot interferometers

From formula (1) one can conclude that the increase in the probability of the photon-axion conversion in a vacuum can be implemented by an increase in the photons optical path length by the use of optical resonators [13], as well as by an increase of the magnetic field B_0. At the same time an axion-photon reconversion can be carried out in a dielectric medium. In this case the group velocity of the corresponding electromagnetic wave can be significantly reduced. In magnetic crystals the internal magnetic field can be created due to the presence of the paramagnetic ions with a large magnetic moment.

2. The dispersion relation of electromagnetic waves in a dielectric medium

In the absence of free charges ρ and currents j for dielectric nonmagnetic medium the electromagnetic waves are given by equation:

$$(\nabla^2 - \frac{\varepsilon\mu}{c_0^2}\frac{\partial^2}{\partial t^2})\vec{E} = 0. \quad (2)$$

Solutions of this equation are relations:

$$i\varepsilon\varepsilon_0 \vec{k}\vec{E}_0 \exp i\left(\vec{k}\vec{r} - \omega t\right) = 0, \quad (3)$$

$$\omega^2 = \frac{c_0^2 k^2}{\varepsilon(\omega)}, \quad (4)$$

where $\varepsilon(\omega)$ is the corresponding dispersion dependence of the dielectric constant. In the simplest case of an isotropic dielectric for one polar mode the equations take place:

$$\ddot{\vec{u}} = -\omega_0^2 \vec{u} + \frac{e\sqrt{F}}{m}\vec{E}; \quad \vec{u} = \vec{u}_0 \exp i\left(\vec{k}\vec{r} - \omega t\right). \quad (5)$$

Here m is an oscillator mass, e is charge of electron, F is the coefficient characterizing the power of an oscillator $(F \sim 1)$, u is an oscillating particle deviation from the equilibrium position. The amplitude of the oscillating deflection from the equilibrium of a charged particle, in accordance with equations (4) and (5) takes the form:

$$\vec{u}_0 = \frac{e\sqrt{F}}{m\left(\omega_0^2 - \omega^2\right)}\vec{E}_0. \quad (6)$$

Respectively for the amplitude of the polarization vector we obtain:

$$\vec{P}_0 = \frac{e_2 F}{mV_0\left(\omega_0^2 - \omega^2\right)}\vec{E}_0. \quad (7)$$

The equation of motion for the polarization vector is represented as:

$$\ddot{\vec{P}} = -\omega_0^2 \vec{P} + \frac{e^2 F}{mV_0}\vec{E}, \ \vec{P} = \frac{e\sqrt{F}\vec{u}}{V_0}\vec{P}_0 \exp i\left(\vec{k}\vec{r} - \omega t\right). \quad (8)$$

Introducing plasma frequency ω_p, we have the relation

$$\ddot{\vec{P}} = -\omega_0^2 \vec{P} + \omega_p^2 \vec{E}; \ \omega_p^2 = \frac{e^2 F}{mV_0}. \quad (9)$$

Accordingly for $\vec{D}_0$ we have:

$$\vec{D}_0 = \varepsilon_0 \vec{E}_0 + \vec{P}_0 = \varepsilon_0 \left[1 + \frac{e^2 F}{mV_0\left(\omega_0^2 - \omega^2\right)}\right]\vec{E}_0 = \varepsilon_0 \varepsilon(\omega)\vec{E}_0 \quad (10)$$

Thus, the dielectric function can be represented as:

$$\begin{cases} \varepsilon(\omega) = 1 + \dfrac{e^2 F}{mV_0\left(\omega_0^2 - \omega^2\right)} = 1 + \dfrac{\omega_p^2}{\omega_0^2 - \omega^2} = \dfrac{\omega_l^2 - \omega^2}{\omega_0^2 - \omega^2}; \\ \omega_l^2 = \omega_0^2 + \omega_p^2; \\ \omega_p^2 = \dfrac{e^2 F}{mV_0}. \end{cases} \quad (11)$$

Account of an electronic oscillator in the elementary volume V_0 leads to factorization of the dielectric function in the form:

$$\varepsilon(\omega)=\varepsilon_\infty \frac{\omega_l^2-\omega^2}{\omega_0^2-\omega^2}; \quad \varepsilon_\infty = n_\infty^2. \tag{12}$$

Respectively the dispersion relation for polaritons in the considered view of the dielectric medium with an electronic oscillator takes the form [14-15]:

$$\omega^2=\frac{c_0^2k^2}{\varepsilon(\omega)}=\frac{c_0^2k^2\left(\omega_0^2-\omega^2\right)}{\varepsilon_\infty\left(\omega_l^2-\omega^2\right)}=\frac{c^2k^2\left(\omega_0^2-\omega^2\right)}{\left(\omega_l^2-\omega^2\right)}; \quad c^2=\frac{c_0^2}{\varepsilon_\infty}. \tag{13}$$

From equation (13) we have the biquadratic equation:

$$\omega^4-\omega^2\left(\omega_l^2+c^2k^2\right)+\omega_0^2c^2k^2=0. \tag{14}$$

The exact solution of this equation defines two polariton branches:

$$\omega_\pm^2=\frac{\left(\omega_l^2+c^2k^2\right)}{2}\left(1\pm\sqrt{1-\frac{4\omega_0^2c^2k^2}{\left(\omega_l^2+c^2k^2\right)^2}}\right). \tag{15}$$

Here indices l, 0 mean longitudinal and transversal modes. Graphically this dependence is shown in Figure 3.

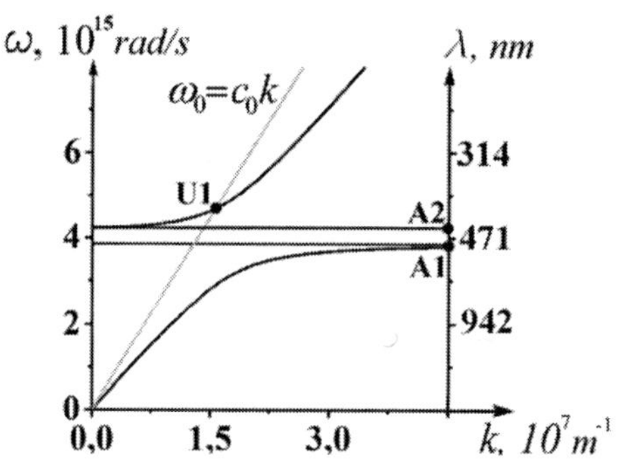

Figure 3. The dispersion law for polariton waves in terbium nitrate hydrate in the visible wavelength range. The corresponding wavelengths are A1 - 488 nm, A2 - 444 nm. The corresponding energy transition is $^5D_4 \rightarrow {}^7F_6$ ($\lambda = 448$ nm)

3. Peculiarities of the rare earth ions spectra

The photoluminescence spectra of terbium ion Tb^{3+} in the optical range is determined by electron transitions between the levels of unfilled inner 4f-shell, which is shielded from external influences of external electrons (usually *p*- and *s*-) shells. Such transition is forbidden in the electron dipole approximation [16].

The disturbance of the parity is possible if a terbium ion is surrounded by lattice or impurities (oxygen, carbon, and nitrogen). In such case, under the influence of the disturbance fields the opposite parity to the 4f-states occurs and the transition becomes permitted [17]. Such screening leads to narrow spectral lines and intense fluorescence.

It is known that the atomic configuration of the rare earth elements including terbium has the form $[Xe]4f^96s^2$ ([Xe] is the configuration corresponding to the electron shell of an xenon atom). In the presence of the crystalline lattice the terbium ion Tr^{3+} assumes the configuration $[Xe]4f^8$.

The splitting of the levels is determined by the symmetry of the nearest environment of the ion Tb^{3+}in the emitting complex [17]. Since the large number of different types of centers of symmetry in the same sample is realized, the emission spectrum has very complex structure.

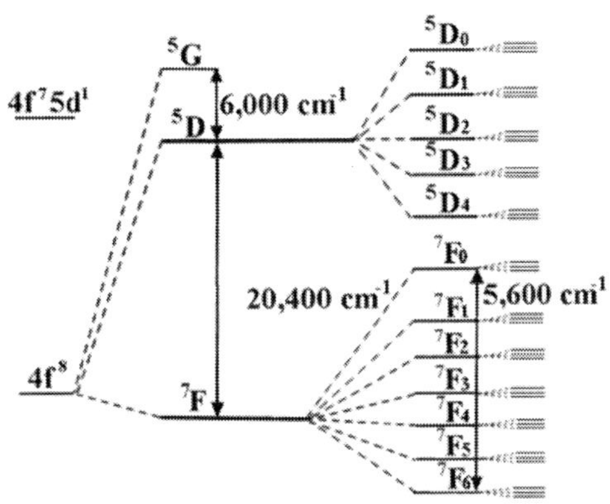

Figure 4. The scheme of the energy levels for terbium ions. The terbium ion energy levels are splitting under the influence of spin-orbit coupling and an external crystalline field

As a result of the spin-orbit interaction between the electrons in the f-shell of Tb^{3+} ^7F ground state splits into a number of terms $^7F_6$, $^7F_5$, $^7F_4$, $^7F_3$, etc. (the left part of Figure 4). Accordingly for excited states term ^5D splits into $^5D_0$, $^5D_1$, $^5D_2$, etc. (see Figure 4). The crystal field of ligands splits the terbium triple ion levels into the sub-levels, thus leading to resolve of degeneracy (the right part of Figure 4).

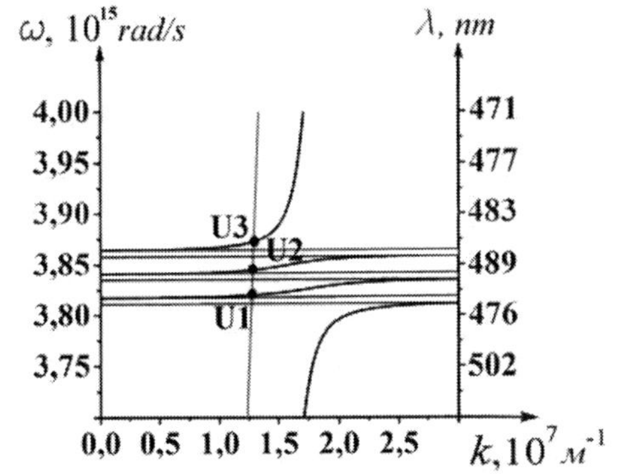

Figure 5. The dispersion relation of polariton waves in terbium nitrate hydrate in the visible wavelength range, taking into account the crystal field splitting.

As we can see from the Figure 5, the tangents to the polariton wave graphs at points U1 and U2 have a small slope. This corresponds to a small value of the group velocity of photons in the medium.

$$V = \frac{d\omega}{dk} = \frac{c^2}{c_0} = \frac{c_0}{\varepsilon_\infty}. \tag{16}$$

Here c_0 is the photon velocity in vacuum, V is the group velocity of polariton wave in the crystal. The slowing down of the group velocity (the value of the permittivity $\varepsilon_\infty \sim 10^3$) determines the increase in the photon-axion interaction time in the medium by a factor of c_0 / V. These effects contribute to the increase in the probability of a photon-axion conversion $\sim 10^3$.

4. Experimental setup

For experimental confirmation the splitting of the levels due to the presence of the crystalline lattice of the terbium ion Tr^{3+} we studied reflection spectra in an installation whose block diagram is shown in Figure 6 (a, b) The installation is consisted of four functional blocks: nitrogen laser ($\lambda = 337$ nm)-source of UV-radiation, samples, receivers and signal processing device. The obtained primary

IOP Conf. Series: Journal of Physics: Conf. Series **1051** (2018) 012015 doi:10.1088/1742-6596/1051/1/012015

excitation (see Figure 6a) was fed along the optical fiber (2) to the holder (4), on which the capillary light guide (5) with the sample (6) was placed. We put in reflected radiation through the optical fiber (3) to the spectrometer (7). The obtained spectra were transmitted to the computer (8) via the interface. The transmission spectra were studied in an installation whose block diagram is shown in Figure 6b. The primary excitation from nitrogen laser was focused into optical fiber (2) and fed to the holder (4).

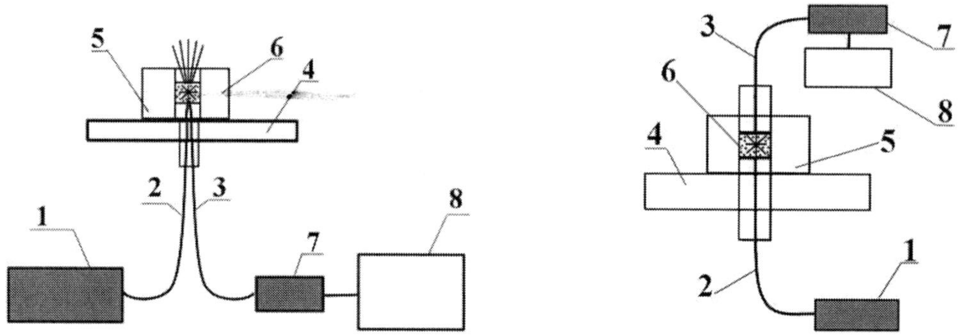

a b

Figure 6. Schematic diagram of the installation for recording the reflection spectra (a) and transmission (b): 1 – nitrogen laser (λ = 337 nm); 2, 3 - light guides; 4 – a holder; 5 - quartz plates; 6 – a sample; 7 -spectrometers; 8 – a computer

The capillary light guide (5) contained the sample (6). The transmitted radiation was fed along the optical fiber (3) to the spectrometer (7). The obtained spectra were transmitted by the interface to the computer (8).

5. An increase of conversion efficiency due to use of paramagnetic crystal

According to the previous experimental results, photon-axion conversion in vacuum requires more than 10 T values of magnetic induction with superconductor elements and cavity length more than 1 m [12]. In this paper we analyze the opportunity of photon-axion conversion efficiency gain by using of paramagnetic crystal $Tb(NO_3)_3{\cdot}6H_2O$ [18].

The unpaired electrons of the lanthanide elements in the 4f shell create a high value of the magnetic moment: $\mu = 11.4\mu_B$. An estimate of the magnetic moment for the f shell electrons is given by dependence $\mu_{eff} = \sqrt{n(n+2)}\mu_B$. Here, n is the number of unpaired electrons on the 4f shell [19, 20]. Crystal $Tb(NO_3)_3{\cdot}6H_2O$ has a low point symmetry (C_s point group) which corresponds to the axial symmetry, resemble to magnetic field symmetry. The presence of a large magnetic moment of theTb^{3+}, located at the site of the crystal structure, gives rise to local strong magnetic field inside the crystalline lattice. The magnitude of the internal magnetic field induction can be estimated from the experimental data on fluorescence spectrum (Figures 7,8) provided by the experimental setup described above. Assuming that the splitting of the spectral lines $\Delta\lambda$ in the spectrum of the Tb^{3+} ion is due to effective the Zeeman effect, we have

$$\Delta E = \hbar\Delta\omega = -\mu_i B. \qquad (17)$$

Here $\mu_i = 11.4\mu_B$ is the magnetic moment of the lanthanide ion, $\mu_B = 0.93\cdot 10^{-23}$ J/T is Bohr magneton, B is the induction of the magnetic field in the crystal lattice. So we obtain for effective magnetic field:

$$B = \hbar\frac{\Delta\omega}{\mu_i} = h\frac{c}{\mu_i\lambda^2}\Delta\lambda. \qquad (18)$$

Using the data of experimental measurements (Figure 8a) and the known constants, we obtain:

IOP Conf. Series: Journal of Physics: Conf. Series **1051** (2018) 012015 doi:10.1088/1742-6596/1051/1/012015

$$B_{\text{int}}^{Tr} = \frac{1.995 \cdot 10^{-25} \cdot 0.9 \cdot 10^{-8}}{11.4 \cdot 0.93 \cdot 10^{-23} (0.488)^2 \cdot 10^{-12}} = 71.7T. \tag{19}$$

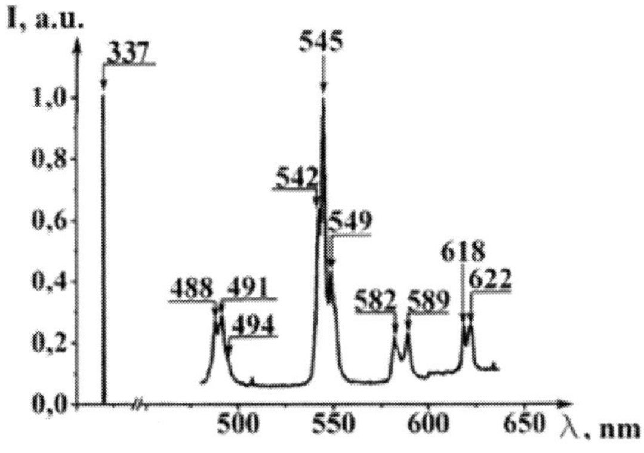

Figure 7. Whole photoluminescence spectrum of terbium nitrate hydrate at room temperature. Excitation by a nitrogen laser ($\lambda = 337$ nm)

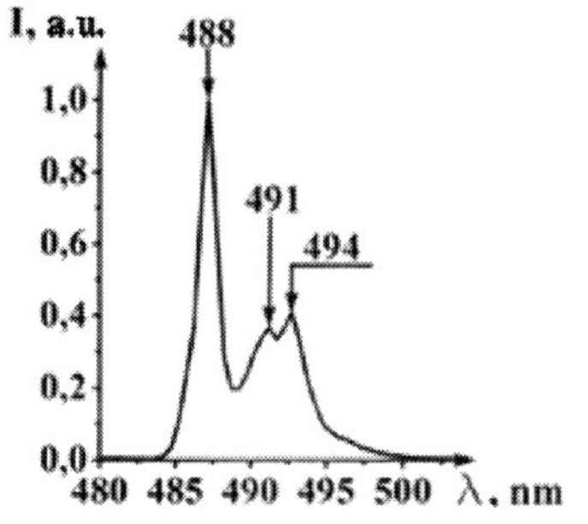

a

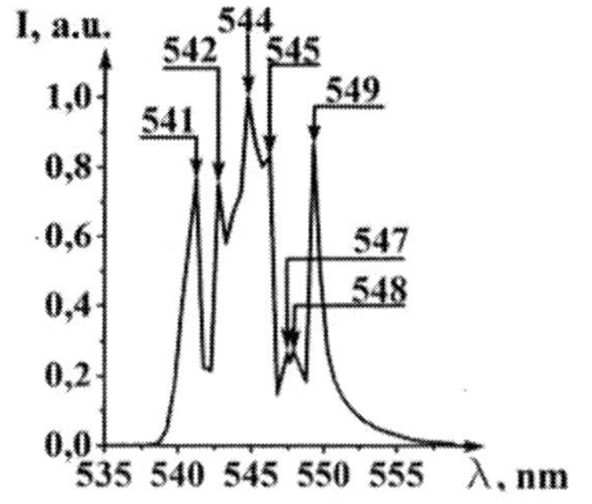

b

Figure 8. Parts of the photoluminescence spectrum of terbium nitrate hydrate for transitions: a) $^5D_4 \rightarrow {}^7F_6$; b) $^5D_4 \rightarrow {}^7F_5$.

Thus, the local internal strong magnetic field produced by a terbium ion in the low symmetrical crystalline lattice (Cs - symmetry) can increase the probability of axion-photon conversion in the paramagnetic ion region by about fifty times compared with the conversion probability in an external superconductor magnetic field of 10 T in vacuum. For approving of momentum conservation law in elemental conversion processes, we should use exciting radiation with frequency, close to the unitary polaritons value (see Figure 5, points U1, U2, U3)

Additional increase in the efficiency of the photon-axion transformation (1) will occur when Fabry-Perot resonators are used, inside which an active medium is placed

$$P_{\gamma \rightarrow a} = \frac{N_a}{N_\gamma} = \frac{1}{4} \left(\frac{Q}{\pi} \right) \left(\frac{\omega}{k_a} \right) \cdot \left(g B_0 l \right)^2 F^2(q). \tag{20}$$

Here Q is the quality factor of the resonator, $F(q) = \frac{2}{ql} \sin \frac{ql}{2}$ is a factor of the order of unity at a resonance. For $g = 2.7 \cdot 10^{-11}$ GeV^{-1}, $B = 10$ T, $l = 1$ m, and $Q \sim 10^5$, the probability of the conversion is $P_{\gamma \rightarrow a} \sim 10^{-14}$. Estimates made for laser $P \sim 10$ W ($\lambda = 0.5$ μm) emitting in a quasicontinuous regime with the number of photons at the output $N_\gamma \sim 10^{20}$ per second determine the number of photons at the output of the second identical resonator (see Figure 2) - $N'_\gamma \sim 10^{-8}$.

Using of a paramagnetic medium placed in a Fabry-Perot resonator, taking into account the local magnetic field $B_{\text{int}} = 71.7\,\text{T}$ and reducing the group photon velocity $c / c_0 \sim 10^3$ within the crystal, will lead to increase in the estimate of the number of photons at the output of the apparatus by a factor of 10^7.

6. Conclusion

Thus in this paper the gain of axion-photon conversion process is proposed by means of paramagnetic crystalline media using instead of vacuum and outer magnetic field, having used before. Low crystalline symmetry of proposed paramagnetic crystal ($Tb(NO_3)_3 \cdot 6H_2O$) proves the strong internal local effective magnetic field, larger than outer superconductor magnetic field, having been used before. For approving of momentum conservation law in elemental conversion processes, it was proposed to use the exciting laser emission with frequency, close to unitary polaritons value (see Figure 5, points U1,U2,U3). In this case the group velocity of exciting light is essentially less with comparing to light velocity in vacuum. Accordingly the efficiency of photon – axion conversion is waited be increased enough for hopeful recording of the secondary emission in Primakoff effect by modern detectors.

References

[1] Sloan J V *et al* 2016 *Physics of the Dark Universe* **14** 95
[2] Peccei R D and Quinn H R 1977 *Phys. Rev. Lett.* **38** 1440
[3] Weinberg S 1978 *Phys. Rev. Lett.* **40** 223
[4] Wilczek F 1978 *Phys. Rev. Lett.* **40** 279
[5] Sikivie P 1983 *Phys. Rev. Lett.* **51** 1415
[6] Ruffelt G G 1990 *Phys.Rept.* **198** 1
[7] Melendez B, Bertolami M M and Althaus L 2012 *Revisiting the Impact of Axions in the Cooling of White Dwarfs (Preprint* arXiv: 1210.0263v1)
[8] Okun L V 1982 *Sov. Phys. JETP* **56** 502
[9] Ringwald A 2015 *The hunt for axions (Preprint* arXiv: 1506.04259v1)
[10] Rosenberg L J 2015 *PNAS* **112** 40 12278
[11] Primakoff H 1951 *Phys. Rev.* **81** 899
[12] Sikivie P, Tanner D and van Bibber K 2007 *Phys. Rev. Lett.* **98** 172002
[13] Gorelik V S and Izmaïlov G N 2011 *Bull. Lebedev Phys. Inst.* **38** 177
[14] Agranovich V M 1968 *Theory of excitons* (M., Nauka, 384 pp.)
[15] Rumyantsev V V and Gumennyk K V 2015 *J. of Photonic Materials and Techn.* **1** 11
[16] Bunker P R and Jensen P 2006 *Molecular Symmetry and Spectroscopy* (NRC Research Press 2nd ed. 748 pp.)
[17] Dieke G H 1968 *Spectra and Energy levels of Rare Earth Ions in Crystal* (Int. Pub.: A division of John Wiley & Sons, N.Y. 401 pp.)
[18] Hoard R *et al* 1985 *IEEE Transactions on Magnetics* **21** 2
[19] Audi G, Bersillon O, Blachot J and Wapstra A H 2013 *Nuclear Physics* A **729** 3
[20] Huray P G, Nave S E and Haire R G 1983 *Journal of the Less Common Metals* **93** 2 p 293

IOP Conf. Series: Journal of Physics: Conf. Series **1051** (2018) 012016 doi:10.1088/1742-6596/1051/1/012016

The possibility of phased network developing for gravitational waves registration by the compact high sensitive system SQUID-magnetostrictor

G N Izmaïlov[1], A M Tshovrebov[2] and L N Zherikhina[2]

[1]Moscow Aviation Institute (National Research University), Moscow, Russia
[2]Lebedev Physical Institute of Russian Academy of Sciences, Moscow, Russia
E-mail: izmailov@mai.ru

Abstract. We consider certain features of registered merging of black holes expressed in gravitation radiation. Some schemes intended to register gravitational waves by utilizing magnetostrictive effect are offered. The magnetic field producing by strain is amplified by SQUID with an additional circuit allows to compress field quantum fluctuations. We bring estimates of proposed sensitivity of a phased network intended for gravitational waves antenna developing.

1. Introduction

Now we have learned some evidences of gravitational waves (GW) registration: GW150914 [1], GW151226 event [2].

Usually we say "postman rings twice" but there was the third event GW170104 [3] recently.

The signals produced by merging of double BHs (with masses ~30 solar masses $M_\odot$) were unexpected for the majority of researchers. As it was detected during the merging the enormous amount of energy $2\div5\ M_\odot c^2$ have been released. Detections provide an estimate of the merger rate of stellar-mass binary BH, and the rate appears to be larger than some models predict. All events are originated very far from Galaxy ($R \sim 1.3 \div 3$ Gly).

Thus, the reality of the GW is established (at the same level of confidence as the registration of the Higgs boson [4-5]), and now efforts are needed to create a GW network of observatories with reliable and inexpensive equipment for the comprehensive study of gravitational waves of a wide range of frequencies.

Moreover two issues are notably important for understanding and classifying generation processes of cosmic GW

- The residual issue is to determine a direction to the source.
- Spins are difficult to measure, but results from GW170104 suggest that the spins may not be aligned with the pair's mutual orbit, a finding that could favor some black hole binary formation mechanisms over others.

2. Field approach for GW parameters estimates

For modeling and mathematical description, Einstein's exact geometrodynamic equation is used. But for estimating GW parameters in regions remoted from the source, a field approach can be applied. Instead of the representation of spacetime as a fabric with special properties, forming distortions of

which are perceived as gravitational waves, we will to use the energy approach to evaluate the results of deformation [6].
Indeed, as is well known, the spatial dependence of the static field potential $\Phi(r)$ on the distance r to the source can be represented in the form of a series

$$\Phi(r)=\frac{\alpha}{r}+\frac{\beta}{r^2}+\frac{\gamma}{r^3}+\frac{\delta}{r^4}+\dots, \quad (1)$$

where the terms of the series determine the contributions of the monopole, dipole, quadrupole, octupole, etc. The numerator of the term symbolizes its contribution to the potential. In the case of gravitational radiation, the laws of conservation of energy and momentum, as well as the equivalence principle, forbid the monopole and dipole radiation (in the electrodynamics variant, a single charge or a dipole made up of an electron and a positron does not emit). In the near zone $r \ll \lambda$, the force produced by the field acting on the unit mass is estimated by the acceleration

$$a_{QP} \propto G\frac{D}{r^4}, \quad (2)$$

but in a far (wave) zone $r \gg \lambda$

$$a'_{QP} \propto G\frac{D}{\lambda^3 r}, \quad (3)$$

here D is quadrupole moment of the system

$$D=\int_V \rho(x_i x_j - \tfrac{1}{3}\delta_{ij}x_i^2)dV, \quad (4)$$

here V is the volume occupied by the system, ρ is the density, x_i is the coordinate along the i or j direction, $i, j = 1,2,3$.
The intensity of the gravitational field in the wave zone is proportional to the square of the acceleration a'_{QP}. Its estimate is

$$I \propto \frac{1}{G}(a'_{QP})^2 \propto G\frac{D^2}{c^6 T^6 r^2}, \quad (5)$$

the magnitude of the gravitational wave Umov-Poynting vector is

$$S_{GW} \propto \frac{G}{c^5}\frac{D^2\omega^6}{r^2}. \quad (6)$$

Therefore, the losses of energy to gravitational radiation by the system is estimated as

$$\frac{d\varepsilon}{dt}=-\frac{G}{c^5}D^2\omega^6. \quad (7)$$

The exact value of radiation losses was obtained by General Relativity. In modern notation it is written as:

$$\frac{d\varepsilon}{dt}=-\frac{32}{5}\frac{G}{c^5}D^2\omega^6. \quad (8)$$

It follows from the last formula that the efficiency of conversion of the merging energy of the system into gravitational radiation is substantially higher with the growth rate of the change in quadrupole moment of the system. On the other hand, gravitational waves should have amplitude, which is denoted by the dimensionless value of h. Its assessment from the energy point of view is

$$h=\frac{\Phi'_{QP}}{c^2} \propto \frac{G}{c^4}\frac{\ddot{D}}{r} \propto \frac{r_s^2}{d \cdot r}, \quad (9)$$

IOP Conf. Series: Journal of Physics: Conf. Series **1051** (2018) 012016 doi:10.1088/1742-6596/1051/1/012016

where the potential Φ'_{QP} corresponds to the acceleration (3), $r_S = \frac{2GM}{c^2}$ is Schwarzschild gravitational radius, d is the characteristic size of the quadrupole, M is its mass and r is the distance to the radiating system. From the last formula it is easy to obtain another estimation of the GW amplitude

$$h = \frac{\Phi_{in}}{c^2} \cdot \frac{\Phi_{out}}{c^2}, \tag{10}$$

here $\Phi_{in} = GM/d$, and $\Phi_{out} = GM/r$.

The upper limit of the GW frequencies is estimated by the time of light travel around the horizon of the events of the black hole. This gives the maximum frequency

$$f \propto \frac{c}{2\pi r_g} = \frac{c^3}{4\pi GM}. \tag{11}$$

An important property of an inhomogeneous gravitational field is its unavoidability when measurements are carried out in a reference frame having finite dimensions. From the standpoint of General Relativity, it was shown that in this case it is convenient to observe the created tidal acceleration in a two-mass system in a reference frame connected with the center of mass of particles and the choice of a special coordinate system (the so-called TT calibration). With this choice of time measurement, t is given by the values of the proper time τ, and the equation of motion of the test masses is described by the forced oscillation equation in ordinary derivatives. Then, if the experimenter takes two test bodies and measures the distance between them, then with the arrival of the GW packet, it is possible to determine the amplitude and frequency of the GW.

The most accurate measurements are carried out interferometrically, so from the 70's of the last century, laser interferometers appeared in various laboratories of the world using the optical schemes of Michelson and Fabry-Perot. Mirrors of interferometers are located on free masses. An independent and earlier proposal for the application of a laser interferometer for recording GW was presented in [7].

3. Approximation for the vortex motion of the merging masses

An substantial feature of the generation of gravitational waves is the spiral motion of merging double black holes. Explain its effect can be so. The body rotating with an angular velocity ω has the energy $E_{rot} = \frac{1}{2} I\omega^2$ where I is the moment of inertia of the body relative to the axis of rotation. This additive to the total energy is proportional to the addition to the mass of the body, since $\Delta m = \frac{E_{rot}}{c^2}$. As the main property of the mass is manifested in the creation of the gravitational field, a new gravitational field is created that has a vortex character.

This addition produces (due to the equivalence principle) the additional gravitational field; therefore a neighboring body must rotate faster in order to compensate a growing attraction force.

So an additional energy of rotation alters the spacetime metric, new effects have arisen.

From radial geodesic equation transcribed in an equatorial plane

$$\frac{d}{ds}\left(g_{rr}\frac{dr}{ds}\right) = \frac{1}{2}\frac{dg_{\mu\nu}}{dr}\frac{dx^\mu}{ds}\frac{dx^\nu}{ds}. \tag{12}$$

For slow rotation and relatively small masses effect of central body rotation is Lense – Thirring field (by analogy with the magnetic field it is often called gravimagnetic). The effect is appeared. But since gravity “pulls”, then Lense-Thirring field "drags" behind itself when a probe body is in orbit. From eq. (13) one gets

IOP Conf. Series: Journal of Physics: Conf. Series **1051** (2018) 012016 doi:10.1088/1742-6596/1051/1/012016

$$ds^2 = -\left(1-\frac{r_s}{r}\right)(cdt)^2 + \frac{1}{\left(1-\frac{r_s}{r}\right)}dr^2 + r^2\left(d\theta^2 + \sin^2\theta d\phi^2\right) - 2\frac{r_s}{r}\frac{I\Omega}{M}\sin^2\theta dt d\phi \tag{13}$$

the change of attractive acceleration is equal to

$$\Delta a = \frac{2GI\Omega_1^2}{R^2c^2}. \tag{14}$$

Here Ω_1 is a new angular velocity. Because there should be a new balance between the new gravitational acceleration and centripetal one, a new angular velocity Ω_1 is and

$$\Delta\Omega_{LT} = \frac{GI\Omega}{R^3c^2}. \tag{15}$$

For the motion of bodies in the field of Lense-Thirring, created by the Earth (the LAGEOS project) and even by the Sun the effect is negligible [8].

However, for a relativistic mass motion of ~ 30 solar masses and at small relative distances the field energy estimate is ~ 10^{23} J, which significantly affects the dynamics of merging. With the implementation of theoretical predictions and the use of the newest metrological standards and technological solutions, the scientific groups working in the joint project LIGO were more successful than others. They already during the engineering run of advanced interferometers and auxiliary equipment fixed the relative displacement of 40-kg mirrors by a distance of a millionth of the proton's diameter. Such a displacement was caused by the release of energy from the confluence of the black holes with an initial total mass of ~ 65 solar masses outside Galaxy.

4. Magnetostrictive detector. Sensitivity and its applications

Above we considered the results obtained with laser interferometers of the LIGO group for the search for gravitational waves of cosmic origin. First of all, it should be noted that there are known ways to further enhance the sensitivity and reliability of the interferometers created. While the achieved sensitivity in the visible range of electromagnetic radiation corresponds to the limiting sensitivity of an eye, but for sound frequencies - the audibility of a whisper at the distance of 1 m.

Along with huge interferometric antennae (LIGO, VIRGO, AIGO, GEO, KARGA, LISA, etc.) we can try to use a net of small but highly sensitive detectors for high frequencies GW. Such detectors and networks of them can be implemented in laboratory experiments and small observatories. As an example let us examine so called magnetostrictors as sensors and SQUIDs as a transducer.

A SQUID (acronym for Superconducting Quantum Interference Device) is a loop of superconducting material interrupted by one or two Josephson junctions. We may sketch a SQUID in the next figure (Figure 1). As it follows from the theory for superconductors [10]

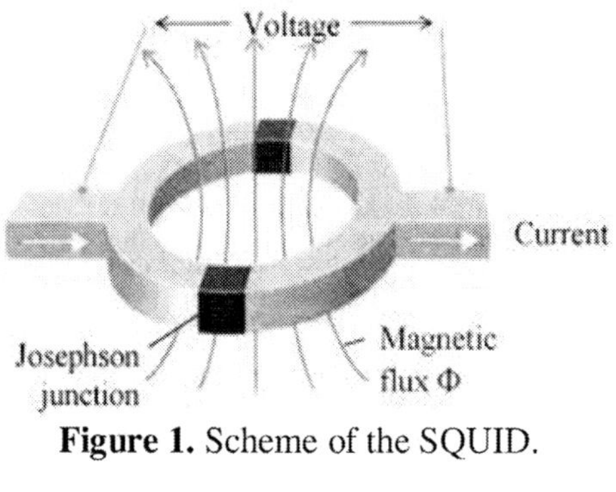

Figure 1. Scheme of the SQUID.

the operation of SQUID on a direct current is described by two Josephson relations: the current through the junction is

$$I = I_0 \sin\phi, \tag{16}$$

here $LI_0 = \Phi_0/2$, where $0 < \varphi < 2\pi$; $\Phi_0 = \frac{\pi\hbar}{e} \approx 2{,}07 \cdot 10^{-15}$ Wb is a magnetic flux quanta for stationary Josephson effect and

$$\hbar\frac{\partial\phi}{\partial t} = 2eV \tag{17}$$

for nonstationary effect. As was shown for intend to compress quantum noise in the device one can include RF-SQUID into a scheme for parametric compression (Figure 2) [11]

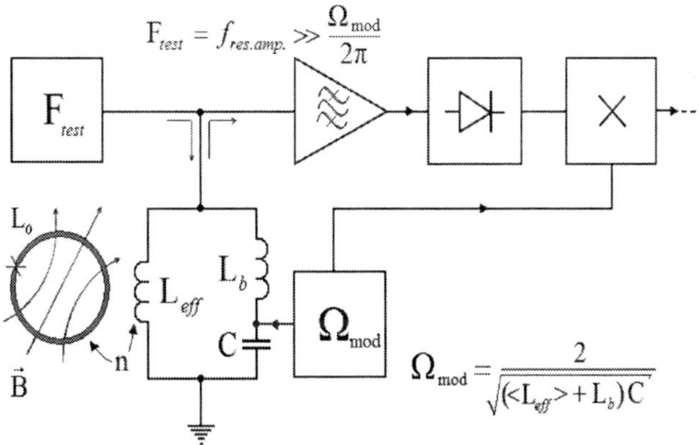

Figure 2. Including of SQUID into a scheme for parametric compression of quantum fluctuations.

When considering a hysteresis-free RF-SQUID in the quantum oscillator mode with compressed coherent states one can expect the growth of sensitivity from $10^{-8} \div 10^{9}$ $\Phi_0 / \sqrt{Hz}$ is determined by noise levels are as low as 3 fT/$\sqrt{\text{Hz}}$.

What kind of materials magnetostrictive detector is suitable for? We can point on 1. Ferromagnetic materials; 2. Piezomagnetic materials and 3. Superparamagnetic materials. For example the alloy 54%Pt46%Fe has $\Lambda^{-1} = \frac{\partial B}{\partial P} \approx 7 \cdot 10^{-6}$ T/Pa where Λ is a magnetostrictive coefficient proportional to the shape change of a ferromagnetic material during magnetization; another alloy is $Tb_xDy_{1-x}Fe_2$.

Below we sketch arrange of a magnetostrictor rod in the field of a planar GW (Figure 3).

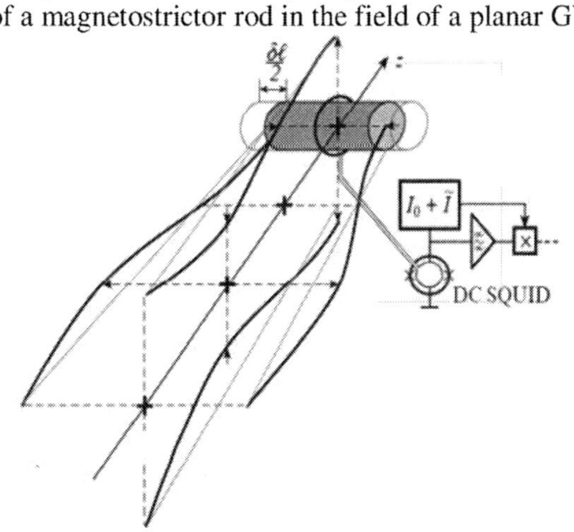

Figure 3. Qualitative relation between the geometry of gravitational wave propagation and the position of a magnetostrictive antenna.

We offer a superconducting concentrator for the magnetic flux to gain effectivity of a device (Figure 4).

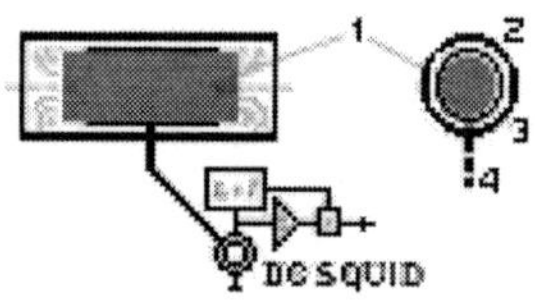

Figure 4. Schematic diagram of a magnetic flux superconducting concentrator.(1) a magnetostrictor; (2) the tubular loop of the superconducting flux transformer; (3) an external superconducting tube (superconducting flux concentrator); (4) a twin-wire connection link of the flux transformer.

The change of the magnetic flux in the last is detected by SQUID. In the foreground the plane-polarized gravitational wave is schematically shown. Arrows point to the regions of decreasing and increasing dimensions of a virtual test body, caused by the influence of the gravitational wave. The proportion between the wavelength λ and the dimension l of the cylindrical antenna is deliberately distorted (really wave length is greater than l); for clarity the variation of geometric dimensions of the test body in the field of the weak gravitational wave is also exaggerated (by 20 orders of magnitude). Here the electric circuit includes a DC- SQUID and elements of service electronics. The crosses in the operating ring of the DC-SQUID mark Josephson tunnel junctions – the principal elements of the superconducting quantum interferometer.

To evaluate the sensitivity of the circuit, we put the sensor response at the level of its noise $\langle\delta\Phi\rangle/\sqrt{Hz} = 10^{-6}\,\Phi/\sqrt{Hz} = 2.07\cdot10^{-21}\dfrac{Wb}{\sqrt{Hz}}$. So the estimation of minimal level of the magnetostrictive rod relative elongation is

$$\langle\delta l/l\rangle/\sqrt{Hz} = \langle\delta F\rangle/S_m\cdot E/\sqrt{Hz} \approx 10^{-24}/\sqrt{Hz}, \qquad (18)$$

if a rod cross-section S_m is 30 sm^2, material strain $E = 100$ GPa.

5. Schemes with magnetostrictors for GW detection

We offer some schemes for GW detection with magnetostrictors (Figure 5 and Figure 6).

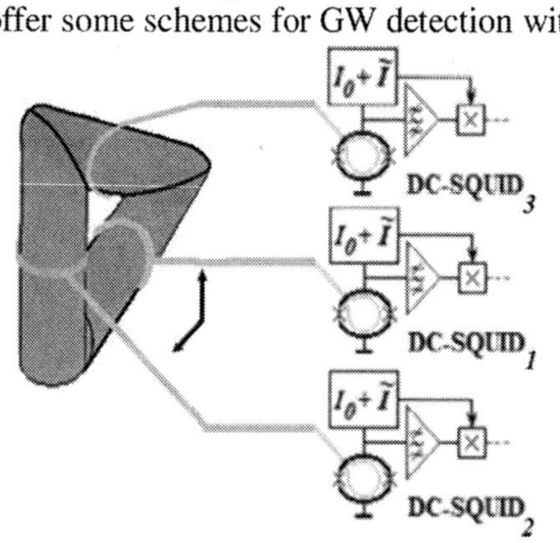

Figure 5. 2D detector scheme.

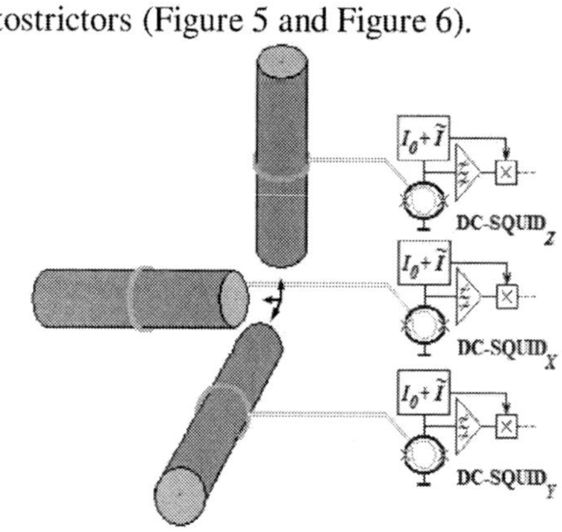

Figure 6. 3D detector scheme.

These schemes on Figures 5 and 6 are basic elements for the GW registration network is destined.

IOP Conf. Series: Journal of Physics: Conf. Series **1051** (2018) 012016 doi:10.1088/1742-6596/1051/1/012016

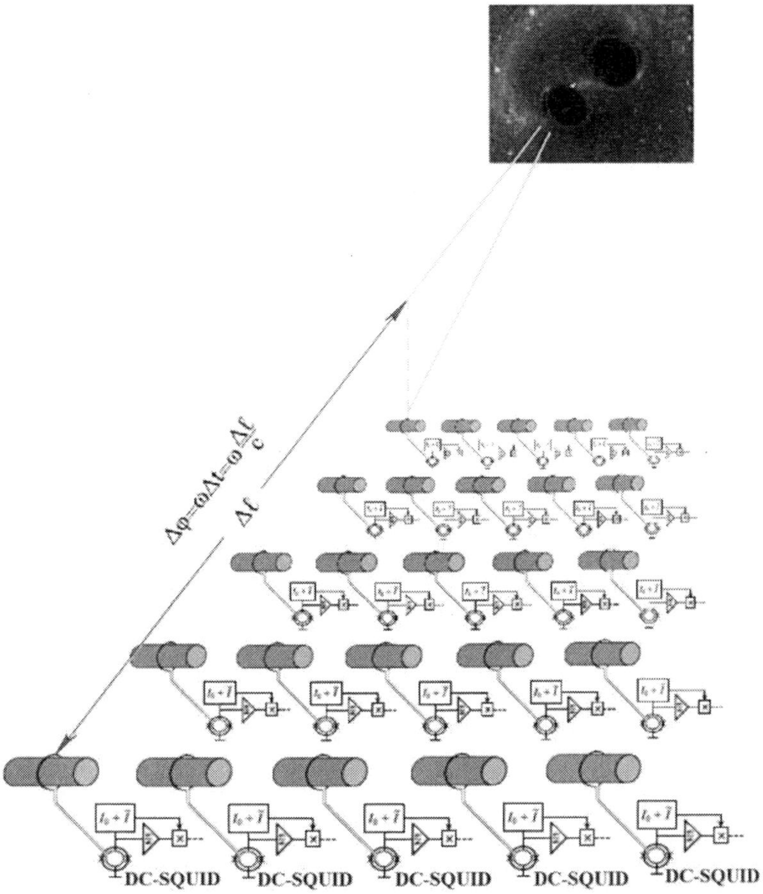

Figure 7. The phased network of SQUID-magnetostrictor detectors for gravitational waves registration. For simplicity the magnetostrictive detectors are schematized.

References

[1] Abbott B P *et al* (LIGO Scientific Collaboration and Virgo Collaboration) 2016 *Phys. Rev. Lett.* **116** 061102

[2] Abbott B P *et al* (LIGO Scientific Collaboration and Virgo Collaboration) 2016 *Phys. Rev. Lett.* **116** 241103

[3] Abbott B P *et al* (LIGO Scientific and Virgo Collaboration) 2017 *Phys. Rev. Lett.* **118** 221101

[4] Chatrchyan S *et al* (CMS Collaboration) 2012 *Phys. Rev. Lett.* **108** 111804

[5] Aad G *et al* (ATLAS Collaboration) 2012 *Phys. Rev. Lett.* **108** 111803

[6] Izmaïlov G N 2016 *Meas. Tech.* **59** 6 560

[7] Gertsenshtein M E and Pustovoit V I 1963 *Soviet Physics-JETP* **16** 2, 433

[8] Golovashkin A I, Izmaïlov G N, Kuleshova G V, Tskhovrebov A M and Zherikhina L N 2009 *Int. Meeting on Physical Interpretations of Relativity Theory "PIRT 2009"* (Moscow) pp. 146-154

[9] Golovashkin A I, Zherikhina L N, Tskhovrebov A M, Izmaïlov G N and Ozolin V V 2010 *JETP* **111** 2 332

[10] Clarke J, Braginski A I eds. 2006 *The SQUID Handbook: Applications of SQUIDs and SQUID Systems*. **2** (Wiley-VCH)

[11] Zherikhina L N, Izmailov G N and Tskhovrebov A M 2016 *Meas. Tech.* **59** 5 460

IOP Conf. Series: Journal of Physics: Conf. Series **1051** (2018) 012017 doi:10.1088/1742-6596/1051/1/012017

Sakharov Curvature in Rowlands Duality Spacetime: Do vacuum 'spacetime forces' curve matter?

S E Karam

Morgan State University, Baltimore, USA
E-mail: sabah.karam@morgan.edu; sabah@dualityScience.com

Abstract. Space and time, inertia and gravity are unseen yet they are the most fundamental concepts needed to understand physics classically, relativistically, and quantumly. Non-Euclidean geometries and relativity theory have yielded new insights in treating the four interrelated metaphysical concepts. Minkowski linked time and space. Einstein linked Newton's first and second laws of motion. Attempts to link electromagnetism / gauge forces to theories of gravity have been somewhat of a challenge. Could discrete *inertia* be the unrecognized link between quantum mechanics and relativity theory? Newtonian mechanics and Maxwell's equations are based on the mathematics of continuous functions and relativistic effects count Newtonian laws as approximations. The classical period of physics, based on ideas of continuity came to an end when discreteness and quantum considerations, like the Planck black-body spectrum, became experimentally and theoretically relevant. General relativity and quantum mechanics have reshaped our way of thinking about classical geometric concepts of space and time as well as inertia and gravity. Unsolvable mathematical frameworks have not provided the inertial / gravitational and space / time superstructures needed to address the unification problem. The first attempts to link gravity to the behavior of subatomic particles originated in papers written by two Russian physicists. Sakharov proposed that Newtonian gravity could be interpreted as a dipolar van der Waals force that emerges from the action of quantum fluctuations of the empty vacuum. Zeldovich gave a new meaning to Einstein's cosmological constant, as a measure of the curvature of space, allowing for new insights into cosmological inflation and dark energy. It will be argued in this paper that zero-point fluctuations which produce inertial reaction forces also produce the elasticity / curvature of spacetime suggested by Sakharov / Zeldovich. Is it possible that dynamic spacetime gives matter its curvature? We will attempt to show the feasibility of this idea by appealing to Rowlands quaternion representation of spacetime and gauge / gravity duality hypothesis. We will then argue that Sakharov's induced continuous gravitational field emerges which balances the discrete inertial *spacetime forces* satisfying a zero-totality condition.

1. Introduction

The "origin and nature of inertia" was described by Vigier [1] as an "unsolved mystery in modern physics" even though it is known that inertia instantaneously opposes acceleration. Newton viewed inertia as an internal property of matter. Mach, on the other hand, thought that inertia was an external property collectively linked to all matter in the universe. General relativity (GR) links spacetime curvature to gravitational fields and inertia is accommodated by Einstein's equivalence principle (EEP). GR cannot, however, accommodate Machian-type inertia and attempts to explain gravity as the source of inertial forces in the geometrical interpretation fails. Any appeal to the equivalence principle leads, rather, to a circular argument. Solutions to Einstein's field equations have been formulated for

empty de Sitter spaces and rotating Godel universes. Vigier's investigations into inertia from a relativistic perspective revealed other conflicts when Mach's principle is considered and he attempted to reconcile GR and Mach's principle proposing that non-Machian forces operated in "local interactions of a Dirac subquanum aether model stemming from Einstein-de Broglie-Bohm causal stochastic quantum mechanics". Also called the Dirac Sea, a theoretical model of the *vacuum* with negative energy, Vigier's aether model combines matter and antimatter, particles and antiparticles. The aether's density is extremely large but cannot be detected because the observable effects cancel each other out. Fluctuations in the vacuum can be understood as the production of a weak dipolar fermion / antifermion pair, a harmonic oscillator creation-annihilation type of mechanism. The same fluctuations are also responsible for the Casimir effect and the Van der Waals force corresponding to the potential for a fluctuating dipole-dipole interaction.
Another model for inertia that uses the electromagnetic part of the Unruh radiation has been suggested by Rueda and Haisch [2-3]. They substitute electromagnetic fields of the quantum vacuum for Mach's principle as the "external causative agent of inertia" providing a "locally-originating reaction force." The "quantum inertia vacuum hypothesis" that inertia originates in interactions between quarks and electrons and the zero-point field of the vacuum was inspired by Sakharov's 1967 conjecture [4]. Cosmological spacetime, surprisingly, can be considered as a phenomenon associated with the physics of particles and fields. Spacetime structure is not directly observed nor measured but is deduced indirectly by looking at the propagation of radiation and matter and has been assimilated into a theory called quantum field theory (QFT). Inertial properties of neutrons and protons arise in the zero-point fluctuations (ZPF) scattering at the level of individual quarks. ZPF, zero-point energy (ZPE), zero-point field, and vacuum energy considerations all appeal to theories in particle physics. The density of the ZPE is linked to Newton's gravitational constant G and the particles may originate in the cosmic microwave background (CMB) which has been experimentally validated. The quantum vacuum stress between conducting plates caused by modifications to the electromagnetic ZPF was predicted by Casimir in 1948. This effect has also been confirmed by experiment. The Zeeman, Aharon-Bohm, Sagnac, and Lamb effects, which are considered as evidence of the Dirac polarized vacuum, are based on classical physical conditions but cannot be replicated in laboratory experiments. The continuity of the filled Dirac vacuum of mass-energy, the Higgs field, and the need for instantaneous correlation of inertial reaction on discrete fermionic or bosonic states requiring an instantaneous and continuous gravitational force to balance it seem to satisfy Rowlands' spacetime zero-totality condition. The following sections will investigate these ideas.

2. Induced gravity

Einstein's theory *about* gravity appeared in 1915. It is also theory *about* space-time, a geometry which is not fixed but governed by a set of field equations. Gravity, rather than being described as a Newtonian force on a fixed space-time manifold, is embedded in the geometry of space-time. Unknown origin of Newtonian-type forces are no longer relevant in this picture of gravity. Sakharov, on the other hand, proposed that vacuum fluctuations, forces, were relevant in gravitational theory. He suggested that gravity was "analogous to the discussion of quantum electrodynamics (QED)." Sakharov's hypothesis "identified the action (in Einstein's theory of gravity) as being the change in quantum fluctuations of vacuum resulting in a distortion of space." Sakharov obtained an expression for Newton's gravitational constant G which is essential in Einstein's field equation. The expression Sakharov derived, $G = \frac{c^3 L^2}{\hbar} = \frac{\hbar c^3}{p_0^2}$, connects Newton's gravitational constant to the elementary Planck length L, the cutoff length of Einstein's action, and a limiting momentum $p_0 = \frac{\hbar}{L}$. If the evolution of the universe in the big bang theory is followed backwards in time it can be shown that the entropy is a consequence of baryon asymmetry. In 1967 Sakharov proposed a general model for the non-zero baryon number in which hypothetical heavy bosons mediated interactions between quarks and leptons. The following year he linked gravitation to spacetime and vacuum emergence treating gravity, not as a fundamental interaction but, as a secondary effect induced by non-gravitational forces. This raises the question about spacetime. Is spacetime really non-Euclidean? Could the measurements of spacetime curvature be nothing more than light propagation through a quantum polarized vacuum? Is it possible that the spacetime vacuum gives curvature to matter? A view supported by the helical molecular structure at the DNA scale in which bipolar van der Waals forces operates. Van der Waal forces are a fundamental aspect of the nature of the weak vacuum which plays a significant role in all material phase transitions. These transitions occur with the creation of nuclear interbaryonic matter through the strong forces between quarks which is observed as a Bose-Einstein condensation.

Today Planck-size forces are collectively called gauge fields and include the electric, weak, and strong interactions. Sakharov's view of gravity as a "metric elasticity of space" arises from particle interactions. He identified the action term of Einstein's geometrodynamics with the "change in the action of quantum fluctuations of the vacuum... and the dependence of the action of the quantum fluctuations on the curvature of space." Sakharov viewed Einstein's action

$$S(R) = -\frac{1}{16\pi G}\int (dx)\sqrt{-gR} \tag{1}$$

of spacetime as a function of curvature. The invariant Ricci curvature tensor, R, determines how matter diverges or converges in time. This leads to Sakharov's idea of an expanding and contracting universe. Sakharov postulated that the "presence of the action (1) leads to a 'metrical elasticity' of space, i.e., to generalized forces which oppose the curving of space." Opposing the curving of space implies that a gravitational field is present. It may also imply, if we appeal to Einstien's equivalence principle, that an inertial field is present. And, at the Planck scale where gauge forces reside, it would also imply a quantized inertia. Sakharov expands the Lagrange density function in powers of the curvature

$$L(R) = L(0) + A\int kdk \cdot R + B\int \frac{dk}{k}R^2 + \ldots, \tag{2}$$

where the first term corresponds to Einstein's cosmological constant and the second term, corresponds to the action (1)

$$G = -\frac{1}{16\pi A\int kdk}, \text{ where } A \sim 1. \tag{3}$$

Integrating (3) and solving for k yields $k_0 \sim 10^{28}$ eV $\sim 10^{+33}$ cm^{-1} which corresponds to the inverse of the Planck length. The third term in (2) "leads to corrections, non-linear in R, to Einstein's equations." Sakharov argues that the "quantity k_0 determines the mass of the heaviest particles existing in nature" and "determines the limit of applicability of present-day notions of space and causality." He argues that the "density of the vacuum Lagrange function in a simplified 'model' of the theory for noninteracting free fields with particles $M \sim k_0$ shows that for fixed ratios of the masses of real particles and 'virtual' particles' (i.e., hypothetical particles which give an opposite contribution to that of the real particles to the R dependent action) *a finite change of action* arises that is proportional to M^2R and which we identify with R/G." The opposite contribution proportional to R is significant as it correlates with a repulsive force. For a modern perspective on Sakharov's induced gravity see [5].

The *finite change of action* identified by Sakharov that is proportional to M^2R can be correlated with an anti-gravitational Hooke-like repulsive term. Newton's second law originally accommodated a discrete Hooke-like term but was rejected several times as not being a natural physical law. Freeing gravity from being a nebulous force does not resolve the problem of continuity and discreteness. A unified picture of the continuous wave / discrete particle nature of light revealed in slit experiments has yet to be resolved. A mathematical attempt at a unified treatment of the discrete point / continuous manifolds has been resolved. The mathematical solution found by Grothendieck is the *topos* which combines topological spaces and the combinatorial structures found in arithmetic. When applied to quantum formalism it provides a solution for the coexistence of discrete and continuous variables. Combining continuous variables with discrete ones appears to be a necessary condition for all physical theories to have universal validity, whether within a physical or mathematical framework. For a full account with historical details see [6-7]. Einstein introduced a cosmological constant to produce a static universe. Recent data collected by the Supernova Cosmology Project and the High-Z Supernove Team provide irrefutable evidence that the universe is expanding. Zeldovich [8] characterized the cosmological constant as the "curvature of empty space." A positive cosmological constant $\Lambda \approx 10^{-52}$ cm^{-2} ostensibly gives the required repulsive force to cause the expansion.

3. Virtual particles

Sakharov's virtual particles produce a repulsive force proportional to R, the second term in (2). Zeldovich regarded the cosmological constant, the first term in (2), as curvature. It seems reasonable to combine these two ideas. The introduction of the cosmological constant into Einstein's field equations yields the following equation

$$R_{\mu\nu} - \frac{1}{2} g_{\mu\nu} R = \frac{8\pi G}{c^4} T_{\mu\nu} - \Lambda g_{\mu\nu}. \tag{4}$$

Equation (4) leads to the Friedmann equation where ρ_M represents matter density and $\rho_\Lambda = \dfrac{\Lambda}{8\pi G}$ is the vacuum energy density

$$R'^2 = \frac{8\pi G}{3}(\rho_M + \rho_\Lambda) R^2 - kc^2. \tag{5}$$

Differentiating (5) and simplifying gives

$$R'' = \left[\frac{8\pi G}{3}(\rho_M + \rho_\Lambda) \frac{\Lambda}{3} \right] R. \tag{6}$$

It follows from (6) that a universe where $R > 0$ is accelerating. The general spherically symmetric vacuum solution of (4) results in the Schwarzschild-de Sitter metric having an asymptotic curvature induced by vacuum energy corresponding to Λ. The weak field limit leads to the potential

$$\varphi = \frac{GM}{r} + \frac{\Lambda c^2}{6} r^2 \tag{7}$$

from which the full-force Newton second law can be derived

$$a = \frac{F}{m} = \nabla \varphi = -\frac{GM}{r^2} + \frac{\Lambda c^2}{3} r^2. \tag{8}$$

Newton's second law of motion (N2L) is clearly visible as the first term in (8). A Hooke-like repulsive second term, identified by Sakharov, also makes an appearance. The repulsive Hooke-like term, originally found by Newton, is derived from the Einstein action, regarded as a repulsive factor by Zeldovich, and examined in astronomical satellite-based studies advanced. Treating F/m as acceleration due to gravity we can set $a = 0$ we can solve for r. We find the distance $r = \sqrt[3]{\dfrac{3GM}{\Lambda c^2}}$ at which a galaxy of mass M, or a star like our Sun, no longer has attractive influence. For a historical overview of the three century rejection of the repulsive Hooke-like term in (8) see [11]. An important thing to note in the two terms in (8) is that *continuous* gravitational attraction is linked to the *discrete* inertial repulsion. This combination is very significant. Gravity and inertia are now seen to be linked by a Machian principle. The Planck mass can be understood to represent an inertial quantum, and an event horizon for the time-delayed inertial force, through which inertial mass appears to be generated. For a fuller discussion of this and the significance of discreteness and continuity problem see Chapters 7, 15, and16 in Rowlands [9].

4. Singularities

Einstein's theory of gravity leads to a QFT which cannot be normalized by standard quantization methods. An increasing number of singularities are encountered when spacetime is curved. Sakharov's hypothesis is that gravitational phenomenon is induced by particle interactions producing a "metrical elasticity of space" which opposes the curving of spacetime. It is known that the quantum vacuum makes a contribution to inertial mass. Part of the inertial force of opposition to acceleration, the inertial reaction force, originates in the zero-point fluctuations (ZPF) of the quantum vacuum. ZPF, zero-point field, vacuum energy, and zero-point energy describe an energy-momentum flux as experienced in a Rindler constant acceleration frame and the Unruh-Davies radiation. Both emerge from event-horizon effects in accelerating reference frames. The force of radiation pressure produced by the Rindler flux is proportional to the acceleration of the reference frame. The hypothesis that the inertia of an object is due to the individual and collective interaction of its quarks and electrons with the energy momentum flux allows for a new interpretation of Sakharov's conjecture about the elasticity of spacetime. If we regard the inertial reaction forces as being emergent from vacuum fluctuations then the elasticity of spacetime, i.e. the resistance to gravity, may be considered as the origin of inertial mass while simultaneously giving rise to curvature in quantum spacetime. Vacuum energy, according to Sakharov, cannot generate a gravitational field but it can generate an inertial field. These considerations naturally lead to the following conclusions (a) Planck length spacetime inertial reaction forces are created by vacuum fluctuations, (b) the inertial forces are responsible for curvature, and (c) the gauge forces are brought into balance by a continuous gravitational field. The third conclusion listed as (c) is an application of Rowlands' zero-totality condition [10] and conjecture that it is discrete inertial forces that are amenable to quantization which result in the emergence of a continuous gravitational field that balances gauge forces. The continuous / discrete tradeoff, similar to the action-reaction of continuous kinetic and discrete potential energy conservation, can also be seen in the two terms (8). The interchange between a continuous attractive gravitational force and a discrete repulsive Hooke-like force can produce a Sakharov contracting / expanding universe. Rowlands' discrete gauge / continuous gravity duality is a conservation law analogous to the kinetic-potential energy conservation law. In classical physics we assume that space is a vacuum; there is nothing in it; it is a complete void. Quantum theory on the other hand postulates that space is full of activity, virtual because it is unobserved, with elementary particles being created and annihilated simultaneously and instantaneously, resulting in a virtual void called zero-point field (ZPF). The random and complex radiation field is considered to be made up of zero-point energy (ZPE) and is still present even at absolute zero. Sakharov suggested that spacetime emerges curved due to the vacuum fluctuations and that gravity corresponds to the "generalized forces which oppose the curving of space." Inertia is a force which opposes gravity. This suggests that spacetime curvature is linked to inertia, not gravity. Einstein's equivalence of gravitational and inertial mass and Rowlands zero-totality condition can then be applied to argue that Sakharov elasticity and emergent curvature, an inertial action force, are intrinsic properties of mass and inherited by matter. Gravity, a dual reaction force to gauge forces, emerges as an inverse equilibrium force satisfying Rowlands' universal zero-totality (ZT) condition. In an earlier paper it was shown how ZT can be regarded as a generalization of Newton's third law [11]. Combining Rowlands' extension of the relativistic Dirac equation to an order 64 Dirac algebra space / anti-space of four fundamental parameters (space, time, mass, charge) and Sakharov's theory of gravitation yields additional insights into the connections between discrete inertial forces and continuous gravitational fields.

5. The search continues

Einstein's search for an extension of the special theory of relativity was to express fundamental physical laws in arbitrary coordinate systems. A satisfactory general theory about gravity was formulated when EEP showed that there is no essential difference between inertia and gravitation. Einstein's thought experiment demonstrated that a person in gravitational free-fall does not feel their own weight. It is as if though the gravitational field did not exist at all. This physical interpretation of the influence of gravity on a human is similar to Galileo's realization, the original principle of relativity, that for a person in uniform motion it seems as if motion does not exist. In any inertial frame of reference, motion is not detected and physical laws are invariant. It is remarkable that a person who does not feel the effects of a first derivative, i.e. uniform motion, is like a person in free fall who does not feel effects of the second derivative, i.e. acceleration. Newton's first law of motion (N1L) is a codification of Galileo's principle of relativity. Although Einstein argued that the general theory of relativity "rests exclusively on this [equivalence] principle" many have challenged its validity. Some argue that accelerometers of arbitrarily small size can detect tidal variations in a non-homogeneous gravitational field. Others argue that small droplets of liquid falling freely in the gravitational field of a spherical body will not be perfectly spherical and tidal effects cause them to be slightly ellipsoidal. Others insist that the equivalence principle is only valid for infinitesimal regions of spacetime and that this restriction renders it meaningless.

A purely geometrical interpretation of gravity would be impossible if gravitational and inertial acceleration were not identical. Gravitation, according to GR, is not something that exists in spacetime, but is considered to be an attribute of spacetime itself; coordinates that curve *are* gravitation. The ontological phenomena called gravity and acceleration are analogous to electric and magnetic fields in the context of special relativity; there are two ways at looking at electric and magnetic phenomena in terms of different coordinate systems. Similarly, in quantum mechanics we have two ways of describing wave-particle duality. We can use special relativity and Heisenberg's quantum mechanics or Lorentzian relativity and Schrödinger's wave mechanics. Special relativity has successfully unified the electric and magnetic fields in free space without trying to describe the nature of the elementary charged particles. Likewise, GR unifies gravity, acceleration, and non-linear coordinates in free space. The question of how gravity is ultimately produced and why the elementary massive particles have the masses they have are still open questions. The purely *geometrical effects* of non-inertial coordinates is associated with the physical phenomenon called *gravitation*. It is argued that the equivalence principle considers only the *passive* response of inertial mass points to a gravitational field. A complete account of gravity surely must include the *active* source of each mass point in the production of the gravitationally curved field. Einstein himself emphasized that both electromagnetism and gravitational theories were incomplete, precisely because they did not include the "source" of the interactions. He said, "*Maxwell's theory of the electric field remained a torso, because it was unable to set up laws for the behavior of electric density, without which there can, of course, be no such thing as an electro-magnetic field. Analogously the general theory of relativity furnished a field theory of gravitation, but no theory of the field-creating masses.*"

Einstein agreed that GR correlates spacetime curvature with gravity but provides no insights nor answers questions related to field-creating masses or inertial forces. There are various terms which describe spacetime deformations. The terms curvature, elasticity, expansion, contraction, and compactification are used interchangeably to describe gravitational phenomena, including gravitational fields, gravitational potentials, time and space dilations, gravitational waves, or simply gravity itself. The geometry of GR is not a suitable framework for describing behaviors where high energies, quantum considerations, and where canonical quantization of the electromagnetic field uses Coulomb gauge not subject to Lorentz invariance. Quantization implies discontinuity; GR equations and its classical Newtonian approximation based on ordinary and partial differential equations are governed by continuous variables. Newton's laws allow for physical interpretations given to the solutions of the ordinary differential equations that describe terrestrial and planetary motion. The focus of special relativity has been on the physics of objects that move at or near the speed of light; it is the speed at which *field-creating masses* move. Quantum effects have undermined the ability to produce valid geometric descriptions of spacetime. Quantum theoretical considerations impose fundamental limitations on the accuracy with which spacetime measurements can be made. It is not possible, for example, to approximate with arbitrary precision free test-point particles in which any spacetime structure is measured according to GR. Time considerations of quantum mechanical clocks also reveal fundamental limitations to measurements of duration as well as distance. The physical meaning given

to spacetime points and manifolds has been seriously questioned. The continuum of spacetime required by GR fails at the quantum mechanical Planck length ($l_p = \sqrt{\frac{Gh}{c^3}} \approx 10^{-33}$ cm) where black holes are theoretically studied. Einstein resisted ideas about origins of the universe, such as the big bang, as being produced from any singularity. In 1939 he published a paper in the Annals of Mathematics proving that black holes could not be formed by the gravitational collapse of a star. In that same year it was clearly shown how spacetime curvature can be strong enough to prevent even light from escaping. According to general relativistic constructs it would appear to outside observers as pitch black. The bending effect inside a black hole is so extreme that not only is spacetime ripped apart but a singularity is formed as curvature becomes infinite. The physics modeled by GR appears not to apply on small scales and requires a new theory which accommodates a so-called *vacuum.*
Historical studies reveal that it was as early as 1916 that Nernst [12] proposed that the vacuum was not empty. Pauli [13], in the 1920s, also became concerned with the gravitational effects of zero-point energy. Dirac, in 1933, speculated about a huge vacuum energy. The ZPF-inertial mass connection agrees with classical representations of the ZPF which have been used to derive a variety of quantum results. These include van der Waals binding force, the ground-state behavior of the quantum-mechanical harmonic oscillator, the Casimir and Davies-Unruh effects, and the spectrum of blackbodies as well. It was Zeldovich, however, who was the first to suggest that vacuum fluctuations contributed to the cosmological constant leading him to conclude that the zero-point energy needed to be considered even when gravity is taken into account. Puthoff [14], like Sakharov, proposed that a cut-off at the Planck mass in the zero-point field could generate a long-range van der Waals type force with an inverse-square property of gravity. Rueda and Haisch [15], expanding on Puthoff's idea, showed that the Davies-Unruh anisotropy of vacuum fluctuations is equivalent to a non-vanishing Poynting vector. This leads to a resistance to acceleration which may be interpreted in the context of Newton's second law (N2L) as inertia or inertial mass, $F = ma$. Haisch and Rueda, like Sakharov, require a Planck length zero-point cut-off. Their use of a Poynting vector within the discrete event horizon is likely equivalent to a gravomagnetic inertial field which generates the inertial mass. Their argument depends on the linearity of the gravitational field being determined by the zero-point cut-off which applies only to the discrete inertial repulsion, and not to the continuous gravitational attraction. The assertion that inertia is time-delayed, while gravity is instantaneous, reinforces the claim that inertia is amenable to quantization and not gravity as Rowlands and McCulloch have shown theoretically and experimentally.

6. Conclusion
Einstein's equivalence principle allows us to write Rowlands **gauge / gravity** duality [10] as a **gauge / inertia** duality which the author has written in the form of a generalized Newton third law (N3L) [11]

gauge forces (+) gravitational force = Z, (9)

where Z represents a generalized zero-totality condition. In this particular case, $Z = 0$. We now rewrite (9) in the form

gauge forces (+) inertial force = 0 (10)

agreeing with Sakharov / Zeldovich that spacetime is produced in the vacuum by particle interactions. These forces, which we might call *spacetime forces*, give us a of view spacetime as being dynamic, producing discrete inertial forces, and simultaneously creating the curvature properties which are inherited by all matter. In equation (10) the inverse force, a gravitational field, balances the gauge forces satisfying Rowlands' zero-totality condition. Spacetime curves matter. This idea also appears to reverse Mach's principle about inertia and that it is matter in the universe that contributes to gravitational field rather than inertia. The work of another British researcher, McCulloch [16-17], proposes such a model in which inertia, rather than gravity, is quantized. McCulloch also explains the connection between an object's inertial mass and the emission of photons in the form of Unruh radiation. Inertial mass is generated by the object's acceleration with respect to surrounding matter; a Modification of Inertia resulting from a Hubble-scale Casimir effect (MiHsC), or Quantised Inertia for

short, follows. The hypothesis of Sakharov and Zeldovich about the effect of vacuum energy on spacetime, viz. its elasticity and curvature, the inertial effects of the vacuum suggested by Haisch and Rueda, the gauge / gravity duality and quantized inertial models of Rowlands, and "quantised inertia" of McCulloch all point to a dynamic spacetime that curves matter while simultaneously producing a reverse Machian ether-type cosmological gravitation field.
Finally, it is important to note that zero-point energy, anticommutativity, and electron spin derive from the harmonic oscillator and can be related to the $\hbar/2$ in the Heisenberg uncertainty relation. Assuming a state is represented by a Poynting vector ψ the expectation value of the variable p_2 becomes $\langle p^2\rangle = \psi P^2\psi^*$ and the mean squared variance $(\Delta p)^2 = \psi\{P-\langle p\rangle I\}^2\psi^* = \psi P'^2\psi^*$ if $P' = P - \langle p\rangle I$ and I is a unit matrix. For operator Q, $(\Delta q)^2 = \psi\{Q-\langle q\rangle I\}^2\psi^* = \psi Q'^2\psi^*$. Since $P'\psi$ and $Q'\psi$ are vectors,

$$(\Delta p)^2(\Delta q)^2 = (\psi P'^2\psi^*)(\psi Q'^2\psi^*) \geq (\psi P'Q'\psi^*)(\psi Q'P'\psi^*) \geq \left|\frac{1}{2}(\psi P'Q'\psi^* - \psi Q'P'\psi^*)\right|^2 \geq$$
$$\geq \frac{1}{4}\left|\psi(P'Q'-Q'P')\psi^*\right|^2 \geq \frac{1}{4}[P,Q]^2.$$

Therefore $(\Delta p)(\Delta q) \geq \frac{1}{2}[P,Q] \geq \frac{\hbar}{2}$ if P and Q do not commute. The significance of this proof is seen in the factor 2. The expression $\hbar/2$ comes from the non-commutation of the P operator. Non-commutativity occurs in the quaternion relationship since $ij + ji = 0$. We can think of i and j as space-operators which are anti-commutative. They also satisfy a zero-totality condition which we may view as N3L action-reaction. We see the same result when applying two rotations in succession to vectors. Reversing the order of the differentiation yields two different results

$$T_{\alpha\mu\nu} - T_{\alpha\nu\mu} = R^{\sigma}_{\alpha\mu\nu}T_{\sigma}, \tag{11}$$

where $R^{\sigma}_{\alpha\mu\nu}$ is the Riemann curvature tensor. The vanishing of (11) is a necessary and sufficient condition for the manifold to be metrically flat which means it is free of intrinsic curvature. This tensor can be regarded as a measure of the degree of non-commutativity of covariant derivative operators in the manifold and can be linked to the Heisenberg uncertainty principle. This idea will be explored in a follow up study.

References

[1] Vigier J-P 1995 *Foundations of Physics* **25** 10 p 1461
[2] Haisch B *et al* 1994 Inertia as a zero-point field Lorentz force *Phys. Rev.* A **49** pp 678-694
[3] Rueda A and Haisch B 2002 The inertial reaction force and its vacuum origin *Gravitation and Cosmology: From the Hubble Radius to the Planck Scale* eds R L Amoroso, G Hunter, M Kafatos and J-P Vigier (Kluwer: Netherlands) pp 447-458
[4] Sakharov A 1967 Vacuum quantum fluctuations in curved space and the theory of gravitation *Sov. Phys. Dokl.* **12** 1040
[5] Visser M 2002 Sakharov's induced gravity: A modern perspective *Mod. Phys. Lett.* A **17** pp 977-992
[6] Connes A 2017 Geometry and the quantum *Preprint* arXiv: 1703.02470v1
[7] Karam S E 2017 Spacetime emergence, inertia, and Rowlands duality *Unified Field Mechanics II* ed R L Amoroso *et al* (World Scientific: Singapore)
[8] Zeldovich Ya B 1967 The cosmological constant and the theory of elementary particles *Usp. Fiz. Nauk* **95** pp 209–230
[9] Rowlands P 2007 *Zero to Infinity: The Foundations of Physics* (World Scientific: Singapore)
[10] Rowlands P 2014 *The Foundations of Physical Law* (World Scientific: Singapore)

[11] Karam S E 2013 Zero-Totality: A generalization of Newton's third law? *The Physics of Reality: Space, Time, Matter, Cosmos* eds R L Amoroso *et al* (World Scientific: Singapore) pp 38-49
[12] Nernst W 1916 Uber einen versuch, von quantentheoretischen betrachtungen zur annahmestetiger energianderungen suruckzukenhren *Verhandlungen der deutche physikalische Gesellschaft* **18** pp 83-116
[13] Enz C P and Thellung A 1960 *Helv. Phys. Acta.* **33** 839
[14] Puthoff H E 1989 Gravity as a zero-point fluctuation force *Phys. Rev.* A **39** p 2333
[15] Rueda A and Haisch B 1998 Contribution to inertial mass by reaction of the vacuum to accelerated motion *Foundations of Physics* **28** p 1057
[16] McCulloch M E 2012 Testing quantised inertia on galactic scales *Astrophysics and Space Science* **342** 2 pp 575-578
[17] McCulloch M E 2013 Inertia from an asymmetric Casimir effect *Europhysics Letters* **101** 5 p 59001

IOP Conf. Series: Journal of Physics: Conf. Series **1051** (2018) 012018 doi:10.1088/1742-6596/1051/1/012018

Quantum Physical Origin of Lorentz Transformations

R Kerner

LPTMC, Sorbonne Université-Université Pierre et Marie Curie, Paris, France
E-mail: richard.kerner@upmc.fr

Abstract. Our aim is to derive the symmetries of the space-time, i.e. the Lorentz transformations, from *discrete* symmetries of the interactions between the most fundamental constituents of matter, in particular quarks and leptons. The role of Pauli's exclusion principle in the derivation of the $SL(2,\mathbb{C})$ symmetry is put forward as the source of the macroscopically observed Lorentz symmetry. Then Pauli's principle is generalized for the case of the Z_3 grading replacing the usual Z_2 grading, leading to ternary commutation relations for quantum operator algebras. In the case of lowest dimension, with two generators only, it is shown how the cubic combinations Z_3-graded elements behave like Lorentz spinors, and the binary product of elements of this algebra with an element of the conjugate algebra behave like Lorentz vectors.

1. Introduction

Since the advent of quantum physics, great care was taken to demonstrate that all quantum phenomena, once averaged over great number of items, lead to the well known classical limits. The existence of such a limit was considered as one of the cornerstones of proper formulation of quantum mechanics, and Bohr made the "correspondence principle" a central point in his construction of quantum mechanics' basic framework [1].

Despite countless attempts, either using the ideas of "hidden parameters", or along the lines of the so-called "geometric quantization", or probabilistic Brownian motion models, a convincing derivation of quantum physics from classical models, no matter how sophisticated, was never found. Therefore it can be stated now without any doubt left, that quantum physics is primordial with respect to other observable phenomena perceived by us on the classical level.

Seen from this angle, the idea to derive the geometric properties of space-time, and perhaps its very existence, from fundamental symmetries and interactions proper to matter's most fundamental building blocks seems quite natural. Many of those properties do not require any mention of space and time on the quantum mechanical level, as was demonstrated by Born and Heisenberg [2] in their version of matrix mechanics, or by von Neumann's formulation of quantum theory in terms of the C^* algebras [3]. The non-commutative geometry is another example of formulation of space-time relationships in purely algebraic terms [4].

In what follows, we shall choose the point of view according to which the space-time relations are a consequence of fundamental *discrete symmetries* which characterize the behavior of matter on the quantum level. In other words, the Lorentz symmetry observed on the macroscopic level, acting on what we perceive as space-time variables, is an averaged version of the symmetry group acting in the Hilbert space of quantum states of fundamental particle systems.

IOP Conf. Series: Journal of Physics: Conf. Series **1051** (2018) 012018 doi:10.1088/1742-6596/1051/1/012018

Quantum Mechanics started as a non-relativistic theory, [2] but very soon its relativistic generalization was created. As an immediate result, the wave functions in the Schrödinger picture were required to belong to one of the linear representations of the Lorentz group, which means that they must satisfy the following *covariance principle*:

$$\tilde{\psi}(\tilde{x})=\tilde{\psi}(\Lambda(x))=S(\Lambda)\psi(x). \tag{1}$$

The nature of the representation $S(\Lambda)$ determines the character of the field considered: spinorial, vectorial, tensorial, etc.... As in many other fundamental relations, the seemingly simple equation

$$\tilde{\psi}(\tilde{x})=\tilde{\psi}(\Lambda(x))=S(\Lambda)\psi(x) \tag{2}$$

creates a bridge between two totally different realms: the space-time accessible via classical macroscopic observations, and the Hilbert space of quantum states. It can be interpreted in two opposite ways, depending on which side we consider as the cause, and which one as the consequence. In other words, is the macroscopically observed Lorentz symmetry imposed on the micro-world of quantum physics, or maybe it is already present as symmetry of quantum states, and then implemented and extended to the macroscopic world in classical limit? In such a case, the covariance principle should be written as follows:

$$j^{\mu'}(S(\psi))=j^{\mu'}(\psi')=\Lambda^{\mu'}_{\mu}(S)j^{\mu}(\psi). \tag{3}$$

In the above formula $j^{\mu}=\bar{\psi}\gamma^{\mu}\psi$ is the Dirac current, ψ is the electron wave function.

In view of the analysis of the causal chain, it seems more appropriate to write the same transformations with Λ depending on S:

$$\psi'(x^{\mu'})=S\psi(x^{\nu})=\psi'(\Lambda^{\mu'}_{\nu}(S)x^{\nu}), \tag{4}$$

$$x^{\mu'}(S\psi,\bar{\psi}\bar{S})=\Lambda^{\mu'}_{\mu}(S)x^{\mu}(\psi,\bar{\psi}). \tag{5}$$

This form of the same relation suggests that the transition from one quantum state to another, represented by the unitary transformation S is the primary cause implying the transformation of observed quantities such as the electric 4-current, and as a consequence, the apparent transformations of time and space intervals measured with classical physical devices.

Although mathematically the two formulations are equivalent, it seems more plausible that the Lorentz group resulting from the averaging of the action of the $SL(2,\mathrm{C})$ in the Hilbert space of states contains less information than the original double-valued representation which is a consequence of the particle-anti-particle symmetry, than the other way round.

In what follows, we shall draw physical consequences from this approach, concerning the strong interactions in the first place. But before considering these, which describe the forces conveyed by gluons and acting among quarks constituting hadrons, let us first see how the Lorentz group appears through the $SL(2,C)$ group action on fermions, in particular, on the electron states.

2. Pauli's exclusion principle

The Pauli exclusion principle, according to which two electrons cannot be in the same state with identical quantum numbers, is one of the most important foundations of quantum physics [5]. Not only does it explain the structure of atoms and the periodic table of elements, but it also guarantees the stability of matter preventing its collapse as suggested by Ehrenfest, and proved later by Dyson [8-9]. The link between the exclusion principle and particle's spin, known as the "spin-and-statistic theorem", is one of the deepest results in quantum field theory.

In the case of the electron, it was discovered that this smallest and undivisible carrier of electric charge is also endowed with a magnetic moment and the intrinsic angular momentum, called "the spin", which can take on exclusively two values. This explains the structure of electronic shells in atoms, where for a given main quantum number n there exist $2n^2$ different states. The n^2 states come from the possibility for the electron with energy state given by n to take on different *magnetic number*

states, labeled by the magnetic number m taking values from $-l$ to $+l$, including 0. Here l is the *azimutal* number, varying from $-n$ to n. Then, for a given choice of n, l and m, there are still *two* possibilities for the electron spin, which can be found in two states, "up" and "down". The result is well known: the first electronic shell corresponds to the spherically symmetric s-state, with $n=0, l=0, m=0$, so that we have no more than two electrons in the lowest energy shell, with opposite spins. The next shell corresponds to $n=1, l=-1,0,+1$ and $m=\pm 1,0$, i.e. three magnetic states. With two possibilities for spin we get the number of states in this subshell equal to 6, and altogether, with the previous shell, 8. The total number of states in all shells will be $2n^2$.
In purely algebraical terms Pauli's exclusion principle amounts to the anti-symmetry of wave functions describing two coexisting particle states. The easiest way to see how the principle works is to apply Dirac's formalism in which wave functions of particles in given state are obtained as products between the "bra" and "ket" vectors [6].
The most important statement concerning the spin of the electron is that two electrons cannot occupy the same state, with all quantum numbers equal including spin. If all quantum numbers except the spin are equal for two electrons, their spins must be opposed. In one word, if the two opposite spin states are created by

$$a_1|0\rangle = |1\rangle,\ a_2|0\rangle = |2\rangle. \quad (6)$$

Both creation operators are supposed to be nilpotent,

$$a_1 a_1|0\rangle = |1,1\rangle = 0,\ a_2 a_2|0\rangle = |2,2\rangle = 0, \quad (7)$$

thus making impossible coexistence of two electrons with the same state of spin.
Let us form an arbitrary linear superposition of two mutually exclusive states.

$$|u\rangle = \lambda|1\rangle + \mu|2\rangle \quad (8)$$

Such a state is created by the corresponding linear combination of creation operators

$$u = \lambda a_1 + \mu a_2. \quad (9)$$

This operator, creating a possible state of an electron, must be also nilpotent, so that

$$u^2|0\rangle = u|u\rangle = |u,u\rangle = 0. \quad (10)$$

Writing explicitly the square of the linear combination $u = \lambda a_1 + \mu a_2$ we get

$$u^2 = (\lambda a_1 + \mu a_2)^2 = \lambda^2 a_1^2 + \lambda\mu a_1 a_2 + \mu\lambda a_2 a_1 + \mu^2 a_2^2.$$

Imposing the nilpotency condition on the operator u makes the above combination vanish. As both operators a_1 and a_2 were supposed to be nilpotent, what remains is

$$\lambda\mu(a_1 a_2 + a_2 a_1) = 0, \rightarrow a_1 a_2 = -a_2 a_1 \quad (11)$$

(the numbers μ and λ commute, so that $\mu\lambda = \lambda\mu$).
Let us introduce a contravariant tensor

$$\varepsilon^{\alpha}\Big|^{\beta},\ \alpha,\beta = 1,2. \quad (12)$$

Contracting with two creation operators, we can write:

$$\varepsilon^{\alpha\beta} a_\alpha a_\beta = \frac{1}{2}\left[\varepsilon^{\alpha\beta} a_\alpha a_\beta + \varepsilon^{\beta\alpha} a_\beta a_\alpha\right].$$

But $a_\alpha a_\beta = -a_\beta a_\alpha$, therefore

$$\varepsilon^{\alpha\beta} a_\alpha a_\beta = \frac{1}{2}\left[\varepsilon^{\alpha\beta} - \varepsilon^{\beta\alpha}\right] a_\alpha a_\beta,$$

from which follows, taking into account that $a_\alpha a_\beta$ are linearly independent,

$$\varepsilon^{\alpha\beta} = -\varepsilon^{\beta\alpha} \quad \rightarrow \quad \varepsilon^{11} = \varepsilon^{22} = 0, \quad \varepsilon^{12} = -\varepsilon^{21}. \quad (13)$$

There exists a natural *dual algebra* of contravariant entities ξ^α, $\alpha,\beta = 1,2$ satisfying the anti-commutation relations

IOP Conf. Series: Journal of Physics: Conf. Series **1051** (2018) 012018 doi:10.1088/1742-6596/1051/1/012018

$$\xi^\alpha \xi^\beta = -\xi^\beta \xi^\alpha$$

with a corresponding two-form

$$\varepsilon_{\alpha\beta} = -\varepsilon_{\beta\alpha}.$$

The 2-form $\varepsilon_{\alpha\beta}$ can be considered as *inverse* to $\varepsilon^{\alpha\beta}$ because one has:

$$\varepsilon_{\alpha\beta}\varepsilon^{\gamma\beta} = \delta^\gamma_\alpha. \qquad (14)$$

Note the order of indices being contracted: with a different choice, we would get the identity matrix with the minus sign on the right.

Now, if we require that Pauli's principle must apply independently of the choice of a basis in Hilbert space, i.e. that after a linear transformation we get

$$\varepsilon^{\alpha'\beta'} = S^{\alpha'}_\alpha S^{\beta'}_\beta \varepsilon^{\alpha\beta} = \det S = 1, = -\varepsilon^{\alpha'\beta'}, \quad \varepsilon^{1'2'} = 1, \qquad (15)$$

then the 2×2 complex matrix $S^{\alpha'}_\alpha$ must have the determinant equal to 1, which defines the $SL(2,\mathrm{C})$ group.

3. From Pauli to Dirac

The existence of two internal degrees of freedom had to be taken into account in fundamental equation defining the relationship between basic operators acting on electron states. At the time when Pauli proposed the simplest equation expressing the relation between the energy, momentum and spin:

$$E\psi = mc^2\psi + c\,\sigma\cdot \mathrm{p}\psi, \qquad (16)$$

the relativistic invariance of electromagnetic interactions was firmly established. However, the linear equation (16) is not invariant under the Lorentz group: once squared, it yields the wrong relation between the energy-momentum E, p and the mass m:

$$E^2 = m^2c^4 + 2mc^3\,\sigma\cdot \mathrm{p} + \mathrm{p}^2.$$

At this point Pauli could have restored the Lorentz invariance by doubling the number of components of the wave function, and by introducing states with negative mass:

$$\begin{aligned} E\psi_+ &= mc^2\psi_+ + c\,\sigma\cdot \mathrm{p}\psi_-, \\ E\psi_- &= -mc^2\psi_- + c\,\sigma\cdot \mathrm{p}\psi_+. \end{aligned} \qquad (17)$$

It is easy to check that now each of the components satisfies the relativistic equation

$$E^2 = m^2c^4 + c^2\mathrm{p}^2 \quad \text{or} \quad \left(\frac{E}{c}\right)^2 - \mathrm{p}^2 = p_\mu p^\mu = m^2c^2. \qquad (18)$$

The existence of *anti-particles* (in this case the positron), suggests the use of the non-equivalent representation of $SL(2,\mathrm{C})$ group by means of complex conjugate matrices. Along with the time reversal, the Dirac equation can be now constructed. It is invariant under the Lorentz group.

$$\begin{aligned} i\hbar\frac{\partial}{\partial t}\psi_+ - mc^2\psi_+ &= i\hbar\sigma\cdot\nabla\psi_-, \\ -i\hbar\frac{\partial}{\partial t}\psi_- - mc^2\psi_- &= -i\hbar\sigma\cdot\nabla\psi_+. \end{aligned} \qquad (19)$$

The two coupled Pauli equations, with two masses, positive and negative, can be represented as follows:

$$E\left[\mathrm{I}_2\otimes \mathrm{I}_2\right]\psi = mc^2\left[\sigma_3\otimes \mathrm{I}_2\right]\psi + c\left[\sigma_1\otimes\sigma\cdot \mathrm{p}\right]\psi. \qquad (20)$$

Here I_2 stands for the 2×2 identity matrix, and the two Pauli matrices are

$$\sigma_3 = \begin{pmatrix}1 & 0\\ 0 & -1\end{pmatrix}, \quad \sigma_1 = \begin{pmatrix}0 & 1\\ 1 & 0\end{pmatrix}$$

and $\psi = \begin{pmatrix} \psi_+ \\ \psi_- \end{pmatrix}$, so that ψ is a column formed with two two-dimensional Pauli spinors.

After multiplying the entire equation from the left by the matrix $\sigma_3 \otimes \mathrm{I}_2$, and moving the momentum term to the left, we get the Dirac equation:

$$E\left[\sigma_3 \otimes \mathrm{I}_2\right]\psi - c\left[i\sigma_2 \otimes \sigma\cdot \mathrm{p}\right]\psi = mc^2\, \mathrm{I}_2 \otimes \mathrm{I}_2\, \psi,$$

which can be written as

$$\gamma^\mu p_\mu \psi = mc^2 \psi$$

with

$$\gamma^0 = \sigma_3 \otimes \mathrm{I}_2, \quad \gamma^k = -i\sigma_2 \otimes \sigma^k .$$

The price to be paid for the recovery of relativistic invariance was the introduction of negative mass, which Pauli was not ready to accept at that time. A few years later Dirac deduced the same system of equations for the electron introducing the operator acting on a four-component spinors as a "square root" of the d'Alembertian operator. The states with negative mass were interpreted as the "Dirac sea", and the holes in that sea were interpreted as positrons - electrons' antiparticles.
The transformation properties of γ-matrices can be now written as follows [10]:

$$S\gamma^\mu S^{-1} = \Lambda^\mu_{\nu'}\left(S^{-1}\right)\gamma^{\nu'}, \tag{21}$$

where $\Lambda^\mu_{\nu'}$ is the Lorentz transformation acting on space-time 4-vectors, while S is a bi-spinor representation of the Lorentz group, composed of two in equivalent matrices of $SL(2,\mathrm{C})$ group.

The apparent Z_2 symmetry ensuring the relativistic invariance of two coupled Pauli equations echoes a similar situation in the description of classical electromagnetic field. The Maxwell equations can be viewed as a coupled system of linear first-order equations for the components of electric and magnetic fields, E and B

$$\frac{1}{c}\frac{\partial \mathrm{E}}{\partial t} = \nabla \wedge \mathrm{B}, \qquad -\frac{1}{c}\frac{\partial \mathrm{B}}{\partial t} = \nabla \wedge \mathrm{E}.$$

These equations can be decoupled by applying the time derivation twice, which in vacuum, where $div\,\mathrm{E} = 0$ and $div\,\mathrm{B} = 0$ leads to the d'Alembert equation for both components separately:

$$\frac{1}{c^2}\frac{\partial^2 \mathrm{E}}{\partial t^2} - \nabla^2\,\mathrm{E} = 0, \qquad \frac{1}{c^2}\frac{\partial^2 \mathrm{B}}{\partial t^2} - \nabla^2\,\mathrm{B} = 0.$$

Nevertheless, neither of the components of the Maxwell tensor, be it E or B, can propagate separately alone; only their combination can travel along the same direction of the common wave vector.

The Dirac equation for the electron displays a similar Z_2 symmetry, with two coupled equations which can be put in the following form:

$$\begin{aligned} i\hbar\frac{\partial}{\partial t}\psi_+ - mc^2\psi_+ &= i\hbar\sigma\cdot\nabla\psi_-, \\ -i\hbar\frac{\partial}{\partial t}\psi_- - mc^2\psi_- &= -i\hbar\sigma\cdot\nabla\psi_+. \end{aligned} \tag{22}$$

where ψ_+ and ψ_- are the positive and negative energy components of the Dirac equation; this is visible even better in the momentum representation:

$$\begin{aligned} \left[E - mc^2\right]\psi_+ &= c\sigma\cdot\mathrm{p}\psi_-, \\ \left[-E - mc^2\right]\psi_- &= -c\sigma\cdot\mathrm{p}\psi_+ \end{aligned} \tag{23}$$

The same effect (negative energy states) can be obtained by changing the direction of time, and putting the minus sign in front of the time derivative, as suggested by Feynman.

Each of the components satisfies the Klein-Gordon equation, obtained by successive application of the two operators and diagonalization:

$$\left[\frac{1}{c^2}\frac{\partial^2}{\partial t^2} - \nabla^2 - m^2\right]\psi_{\pm} = 0.$$

As in the electromagnetic case, neither of the components of this complex entity can propagate by itself; only all the components can.

4. Ternary exclusion principle

Our next goal is to find how the Lorentz invariance can be derived from quantum symmetries of strong interactions. According to the present knowledge, the hadrons, behaving like fermions, are in fact composite particles. The true elementary particles carrying both baryonic and fractional electric charges are quarks. The experimental data obtained via deep inelastic scattering reveal that quarks behave like almost point-like objects as compared with proton's or neutron's size: in fact, their dimension is about 1000 times smaller than that of a hadron - about the same ratio as the dimension of atomic nuclei compared with dimension of an atom [11-13].

It seems that at this scale the usual space-time relations are hard to be implemented from outside, especially taking into account that due to the strength of interactions the uncertainty principle should be at work, thus making it even harder to speak of distances and time delays. On the other hand, strict selection rules are apparently at work, because only three-quark states lead to observable fermionic configurations, and quark-antiquark states produce strongly interacting π-mesons. There are two fundamental quark states, called u and d ; the u quark is endowed with fractional electric charge $+2/3$, while the d quark has the charge equal to $-1/3$. The anti-quarks $\bar{u}$ and $\bar{d}$ gave the same charges with opposite signs, $-2/3$ for $\bar{u}$ and $+1/3$ for $\bar{d}$. Thus, a proton is constructed as a product state uud, and neutron corresponds to the product state udd. The three π-mesons are generated by the quark-antiquark pairs according to the scheme

$$\pi^+ \sim u\bar{d}, \quad \pi^- \sim d\bar{u}, \quad \pi^0 \sim \frac{1}{\sqrt{2}}\left(u\bar{u} - d\bar{d}\right).$$

Besides, there is another important super-selection rule: three colors are needed to combine three quarks into a hadron, anti-colors for an anti-hadron, and colorless combinations of colors and anti-color are needed to form a meson [14-15].

Although in Quantum Chromodynamics quarks are treated as fermions, there is no direct proof of such assertions, because quarks are never observed as freely propagating particles outside the hadrons where they interact by means of gluon exchange. The fact that *three* quarks combine into stable states, with *two* identical states coexisting, but not three, strongly suggests that a ternary analogue of Pauli's principle may be involved, based on the representations of the cyclic group Z_3 instead of the cyclic group Z_2. We shall show how certain unusual representations of the $SL(2,\mathrm{C})$ group can be constructed as invariance groups of ternary generalization of Pauli's exclusion principle. This suggests that a convenient generalization of Pauli's exclusion principle would be that no *three* quarks in the same state can be present in a nucleon.

Let us require then the vanishing of wave functions representing the tensor product of *three* (but not necessarily two) identical states. That is, we require that $\hat{O}\left(x,x,x\right)=0$ for any quantum state $|x\rangle$. As in the former case, consider an arbitrary superposition of three different states, $|x\rangle$, $|y\rangle$ and $|z\rangle$,

$$|\omega\rangle = \alpha|x\rangle + \beta|y\rangle + \gamma|z\rangle$$

and apply the same criterion, $\hat{O}\left(\omega,\omega,\omega\right)=0$.

We get then, after developing the tensor products,

$$\begin{aligned}
&\Phi(\omega,\omega,\omega)=\alpha^3\Phi(x,x,x)+\beta^3\Phi(y,y,y)+\gamma^3\Phi(z,z,z)+\\
&+\alpha^2\beta\left[\Phi(x,x,y)+\Phi(x,y,z)+\Phi(y,x,x)\right]+\gamma\alpha^2\left[\Phi(x,x,z)+\Phi(x,z,x)+\Phi(z,x,x)\right]+\\
&+\alpha\beta^2\left[\Phi(y,y,x)+\Phi(y,x,y)+\Phi(x,y,y)\right]+\beta^2\gamma\left[\Phi(y,y,z)+\Phi(y,z,y)+\Phi(z,y,y)\right]+\\
&+\beta\gamma^2\left[\Phi(y,z,z)+\Phi(z,z,y)+\Phi(z,y,x)\right]+\gamma^2\alpha\left[\Phi(z,z,x)+\Phi(z,x,z)+\Phi(x,z,z)\right]+\\
&+\alpha\beta\gamma\left[\Phi(x,y,z)+\Phi(y,z,x)+\Phi(z,x,y)+\Phi(z,y,x)+\Phi(y,x,z)+\Phi(x,z,y)\right]=0
\end{aligned}$$

The terms $\hat{O}(x,x,x)$, $\hat{O}(y,y,y)$ and $\hat{O}(z,z,z)$ do vanish by virtue of the original assumption; in what remains, combinations preceded by various powers of independent numerical coefficients α, β and γ, must vanish separately. This is achieved if the following Z_3 symmetry is imposed on our wave functions:

$$\hat{O}(x,y,z)=j\hat{O}(y,z,x)=j^2\hat{O}(z,x,y)$$

with $j=\exp\left[\frac{2\pi i}{3}\right]$, $j^3=1$, $j+j^2+1=0$.

Note that the complex conjugates of wave functions $\hat{O}(x,y,z)$ transform under cyclic permutations of their arguments with $j^2=\bar{j}$ replacing j in the above formula

$$\Psi(x,y,z)=j^2\Psi(y,z,x)=j\Psi(z,x,y).$$

It is quite easy to imagine simple - and unique - cubic generalization of anti-commutation relations defining the usual Pauli's exclusion principle. The Z_2 group is generated by an idempotent element, represented on the complex plane by multiplication by -1. The anti-commutation relations are just the faithful representation of the cyclic permutation: $ab=(-1)ba$. Now, there are two different representations of the cyclic group Z_3 in the complex plane, both generated by cubic roots of unity. Let us denote

$$j=\exp\left[\frac{2\pi i}{3}\right],\ j^2=\exp\left[\frac{4\pi i}{3}\right],\ j^3=1,\ j+j^2+1=0. \tag{24}$$

Let us introduce N generators spanning a linear space over complex numbers, satisfying the following cubic relations [16-17]:

$$\theta^A\theta^B\theta^C=j\theta^B\theta^C\theta^A=j^2\theta^C\theta^A\theta^B,\qquad A,B=1,2,...N. \tag{25}$$

We shall also introduce a similar set of *conjugate* generators, $\bar{\theta}^{\dot{A}}$, $\dot{A},\dot{B},...=1,2,...,N$, satisfying similar condition with j^2 replacing j:

$$\bar{\theta}^{\dot{A}}\bar{\theta}^{\dot{B}}\bar{\theta}^{\dot{C}}=j^2\bar{\theta}^{\dot{B}}\bar{\theta}^{\dot{C}}\bar{\theta}^{\dot{A}}=j\bar{\theta}^{\dot{C}}\bar{\theta}^{\dot{A}}\bar{\theta}^{\dot{B}}. \tag{26}$$

Combined with the associatively, these cubic relations impose finite dimension on the algebra generated by the Z_3-graded generators. As a matter of fact, cubic expressions are the highest order that does not vanish identically. The proof is immediate:

$$\theta^A\theta^B\theta^C\theta^D=j\theta^B\theta^C\theta^A\theta^D=j^2\theta^B\theta^A\theta^D\theta^C=j^3\theta^A\theta^D\theta^B\theta^C=j^4\theta^A\theta^B\theta^C\theta^D \tag{27}$$

and because $j^4=j\neq 1$, the only solution is

$$\theta^A\theta^B\theta^C\theta^D=0. \tag{28}$$

Similar conclusion concerns the conjugate generators, for which the highest monomials are also cubic. To complete the algebra, we should impose commutation relations between ordinary and conjugate generators. Our choice is the following relation:

$$\theta^A\bar{\theta}^{\dot{B}}=-j\bar{\theta}^{\dot{B}}\theta^A,\qquad \bar{\theta}^{\dot{B}}\theta^A=-j^2\theta^A\bar{\theta}^{\dot{B}}. \tag{29}$$

Such an algebra is naturally Z_3-graded: the generators θ^A are of Z_3-grade 1, their binary products $\theta^A\theta^B$ of Z_3-grade 2, and cubic products $\theta^A\theta^B\theta^C$ are of Z_3-grade 3 equivalent to 0. The conjugate operators $\bar{\theta}^{\dot{A}}$ are endowed with grades opposite to that of θ^A's, namely grade 2 for $\bar{\theta}^{\dot{A}}$, Z_3-grade 1 (which is 4 modulo 3) for products $\bar{\theta}^{\dot{A}}\bar{\theta}^{\dot{B}}$, and grade 0 for cubic products. (corresponding to 6 modulo 3). Under multiplication the grades add up modulo 3. Such algebras were considered in [18-19].
In principle, one should ask the question what is the effect of non-cyclic (odd) permutations; however, in the case of two generators only, all permutations are equivalent with cyclic ones. From now on, we shall assume that the generalized ternary Pauli's principle is applied to the set of only two quantum operators. Let us symbolize the operator creating an u-quark by θ^1, and the operator creating a d-quark by θ^2. Then we have only two linearly independent three-quark states:

$$\theta^1\theta^2\theta^1 = j\theta^2\theta^1\theta^1 = j^2\theta^1\theta^1\theta^2 \quad \text{and} \quad \theta^2\theta^1\theta^2 = j\theta^1\theta^2\theta^2 = j^2\theta^2\theta^2\theta^1 .$$

According to (27), no observable states of four or more quarks can be produced. Also the states with three u or three d quarks are prohibited, which agrees with the experiment.
Let us show now how ternary generalization of Pauli's exclusion principle leads to some special representation of the $SL(2,C)$ group. As in the case of the Z_2 symmetry, we shall introduce an invariant *three-form* ρ^{α}_{ABC}, $\alpha,\beta = 1,2...K,\ A,B,... = 1,2,...N$. The upper indices α,β run from 1 to K, where $K = \dfrac{N^3 - N}{3}$, the number of linearly independent 3-forms satisfying the imposed Z_3 symmetry. The invariant 3-forms are then defined as follows:

$$\rho^{\alpha}_{ABC}\theta^A\theta^B\theta^C = \frac{1}{3}\left[\rho^{\alpha}_{ABC}\theta^A\theta^B\theta^C + \rho^{\alpha}_{BCA}\theta^B\theta^C\theta^A + \rho^{\alpha}_{CAB}\theta^C\theta^A\theta^B\right] =$$

$$= \frac{1}{3}\left[\rho^{\alpha}_{ABC}\theta^A\theta^B\theta^C + \rho^{\alpha}_{BCA}\left(j^2\theta^A\theta^B\theta^C\right) + \rho^{\alpha}_{CAB}j\left(\theta^A\theta^B\theta^C\right)\right],$$

$$\rho^{\alpha}_{ABC}\theta^A\theta^B\theta^C = \frac{1}{3}\left[\rho^{\alpha}_{ABC} + \rho^{\alpha}_{BCA}j^2 + \rho^{\alpha}_{CAB}j\right]\theta^A\theta^B\theta^C, \tag{30}$$

from which we get the following properties of the cubic ρ-matrices:

$$\rho^{\alpha}_{ABC} = \rho^{\alpha}_{BCA}j^2 = \rho^{\alpha}_{CAB} . \tag{31}$$

Even in this minimal and discrete case, there are covariant and contravariant indices: the lower and the upper indices display the inverse transformation property. If a given cyclic permutation is represented by a multiplication by j for the upper indices, the same permutation performed on the lower indices is represented by multiplication by the inverse, i.e. j^2, so that they compensate each other. The upper index α can take on two values because there are only two independent combinations of three generators, 121 and 212. Hopefully enough, they may coincide with the index of a Pauli spinor.
Similar reasoning leads to the definition of the conjugate forms $\bar{\rho}^{\dot{\alpha}}_{\dot{A}\dot{B}\dot{C}}$ satisfying the relations similar to (31) with j replaced by its conjugate, j^2:

$$\bar{\rho}^{\dot{\alpha}}_{\dot{A}\dot{B}\dot{C}} = j\bar{\rho}^{\dot{\alpha}}_{\dot{B}\dot{C}\dot{A}} = j^2\bar{\rho}^{\dot{\alpha}}_{\dot{C}\dot{A}\dot{B}} . \tag{32}$$

In the simplest case of two generators, the j-skew-invariant forms have only two independent components:

$$\rho^1_{121} = j\rho^1_{211} = j^2\rho^1_{112} ,$$

$$\rho^2_{212} = j\rho^2_{122} = j^2\rho^2_{221}$$

and we can set

$$\rho^1_{121} = 1,\ \rho^1_{211} = j^2,\ \rho^1_{112} = j ,$$

$$\rho^2_{212} = 1,\ \rho^2_{122} = j^2,\ \rho^2_{221} = j .$$

5. Lorentz symmetry on quark states

The constitutive cubic relations between the generators of the Z_3 graded algebra can be considered as intrinsic if they are conserved after linear transformations with commuting (pure number) coefficients, i.e. if they are independent of the choice of the basis. Let $U^{A'}_A$ denote a non-singular $N \times N$ matrix, transforming the generators θ^A into another set of generators, $\theta^{B'} = U^{B'}_B \theta^B$. We are looking for the solution of the covariance condition for the ρ-matrices:

$$S^{\alpha'}_\beta \rho^\beta_{ABC} = U^{A'}_A U^{B'}_B U^{C'}_C \rho^{\alpha'}_{A'B'C'} . \tag{33}$$

Now, $\rho^1_{121} = 1$, and we have two equations corresponding to the choice of values of the index α' equal to 1 or 2. For $\alpha' = 1'$ the ρ-matrix on the right-hand side is $\rho^{1'}_{A'B'C'}$, which has only three components,

$$\rho^{1'}_{1'2'1'} = 1,\ \rho^{1'}_{2'1'1'} = j^2,\ \rho^{1'}_{1'1'2'} = j ,$$

which leads to the following equation:

$$S^{1'}_1 = U^{1'}_1 U^{2'}_2 U^{1'}_1 + j^2 U^{2'}_1 U^{1'}_2 U^{1'}_1 + j U^{1'}_1 U^{1'}_2 U^{2'}_1 = U^{1'}_1 \left(U^{2'}_2 U^{1'}_1 - U^{2'}_1 U^{1'}_2 \right),$$

because $j^2 + j = -1$.

For the alternative choice $\alpha' = 2'$ the ρ-matrix on the right-hand side is $\rho^{2'}_{A'B'C'}$, whose three non-vanishing components are

$$\rho^{2'}_{2'1'2'} = 1,\ \rho^{2'}_{1'2'2'} = j^2,\ \rho^{2'}_{2'2'1'} = j .$$

The corresponding equation becomes now:

$$S^{2'}_1 = U^{2'}_1 U^{1'}_2 U^{2'}_1 + j^2 U^{1'}_1 U^{2'}_2 U^{2'}_1 + j U^{2'}_1 U^{2'}_2 U^{1'}_1 = U^{2'}_1 \left(U^{1'}_2 U^{2'}_1 - U^{1'}_1 U^{2'}_2 \right).$$

The remaining two equations are obtained in a similar manner. We choose now the three lower indices on the left-hand side equal to another independent combination, (212). Then the ρ-matrix on the left hand side must be ρ^2 whose component ρ^2_{212} is equal to 1. This leads to the following equation when $\alpha' = 1'$:

$$S^{1'}_2 = U^{1'}_2 U^{2'}_1 U^{1'}_2 + j^2 U^{2'}_2 U^{1'}_1 U^{1'}_2 + j U^{1'}_2 U^{1'}_1 U^{2'}_2 = U^{2'}_1 \left(U^{1'}_2 U^{2'}_1 - U^{1'}_1 U^{2'}_2 \right)$$

and the fourth equation corresponding to $\alpha' = 2'$ is:

$$S^{2'}_2 = U^{2'}_2 U^{1'}_1 U^{2'}_2 + j^2 U^{1'}_2 U^{2'}_1 U^{2'}_2 + j U^{2'}_2 U^{2'}_1 U^{1'}_2 = U^{2'}_2 \left(U^{1'}_1 U^{2'}_2 - U^{2'}_1 U^{1'}_2 \right).$$

The determinant of the 2×2 complex matrix $U^{A'}_B$ appears on the right-hand side:

$$S^{2'}_1 = -U^{2'}_1 \left[\det(U) \right]. \tag{34}$$

The remaining two equations are obtained in a similar manner, resulting in the following:

$$S^{1'}_2 = -U^{1'}_2 \left[\det(U) \right], \qquad S^{2'}_2 = -U^{2'}_2 \left[\det(U) \right]. \tag{35}$$

The determinant of the 2×2 complex matrix $U^{A'}_B$ appears everywhere on the right-hand side. Taking the determinant of the matrix $S^{\alpha'}_\beta$ one gets immediately

$$\det(S) = \left[\det(U) \right]^3 . \tag{36}$$

However, the U-matrices on the right-hand side are defined only up to the phase, which due to the cubic character of the covariance relations and they can take on three different values: $1, j$ or j^2, i.e. the matrices $jU^{A'}_B$ or $j^2 U^{A'}_B$ satisfy the same relations as the matrices $U^{A'}_B$ defined above. The determinant of U can take on the values $1, j$ or j^2 if $\det(S) = 1$. This will be true if we admit that the indices $\alpha, \beta ...$ relate to the usual Pauli fermions, and that matrices S represent the $SL(2,\mathrm{C})$ group.

A similar covariance requirement can be formulated with respect to the set of 2-forms mapping the quadratic quark-anti-quark combinations into a four-dimensional linear real space. As we already saw, the symmetry (29) imposed on these expressions reduces their number to four. Let us define two quadratic forms, π^{μ}_{AB} and its conjugate $\bar{\pi}^{\mu}_{\dot{A}\dot{B}}$

$$\pi^{\mu}_{A\dot{B}}\theta^{A}\bar{\theta}^{\dot{B}} \quad \text{and} \quad \bar{\pi}^{\mu}_{\dot{B}A}\bar{\theta}^{\dot{B}}\theta^{A}. \tag{37}$$

The Greek indices $\mu, \nu ...$ take on four values, and we shall label them $0,1,2,3$. The four tensors $\pi^{\mu}_{A\dot{B}}$ and their hermitian conjugates $\bar{\pi}^{\mu}_{\dot{B}A}$ define a bi-linear mapping from the product of quark and anti-quark cubic algebras into a linear four-dimensional vector space, whose structure is not yet defined. Let us impose the following invariance condition:

$$\pi^{\mu}_{A\dot{B}}\theta^{A}\bar{\theta}^{\dot{B}} = \bar{\pi}^{\mu}_{\dot{B}A}\bar{\theta}^{\dot{B}}\theta^{A}. \tag{38}$$

It follows immediately from (29) that

$$\pi^{\mu}_{A\dot{B}} = -j^{2}\bar{\pi}^{\mu}_{\dot{B}A}. \tag{39}$$

Such matrices are non-hermitian, and they can be realized by the following substitution:

$$\pi^{\mu}_{A\dot{B}} = j^{2}i\sigma^{\mu}_{A\dot{B}}, \qquad \bar{\pi}^{\mu}_{\dot{B}A} = -ji\sigma^{\mu}_{\dot{B}A}, \tag{40}$$

where $\sigma^{\mu}_{A\dot{B}}$ are the unit 2 matrix for $\mu = 0$, and the three hermitian Pauli matrices for $\mu = 1,2,3$. Again, we want to get the same form of these four matrices in another basis. Knowing that the lower indices A and $\dot{B}$ undergo the transformation with matrices $U^{A'}_{B}$ and $\bar{U}^{\dot{A}'}_{\dot{B}}$, we demand that there exist some 4×4 matrices $\Lambda^{\mu'}_{\nu}$ representing the transformation of lower indices by the matrices U and $\bar{U}$

$$\Lambda^{\mu'}_{\nu}\pi^{\nu}_{A\dot{B}} = U^{A'}_{A}\bar{U}^{\dot{B}'}_{\dot{B}}\pi^{\mu'}_{A'\dot{B}'}, \tag{41}$$

This defines the vector (4×4) representation of the Lorentz group.

The first four equations relating the 4×4 real matrices $\Lambda^{\mu'}_{\nu}$ with the 2×2 complex matrices $U^{A'}_{B}$ and $\bar{U}^{\dot{A}'}_{\dot{B}}$ are as follows:

$$\begin{aligned}
\Lambda^{0'}_{0} + \Lambda^{0'}_{3} &= U^{1'}_{1}\bar{U}^{\dot{1}'}_{\dot{1}} + U^{2'}_{1}\bar{U}^{\dot{2}'}_{\dot{1}}\\
\Lambda^{0'}_{0} - \Lambda^{0'}_{3} &= U^{1'}_{2}\bar{U}^{\dot{1}'}_{\dot{2}} + U^{2'}_{2}\bar{U}^{\dot{2}'}_{\dot{2}}\\
\Lambda^{0'}_{0} - i\Lambda^{0'}_{2} &= U^{1'}_{1}\bar{U}^{\dot{1}'}_{\dot{2}} + U^{2'}_{1}\bar{U}^{\dot{2}'}_{\dot{2}}\\
\Lambda^{0'}_{0} + i\Lambda^{0'}_{2} &= U^{1'}_{2}\bar{U}^{\dot{1}'}_{\dot{1}} + U^{2'}_{2}\bar{U}^{\dot{2}'}_{\dot{1}}
\end{aligned}$$

There are three other sets of four equations similar to the one displayed above, corresponding to three alternative choices of values of the upper primed index $1'$, $2'$ and $3'$. We do not display them here explicitly; the resulting matrices $\Lambda^{\mu'}_{\nu}$ span a vector representation of the Lorentz group, although with special complex coefficients.

The metric tensor $g_{\mu\nu}$ can be defined in the following manner.

With the invariant "spinorial metric" in two complex dimensions, ε^{AB} and $\varepsilon^{\dot{A}\dot{B}}$ such that $\varepsilon^{12} = -\varepsilon^{21} = 1$ and $\varepsilon^{\dot{1}\dot{2}} = -\varepsilon^{\dot{2}\dot{1}}$, we can define the contravariant components $\pi^{\nu A\dot{B}}$. It is easy to show that the Minkowskian space-time metric, invariant under the Lorentz transformations, can be defined as

$$g^{\mu\nu} = \frac{1}{2}\left[\pi^{\mu}_{A\dot{B}}\pi^{\nu A\dot{B}}\right] = \mathrm{diag}(+,-,-,-) \tag{42}$$

According to this picture, the Lorentz symmetry observed on macroscopic scale is the result of the action of the $SL(2,\mathrm{C})$ group on the Hilbert space of quantum states of elementary fermions (leptons), i.e. electrons, μ-mesons and neutrinos, all of which satisfy the Dirac equation and Pauli's exclusion

principle. The essential ingredient is the appearance of *two* Z_2 symmetries, one due to the dichotomic spin states, another one to the particle-antiparticle symmetry. Due to the last symmetry, Dirac spinors ψ are composed out of two in equivalent Pauli spinors. The Dirac spinors, in turn, produce Lorentz scalars or vectors due to the introduction of charge conjugation, $\bar{\psi} = \psi^+ \gamma^0$. Then Lorentz scalars and Lorentz 4-vectors are produced out of binary products $\bar{\psi}\psi = \psi^+\gamma^0\psi$ and $\bar{\psi}\gamma^k\psi = \psi^+\gamma^0\gamma^k\psi$.
For hadrons and strongly interacting mesons (here we consider only protons and neutrons, plus the π-mesons) the apparent Lorentz symmetry stems from the Z_3 generalization of Pauli's principle implemented in quarks, so that only two three-quark configurations can be observed, uud and udd, as well as three independent quark-antiquark pairs $u\bar{d}$, $d\bar{u}$ and $\frac{1}{\sqrt{2}}\left(u\bar{u} - d\bar{d}\right)$. The Z_3 graded ternary algebra quark generators are preserved by the Z_3-twisted representation of the $SL(2,\mathrm{C})$ group. Cubic combinations of these operators behave like Pauli spinors, transforming under usual representations of $SL(2,\mathrm{C})$. Their quadratic combinations transform as Lorentz 4-vectors or Lorentz scalars.
The dynamics of quarks is very well described by Quantum Chromodynamics [14-15], where quarks are treated as usual Dirac fermions interacting via vector particles called gluons. All these fields are endowed with an extra property called *color*, and become observable only in “colorless” combinations, with three different colors (red, green and blue) for the three-quark states producing nucleons, and eight color-anticolor combinations for gluons, spanning an octet representation of the $SU(3)$ algebra. On the other hand, the possibility of deriving a new Z_3-twisted representation of the Lorentz group suggests that a generalization of the Dirac equation incorporating this symmetry coexisting with the $Z_2 \times Z_2$ symmetry in the usual case should be possible. The color dynamics based on this generalization is beyond the scope of the present paper. The generalized Dirac equation mixing three colors, based on the $Z_2 \times Z_3 \times Z_2$ symmetry is constructed and discussed in papers [20-21].

References

[1] Nielsen J R 1976 Niels Bohr Collected Works, Vol. 3: The Correspondence Principle (1918-1923), (Amsterdam: North-Holland Publishing)
[2] Born N and Jordan P 1925 *Zeitschrift fur Physik* **34** pp 858-878
[3] von Neumann J 1996 *Mathematical Foundations of Quantum Mechanics* (Princeton Univ. Press)
[4] Dubois-Violette M, Kerner R and Madore J 1990 *Journ. Math. Phys.* **31** pp 316-322
[5] Pauli W 1927 *Zeitschrift fur Physik* **43** pp 601-623
[6] Dirac P A M 1928 *Proc. Royal. Soc.* (London) A **117** pp 610-624
[7] Pauli W 1940 *Physical Review* **58** pp 716-722
[8] Dyson F J 1967 *Journ. Math. Phys.* **8** pp 1538-1545
[9] Dyson F J and Lenard A 1968 *Journ. of Math. Physics* **8** pp 423-434
[10] Bjorken J D and Drell S D 1964 *Relativistic Quantum Mechanics* (McGraw-Hill, New York)
[11] Gell-Mann M and Ne'eman Y 1964 *The Eightfold Way* (Benjamin, New York)
[12] Lipkin H J 1987 Frontiers of the Quark Model *Preprint* WIS- 87-47-PH (Weizmann Institite)
[13] Okubo S 1993 *Journ. of Math. Physics* **34** p 3273
[14] Gross D, Wilczek F and Politzer H 1973 *Phys. Rev. Letters.* **30** pp 1343-1347
[15] Wilczek F 2000 *Physics Today* **53** pp 22-28
[16] Kerner R 1991 *Comptes Rendus Acad. Sci. Paris.* **10** 1237
[17] Kerner R 1992 *Journ. of Math. Phys.* **33** pp 403-411
[18] Abramov V, Kerner R and Le Roy B 1997 *Journ. of Math. Phys.* **38** pp 1650-1669
[19] Kerner R 1997 Class. *and Quantum Gravity* **14** A203-A225

[20] Kerner R and Suzuki O 2012 *Int. Journ. for Geom. Methods of Modern Physics* **9**
[21] Kerner R 2017 *Physcis of Atomic Nuclei* **80** 3 pp 522-534

Macroscopic entanglement and time reversal causality by data of the Baikal Experiment

S Korotaev[1,2], N Budnev[3], V Serdyuk[2], E Kiktenko[1,2], J Gorohov[4] and V Zurbanov[3]

[1] Bauman Moscow State Technical University, Moscow, Russia
[2] Geoelectromagnetic Research Centre, Schmidt Institute of Physics of the Earth, Russian Academy of Sciences, Moscow, Russia
[3] Institute of Applied Physics, Irkutsk State University, Irkutsk, Russia
[4] Pushkov Institute of Terrestrial Magnetism, Ionosphere and Radio Wave Propagation, Russian Academy of Sciences, Troitsk, Moscow, Russia
E-mail: korotaev@igemi.troitsk.ru

Abstract. Although the general theory macroscopic quantum entanglement is still in its infancy, consideration of the matter in the framework of action-at-a-distance electrodynamics predicted, for the random dissipative processes, observability of the advanced nonlocal correlations (time reversal causality). These correlations were really revealed in our previous experiments with some large-scale random heliogeophysical processes as the source ones and the lab detectors as the probe ones. However, the strongest macroscopic nonlocal correlations are observed at extremely low frequencies; therefore, the long-terms experiments therewith under very stable conditions, which are difficult to achieve in a usual laboratory, are necessary. To overcome this difficulty, a new experiment has been conducted on the base of Baikal Deep Sea Neutrino Observatory since 2012. Baikal thick water layer is an excellent shield against any local impacts on the detectors. The long-term series of measurements demonstrates that detector signals respond to the random global heliogeophysical processes, but this nonlocal causal connection proves to contain considerable time reversal component. Nonlocal nature of this connection is confirmed by violation of the steering inequality. In addition, advanced nonlocal correlation of the detector signal with a regional source-process – the random component of hydrological activity (macroturbulence) in the upper layer is revealed. The possibilities of the random processes forecast on nonlocal correlations are demonstrated.

1. Introduction

The general theory macroscopic quantum entanglement is still in its infancy, but consideration of the matter in the framework of action-at-a-distance electrodynamics [1] predicts for the random dissipative processes observability of the advanced nonlocal correlations through timelike interval (that is time reversal causality, which is allowed by the quantum principle of weak causality in the uncontrolled entangled states [2]). Moreover, action-at-a-distance electrodynamics predicts that the advanced component of nonlocal correlation can exceed the retarded one [3]. The propagation velocity for diffusion entanglement swapping through a medium can be very small. Accordingly, the retardation and advancement can be very large. Such correlations were revealed in our previous experiments with some large-scale random heliogeophysical processes as the source ones and the lab detectors as the probe ones [1].

To overcome difficulty with detector protection against long-period local impacts under usual laboratory conditions, a new experiment is conducting on the base of Baikal Deep Sea Neutrino Observatory [4-5]. Baikal thick water layer is an excellent shield against any local impacts on the detectors. The Baikal experiment aims, first, study of nonlocal correlations between the electrode detectors at different horizons in the lake and spaced at 4200 km one in the land laboratory, and second, study of correlations between the detector signals and large-scale natural dissipative processes with big random components. Data are processed by the methods of spectral, correlation and causal analysis. The several annual series of measurements, obtained since 2012, demonstrated that the detector signals respond to the main global source-process (which is, by data of all the previous lab experiment, the random component of solar activity and, to a lesser extent, of geomagnetic one). The causal connection of the detector signals is directed downwards: from the Earth surface to the Baikal floor. This nonlocal causal connection proved to contain considerable time reversal component, in all cases exceeding the retarded one. In addition, advanced nonlocal correlations of the detector signal with some regional source-processes were revealed. In particular, it was macroturbulence in the upper layer of the lake. The possibility of its forecast on nonlocal correlations was demonstrated. An important result was displaying of the advanced response of nonlocal correlation detector to the earthquake.

However, the fact that the main source-process is solar activity should be confirmed explicitly by data of a deep-sea detector. On the other hand, although the earthquakes are powerful and interesting events, they occur very seldom to obtain statistically reliable conclusions, therefore weaker, but continuous process of hydrological fluctuations is more convenient in this respect.

In the current paper we consider nonlocal correlations and phenomenon of time reversal causality with the main external global source-process (solar activity) and with an internal local source-process (macroturbulence in the upper Baikal layer), using the detector with the highest signal-to-noise ratio with respect to these processes. To prove nonlocal nature of advanced correlations by violation of steering inequality, we use also one more source: global geomagnetic activity. We demonstrate forecasting capability of the random processes on macroscopic nonlocal correlations.

2. Experiment and data

The Baikal deep-sea experimental setup is described in detail in Refs. [4] (primary configuration) and [25] (final configuration); so we present here only short description of the latter one.

The experimental setup is deployed in the Southwestern part of the lake. The site depth is 1367 m. The bottom detector *Ub* is set at the depth 1337 m, the top one *Ut* is set at the depth 47 m. For comparison, synchronous measurements with the detector on the Earth surface *Ul* (located in the laboratory in Troitsk spaced at 4200 km) are conducted. All the detectors are of electrode type, which by previous experience turned out the most reliable among other types (e.g. [1]). This type of detector is based on measuring of spontaneous variations of self-potentials of weakly polarized electrodes in the electrolyte (under suppressing of all classical local impacts). Every detector represents a couple of weakly polarized AgClAg electrodes HD-5.519.00 with practically zero separation. These electrodes are best in the World by their self-potential insensitivity to the environmental conditions.

The detector signals are measured and stored in the electronics unit set at the depth 20 m. The sampling rate is 10 s. The relative error of measurements is less than 0.01%. In addition, the electronics unit contains the temperature and acceleration sensors. The setup is fixed by the heavy anchor on the floor and by the submersed buoy at the depth 15 m. At a few kilometers from the setup, auxiliary measurements of the sea currents and magnetic field are conducted.

The deep-sea setup is designed to be operated autonomically for a year. Every March the electronic unit is lifted on the ice for data reading and battery changing and then it is installed again for the next year.

By some technical reasons, from 2015 the *Ub* detector channel ceased to withstand annual operation. But by previous data the detector *Ut* proved to be optimal for the signal-to-noise ratio in the study of external heliogeophysical processes (the greatest noise is contained in the detector on the Earth surface

Ul, the smallest – in the bottom one *Ub*, but in the latter the signal from the external processes have significantly shielded the water column). It stands to reason that *Ut* is optimal also in study of internal hydro-thermodynamic processes (macroturbulence), which occur practically only in rather thin upper layer of the Baikal Lake. Therefore, for the main purposes of current work we use data of the detector *Ut*.

According to entanglement monohamity principle a probe process in the detector can be entangled with only a few processes in the environment. Nonetheless, there is a problem of the correlation separation if (as in our case) the spectra of the different sources are overlapped. In addition, the larger space scale of the source, the larger expected advancement of macroscopic nonlocal correlation [1] and the longer time series is needed. For this reason, for study of correlation with solar activity in this work, we use biannual time series 2013/2015, when this activity was rather high, while for study of correlation with hydro-thermodynamic activity we use annual series (2016/2017), when solar one was almost minimal.

The macroscopic entanglement equation [1, 4] relates the entropy productions in the probe and source processes. In turn, the theory of the detector relates probe entropy variation with its signal. The same can easily be done for a simple source. However, since our source are not too simple, and not to contaminate the results by any model assumptions, we relate detector signal with source activity indexes qualitatively reflected the entropy variations. As the solar activity indexes we use the radio wave flux *R* at 2800 MHz and *X*-ray flux at 0.5 – 4 Å. The radio frequency 2800 MHz corresponds to emission from the level of maximal dissipation in the solar atmosphere (upper chromosphere – lower corona, where the magneto-sound waves dissipate). The *X*-rays emit approximately from the same level. As the index of macroturbulence, we use random long-period temperature fluctuations at the depth 20 m.

3. Data treatment

Data were processed by the methods of spectral, correlation and causal analysis. Although we consider quantum phenomena, in the experiment we deal with only classical output of measuring device and may use simpler classical version (just as in the experimental verification of Bell inequality using classical correlations or Shannon entropies).

Recall the essence of classical causal analysis. For any variables *X* and *Y*, in terms of Shannon marginal $S(X)$, $S(Y)$ and conditional $S(X|Y)$, $S(Y|X)$ entropies the independence functions can be defined:

$$i_{Y|X} = \frac{S(Y\,|\,X)}{S(Y)},\ i_{X|Y} = \frac{S(X\,|\,Y)}{S(X)},\ 0 \le i \le 1. \tag{1}$$

Roughly saying, the independence functions behave inversely to module of correlation one. However, they characterize one-way correlations, which are asymmetric for causally related variables. In addition, they are well suited for any linear and nonlinear relationships. Next, the causality function γ is considered:

$$\gamma = \frac{i_{Y|X}}{i_{X|Y}},\ 0 \le \gamma < \infty. \tag{2}$$

By definition *X* is the cause and *Y* is the effect if $\gamma < 1$. And inversely, *Y* is the cause and *X* is the effect if $\gamma > 1$. On theoretical and plenty of experimental examples, it had been shown that such a formal approach to causality did not contradict its intuitive understanding.

In terms of γ the principle of classical causality is formulated as follows:

$$\gamma < 1 \Rightarrow \tau > 0,\ \gamma > 1 \Rightarrow \tau < 0,\ \gamma \to 1 \Rightarrow \tau \to 0, \tag{3}$$

where τ is time shift of *Y* relative to *X*. Only in case of nonlocal correlation, one can observe violation of this principle. It is just the case of weak causality, which does not obey the combination of inequalities of axiom (3).

IOP Conf. Series: Journal of Physics: Conf. Series **1051** (2018) 012019 doi:10.1088/1742-6596/1051/1/012019

The strict dichotomy of the weak causality (violating chronology) and its counterpart, the strong one (respecting chronology), can be done only in the framework of quantum causal analysis, where another, more complicated measure of causality is used (the course of time c_2 instead of the causality function γ, because the latter loses sense at a negative conditional entropy) [1]. However, if both conditional entropies are non-negative, in particular for device classical output (even measuring an entangled state), both principles, the classical and strong causality are equivalent. Finite causality, which does not obey these principles, is weak causality. It may occur only in the entangled states. Thus violation of Ineqs. (3), say, $\gamma > 1$ at $\tau > 0$, is entanglement evidence that is a sufficient condition of nonlocality.

There is a way to test a wider sufficient condition of nonlocality in usual manner, not appealing to time shifts, as it is done through the violation of Bell-like and steering inequalities. Suppose some process C acts upon a distant process A by means of any local interaction along the causal chain $C \to B \to A$. The intermediate process B is situated so that local carriers of interaction cannot come A avoiding B (e.g. B occupies a spherical layer around A). Then the claim of locality implies the steering inequality (derived, e.g. in Ref. [1]):

$$i_{A|C} \geq \max\left(i_{A|B}, i_{B|C}\right). \tag{4}$$

Violation of Ineq. (4) is a sufficient condition of entanglement.

4. Results

The detailed study of nonlocal correlations between all three detectors: bottom one *Ub*, top one *Ut* and land (laboratory) one *Ul* was undertaken before during the year of almost maximal solar activity 2013/2014 [5]. The nonlocal causal connections of their signals proved to be directed downwards: *Ul*→*Ut*, *Ut*→*Ub*, *Ul*→*Ub*, but they contain the time reversal component, in all cases exceeding the retarded one. This excess depends on the mass of the absorbing medium separating the nonlocally correlating processes; the ratios of the maximal time reversal and time respecting (normal) causality functions reached 1.03, 1.12, 2.05 respectively.

Now we are concentrated on nonlocal correlation with solar (and partly geomagnetic) activity by data of biannual 2013/2015 time series with the detector *Ut*, optimal for the signal-to-noise ratio. The main source-process is solar activity. According to entanglement monogamy property, great nonlocal correlation with one subsystem implies only small correlations with other ones. Indeed, nonlocal correlations of detector signal with other natural (geophysical) random dissipative processes by data of all previous experiments turned out smaller. The dominance of solar activity is less in 2016/2017 series, we will confirm this fact below.

In Figure 1 the normalized amplitude spectra of the detector *Ut*, solar radio wave flux *R* (2800 MHz) and *X*-ray flux (0.5-4 Å) at the period range is from 10 to 460 days (d). As expected, *Ut* practically does not respond to deterministic variations of the solar activity, which is represent in *R* and *X* by split 27-day variation and its harmonics. Over longer periods, where the random component (in particular, the intermittent variation [6]) is dominated, the response in *Ut* is clearly visible. Detailed analysis showed the greatest similarity of the spectra of *Ut* with *R* in the band of periods 365>T>59 d, and with *X* in the band 365>*T*>77 d.

Next, to causal and correlation analysis data from such a band-pass filtration were used. The computation results are presented in Figures 2 and 3. Their stability was tested by alternate noising of the time series by the flicker noise (21% power).

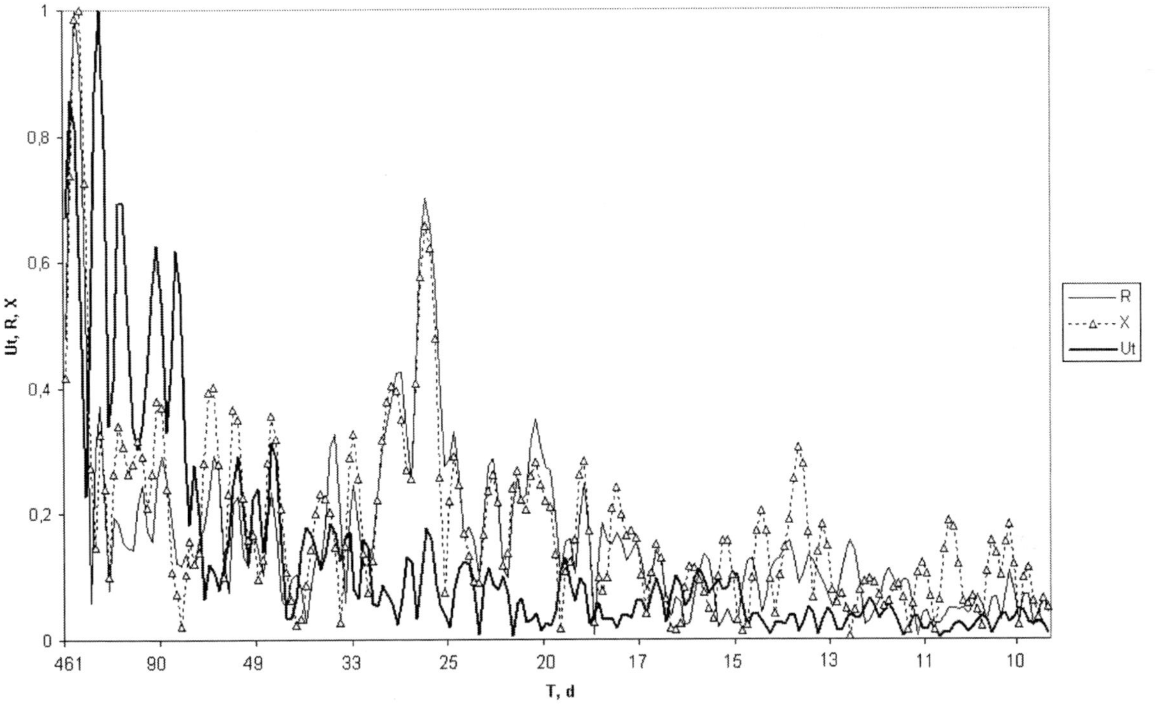

Figure 1. Normalized amplitude spectra of the signals of detector *Ut*, solar radio wave flux *R* and *X*-ray flux.

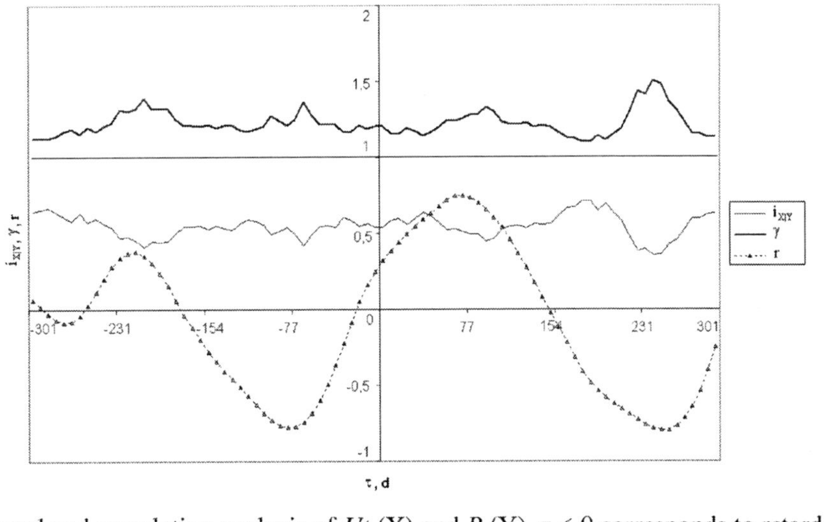

Figure 2. Causal and correlation analysis of *Ut* (X) and *R* (Y). $\tau < 0$ corresponds to retardation of *Ut* relative to *R*, $\tau > 0$ – to advancement.

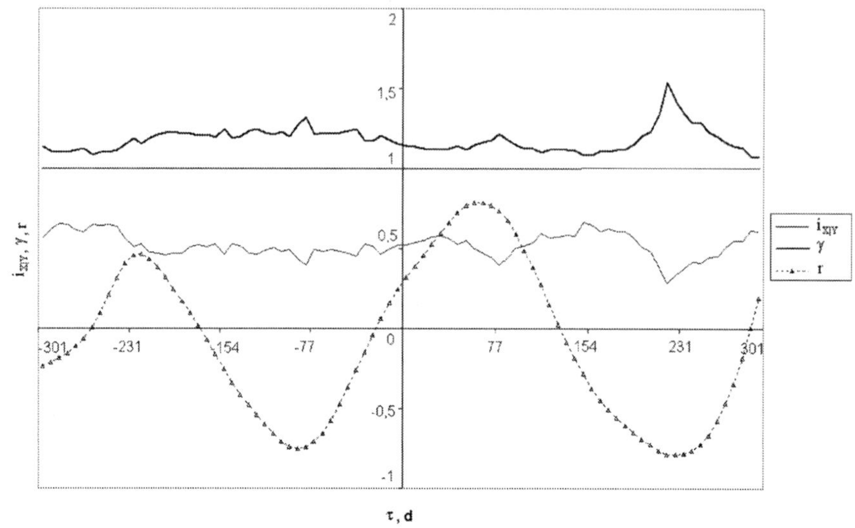

Figure 3. Causal and correlation analysis of *Ut* (X) and *X* (Y). τ < 0 corresponds to retardation of *Ut* relative to *X*, $\tau > 0$ – to advancement.

In a couple of $Ut - R$ it is seen that solar activity is a cause with respect to detector signal ($\gamma > 1$) and the highest maximum of causality $\gamma = 1.5^{+0.0}_{-0.1}$ is observed at advancement of *Ut* with respect to *R* for 250 days. At the same advancement the deepest minimum of independence $i_{Ut|R} = 0.35^{+0.05}_{-0.00}$ and maximal correlation $r = -0.79^{+0.02}_{-0.01}$ are observed. In a couple of Ut – X the results are close: the extreme $\gamma = 1.5^{+0.1}_{-0.0}$, $i_{Ut|R} = 0.28^{+0.02}_{-0.00}$, $r = -0.79^{+0.02}_{-0.01}$ are at advancement 230 days. Thus, the characteristic for nonlocal connection of the random processes time reversal causality is observed.

The level of advanced correlation is sufficient to forecast solar activity with optimal advancement about 250 days. However, the large value of optimal advancement with dramatically reduced, due to the above-mentioned filtration, the length of *Ut* series make impossible yet to use forecasting algorithms based on computation of impulse transient response [1], or even simpler one based on sliding regression (as it was done in the similar problem in Ref. [5]), because they need long training interval. However, the very possibility of solar activity forecast is easily demonstrated by the series shift. It is shown in Figure 4 by the example of *R*.

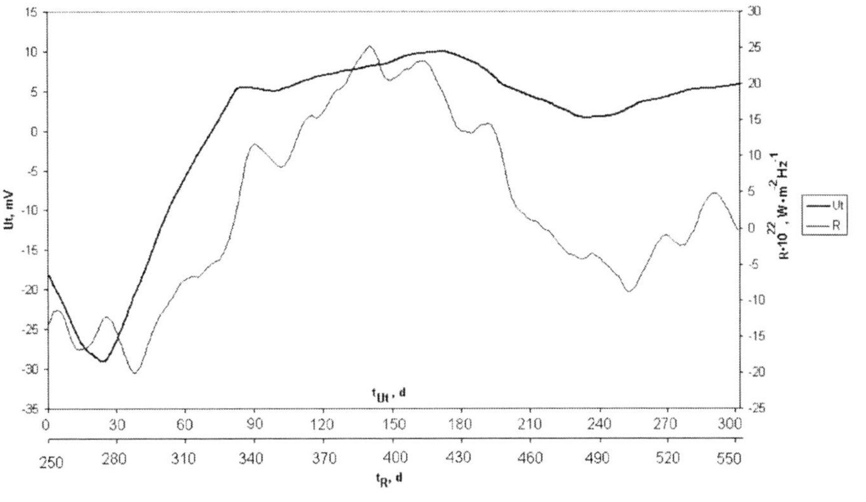

Figure 4. *Ut* approximately forecasts *R* with advancement 250 days.

In agreement with the results of all previous experiments of this kind, there is also a weaker advanced connection of the detector signal with the global geomagnetic activity characterized by the *Dst*-index. With filtration 365>*T*>59 days in the couple *Ut* – *Dst* at advancement of *Ut* with respect to *Dst* for 110 days the extreme values $\gamma = 1.3^{+0.0}_{-0.1}$, $i_{Ut|Dst} = 0.51^{+0.08}_{-0.00}$, $r = -0.65^{+0.01}_{-0.00}$ are observed; with filtration 365>*T*>77 days the extreme values are observed at the same advancement and amount to $\gamma = 1.2^{+0.1}_{-0.0}$, $i_{Ut|Dst} = 0.47^{+0.00}_{-0.01}$, $r = -0.68^{+0.03}_{-0.00}$. The smaller value of advancement compared to solar activity is also consistent with (observed in all previous experiments) a trend to a direct relation of this quantity with the scale of the source.

Although the nonlocal nature of the correlations follows from the observed violation of the principle of strong causality ($\gamma > 1$ at $\tau > 0$), the combination of connections of *Ut* with geomagnetic and solar activities presents the opportunity to another, independent proof of nonlocality, not appealing to time shifts, through violation of Ineq. (4).

In our case $A = Ut$, $B = Dst$, $C = R$ or X. Classical local influence of solar activity, indexed by *R* or *X* on geomagnetic one, indexed by Dst, is well known. However the electrode detector is insensitive to the local magnetic field (at least, up to 10^{-3} T), as well as it is insensitive to classical local impact of solar activity mediated by the cosmic ray flux [1]. However, even assuming that these ideas are incomplete and some classical correlations *R* or *X* with *Ut* are available, Ineq. (4) must be satisfied. Let us check it by substitution of experimental values of independence functions under the same filtration of the series $A = Ut$, $B = Dst$, $C = R$ or X.

For $C = R$ with filtration 365>*T*>59 days: $i_{Ut|R} = 0.35^{+0.05}_{-0.00}$, $i_{Ut|Dst} = 0.51^{+0.08}_{-0.00}$, $i_{Dst|R} = 0.57^{+0.00}_{-0.01}$; with filtration 365>*T*>77 days: $i_{Ut|R} = 0.34^{+0.00}_{-0.05}$, $i_{Ut|Dst} = 0.47^{+0.00}_{-0.01}$, $i_{Dst|R} = 0.54^{+0.00}_{-0.00}$. For $C = X$ with filtration 365>*T*>59 days: $i_{Ut|X} = 0.36^{+0.14}_{-0.00}$, $i_{Ut|Dst} = 0.51^{+0.08}_{-0.00}$, $i_{Dst|X} = 0.55^{+0.00}_{-0.00}$; with filtration 365>*T*>77 days: $i_{Ut|X} = 0.28^{+0.02}_{-0.00}$, $i_{Ut|Dst} = 0.47^{+0.00}_{-0.01}$, $i_{Dst|X} = 0.55^{+0.01}_{-0.01}$.

In all four variants Ineq. (4) is reliably violated. Thus, the nonlocal (quantum) nature of the observed in the Baikal experiment correlations is proved also through the violation of the steering inequality without resorting to arguments of time shifts.

Before to turn to analysis of the 2016/2017 series for study of detector correlation with internal (hydrological) processes, we note, that correlation with external (heliogeophysical) processes in spite of low solar activity certainly must remain. Consider the result of causal analysis of the detectors *Ut* and *Ul* (Figure 5) and compare it with the one obtained at high solar activity 2013/2014 [5]. In Figure 5 we observe $\gamma > 1$, that is there a causal connection *Ul*→*Ut* (as well as in 2013/2014); the highest maximum of causality $\gamma \approx 1.3$ and the deepest minimum of independence $i_{Ut|Ul} \approx 0.49$ are at advancement 24 days (1.4 and 0.34 respectively just at the same advancement in 2013/2014); the ratio of the maximal time reversal and time respecting (normal) causality functions is 1.02 (1.03 in 2013/2014). Causal connection remained similar, but weaker.

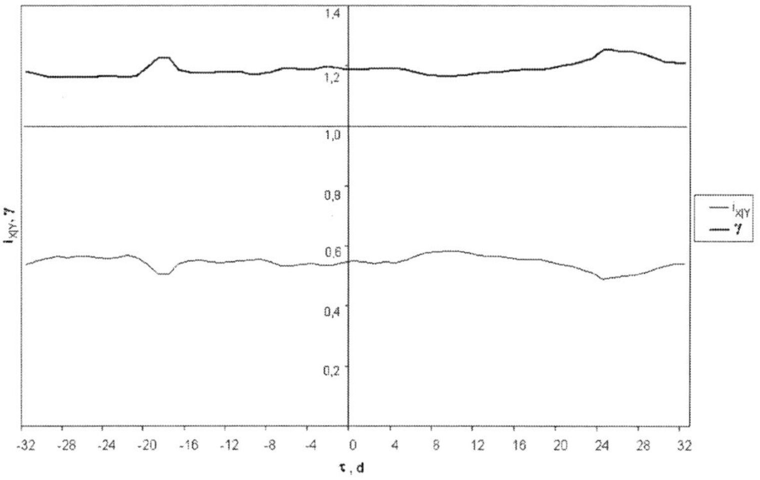

Figure 5. Causal analysis of *Ut* (X) and *Ul* (Y). τ < 0 corresponds to retardation of *Ut* relative to *Ul*, $\tau > 0$ – to advancement.

In Figure 6 the amplitude spectra (in absolute units) of the detector *Ut* and temperature *t* in the active layer are presented. It is seen that at the long periods there is a certain similarity of the spectra *Ut* and *t*. Therewith their amplitude ratio proves to be much greater (of order 1 mV/K) than upper bound of a local influence (electrode temperature coefficient is 0.04 mV/K). It may be explained only by some nonlocal correlation.

IOP Conf. Series: Journal of Physics: Conf. Series **1051** (2018) 012019 doi:10.1088/1742-6596/1051/1/012019

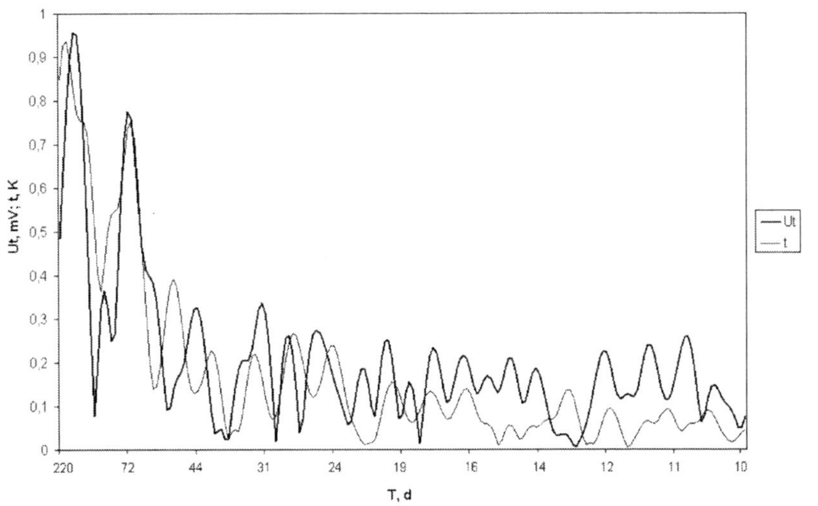

Figure 6. Amplitude spectra of the detector *Ut* (at horizon 47 m) and temperature *t* (at horizon 20 m).

Consider the results of causal and correlation analysis of *Ut* and *t* (Figure 7) with the broadband filtration, stemming from the spectra and corresponding to low-frequency macroturbulence, 156>*T*>48 days. $\gamma > 1$ that is *t* is a cause with respect to *Ut*, the advanced maxima of the causality function are greater than the retarded ones. We observe two, both advanced highest $\max \gamma = 1.5^{+0.1}_{-0.0}$, but only one of them, at τ = 25 days corresponds to the deepest $\min i_{Ut|t} = 0.19^{+0.1}_{-0.0}$ and the greatest value of correlation function $\max r = 0.99^{+0.00}_{-0.03}$. It is just manifestation of advanced nonlocal connection *t*→*Ut*.

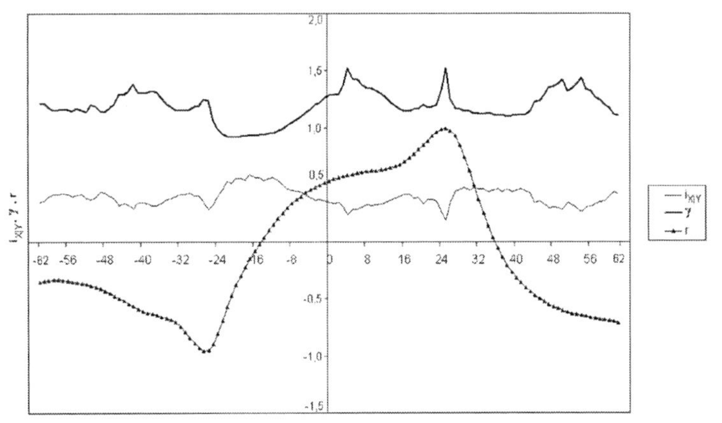

Figure 7. Causal and correlation analysis of U_t (*X*) and *t* (*Y*). $\tau < 0$ corresponds to retardation of U_t relative *t*, $\tau > 0$ – to advancement.

We can use both forecasting algorithms, mentioned above, to forecast the random temperature variation Δ*t* by the filtered signal of *Ut* with fixed advancement 25 days. However, the length of the possible forecasted Δ*t* series is short, because, first, band-pass filtering strongly shortens the length of *Ut* series and, second, the need for a primary training interval takes in given case at least 75-80 first days. In addition, with algorithm based on impulse transient response, third, the need of the solution regularization (postfiltering) shortens the length of forecasted Δt series again. Note also, that in any classical forecast, e.g. weather one, a forward problem is solved (computation of the effects by the causes). In contrast, we forecast the causes by the effects that are we solve, much more complicated inverse problem. Our purpose for now is only demonstration of a real forecast based on time reversal causality.

In the regression algorithm the regression coefficient on the training interval is computed, which then used for the computation of the first forecasted value. On the next step (hour), the training interval is moved forward, the procedure is repeated and the next value with the same advancement is forecasted, and so on. In the transient response algorithm [1] the long impulse transient response on the training interval is computed (by solving of the integral equation), which then is used, as its convolution with multitude of the preceding detector signal values, for the computation of the first forecasted value. Next, the procedure is repeated as well as in the former algorithm. Advantage of the latter one is taking into account non-Markovian nature of the process relation. Since the both methods are linear (although taking into account non-stationarity), the optimal advancement of forecasting is determined by the position of the main correlation function maximum.

The results are presented in Figure 8. The forecast on the sliding regression (Figure 8a), in spite of some unfit, demonstrates the mean accuracy acceptable for the practical purposes. The forecast series on the sliding impulse transient response (Figure 8b) is short to look impressive, but its mean accuracy is much better, as it was expected.

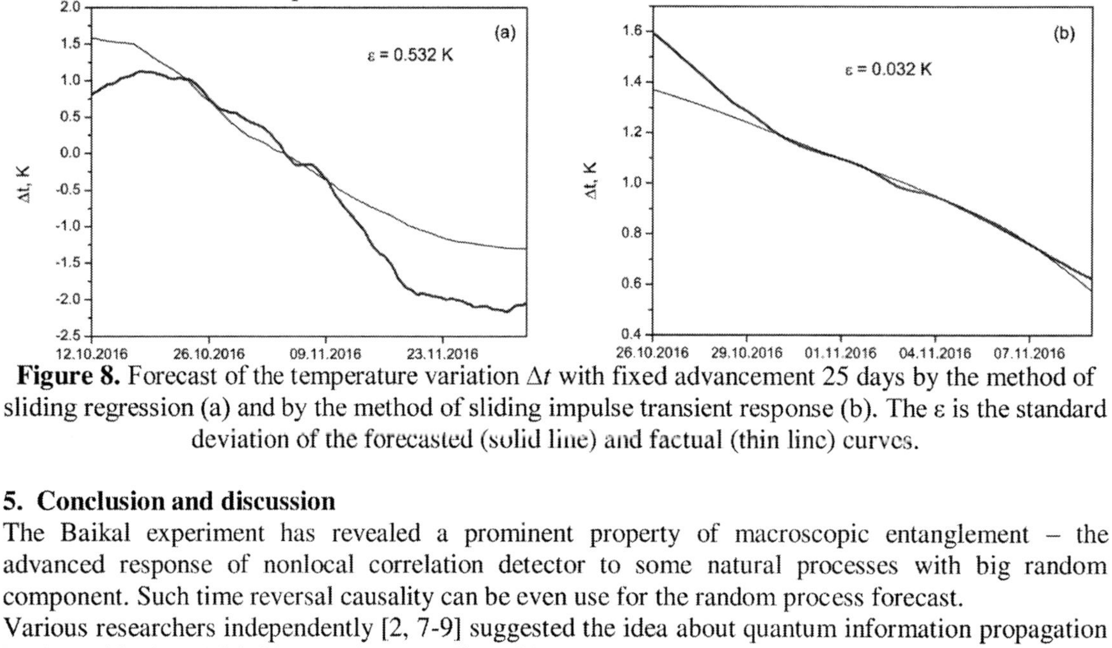

Figure 8. Forecast of the temperature variation Δt with fixed advancement 25 days by the method of sliding regression (a) and by the method of sliding impulse transient response (b). The ε is the standard deviation of the forecasted (solid line) and factual (thin line) curves.

5. Conclusion and discussion

The Baikal experiment has revealed a prominent property of macroscopic entanglement – the advanced response of nonlocal correlation detector to some natural processes with big random component. Such time reversal causality can be even use for the random process forecast.

Various researchers independently [2, 7-9] suggested the idea about quantum information propagation backward in time. This idea has been confirmed by observations of advanced nonlocal correlations in the experiments with low-dimensional systems (including a few particles) [8-11]. In these experiments, the advanced correlations were revealed *post factum*, in agreement with common view that quantum information can be decoded only with ancillary classical one, propagating forward in time. However, the latter is true if we aim to get decoded quantum information from the future (*a posteriory*) with unit probability. If we aim to get such information with finite probability, it becomes possible without ancillary classical channel [12]. It can be illustrated by the simplest example of qubit teleportation protocol, where Bob do not use classical communication with Alice, but instead measure state of his particle (one from the entangled pair emitted by EPR-source) with result $|0\rangle_B$ or $|1\rangle_B$. Suppose Alice teleports a definite state $|0\rangle$. Performing its joint measurement with her particle from the entangled pair, she obtain randomly one of four results (Bell states) $\left|\Psi^{\pm}\right\rangle_A$, $\left|\Phi^{\pm}\right\rangle_A$, and one can guess which one with probability 1/4. The joint three-particle state at any moment of Bob measurement, including a moment before (!) Alice measurement (Bob can be closer to the source than Alice) is $1/2\left(-\left|\Psi^{-}\right\rangle_A\left|1\right\rangle_B+\left|\Psi^{+}\right\rangle_A\left|1\right\rangle_B+\left|\Phi^{-}\right\rangle_A\left|0\right\rangle_B-\left|\Phi^{+}\right\rangle_A\left|0\right\rangle_B\right)$. Hence, Bob can guess Alice result with probability 1/2. The probability of his successful forecast is twice higher as compared to random guess. Physically, such a possibility to observe the random future is explained in this example by propagation of quantum information backward in time from Alice setup to EPR-source [2, 7-9].

Quantum information by itself can only partly be decoded. Therefore, the forecast of the random processes on nonlocal correlations is doomed to be probabilistic. However, for high-dimensional (macroscopic) system the probability of success can be much greater than in above qubit example. On the other hand, owing to decoherence, the entanglement degree of the macroscopic states is unavoidable very low and that prevents the success. Most of contemporary experimental studies of macroscopic entanglement (e.g. [13-14]) does not concern time problem yet. Our investigation was inspired by the prominent experimental results of Nikolay Kozyrev who was the first to discover advanced nonlocal correlations on interstellar scale, although he interpreted them in semiclassical terms [15].

The problem of time reversal propagation of quantum is associated with the problem of closed timelike curves (CTC). David Deutsch in his seminal work [16] stressed that his quantum-mechanical proof of the time travel possibility along CTC is applicable without any change to a macroscopic body. CTCs can exist not only in the curved space-time of the wormholes, but also in the flat space-time with some quantum circuits, although their properties are rather different [17]. Self-consistency of CTC allows a probabilistic forecast of the random future, as in our case.

References

[1] Korotaev S M 2011 *Causality and Reversibility in Irreversible Time* (Irvine, CA: Scientific Research Publishing)

[2] Cramer J G 1980 *Phys. Rev.* D **22** 362

[3] Hoyle F and Narlikar J V 1995 *Rev. Mod. Phys.* **67** 113

[4] Korotaev S M, Budnev N M, Serdyuk V O, Gorohov J V, Kiktenko E O, Zurbanov V L, Mirgazov R R, Buzin V B and Novysh A V 2013 *Physical Interpretations of Relativity Theory* ed V O. Gladyshev, A N Morozov and P Rowlands (Moscow: BMSTU PH), pp 141-151

[5] Korotaev S M, Serdyuk V O, Kiktenko E O, Budnev N M and Gorohov J V 2015 *Unified Field Mechanics* ed R L Amoroso, L H Kauffman and P Rowlands (Singapore: World Scientific) pp 366–373

[6] Lean J L and Brueckner G E 1989 *Astrophys. J.* **337** 568

[7] Penrose R 1998 *Soc. of London Phil. Tr.* A **1743** 1927

[8] Laforest M, Baugh J and Laflamme R 2006 *Phys. Rev.* A **73** 032323

[9] Lloyd S, Maccone L, Garcia-Patron R, Giovannetti V, Shikano Y, Pirandola S, Rozema L A, Darabi A, Soudagar Y, Shalm L K and Steinberg A M 2011 *Phys. Rev. Lett.* **106** 040403

[10] Ma X-S, Zotter S, Kofler J, Ursin R, Jennewien T, Brukner Č and Zeilinger A 2012 *Nature Physics* **8** 479

[11] Megidish E, Halevy A, Shacham T, Dvir T, Dovrat L and Eisenberg H S 2013 *Phys. Rev. Lett.* **110** 210403

[12] Kiktenko E O and Korotaev S M 2014 *Physics Essays* **27** 548

[13] Paternostro M, Vitali D, Gigan S, Kim M S, Brukner Č, Eisert J and Aspelmeyer M 2007 *Phys. Rev. Lett.* **99** 250401

[14] Lee S-S B., Park J and Sim H-S 2015 *Phys. Rev. Lett.* **114** 057203

[15] Kozyrev N A and Nasonov V V 1980 in *Manifestation of Cosmic Factors on the Earth and Stars* ed A A Efimov (Moscow: VAGO Press,) pp 76-84

[16] Deutsch D 1991 *Phys. Rev.* D **44** 3197
[17] Korotaev S M and Kiktenko E O 2015 *Physica Scripta* **90** 085101

Non-stationary Noise Sources in an Optoacoustical Gravitational Antenna

V A Krysanov

Senior Researcher, Institute for Nuclear Research of Russian Academy of Sciences, Russia
E-mail: krysanov@mail.ru

Abstract. The optoelectronic Pound-Drever-Hall scheme is considered as a registration system of small variations of length and eigenfrequency of reference Fabry-Perot resonator with base of 2 m fixed on the aluminum test body of the GW antenna. In uniform sensitivity analysis shot noise of photoelectrons, intrinsic technical LF laser stochastic frequency deviations and power fluctuations (PF) are considered. Mechanism of PF penetration into the precise feedback chain is pointed out. It is shown that the carrier of PF noise is the quasistatic error signal of the laser frequency tracking system following the drifting frequency of the resonator, since this voltage is formed by synchronous detector from the photodiode current. The frequency discriminator channel of the installation OGRAN can also introduce PF noise, since the output of it's synchronous detector contains significant LF noise (10÷300 Hz) voltage created by industrial seismic and vibration disturbances.

1. Introduction

The basis of a bar gravitational wave (GW) antenna [1] is a cylindrical aluminum acoustic resonator as a test body. Gravitational field metric perturbations $h(t)$ are assumed in form $h = h_m \cos \omega t$ with limited exposure time τ [2-4]. Frequency ω should be fairly close to bar eigen frequency ω_μ. Effect of GW is reduced to equivalent harmonic signal force with amplitude [2]

$$F_{Sm} = h_m M \omega_\mu^2 L_G / 2, \tag{1}$$

where M is an equivalent oscillator mass and L_G is a bar length.

Small variations in bar length are registered by a sensitive measuring system (measurer). During the development of the project OGRAN, progress is achieved by reducing the threshold signal (resolution) $x_{\min}$ of the measurer ($S / N = 1$) as it's single relevant parameter.

In absence of a signal, the measurer registers thermal vibrations of the test body. At the antenna's output they look as a harmonic process with stochastically slowly varying amplitude. The corresponding spectrogram is presented in article [1] in Figure 7. Registration of this process allows us to control achieved measurer resolution as a result of modifications and adjustments. Spectrogram shows noises of the test body and the measurer in common. And anyone can see noise "pedestal" at level

$\sim 3\cdot 10^{-15}$ cm/Hz$^{1/2}$ ($\sim 4\cdot 10^{-3}$ Hz/(Hz)$^{1/2}$). The aim is to show that this noise level determines uniquely the numerical value of the new parameter introduced for the first time in the article [1]: bands of 4 Hz for $h\approx 10^{-19}$ Hz$^{-1/2}$. Accordingly, this value should be considered as the achieved resolution of the measurer. An explanation of noise origin is offered.

2. Resolution of the measurer; phenomenology

The main result of article [1] is presented in the form of a graph of the spectral density (SD) of the metric in Figure 8. Certain numerical characteristic of this dependence is represented.

To compose the theoretical forecast of this spectrogram we should to describe noise and dynamic parameters of the test body and to write out the equation of an equivalent oscillator

$$M\ddot{x}+H_{\mu}\dot{x}+k_{\mu}x=F_S+F_T. \tag{2}$$

The Nyquist force F_N has spectral density [1,3]

$$G_T(f)=2\pi G_T(\omega)=4kTH_{\mu}=\frac{4kTM\omega_{\mu}}{Q}. \tag{3}$$

Near the resonance there is a dynamic transfer characteristic

$$\frac{x_m}{F_m}\equiv K_{FX}(\omega)=\frac{Q/M\omega_{\mu}^2}{1\sqrt{\left[1+(\omega-\omega_{\mu})^2/\delta_{\mu}^2\right]}}. \tag{4}$$

Spectral densities of force and displacements are related: $G_{XT}(f)=F_{FX}^2G_T(f)$.

The expression of bar thermal noise follows from expressions (3), (4). With the addition of the phenomenological noise of measurer (sensor) $G_S(f)$ (as a "pedestal"), we obtain forecast

$$G_X(f)=G_{XT}(f)+G_S(f)=G_{T0}\left[1+\left(\frac{\Delta\omega}{\delta_{\mu}}\right)^2\right]^{-1}+G_S(f), \tag{5}$$

$$G_{T0}=\frac{4kTQ}{M\omega_{\mu}^3}. \tag{6}$$

Expression (5) describes the experimental spectrogram in Figure 7 in ref. [1].

We assert further that spectral density in the Figure 8 has been obtained completely by recalculating of the experimental dependence in the Figure 7:

$$G_h(f_0+\Delta f)=G_X(f_0+\Delta f)/K_{hX}^2(f_0+\Delta f). \tag{7}$$

Here the transfer function K_{hX} is defined by expressions (1) and (4):

$$K_{hX}(f_{\mu}+\Delta f)=x_m(f_{\mu}+\Delta f)/h_m=Q(L_G/2)k(\Delta f)=K_{hX0}k(\Delta f).$$

Here we had considered $\Delta\omega=\omega-\omega_{\mu}$, $|\Delta\omega|\geq 3\delta_{\mu}$, $\Delta f=f-f_{\mu}$, $k(\Delta f)=\dfrac{\Delta f_{\mu 0}/2}{|\Delta f|}$.

The meaning of the new entered band of 4 Hz [1] is a difference between two frequency values as points on the abscissa corresponding to the ordinate value $(G_h)^{1/2}=10^{-19}$ Hz$^{-1/2}$.

Let's prepare relevant numerical parameters. For the values of $Q=10^5$, $\dfrac{L_G}{2}=100\,\text{cm}$, $f_0=1.3\cdot 10^3\,\text{Hz}$ [1] we have $K_{hX0}=10^7$ cm, $\Delta f_0\equiv\dfrac{f_0}{Q}=1.3\cdot 10^{-2}\,\text{Hz}$. Then for the value $2\Delta f^*=4Hz$ of frequency deviation in

each branch, we find $k(\Delta f)=3.25\cdot 10^{-3}$. From formula (7), substituting also $(G_h)^{1/2}=10^{-19}$ Hz$^{-1/2}$, we find $(G_X)^{1/2}\cong 3\cdot 10^{-15}$ cm/Hz$^{1/2}$. This is just the value that can be seen as the "pedestal" $(G_S)^{1/2}$ of the spectrogram in Figure 7 and is the main real registered resolution value of the antenna's laser measurer.
To confirm this approval, we ascertain the fact that in Figure 7 [1] one may discover $(G_{T0})^{1/2}\cong 10^{-13}\,cm/Hz^{1/2}$. For $\Delta f^{*}=2$ Hz we find forecast estimate: $(G_{XT})^{1/2}=(G_{T0})^{1/2}k(\Delta f)\cong 3\cdot 10^{-16}\,cm/Hz^{1/2}<<(G_S)^{1/2}$. It means $G_X=G_S$. Those two points corresponds to area of the "pedestal".
The condition $G_{XT}=G_S$ is satisfied with the value $\Delta f^{**}=0.2$ Hz. Within this band $(|\Delta f|<0.2Hz)$ bar thermal noise dominates. Here the relation (7) gives the analytical expression $G_h(f)=\left(\frac{4}{L}\right)^2\frac{kT}{MQ\omega_\mu^3}$.
This coincides with the resultant expression of the antenna's theory [1, 3]:

$$h_{\min}\cong(4/L_G)\sqrt{(kT/M\omega_\mu^3 Q)\Delta f}\sqrt{F}\approx 10^{-20}\sqrt{F\Delta f}\,Hz^{-1/2}. \quad (8)$$

In addition to confirmation, for the peak of the bar thermal spectrum, at the frequency f_0, where $k(\Delta f)=1$, we fine $(G_{h0})^{1/2}=\frac{(G_{T0})^{1/2}}{K_{hX0}}=10^{-20}$ Hz$^{-1/2}$; it exactly coincides to the numerical estimate in expression (8).

3. Classic basics of antenna sensitivity analysis

In the basic formula (8), the contribution of the optoelectronic measurer is introduced by the noise factor $F=\frac{2M}{\tau}\left(\frac{G_E}{G_T}\right)^{1/2}$. Parameters of a conceptual schema are presented [1,3]:

$$G_E=B\omega_\mu^2\left[2h\nu/(\eta P)\right]\left[\lambda/(2\pi N)\right]^2,$$

where λ is wavelength of laser radiation, $\eta\sim 0.7$ is the quantum yield of the photodiode, N is an "effective number of reflections" in the FP resonator, P is the optical power and "$B=1-10^3$ is the phenomenological factor that indicates by how many times exceeds the Poisson level" [1],
To highlight the characteristics of this meter, consider the formula in ref. [3]:

$$(S/N)_{out}=\pi^{-1}\int_0^\infty\frac{|f_\omega(j\omega)|^2\,d\omega}{G_T+|Z_\mu(j\omega)|^2\,G_E}.$$

It contains expression for SD of equivalent total stochastic force F_N in the equation (2): $G_F=G_T+|Z_\mu|^2 G_E$. Applying it, as above, to the dynamic system (4), we obtain

$$G_X=\frac{G_T}{\omega^2}|Z_\mu|^2+\omega^{-2}G_v=G_{XT}(\omega)+G_S(\omega).$$

And analytical expression for meter resolution in expression (5) is extracted: $G_S=\omega^{-2}G_E$,

$$G_S^{1/2}(\omega)=\frac{\lambda}{\pi N}\sqrt{\frac{h\nu}{2\eta P}}B=\frac{\lambda}{\sqrt{2}\mathrm{F}}\sqrt{\frac{h\nu}{\eta P}}B. \quad (9)$$

In the last expression are taken into account relations: $N = \dfrac{1}{1-R_m}$ and $F = \dfrac{\pi}{1-R_m}$ as finesse [5], where R_m is an energy reflection coefficient in mirrors of Fabry-Perot (FP) resonator.

On the other hand, there is an article [4], in which a meter on an FP interferometer had been presented and described briefly; it's threshold signal for $\eta = 1$ is represented by the formula:

$$x_{\min} = \frac{(1-R_m)c}{2\pi}\sqrt{\frac{2h}{N_0 \nu_0 \tau}}A = \frac{\lambda}{2\mathrm{F}}\sqrt{\frac{2h\nu}{P}A\Delta f}\,. \tag{10}$$

Here laser noise is represented by a numerical factor A too.

In this article the FP resonator mirrors are attached to ends a bar; it's length variation δx cause locked mode natural frequency variations of an interferometer $\delta \nu_R$. For enough hard mount mirrors the relation is: $\dfrac{\delta x}{L_G} = \dfrac{\delta \nu_R}{\nu_R}$. Registered variations of frequency allow us to determine variations in length. In article [1] this relationship is expressed in the left and right scales of noise spectrograms. For $L_G = 200$ cm and $\nu_R = 2.8 \cdot 10^{14}$ Hz we have for the scales the conversion factor value $\dfrac{\delta x}{\delta \nu_R} = 0.7 \cdot 10^{-12}$ cm/Hz.

Condition $A = B = 1$ in formulas (9) and (10) establishes the resolution limit, determined by shot noise of photoelectrons in a photodiode.

4. Laser noises

To remove some uncertainty in resolution sensitivity forecast ($B = 1 \div 10^3$) an additional study of laser noise characteristics is actual. So, this noise also requires consideration in heterodyne (homodyne) receiving of optical signals. In articles [6,7] and in the technical essay [8] there had been studied fluctuations as "amplitude chaotic modulation of the laser radiation: $P(t) = P_0\left[1+\xi(t)\right]$, where $\xi(t)$ is a dimensionless modulating process" having the spectral density $m_L(f)$. LF power fluctuation had been measured for the laser used in OGRAN pilot model [9], represented and applied for analysis [10].

Pound-Drever-Hall (PDH) technique using modulation-demodulation is applied to measure displacements in GW detector AURIGA [11]. Low frequency (LF) laser power fluctuations (LPF) in the noise budget are replaced by negligible RF LPF.

In reports [12,13] there are introduced into consideration phenomenological LF stochastic laser frequency deviations (SLFD). In AURIGA project their influence is negligible in consequence of smallness of the interferometer base – $L_{R*} = 0.1$ mm. On the other hand, the detector AURIGA is sensitive to LPF. They are under control because they create fluctuations in the light pressure force in the resonant displacement transformer.

Spectral density of SLFD had been measured approximately: $\left(S_\nu\right)^{1/2} \approx 10$ Hz/(Hz)$^{1/2}$ at the antenna operating frequency of 1.3 kHz [13]. In ref. [9] on the graph of noise spectra one can see value $\left(S_\nu\right)^{1/2} \approx 4$ Hz/(Hz)$^{1/2}$.

This noise influence is suppressed by stabilizing laser frequency applying a reference FP interferometer; the GWA resonator just performs this function. For this reason the enlarged gain factor K_{0S} has required in locked feedback circuit [13]; the relevant theory is represented below. The value $K_{0S} = 10^3$ at 1.3 kHz is implemented in the OGRAN installation [1,9,13], and calculated noise component in laser frequency deviations is $\left(S_{\nu l}\right)^{1/2} = \frac{\left(S_{\nu}\right)^{1/2}}{K_{0S}} \approx 4\cdot10^{-3}$ Hz/(Hz)$^{1/2}$. The conversion factor $0.7\cdot10^{-12}$ cm/Hz gives the resolution forecast for the OGRAN measurer: $x_{\min}\approx 3\cdot10^{-15}$ cm$\left(\Delta f\right)^{1/2}$; this satisfactory corresponds to the test result, which had been published implicitly, without interpretation [1, Figure 7]. This result has been revealed above.

Further efforts on suppression of SLFD had not succeeded yet.

To discover an additional noise source there was revealed an effect of residual amplitude modulation (RAM) arising in an imperfect RF modulator [10,14]. Certain efforts had been made to suppress it [14]. As a result (see below), RAM noise is assessed as not primary.

Identifying a second powerful source of technical noise is required.

5. PDH scheme sensitivity analysis

In ref.1 RAM noise is pointed out as the primary technical noise. Meanwhile, it's theory and calculation has not presented. The theory is relevant. Also, introduced non-stationary sources of noise are similar in nature to RAM noise. Therefore, it is relevant to present and apply a universal instrument of sensitivity analysis. It includes conjoint consideration of photoelectron shot noise $(B=1)$ and mechanisms of penetration of the laser noise into the precise chain.

This is an improved version of the analysis used in the report [12].

The Pound-Drever-Hall (PDH) scheme [5] is presented at Figure 1. It is a laser frequency retuning and stabilization system.

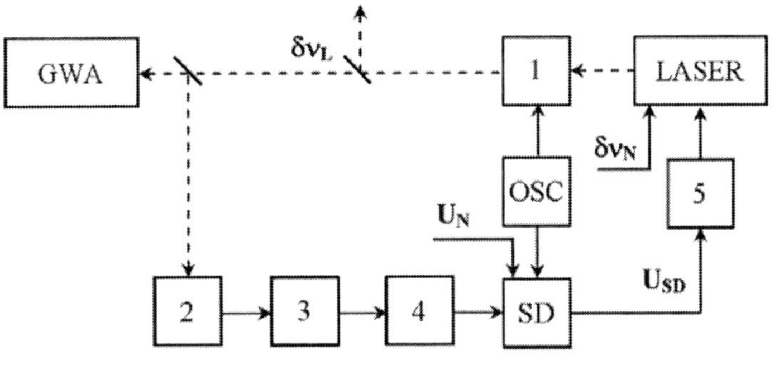

Figure1. 1 – Electro-optical modulator, 2 – photodiode, 3 – current preamplifier, 4 – RF selective filter-amplifier, 5 – servo DC amplifier, OSC – bearing RF generator, SD – synchronous detector.

It uses RF phase modulation of laser radiation and forms a discrimination feature at the output of a synchronous detector (SD); laser frequency $\nu_L(t)$ follows promptly FP resonator intrinsic frequency $\nu_R(t)$. Frequency variations $\delta\nu_L$ containing signal and noise are formed.

Further the electro-optical discriminatory channel converts laser frequency variations into output voltage variations of antennae's measurer on the slope of it's discrimination feature.

A beam with power P_{IN} falls on the interferometer. After electro-optical RF phase modulator 1 we have three spectral components [5]:

$$E_{inc} \cong E_0\left[J_0(\beta)e^{i\omega t} + J_1(\beta)e^{i(\omega+\Omega_0)t} - J_1(\beta)e^{i(\omega-\Omega_0)t}\right].$$

Their powers are related: $P_C + 2P_S \cong P_{IN}$. In a simplified treatment [5], the condition is $R_m + A_m = 1$, where A_m is an energy penetration coefficient in mirrors.

The power in the beam reflected from the interferometer and fallen on photodiode (PD) 2 is given by the expression [5]:

$$P_{REF} = P_C \left(\frac{\Delta \nu_{RL}}{\Delta \nu_{R0}} \right)^2 + 2P_S + D_{PG} \Delta \nu_{RL} \cos \Omega_0 t. \qquad (11)$$

Here $D_{PG} = 8 \frac{(P_C P_S)^{1/2}}{\Delta \nu_{R0}}$ – power decrement, $\Delta \nu_{RL} = \nu_R - \nu_L$, $\Delta \nu_{R0}$ is bandwidth of the resonator at the level -3 dB, $\Delta \nu_{R0} = \frac{\Delta \nu_{FSR}}{F}$, $\Delta \nu_{FSR} = c/2L$ - free spectral range. Sidebands P_S are reflected in 100%. Assuming the first term in (11) to be small, for the persistent component of the power on the photodiode we have the relation: $P_- = 2P_S$.

In reality we have laser power fluctuation and residual amplitude modulation:

$$P_{IN} = P_{IN.0} \left[1 + \xi(t) \right] \left(1 + m_{RC} \cos \Omega_0 t + m_{RS} \sin \Omega_0 t \right). \qquad (12)$$

The typical value of the modulation depth is [10]:

$$\left(m_{RC}^2 + m_{RS}^2 \right)^{1/2} \cong (1 \div 2)\%.$$

On condition $\Delta \nu_{RL} << \Delta \nu_{R0}$ from (11) we find

$$P_{REF} = P_{PH} = 2P_{S0} \left[1 + \xi(t) \right] \left(1 + m_{RC} \cos \Omega_0 t + m_{RS} \sin \Omega_0 t \right) + D_{PG} \Delta \nu_{RL} \cos \Omega_0 t.$$

At PD variation components of power $P_\sim$ and current $I_\sim$ are

$$P_\sim = D_{PG} \Delta \nu_{RL} \cos \Omega_0 t + 2P_{S0} \left[1 + \xi \right] \left(m_{RC} \cos \Omega_0 t + m_{RS} \sin \Omega_0 t \right),$$

$$I_\sim = \frac{\eta e}{h \nu_L} P_\sim + I_{SHOT}.$$

Expression for persistent current components of PD has the form: $I_0 = 2 \frac{\eta e}{h \nu_L} P_{S0}$, where $h\nu_L$ is quantum energy, e is an electron charge.

After low noise current amplifier 3 with gain R_A broadband shot noise I_{SHOT} arrives at a narrowband filter-amplifier. At it's output, the noise voltage becomes a narrow-band stochastic process: $U_{SHOT}(t) = K_F(\Omega) R_A I_{SHOT} \cong U_{mC}(t) \cos \Omega_0 t + U_{mS}(t) \sin \Omega_0 t$, where quadrature components of the complex amplitude are contained.

Input voltage of the SD 4 is given by

$$U_{SD.IN} = K_F R_A \left\{ \frac{\eta e}{h \nu_L} D_{PG} \Delta \nu_{RL} \cos \Omega_0 t + I_0 \left[1 + \xi(t) \right] \left(m_{RC} \cos \Omega_0 t + m_{RS} \sin \Omega_0 t \right) \right\} + U_{mC} \cos \Omega_0 t + U_{mS} \sin \Omega_0 t$$

The synchronous detector converts the harmonic voltage amplitude into persistent and LF voltage: $U_m \cos \Omega_0 t \to K_{SD} U_m$; $U_m \sin \Omega_0 t \to 0$.

Then at the SD output we have

$$U_{SD.OUT} = K_{SD} K_F K_A \left[\frac{\eta e}{h \nu_L} D_{PG} \Delta \nu_{RL} + I_0 m_{RC} (1 + \xi) + I_{mC} \right] = S + N. \qquad (13)$$

Here $I_{mC} = U_{mC}/K_F R_A$ is reduced quadrature component of noise current. Expression (13) is an error signal [5] containing noises.
The theory of narrow-band stochastic processes gives expression for the spectral density of a given quadrature: $S_{SII}(f) = 4eI_0$.
First term in (13) determines the slope of discrimination feature; it is a decrement too. In ref. [9] the measured numerical expression of the slope is given; it is 5.8 kHz/V.
Second term in (13) shows how the persistent voltage carries RAM noise.
To start feedback sensitivity analysis, we separate voltage variational components. In general $\Delta v_R = \Delta v_{RT} + \delta v_R$; here Δv_{RT} is slow thermal drift.

Put at first $\Delta v_{RT} = 0$; In (13) it means $\Delta v_{RL} = \delta v_R = \dfrac{v_L}{L_G}\delta x$.

We introduce denotation: $K_1 = K_{SD} K_F R_A$, $D_{IG} = \dfrac{\eta e}{h v_L} D_{PG}$,

$$I_N = I_0 m_{RC}\xi + I_{mC},\ U_N = K_1 I_N.$$

When the feedback is locked, a variational component of laser frequency δv_L arises. The expression (13) takes the relevant form:

$$U_{SD} = K_1\left[D_{IG}\left(\delta v_R - \delta v_L\right) + I_N\right]. \tag{14a}$$

Here the term from the persistent component of the SD voltage $K_1 I_0 m_{RC}$ is eliminated. It introduces the imbalance of the equation for quasi-static. The special adjustment is used in the scheme to compensate it.
The second equation locks the PDH feedback circuit:

$$\delta v_L = \delta v_N + \beta_L K_2 U_{SD}. \tag{14b}$$

Here K_2 is a gain of servo amplifier 5, β_L is parameter of laser retuning driver, $\delta v_N(t)$ is chaotic process, introduced phenomenologicaly into analytical consideration to characterize stochastic laser frequency deviations [12,13].
The solution of the equation system (14) has the form

$$\delta v_L = \frac{K_{0S}}{K_{0S}+1}\delta v_R + \frac{\delta v_N}{K_{0S}+1} + \frac{K_1 K_2 \beta_L}{K_{0S}+1} I_N = S + N.$$

Here $K_{0S} = K_1 K_2 \beta_L D_{IG}$ is entire gain of the feedback loop.
An expression for the minimum detectable frequency variation ($S/N = 1$) is

$$\left(\delta v_L\right)_{\min} = \frac{\delta v_N}{K_{0S}} + \frac{I_N}{D_{IG}}. \tag{15}$$

The first term determines the suppression of laser frequency fluctuations. This is stabilization of the frequency by means of a reference resonator. The second term shows that for the noise introduced standardly, the presence and depth of feedback does not affect the threshold signal (sensitivity). I.e., one can determine resolution by unlocking feedback speculatively.

Eliminating SLFD and taking into account the relation $\delta v_R = \dfrac{v_R}{L_G} x$ in (15), we find general expressions for small displacement resolution:

$$x_{\min} = \left(\frac{L_G}{v_R}\right)\frac{I_0 m_{RC}\xi + I_{mC}}{D_{IG}}, \tag{16}$$

$$G_S(f)=\frac{m_L I_0^2 m_{RC}^2+4eI_0}{\alpha_{xI}^2}=\frac{S_{IR}+S_{SH}}{\alpha_{xI}^2},$$

where α_{xI} is a parameter of "displacement-photocurrent" conversion, $\alpha_{xI}=\frac{\delta I_S}{x}=\frac{\nu_R}{L_G}D_{IG}$.

To carry out actual numerical estimates, we take into account the additional phenomenological parameter K_G of a real interferometer with energy loss in reflections ($R_m+A_m<1$). This parameter was introduced in [9], and article [10] shows that it reduces the decrement as $K_G/2$, if $K_G\le\frac{1}{3}$. Than we obtain the actual calculation expression

$$\alpha\cong 8\frac{K_G \mathrm{F}_G}{\lambda}\left(\frac{\eta e}{h\nu_L}\right)\sqrt{P_C P_S}\ .$$

To present numerical estimates we employ the values: $\mathrm{F}_G=3000$, $K_G=0.2$, $\lambda=1.06\ \mu\mathrm{m}$, $\eta=0.7$, $P_{PH}=50\ \mathrm{mW}$ [1]. The optimal value of the phase modulation index $\beta=1.15$ rad [5] corresponds to the ratio $\frac{P_S}{P_C}\cong 0.4$. The persistent component of the radiation power on the PD is given by the expression: $P_{PH}=2P_S+(1-K_G)P_C$. Using it we find $P_C=30\ \mathrm{mW}$, $P_S=12\ \mathrm{mW}$ and $\alpha\cong 5\cdot 10^7$ A/m. Spectral density of RAM noise current in (16) for values $I_0=30$ mA и $m_L=10^{-12}\ \mathrm{Hz}^{-1}$ [10] has the value $(S_{IR})^{1/2}=3\cdot 10^{-10}\ \mathrm{A/Hz}^{1/2}$. Value $(G_S)^{1/2}\cong 3\cdot 10^{-16}\ \mathrm{cm/Hz}^{1/2}$ [1] as the reported RAM noise contribution estimate defines value $m_{RC}=0.5\%$. And a real value $m_{RC}=1.5\%$ defines contribution estimate of $10^{-15}\ \mathrm{cm/Hz}^{1/2}$. Thus, there is shown that in installation [1] suppression of RAM is not important.

The uniform noise analysis makes it possible to estimate the shot current noise: $(S_{SH})^{1/2}=1.4\cdot 10^{-10}\ \mathrm{A/Hz}^{1/2}$ and the forecast of the OGRAN measurer resolution limit as $3\cdot 10^{-16}\ \mathrm{cm/Hz}^{1/2}$.

The initial formula for the limiting sensitivity of the circuit is derived in article [5] and report [12]. Here it is obtained by substituting the relation $4eI_0=\frac{8\eta e^2 P_S}{h\nu_L}$ into the expression (16). The expression for the limit resolution of the PDH measurer takes the form:

$$x_{\min}=\frac{\lambda}{8\mathrm{F}}\sqrt{\frac{2h\nu_L}{\eta P_C}\Delta f}\ .$$

There is a small numerical difference from the known formulas [5,12], determined by the conversion of RF noise into LF noise when implementing synchronous detection.

6. Noise of a quasi-static error signal

A different path of low frequency laser power fluctuation penetration into the precision measuring chain may be pointed out. There is significant temperature drift in the aluminum acoustic resonator length (2 m) causing frequency changes of interferometer $\Delta\nu_{RT}$. The laser frequency variation $\Delta\nu_{LT}$ follows it; voltage on the thick piezoelectric block of slow laser tuning is varied. At the moment of tracking capture this voltage is close to zero. Then the voltage rapidly increases as temperature in the MSU boathouse is very unstable. When the voltage reaches a value of 1000 volts the failure of tracking happens. This temperature drift significantly limits time of data accumulation for high resolution spectrogram (Figure 7 [1]). While tracking, quasi-static error signal is changed. Feedback equation for quasi-static variations gives the solution: $\Delta\nu_{LT} = \Delta\nu_{RT}\dfrac{K_{00}}{K_{00}+1}$. Here $K_{00} >> K_{0S} >> 1$ is a gain in the feedback loop at zero frequency.

Operating point at the FP resonant transmittance curve drifts from it's maximum:

$$\Delta\nu_{RLT} = \Delta\nu_{RT} - \Delta\nu_{LT} \cong \frac{\Delta\nu_{LT}}{K_{00}}. \tag{17}$$

This deviation causes variable power component $P_{\sim}$ falling to a photodiode: $P_{\sim} = D_{PG}\Delta\nu_{RLT}\cos\Omega_0 t$. For ideal optical resonator [5] $D_{PG} = 8\dfrac{(P_C P_S)^{1/2}}{\Delta\nu_{RG0}}$.

Here $\Delta\nu_{RG0} = \Delta\nu_{FSRG}/\mathrm{F}_G$, $\Delta\nu_{FSRG}$ - free spectral range, $\Delta\nu_{FSRG} = c/2L_G$ = 75 MHz for $L_G = 2$ m and $\Delta\nu_{RG0} = 25$ kHz for $\mathrm{F}_G = 3000$ [1]. Put ratio $P_S/P_C = 0.25$; it is close to the optimum value of 0.42. Then decrement is $D_{PG} = 16\dfrac{P_S}{\Delta\nu_{RG0}}$. We can consider the modulation depth of radiation power and photodiode current $m_T = P_m/P_{\_} = D_{PG}\Delta\nu_{RLT}/P_{\_}$,

$$m_T = 8\frac{\Delta\nu_{RLT}}{\Delta\nu_{RG0}}.$$

For very arbitrary example values $\Delta\nu_R = 2.5\cdot 10^9$ Hz and $K_{00} = 10^7$ we find $\Delta\nu_{RLT} = 250$ Hz and $m_T = 8\%$. This estimate is approximate also because the actual contrast value $K_G = 0.2$ has not taken into account; it substantially reduces the estimate to 0.5%.

In formalized analyzing this effect, while feedback for variable components is unlocked speculatively, the formula (13) is transformed to the form

$$U_{SD.OUT} = K_{SD}K_F R_A\left[\left(\frac{h\nu_L}{\eta e}\right)8\sqrt{P_{C0}P_{S0}}\frac{\Delta\nu_{RLT}+\delta\nu_R}{\Delta\nu_{RG0}}\left[1+\xi(t)\right]+I_{mC}(t)\right].$$

The expression for the aggregated threshold signal follows from it

$$x_{\min} = \left(\frac{L_G}{\nu_L}\right)\Delta\nu_{RLT}\xi + \frac{I_{mC}}{\alpha_{Ix}}.$$

Here the first new term is independent of the decrement expression.

As a result, we obtain the formula for additional noise contribution

$$G_{S\Delta T}^{1/2} = (L_G/\nu_L)\Delta\nu_{RLT}m_L^{1/2}. \tag{18}$$

For the above estimate $\Delta\nu_{RLT} = 250$ Hz, we obtain $(S_{\nu\Delta T})^{1/2} \cong 2.5\cdot 10^{-4}$ Hz/Hz$^{1/2}$ and $(G_{S\Delta T})^{1/2} \cong 2\cdot 10^{-16}$ cm/Hz$^{1/2}$. Thus, the expressions (17) and (18) make it possible to calculate numerically the threshold signal contribution of noise conditioned by ambient temperature drift.

7. Discriminator vibration noise

Another way of LF laser power fluctuations penetration may be proposed to attention. It occurs in the discriminator channel [1,9,11,13].

IOP Conf. Series: Journal of Physics: Conf. Series **1051** (2018) 012020 doi:10.1088/1742-6596/1051/1/012020

Discriminator channel contains an additional FP resonator and second PDH scheme. It converts laser frequency deviations into antenna's output voltage variations: $U_{SDd} = D_{DUd}\delta\nu_L$. Here D_{DUd} is discriminator decrement or a slope of the PDH discriminatory curve close to it's zero. The natural (eigen) frequency of the discriminator resonator is tuned by the piezoelectric element, to which one mirror is attached. By means of feedback it's eigen frequency follows the frequencies of the laser $\Delta\nu_{LT}$ and base sensor resonator $\Delta\nu_{RT}$. The feedback in the second channel is realized at low frequencies (< 100 Hz) [1]. This provides its function.

In general, this curve has a well-known nonlinear form [5,11]; when the amplitude of the laser frequency deviations increases (in the range $f > 10$ Hz), the transformation ceases to be linear and further reaches full limitation.

There is an observable phenomenon: industrial vibration and seismic noise with frequency range (10÷300) Hz is available at the output of the OGRAN discriminator channel. It's maximum scope is commensurate with the scope of discriminatory curve. Last value can be estimated. So, the standard bandwidth of the resonator is determined by the relation: $\Delta\nu_{RD0} = \dfrac{c/2L_D}{F_D}$. For values $L_D = 0.45$ m and $F_D = 1.2\cdot 10^4$ [1] one can find $\Delta\nu_{RD0} = 28$ kHz , and then the scope can be estimated as (40÷50) kHz.

The fact of influence of significant Moscow industrial seismic background has been noticed. While amplification is nonlinear and significant relatively low-frequency interference is present, the effective gain of a weak signal decreases. For this reason it had been proposed a modified version of a GW detector displacement measurer with the second laser and heterodyning to extend the linearity of frequency demodulation range by an order of magnitude [PIRT'2013]. In OGRAN installation discriminator channel [1] has simple vibration isolation. This second PDH scheme of antenna with slow frequency tracking (10 Hz – 100 Hz) weakly suppresses LF interference. Usually such precise measurements are made at night.

To estimate the effect we impede the interference: $U_{SDd} = D_{DUd}\Delta\nu_m \sin\Omega_{\text{In}} t$, where $\Delta\nu_m$ is amplitude of interference deviation. It is very likely that vibrational interference arrives at the synchronous detector from the photodiode. Then one can refine analytically

$$U_{SDd} = D_{DU}\Delta\nu_m\left[1+\xi(\mathrm{t})\right]\sin\Omega_1 t.$$

A noise component $U_{OUT,N} = \xi D_{DU}\Delta\nu_m \sin\Omega_1 t$ has occurred. In this way voltage carries laser power fluctuations; selected narrowband antenna noise is modulated by function $|\sin\Omega_1 t|$. Envelope averaging gives the equivalent constant voltage $U_{SD-} = \dfrac{2}{\pi} D_D \Delta\nu_m$. We put $\Delta\nu_m = \dfrac{\Delta\nu_{RD0}}{4}$. Then $(S_\nu)^{1/2} = \dfrac{2}{\pi}\Delta\nu_m (m_L)^{1/2} = 4.5\cdot 10^{-3}\ \mathrm{Hz}/(\mathrm{Hz})^{1/2}$. It determines measurer resolution estimation $3\cdot 10^{-15}\ \mathrm{cm}/(\mathrm{Hz})^{1/2}$ coinciding in level with above main source of noise.

Here, as above, the contribution of the noise source is determined quantitatively by the product $\Delta\nu_m\xi$. So through the voltage at the output of the detector laser noise penetrate into the precise circuit. Thus, there has been presented an explanation version of excessive technical noise level had been registered in the installation OGRAN in conditions of increased Moscow seismic background. In conditions of Baksan Neutrino Observatory of INR RAS one should expect an essential decrease in vibration background.

References

[1] Bagaev S N *et al* 2015 *Instruments and Experimental Techniques* **58** 257
[2] Braginsky V B 1993 *Vistas in Astronomy* **37** 341
[3] Gusev A V, Kulagin V V and Rudenko V N 1996 *Gravitation and Cosmology* **5** 68
[4] Braginskii V B 1968 *Soviet physics JETP* **26** 831
[5] Black E D 2001 *Am. J. Phys.* **69** 78
[6] Khaikin N Sh and Yurist B V 1971 *Radiotekhnika I Elektronika* **16** 373
[7] Khaikin N Sh and Yurist B V 1974 *Radiotekhnika I Elektronika* **19** 1952
[8] Egorov-Kuz'min A S, Krysanov V A and Kuruliov B S 1977 *Bulletin of Inventions, USSR* **15** Cert. 555551
[9] Bezrukov L B *et al* 2010 *Instruments and Experimental Techniques* **53** 423
[10] Krysanov V A 2011 *Gravitation and Cosmology* **17** 97
[11] Conti L *et al* 1998 *Rev. Sci. Instrum.* **69** 554
[12] Krysanov V A 2008 *Proc. Int. meeting "PIRT-2007"* (Moscow: BMSTU) 55
[13] Krysanov V A 2009 *Proc. Int. meeting "PIRT-2009"* (Moscow: BMSTU) 242
[14] Vishnyakov V I *et al* 2013 *Tech. Digest Int. Symp. "MPLP-2013"* (Novosibirsk Russia) 179

The variable Planck's constant due to imaginary gravitational temperature

A K M Masood-ul-Alam

Tsinghua University, Beijing, China
E-mail: abulm@math.tsinghua.edu.cn

Abstract. In a talk in PIRT-2015 we described how the Tolman-Thorne temperature of a general relativistic spinning star can be used to define a variable Planck-type constant where the temperature becomes imaginary and how this variable has a negligible value except in some region near a stationary gravitating particle of given mass and spin. In the present talk we recover the Lamb shift of the proton-electron system at thousands' place in MHz from the spread of this region. While this may suggest that the variable Planck's constant could be used with linear Fourier analysis to approximate gravitational effects in coupling gravity with non-gravitational interactions quite effectively, we also consider Einstein-Yang-Mills coupled equations for possible solutions that could be stable because the variable coupling implied by a variable Planck's constant could stop the solutions from blowing up. While no new significant result for these coupled equations has been found yet, we provide a progress report of an ongoing program in which eventually we would like to explain at least qualitatively every valid result in QM and QFT with the help of the variable Planck's constant suggested by GR (by which we mean here Einstein equations) and without using renormalization. We elaborate the first problem in more details. The remarkable coincidence in the earlier talk gave us a variable Planck's constant h_g of gravitational origin roughly at 0.2 Bohr radius for a gravitating particle having the mass and spin of a proton. In this talk we visit Welton's argument for the Lamb shift to see how far the Lamb shift can be explained with this h_g. The fluctuations of the quantized EM field used in Welton's argument would take place only in the region where h_g is meaningful. If we consider the Fourier transform of a real-valued function of a single variable having nonzero value 1 only on an interval of the positive axis, we see that the outer endpoint is roughly inversely proportional to the lowest significant wave number and inner endpoint is roughly inversely proportional to the highest significant wave number. The omitted frequencies produce the Gibbs' phenomenon. Thus the ratio of the cutoff frequencies used in Welton's argument $\frac{k_{\max}}{k_{\min}}$ becomes in our discussion $\frac{r_{\max}}{r_{\min}}$. If we take $r_{\min}$ to be the distance from the center of mass of the proton-electron system where the gravitational temperature becomes imaginary and $r_{\max}$ to be the distance where $h_g \approx h$ (h_g decreasing as inverse square of r thereafter), we get values of our Lamb shift in the range 1019 to 1249 MHz. We do not expect to get the experimental value of 1057 MHz following our crude qualitative arguments. The discussion on the variation of values based on Welton's argument namely 667 to 1394 MHz (see for example the textbook of F. Schwabl) makes one appreciative of our values and hence of our suspicion that gravity is behind the significant part of the Lamb shift. The next task is to look closely wherever Planck's constant is used in the equations of quantum

IOP Conf. Series: Journal of Physics: Conf. Series **1051** (2018) 012021 doi:10.1088/1742-6596/1051/1/012021

physics. Just naively replacing this Planck's constant by h_g will not give the full or a consistent story. In particular one would like to know how the Dirac equation and Einstein-Yang-Mills coupled equations get modified with h_g. A variable Planck's constant will make the coupling constant of the Einstein-Yang-Mills coupled system a varying function and the scale invariance of the system will be lost. We have done some initial calculation for the case of static Einstein-Yang-Mills equations. Spinning has not been added yet to keep the field equations manageable and so the justification behind using h_g is somewhat missing. However the indications are that we cannot have SU(2) solutions, and we cannot have spherically symmetric solutions unless possibly when the gauge group is more complicated. Otherwise the coupling is forced to become constant and we get the known unstable solutions. A sufficiently complicated gauge group gives more equations. Looking at the phase space of the known solutions we then expect that the varying coupling field may stop the blowing up of some otherwise singular solutions. The global solution is expected to be a stable solution because its constant coupling analog is not the unstable separator of the classes of YM-potentials going to $\pm\infty$. Some related issues and how our results can be improved will be discussed.

1. Introduction

In [1] we have a function h_g that plays the role of the Planck's constant in a Schrödinger-type equation guessed from the imaginary values of a gravitational temperature obtained by extrapolating the Tolman-Thorne temperature in some way in the exterior of a massive stationary spinning star or a star-like particle. This h_g has values close to the Planck's constant only in a narrow region outside the particle. In other region h_g either does not exist because the gravitational temperature (g-temperature) is real, or h_g is small decaying rapidly far away from the axis. We suppose that the quantum phenomena will be observed only in the volume of importance where h_g is defined and significant. We recall from [1]

$$\hbar_g \approx \frac{1{,}6R^{2+1{,}5\alpha}k_B}{M^{\alpha/2}r^{1+\alpha}}. \tag{1}$$

We assume that the Newtonian approximations used in deriving this equation crudely apply to a proton and take $R =$ proton radius $0{,}88\cdot10^{-13}$ cm and $M =$ proton mass $=1{,}24\cdot10^{-52}$cm. We had another equation estimating the distance at which g-temperature becomes imaginary. This distance is of the order of Ω^{-1} (in geometrized unit) where Ω is the angular velocity of the star. The angular velocity (angular frequency) of the proton is related to its spin $1/2$ by $\Omega^{-1}=2\hbar M^{-1}R^{-2}$ which equation we use to estimate this distance.

In this formula $1\text{s}=3\cdot10^{10}$ cm. Thus for a proton the imaginary temperature comes out at about a distance of $1{,}8\cdot10^{-13}$ cm. With $\alpha=1$, Eq.(1) gives $h_g\approx2{,}87\cdot10^{-85}r^{-2}\,\text{cm}^4$ at $r\approx0{,}33\times10^{-9}$ cm.

We shall revisit Welton's argument in view of the variations in h_g. In [1] we suspected that h_g has something to do with the Planck's constant h because for a nucleon we got $h_g\approx h$ at $r\approx10^{-9}$ cm $\approx0{,}2$ Bohr radius, while for a star it is at an enormously great distance. In this paper we show that the size of the region of importance gives the major part of the Lamb shift frequency for the hydrogen atom matching at thousands' place in MHz. Thus we get another strong suggestion that the Planck's constant is of gravitational origin.

2. Major part of the Lamb shift

We suppose that the fluctuations of the quantized EM field would take place only in the region where h_g is defined and appreciably nonzero. We are neglecting h_g for electron itself. An electron will be out of the region where h_g due to the electron is important. This is because of the inner cutoff and hence its effect could only be indirect. So we shall neglect it. For a proton the region of interest is from $r \sim 1,8 \cdot 10^{-13}$ cm to $r \sim 0,33 \cdot 10^{-9}$ cm as described in the previous two paragraphs.

2.1. Computation

We approximate h_g with $\hat{h}\chi$ where χ is the characteristic function of the interval $(r_{\min}, r_{\max})$ and $\hat{h}$ is some suitable average height. We write $\hat{h}\chi$ as a difference of two rectangular functions on intervals $(0, r_{\min})$ and $(0, r_{\max})$. We approximate the doubles of these two functions formed by reflecting the graphs on the vertical axis (the doubles being even functions) by their Fourier integrals. Let k_1, k_2 denote the maximum wave numbers we would like to keep for each of the approximations. Such cutoffs produce some mismatch due to Gibb's phenomena but little mismatch will not matter much because we are dealing with approximations of h_g anyway. Now k_1 is inversely proportional to $r_{\min}$ (k_2 is inversely proportional to $r_{\max}$). This is because if we approximate the even function $f(r) = \hat{h}\chi_{(-r_0, r_0)}(r)$ for $r_0 > 0$, where $\chi_{(-r_0, r_0)}(r)$ is the characteristic function of the interval $(-r_0, r_0)$, by its Fourier cosine integral with frequency cutoff ω_0 then ω_0 is roughly inversely proportional to r_0:

$$f(r) = \frac{2\hat{h}}{\pi}\int_0^\infty \frac{\cos(\omega r)\sin(\omega r_0)}{\omega} d\omega \approx \frac{\hat{h}}{\pi}\int_0^{\omega_0} \frac{\sin(\omega(r + r_0))}{\omega} d\omega - \frac{\hat{h}}{\pi}\int_0^{\omega_0} \frac{\sin(\omega(r - r_0))}{\omega} d\omega.$$

The graph of $\omega^{-1}\sin(\omega(r + r_0))$ as a function of ω has the maximum height of $(r + r_0)$ at $\omega = 0$ falling to zero at $\omega = \pm\frac{\pi}{r + r_0}$. $\omega^{-1}\sin(\omega(r + r_0))$ has appreciable value only for $|\omega| < \pm\frac{\pi}{r + r_0}$. We see that the cutoff frequency ω_0 is roughly inversely proportional to r_0. Hence we denote k_1 by $k_{\max}$ and k_2 by $k_{\min}$. Then we have $\frac{k_{\max}}{k_{\min}} \approx \frac{r_{\max}}{r_{\min}}$.

Thus if we compute the energy due to fluctuations of the relative position of an electron caused by the fluctuations of the quantized EM field following Welton [2] (see also Bethe [3]) without using his argument for the upper and lower cutoff of frequencies we find the frequency shift

$$E \approx 135,64 \ln \frac{k_{\max}}{k_{\min}} \approx 135,64 \ln \frac{0,33 \cdot 10^{-9}}{1,8 \cdot 10^{-13}} = 1019 \text{ MHz} \tag{2}$$

or possibly

$$E \approx 135,64 \ln \frac{k_{\max}}{k_{\min}} \approx 135,64 \ln \frac{10^{-9}}{10^{-13}} = 1249 \text{ MHz}. \tag{3}$$

2.2. Discussion

The experimentally observed shift $2S_{1/2} - 2P_{1/2}$ in the hydrogen atom is approximately 1057,86 MHz. Actually the estimate by Welton's method vary from about 667 MHz (see Schwabl [4] page 193) to 1394 (or 1413) MHz depending on the choice of the lower limit $k_{\min}$. Welton's value was roughly the average though Welton did not explicitly stated so in [2]. Schwabl notes it out of curiosity. Theoretical argument based on standard theory gives roughly about 1052. Our value E is supposed to be for the state $2S_{1/2}$. Contribution of $2P_{1/2}$ is not much. So we shall not worry about it. Theoretical values in

IOP Conf. Series: Journal of Physics: Conf. Series **1051** (2018) 012021 doi:10.1088/1742-6596/1051/1/012021

text books may also differ because some other important corrections are added affecting up to tens' place. The factor of 135.64 in Eq.(2) was derived using several simplifying assumptions (see, for example, page 482 in [5]). One of these assumptions may need a closer look in view of our method. This factor has the value of the wave function at the origin at the center of mass of the proton-electron system and it comes out because of the convolution of the wave function with the delta function singularity of the Coulomb potential. Center of the proton charge will be outside the domain of definition of h_g Thus this factor could be smaller and in that case the second estimate given in Eq.(3) which is more realistic in derivation but which is an overestimate will also be smaller.

2.3 Work remains to be done

Now we discuss the work remains to be done. Welton's method is based on the non-relativistic Schrödinger equation. The fluctuation of the EM potential is calculated only up to second order in the fluctuation of the position of the electron. We hope somebody recalculates what we did using the Dirac equation. At present we did not sort out what $e^{-imc^2t/\hbar_g}$ means. Our values of h_g in the spacetime show that an expression such as $\frac{e^2}{\hbar c}$ (here e is the charge of the electron), should be avoided at both low and high frequency ends. In certain quantum calculation an expansion in powers of $\frac{e^2}{\hbar c}$ is used. Bloch and Nordsieck [6] removed the divergence in the low frequency limit by considering alternative expansion procedure in which ħ does not occur in the denominator. It is natural to demand that transition to classical limit $\hbar \to 0$ is possible in the low frequency limit. High frequency limit is considered far from the classical regime and the need to eliminate high frequency divergence did not get similar attention.

Two other potential sources of corrections are as follows. One is the process involving the emission of virtual quanta by the field and their absorption by the electron. The other is the Uehling's vacuum polarization. For the Lamb shift $2S_{1/2} - 2P_{1/2}$ in the hydrogen atom these corrections may affect only up to the tens' place in MHz while our heuristic method is crude at hundreds' place. So at present it is meaningless to include them in our calculation in terms of the renormalized charge or mass of the electron (see pages 157-159 in [7]). Assuming that these activities are important only in the region of the physical spacetime where the variable Planck's constant is important (and neglecting the completely disconnected Feynman graphs), one suspects that only h_g produced by the nucleus of the hydrogen atom is important. For this contribution to h_g both low and high frequency cutoffs as mentioned above would possibly remove divergences. As for the contribution to h_g from a quantum we remember the discussion of Feynman [8]. Feynman suggests that the electron wave function should be kept away from the light cone of the quanta to avoid a divergence. However for our purpose the light cone of the quanta will be inside the light cone of the radius where the g-temperature due to the quanta becomes imaginary. Hence one expects that the net effect of the contribution to h_g from the field quanta may not have much significance. How to calculate this contribution? Some naive calculation may be possible as in the case of the WKB analysis of electromagnetic waves giving the photon number conservation as well as considering a photon as a star-like particle of limiting zero rest mass in a co-moving frame of limiting velocity c, the star-like particle being a small region about the peak of the electromagnetic field where the energy density is momentarily maximum. Such calculation may also explain the meaning of the analog of the Einstein relation namely $mc^2 = \omega\hbar_g$ and could be useful in clarifying the meaning of e^{-imc^2t/h_g} as $h_g \to 0$ if the expression occurs in a solution of the modified Dirac equation.

3. EYM equations

We now discuss the possible effect of the varying h_g on Einstein-Yang-Mills coupled equations. A variable Planck's constant possibly indicates that the coupling constant of the EYM equations may not be actually constant. While it is not totally clear how a varying h_g would affect the coupling we now report on an ongoing investigation and some ideas involving variable coupling for EYM equations.

3.1. Variable coupling

For the electromagnetism (in short EM) the coupling constant is $l_{em} = \frac{1}{4}\varepsilon_0 c$ so that we get the Maxwell equations in SI units. We recall the Einstein equations $R_{ab} - \frac{1}{2}Rg_{ab} = \left(\frac{8\pi G}{c^4}\right)T_{ab}$ coupled to EM field having the energy-momentum tensor

$$T_{ab} = \frac{1}{4}\varepsilon_0 c^2\left(F_{ac}F_{bd}g^{cd} - \frac{1}{4}g_{ab}F_{ij}F_{kl}g^{ik}g^{jl}\right).$$

There is no Planck's constant in the Lagrangian density of the Einstein-Maxwell coupled equations. The Lagrangian density is $L = -\kappa R - l_{em}\|F\|^2$ where $\kappa = \frac{c^3}{16\pi G}$. Planck's constant naturally enters in minimal coupling to make the Schrödinger equation or the Klein-Gordon equation for an electron in an EM field invariant under spacetime point-dependent U(1) gauge transformation. For example when introducing EM interaction to a free particle one takes

$$\partial_t \to \partial_t - i\left(\frac{e}{\hbar}\right)A^0,\ \vec{\nabla} \to \vec{\nabla} + \frac{ie}{\hbar c}\vec{A}.$$

Since the coupling constant also enters in front of the gauge covariant derivative we have, with $\beta = \frac{ie}{\hbar c}$, curvature $F = dA + \beta A \wedge A$. For a general function β, usually for Yang-Mills (YM) potential one defines $\tilde{A} = \beta A$ and $\tilde{F} = \beta F$. The Lagrangian density for $\tilde{A}$ then modifies to

$$L = -\kappa R - l\beta^{-2}\|\tilde{F}\|^2$$

for some constant l which we shall henceforth absorb in β^{-2}. When β is constant, $\tilde{F}$ satisfies $\tilde{F} = d\tilde{A} + \tilde{A} \wedge \tilde{A}$. It is well-known that the EYM equations is scale invariant when β is constant. However in case β is not constant $\tilde{F}$ satisfies

$$\tilde{F} = d\tilde{A} + \tilde{A} \wedge \tilde{A} + A \wedge d\beta \quad (4)$$

and the scale invariance is gone. In view of our variable Planck's constant h_g and in case the coupling constant depends on the Planck constant, it is likely that β will not be constant. h_g suggests one possibility that β^{-1} is bounded going to zero at infinity and is not defined in an interior region. This region is actually cylindrical (see [1]) but for simplicity in this paper we assume it to be spherical. Without spinning we may not get stable solutions but still it could give some insight for modeling composite particles having spin zero. We therefore consider β to be an unknown function determined from the field equations.

3.2. Coupling derived from an equation

EYM coupled equations with variable coupling has been considered by many authors. References can be found in Bekenstein [9] and in the volume edited by Solà [10]. Although the title of the latter work says time variations, the volume also contains some discussion of space variations. Bekenstein [9] discussed the problem of finding a "proper" dynamical equation for the coupling from a separate action for it. We did not worry on this matter. Our variable coupling field is the result of the imaginary gravitational temperature of an EYM system. We do not even know that the equations we seek are derivable from an action integral in the usual way. We are interested only in field equations and we keep the option for matching constant coupling equations at two ends. We also do not know whether we need to add Higgs field because we already have a gravitational mass, and a non-constant coupling will behave like a scalar field. In general there would be more than one coupling for various components of the gauge group. If a Higgs-type field can be derived from the coupling by adding a separate action for the varying coupling or by some other modifications that is fine. There has been effort to derive Higgs potential from extended Brans–Dicke theory (see Solà, Karimkhani and Khodam-Mohammadi [11]). One also recalls the suggestive remark by Bartnik, Fisher and Oliynyk [12] in the introduction of their paper. Later we may add some structure mimicking the cosmological constant of appropriate sign. Variable coupling has some similarities with certain non-minimal coupling EYM solutions (see for example Müller-Hoissen [13], Balakin and Zayats [14-15]).

3.3. Forms of the metric and gauge potential considered

Now we specify the forms of the spacetime metric and the YM potential we use. Static spherically symmetric field will not produce imaginary g-temperature. However for the sake of simplicity we shall not consider stationary axisymmetric solution in this paper. Our spacetime is however a little bit more complicated than a spherically symmetric spacetime. This is an artificial situation but equations are much simpler than the stationary axisymmetric case. Our aim is to seek nontrivial solutions displaying qualitative features that may provide hope before we start a complicated program. We take a static spacetime metric of the form

$$g=-V^2dt^2+\gamma=-V^2dt^2+N^{-1}dr^2+r^2(d\theta^2+\sin^2\theta\phi^2),$$

where $V=V(r,\theta), N=N(r)$. Here we consider only $SU(2)$ YM fields. We assume that we can choose a basis of the Lie algebra so that the (modified) YM potential $\tilde{A}$ is of the form

$$\tilde{A}=a\tau_3dt+b\tau_3dr+(C\tau_1+D\tau_2)d\theta+(\cot\theta\tau_3+C\tau_2-D\tau_1)\sin\theta d\phi$$

where τ_i are the Pauli spin matrices corresponding to the Lie algebra basis. Initially we suppose C,D are functions of r, a, b are functions of r and θ. Later we shall take $D=0=b$. The coupling β is assumed to be a function of r and θ. We use bold indices as Lie algebra indices. Latin indices are spacetime indices. The components of the YM potential are

$$\begin{pmatrix} \tilde{A}_0^{\mathbf{1}} & \tilde{A}_r^{\mathbf{1}} & \tilde{A}_\theta^{\mathbf{1}} & \tilde{A}_\varphi^{\mathbf{1}} \\ \tilde{A}_0^{\mathbf{2}} & \tilde{A}_r^{\mathbf{2}} & \tilde{A}_\theta^{\mathbf{2}} & \tilde{A}_\varphi^{\mathbf{2}} \\ \tilde{A}_0^{\mathbf{3}} & \tilde{A}_r^{\mathbf{3}} & \tilde{A}_\theta^{\mathbf{3}} & \tilde{A}_\varphi^{\mathbf{3}} \end{pmatrix}=\begin{pmatrix} 0 & 0 & C & -D\sin\theta \\ 0 & 0 & D & C\sin\theta \\ a & b & 0 & \cos\theta \end{pmatrix}.$$

Components of the YM curvature are (using Eq.(4))

$$\tilde{F}_{ab}^{\mathbf{c}}=\partial_a\tilde{A}_b^{\mathbf{c}}-\partial_b\tilde{A}_a^{\mathbf{c}}+f_{\mathbf{mn}}^{\mathbf{c}}\tilde{A}_a^{\mathbf{m}}\tilde{A}_b^{\mathbf{n}}+A_a^{\mathbf{c}}\partial_b\beta-A_b^{\mathbf{c}}\partial_a\beta.$$

Here $f_{\mathbf{mn}}^{\mathbf{c}}$ is the totally antisymmetric tensor $\varepsilon^{\mathbf{cmn}}$. Lie algebra indices are written in bold. A review of the EYM system can be found in the report by Volkov and Gal'tsov [16].

We shall now count the nontrivial EYM equations for the case $D=0=b$. In the following list the Greek index stands for r, θ or φ.

For **c=3**, $\alpha=r$, the YM equation is trivial if we choose $D=0=b$.

For **c=3**, $\alpha=\theta$, the YM equation is trivially satisfied.

For **c=3**, $\alpha=\varphi$, the YM equation gives the 2-Laplacian of $\beta(r,\theta)$.

For **c=1**, $\alpha=r$, the YM equation gives an equation containing the mixed partial of β:

$$\partial_\theta\partial_r\beta^{-1}+\partial_r\beta^{-1}\cot\theta+(\ln C)'\partial_\theta\beta^{-1}+(\ln C)'\beta^{-1}\partial_\theta\ln V+\partial_r\beta^{-1}\partial_\theta\ln V=0.$$

For **c=1**, $\alpha=\theta$, the YM equation gives an equation for $C(r)$.

For **c=1**, $\alpha = \varphi$, the YM equation is trivially satisfied.
For **c=2**, $\alpha = r$, the YM equation is trivially satisfied.
For **c=2**, $\alpha = \theta$, the YM equation is trivially satisfied.
For **c=2**, $\alpha = \varphi$, the YM equation gives an equation containing C and β.
With $a \neq 0$ there are two more equations:
For **c=2**, spacetime index 0, the YM equation gives $a^2 = V\beta^2\xi(r)$ for some ξ.
For **c=3**, spacetime index 0, the YM equation gives an equation for $\xi(r)$.
For **c=1**, spacetime index 0, the YM equation is trivial.

With $a = 0$ there are four nontrivial YM equations when β is not constant. Of these the following three equations threaten to render β constant. This is certainly the case if V is independent of θ (spherically symmetric case) because of the $\theta\theta$ and $\varphi\varphi$ components of the Ricci tensor in the Einstein equations. $\theta\theta$ and $\varphi\varphi$ components of the Ricci tensor give two equations that are slightly different when β is not constant. The three equations are as follows: The equation for **c=3**, $\alpha = \varphi$. The equation for **c=2**, $\alpha = \varphi$. The equation for **c=1**, $\alpha = r$. The equation for **c=1**, $\alpha = \theta$ gives the only surviving equation in the constant β case.
Einstein equations give six nontrivial equations. In the following 3-Ricci refers to the Ricci curvature of the 3-metric γ. Δ_γ is the Laplacian of γ. The equations come from 0φ-component for 4-Ricci, rr-component of 3-Ricci, $\theta\theta$-component of 3-Ricci, $\varphi\varphi$-component of 3-Ricci, $r\theta$-component of 3-Ricci and 00-component for 4-Ricci. 00-component for 4-Ricci gives an equation for $\partial_r^2 V$ coming from $\Delta_\gamma V$.
Equations for $r\varphi$-component and $\theta\varphi$-component of 3-Ricci, and $0r$-component and 0θ-component of 4-Ricci are trivial. So for $SU(2)$ EYM system in the spacetime setting described above and for the chosen form of the YM potential, we have ten equations for four variables namely β, C, g_{rr} and V. ξ satisfies a second order ODE involving C, g_{rr} and V and we need not worry about it now. This is a very tight situation. The work is incomplete.

4. Conclusion

Our concluding remarks on the work on the Lamb's shift is as follows. If our result is not due to mere coincidence, then it suggests that using the variable Planck's constant with the linear theory (Fourier analysis) one can get a way of calculating the gravitational effects on non-gravitational interaction that reflects the already unified nature of Einstein equations. It is likely that approximating this way one will not need renormalization because the divergence integrals will disappear. It seems we have gravitational justification for the cutoffs for both high and low frequencies. One therefore needs to study the cutoffs considered in the case of emission and absorption of virtual quanta in Feynman's 1948 model (or those in Bopp-Podolsky model). Our selective use of the results of QM and QFT in essentially classical arguments is not arbitrary. At present we are free to use any experimentally verified conclusion of QM and QFT in our analysis. In fact we can use them to get a better equation giving h_g from the rigorously defined Tolman-Thorne temperature in stationary spacetimes. Then we can guess how to extend the relevant concepts to the local charts of non-stationary spacetimes. Our eventual aim is to do the reverse, that is, use h_g to explain all the valid results of quantum physics. As for the second ongoing problem discussed at the end of the last subsection, one should not be disappointed if there are no nontrivial solutions for static $SU(2)$ EYM system in the non-constant coupling case for the form of the metric and YM potential we considered. Because it would possibly suggest that the gauge group should be more complicated for nontrivial static solutions. In that case there is also the added opportunity to use more than one coupling parameter. Still we expect that the pursuit would provide some useful hints for the search of static solutions modeling a spin zero composite particle with somewhat more complicated but realistic action. Ultimately one would like to explain the origin of the gauge group from the local symmetry of the Minkowski spacetime and hydrodynamic and thermodynamic considerations of Einstein equations.

References

[1] Masood-ul-Alam A K M 2015 An imaginary temperature far away from a stationary spinning star *Int. Meeting on Physical Interpretations of Relativity Theory "PIRT-2015"* (Moscow)
[2] Welton T A 1948 Observable effects of the quantum-mechanical fluctuations of the electromagnetic field *Phys. Rev.* **74** pp 1157-1167
[3] Bethe H A 1947 The electromagnetic shift of energy levels *Phys. Rev.* **72** pp 339-341
[4] Schwabl F 2005 *Advanced Quantum Mechanics* (Springer)
[5] Park D 1992 *Introduction to the Quantum Theory* (Dover)
[6] Bloch F and Nordsieck A 1937 Note on the radiation field of the electron *Phys. Rev.* **52** pp 54-59
[7] Bjorken J D and Drell S D 1964 *Relativistic Quantum Mechanics* (McGraw-Hill)
[8] Feynman R P 1948 Relativistic cut-off for quantum electrodynamics *Phys. Rev.* **74** pp 1430-1438
[9] Bekenstein J D 1982 Fine-structure constant: Is it really a constant? *Phys. Rev.* D **25** pp 1527-1539
[10] Solà J 2015 *Fundamental Constants in Physics and Their Time Variation Mod. Phys. Lett.* A **30**
[11] Solà J, Karimkhani E and Khodam-Mohammadi A 2017 Higgs potential from extended Brans–Dicke theory and the time-evolution of the fundamental constants *Class. Quantum Grav.* **34** 025006
[12] Bartnik R, Fisher M and Oliynyk T A 2010 Static Spherically Symmetric Solutions of the SO(5) Einstein Yang-Mills Equations *J. Math. Phys.* **51** 032504
[13] Muller-Hoissen F 1988 Modification of Einstein-Yang-Mills theory from dimensional reduction of the Gauss-Bonnet action *Class. Quantum Grav.* **5** L35-L40
[14] Balakin A B and Zayats A E 2007 Non-minimal Wu-Yang monopole *Physics Letters* B **644** pp 294-298
[15] Balakin A B and Zayats A E 2006 Curvature coupling in Einstein-Yang-Mills theory and non-minimal self-duality *Grav. Cosmol.* **12** pp 302-306
[16] Volkov M S and Gal'tsov D V 1999 Gravitating non-Abelian solitons and black holes with Yang-Mills fields *Physics Reports* **319** pp 1-83

Acknowledgements
Author's attendance to PIRT-2017 was supported by a grant from the Tsinghua University, Beijing, P. R. China.

Commutative Fuzzy Geometry and Geometric Quantum Mechanics

S N Mayburov

P.N. Lebedev Physical Institute of Russian Academy of Sciences, Moscow, Russia
E-mail: mayburov@sci.lebedev.ru

Abstract. Commutative fuzzy geometry considered as the possible mathematical framework for reformulation of quantum-mechanical formalism in strictly geometric terms. In that approach the states of massive particle m correspond to elements of fuzzy manifold called fuzzy points. Due to manifold weak topology, in 1-dimensional case m space coordinate x acquires principal uncertainty σ_x and described by positive, normalized density $\omega(x,t)$, the analogous uncertainties appear for 3-dimensional manifold. It's shown that m states on such manifold are equivalent to Dirac vectors of complex Hilbert space, their evolution correspond to Shroedinger dynamics of massive quantum particle.

1. Introduction

It's well known that quantum mechanics (QM) can be consistently described by several alternative mathematical formalisms such as Shroedinger (standard) one, algebraic QM, functional integral, etc. [1]. In last years, possible reformulation of QM formalism in geometric terms attracted significant attention [2-3]. Really, in some fields of physics, like optics and general relativity, geometric ideas have been very useful, hence one can hope that they also can help to study some important QM problems, first of all, quantum theory of gravity and theory of gauge fields. Up to now, several formalisms were exploited as the basis of QM geometrization such as symplectic geometry, Hilbert manifolds, Kahler bundles, etc. ([3] and refs. therein). In this paper, we'll study as possible basis for QM geometric reformulation the formalism based on fuzzy calculus [4-5]. During last 50 years, the fuzzy set theory and other branches of fuzzy mathematics were applied in a wide range of scientific areas such as biology, economics and computer science. From the early days of its development, the significant similarity of that theory and QM was noticed [6-8]. It was argued, in particular, that the parameter and proposition uncertainty (fuzziness) which is generic for fuzzy mathematics can be, in fact, equivalent to QM uncertainty of particle coordinate, momentum and other QM observables [6-7]. Its worth to notice that in modern QM the fuzzy (unsharp) observables became the important part of QM theory of measurements [9], hence QM formalism based on fuzzy methods can be interesting both from fundamental and applied angles. However, until now such studies were performed only in the context of fuzzy (multivalued) logics ([10] and refs. therein). In distinction, our approach deals with standard logics and based on the results of fuzzy topology and geometry [7-8, 11]. In particular, this formalism exploits the system phase space equipped with fuzzy topology, it will be shown that the corresponding structure of state space is equivalent to QM Hilbert space [12]. In its essence, the proposed formalism is geometric; it exploits, in fact, the phase space geometry with specific weak topology. As the result, it permits to derive the quantum system evolution from geometric arguments. It will be considered here for nonrelativistic particle, such system traditionally exploited as testing

ground for QM formalism construction [13]. In mathematics the formalism exploited here called fuzzy geometry [8,14], but in modern physics such term ascribed to some noncommutative field theories [15], so to avoid the confusion, it will be called here commutative fuzzy geometry (CFG).
Our paper organized as follows. In sect. 2 basic features of fuzzy topology and geometry are reviewed and their relations with set theory and topology are discussed. In sect. 3 1-dimensional model of particle evolution on CFG manifold considered and resulting space of particle states is defined. In sect. 4 particle evolution equation in this model derived. In sect. 5 general formalism for 3-dimensional case constructed. Sect. 6 presents our concluding remarks.

2. Geometric Fuzzy Structures

Here we'll consider the fuzzy structures important for formalism construction, for the detailed review on CFG and related topics see [8,11,14]. The properties of fuzzy objects can be introduced *ad hoc*, but its worth to start from consideration of their analogs in set theory which provides the useful link to the realm of fuzzy structures and illustrates their physical meaning.
Remind that in set theory the sets can be classified according their ordering structure; the simplest case presents the totally ordered set, for all its element pairs a_i, a_j the ordering relations $a_k \le a_l$ (or vice versa) fulfilled. In distinction, for partially ordered set (Poset), some its element pairs can obey to incomparability relations (IR) between them: $a_j \sim a_k$. In this case, both $a_j \le a_k$ and $a_k \le a_j$ propositions are false [16]; in some aspects IR is discrete analogue of fuzzy relations. To illustrate this analogy, consider poset $S^p = A^p \bigcup B$, which includes the subset of 'incomparable' elements $A^p = \{a_j\}$, and ordered subset $B = \{b_i\}$. For the simplicity suppose that in B the element indexes grow correspondingly to their ordering, so that $\forall i,\ b_i \le b_{i+1}$; relations between an arbitrary pairs a_j, a_i or a_i, b_l can be ordered, as well as incomparable. Let's consider B interval $\{b_l, b_n\}$ and suppose that A^p element a_j is confined in $\{b_l, b_n\}$, i.e. $b_l \le a_j;\ a_j \le b_n$, and simultaneously a_j is incomparable with all internal $\{b_l, b_n\}$ elements: $b_i \sim a_j;\ \forall i;\ l+1 \le i \le n-1$. In this case, a_j is, in a sense, smeared over such interval, which is the rough analogue of a_j coordinate uncertainty, if to treat the sequence of B elements $\{b_i\}$ as the coordinate axe.
The next step to fuzzy structures is to substitute described set-theoretical relations between S^p elements by fuzzy relations. Let's put in correspondence to each a_j, b_i pair of S^p set the nonnegative weight $w_i^j \ge 0$ with norm $\sum_i w_i^j = 1$, resulting w_i^j characterizes the rate of closeness (membership) between a_j, b_i [4-5]. In particular, $w_i^j = 0$ means that a_j, b_i are ordered relative to each other. For example considered above, one can ascribe $w_i^j = (n-l-2)^{-1}$ to all b_i inside $\{b_l, b_n\}$ interval, $w_i^j = 0$ for other b_i. The similar structure can be defined for continuum, as the example, consider the continuous set S^f such that $S^f = A^p \bigcup X$ where A^p is the same discrete subset, X is the continuous ordered subset. If the flat metrics $M(x, x')$ is defined on X, then it's equivalent to R^1 real number axe, and fuzzy relations between elements a_j, x are described by real, nonnegative functions $w^j(x) \ge 0$ with the norm $\int w^j dx = 1$ [5]. In this case, $\{a_j\}$ called the fuzzy numbers $\tilde{x}_j$ or in geometric framework, 1-dimensional fuzzy points (FPs) [4-5,14]. Then, ordered point $x_i \in X$ is characterized by $w^a(x) = \delta(x - x_i)$, hence the ordered points and FPs can be regarded formally on the same ground.

In 3-dimensional case one can consider the fundamental set: $C^f = A^p \bigcup X'$ where A^p defined above, X' is the continuous set. Suppose that on X' 3-dimensional linear space R^3 with flat metrics M_{ij} is defined. Then FP a_j is described by the nonnegative function $w^j(\vec{r})$ with norm $\int w^j d^3x = 1$, such C^f is called fuzzy manifold, in our case, it denoted as $\tilde{R}^3$ [5,14].

3. Particle States on Fuzzy Manifold

Here we shall consider the dynamical model on fuzzy manifold called fuzzy mechanics (FM); it will be constructed as the minimal theory, i.e. at every step we'll choose ansatz with minimal number of theory parameters (physical constants). In geometric framework, classical particle is material point $\tilde{r}_a(t) \in R^3$, in FM formalism, the particle m corresponds to FP $a(t)$ on fuzzy manifold $\tilde{R}^3$ and characterized by normalized positive density $w(\vec{r},t)$ on R^3. Beside w, m state $|\zeta\rangle$ called the fuzzy state can depend, in principle, on other m degrees of freedom (DFs). We shall start FM construction from 1-dimensional case, because for it the theory premises are most simple and transparent. Let's consider in this framework m average velocity

$$\bar{\upsilon}(t) = \frac{\partial}{\partial t}\int_{-\infty}^{\infty} xw(x,t)dx = \int_{-\infty}^{\infty} x\frac{\partial w}{\partial t}dx. \tag{1}$$

It's reasonable to assume that in general $\bar{\upsilon}(t)$ can be independent of $w(x,t)$; we shall look for additional m DFs in form of real functions $q_j(x,t);\ j = 1,\dots,k$. Let's suppose that in FM m state evolution is local, in particular,

$$\frac{\partial w}{\partial t}(x,t) = \Phi\left[w(x,t), q_1(x,t),\dots,q_k(x,t)\right], \tag{2}$$

where Φ is an arbitrary function. From w norm conservation it follows that

$$\int_{-\infty}^{\infty}\Phi(x,t)dx = \int_{-\infty}^{\infty}\frac{\partial w}{\partial t}(x,t)dx = \frac{\partial}{\partial t}\int_{-\infty}^{\infty} w(x,t)dx = 0. \tag{3}$$

One can substitute: $\Phi = -\frac{\partial J}{\partial x}$ where $J(x)$ is differentiable function, which obeys to condition:

$$J(\infty,t) \quad J(-\infty,t) = 0, \tag{4}$$

i.e. there is no influence on w norm from infinity. J can be decomposed formally as: $J(x) = w(x)\upsilon(x)$, in these terms eq. (2) can be rewritten in form of flow continuity equation [17]

$$\frac{\partial w}{\partial t} = -\upsilon\frac{\partial w}{\partial x} - \frac{\partial \upsilon}{\partial x}w. \tag{5}$$

By the analogy with fluid dynamics, we shall assume that flow velocity $\upsilon(x)$ can be considered as independent m DF; the double $\rho^0 = \{w,\upsilon\}$ called observational $|\zeta\rangle$ representation. Plainly, to make the formalism consistent, beside eq. (5) for w evolution, it's necessary to find also υ evolution equation. Yet, ρ^0 ansatz isn't optimal for that purpose, so it's instructive to look for some dynamical $|\zeta\rangle$ representation η which will be more convenient. We'll suppose that η corresponds to the set of real functions $\{\eta_i(x)\}$, $i = 1, n_a$, plus some η_i algebra; each η_i described by 1-st time order evolution equation. Then, the most general η ansatz is $\eta_j(x) = Y_x^j(w,\upsilon)$ where Y_x^j are some w, υ functionals and x is their parameter. For $|\zeta\rangle$ characterized by two DFs w, υ, one can start η ansatz search from complex $\eta(x)$, not assuming yet, that η is L^2-normalized. Hence $n_a = 2$ and $\eta(x)$ can be expressed as

$$\eta(x) = Y_x^1 + iY_x^2 = \Omega_x(w,\upsilon)e^{i\lambda_x(w,\upsilon)}, \tag{6}$$

where Ω_x, λ_x are real functionals. It's natural to assume also that if $w(x_a,t)=0$ for some x_a, then $\eta(x_a,t)=0$ and vice versa. It can be easily shown that the minimal Ω_x ansatz for that is $\Omega_x = f\left[w(x)\right]$, where $f \to 0$ for $w \to 0$.

We'll suppose also that analogously to classical mechanics, in FM particle evolution can be characterized by its velocity 'as the whole' $u(t)$ described by the corresponding normalized distribution $w_u(u,t)$ for which

$$\bar{u} = \bar{\upsilon} = \int_{-\infty}^{\infty} \upsilon(x)w(x)dx, \tag{7}$$

w_u supposedly can be defined in u measurements, however, u isn't assumed beforehand to be independent $|\zeta\rangle$ DF. In place of u, below it will be convenient to use the variable $p=\mu u$ where μ is the theory constant; for the corresponding distribution $w_p(p)$ it gives $w_p(p)=\mu^{-1}w_u(\mu^{-1}p)$. m state $|\zeta\rangle$ presumably contains the information on the expectation value of any m observable Q in form of some η functional. In particular, $w_p(p)=F_p(\eta)$, and it can be shown that F_p functional is related to $\eta(x)$ fourier transform. To prove it and calculate η, w_p, let's introduce the auxiliary form: $\varphi(p)=w_p^{1/2}e^{i\beta(p)}$, here $\beta(p)$ is the auxilary real function on which final w_p ansatz wouldn't depend. We shall look for w_p, β such that φ fourier decomposition on X is equal to:

$$\varphi(p) = \int_{-\infty}^{\infty} \eta(x)e^{-ipx}dx = \int_{-\infty}^{\infty} fe^{i\lambda_x - ipx}dx, \tag{8}$$

w_p is normalized, so the application of Plancherele formulae to that norm gives [19]:

$$\int_{-\infty}^{\infty} w_p(p)dp = \int_{-\infty}^{\infty} \varphi(p)\varphi^*(p)dp = \int_{-\infty}^{\infty} f^2(w)dx = 1. \tag{9}$$

To calculate $f(w)$, let's denote $\Theta = f^2$ and consider its variation $\delta\Theta$. As follows from eq. (9):

$$\int_{-\infty}^{\infty} \delta\Theta dx = \int_{-\infty}^{\infty} \frac{\partial(f^2)}{\partial w}\delta w dx = 0 \tag{10}$$

with additional δw constraint: $\int_{-\infty}^{\infty} \delta w dx = 0$. Let's substitute $\delta w = \frac{\partial \varpi}{\partial x}$; under these conditions its possible to choose $\varpi(\pm\infty)=0$, then such ϖ satisfies to the conditions of Du Bois-Reymond lemma considered in Appendix [18]. Its application gives: $\Theta = w$, so that $f = \pm\sqrt{w}$. Now $\bar{p}$ can be calculated anew from 2-nd Plancherele formulae :

$$\bar{p} = \int_{-\infty}^{\infty} p\varphi(p)\varphi^*(p)dp = \int_{-\infty}^{\infty} \frac{\partial \lambda_x}{\partial x} f^2 dx = \int_{-\infty}^{\infty} \frac{\partial \lambda_x}{\partial x} w(x)dx. \tag{11}$$

From its comparison with eq. (7) and equality $\bar{p} = \mu\bar{u}$, it follows: $\lambda_x = \gamma(x) + \chi(w)$ where γ is the functional:

$$\gamma(x)=\mu\int_{-\infty}^{x}\upsilon(\xi)d\xi+c_{\gamma}, \tag{12}$$

here c_γ is an arbitrary real number; $\chi(w)$ is an arbitrary real function which obeys to the condition:

$$\int_{-\infty}^{\infty}\chi(w)\frac{\partial w}{\partial x}dx=0. \tag{13}$$

The resulting m state $|\zeta\rangle$ in x-representation is equal to:

$$\eta(x)=w^{1/2}(x)e^{i\gamma+i\chi}, \tag{14}$$

so $\eta(x)$ is the vector (ray) of L^2-complex Hilbert space H. $w_p(p)$ and $\beta(p)$ can be calculated from eq. (8) as functions of χ, in particular

$$w_p(p)=\left|\int_{-\infty}^{\infty}\sqrt{w}e^{i\gamma+i\chi-ipx}dx\right|^2 \tag{15}$$

is independent of $\beta(p)$, so w_p is just η, i.e. w, υ functional, the same is true for β_p.

4. Linear Model of Fuzzy Dynamics

Obtained m state can be rewritten as: $\eta=e^{i\chi}g$, where $g(x,t)$ corresponds to standard QM wave function. If $\chi(w)\neq 0$, it demands to introduce at least two additional theory parameters, hence we shall consider first $g(x,t)$ evolution, η evolution for $\chi(w)\neq 0$ will be analyzed afterwards. The evolution equation for g presumably is of the first order in time, i.e.

$$i\frac{\partial g}{\partial t}=\hat{H}g. \tag{16}$$

In general, $\hat{H}$ can be nonlinear operator, for the simplicity we shall start with the linear $\hat{H}$ and turn to nonlinear case later. Free m evolution is invariant relative to x space shift on arbitrary x_0 performed by the operator $\hat{W}(x_0)=\exp\left(x_0\frac{\partial}{\partial x}\right)$. Because of it, corresponding operator $\hat{H}_0$ should commute with $\hat{W}(x_0)$ for the arbitrary x_0, i.e. $\left[\hat{H}_0,\frac{\partial}{\partial x}\right]=0$. It holds only if $\hat{H}_0$ is differential polinom of the form

$$\hat{H}_0=-\sum_{l=1}^{n}b_{2l}\frac{\partial^{2l}}{\partial x^{2l}}, \tag{17}$$

where b_{2l} are arbitrary complex values, n is arbitrary number. From correspondence to classical mechanics, it supposed that the influence of potential field U on m evolution can be accounted in $\hat{H}$ additively: $\hat{H}=\hat{H}_0+U(x,t)$, where U is real, nonsingular function. Then, eq. (16) can be rewritten as

$$i\frac{\partial g}{\partial t}=\left(i\frac{\partial\sqrt{w}}{\partial t}-\sqrt{w}\frac{\partial\gamma}{\partial t}\right)e^{i\gamma}=e^{i\gamma}\hat{Z}g, \tag{18}$$

where $\hat{Z}=e^{-i\gamma}\hat{H}$. Hence

$$\frac{\partial\sqrt{w}}{\partial t}=\mathrm{Im}\left(\hat{Z}g\right). \tag{19}$$

If to substitute $\upsilon(x)$ by $\gamma(x)$ in eq. (5) and transform it to $\sqrt{w}$ time derivative, it gives

IOP Conf. Series: Journal of Physics: Conf. Series **1051** (2018) 012022 doi:10.1088/1742-6596/1051/1/012022

$$\frac{\partial\sqrt{w}}{\partial t}=-\frac{1}{\mu}\frac{\partial\sqrt{w}}{\partial t}\frac{\partial\gamma}{\partial x}-\frac{1}{2\mu}\sqrt{w}\frac{\partial^2\gamma}{\partial x^2}. \tag{20}$$

Plainly, the right parts of equations (19) and (20) should coincide for arbitrary w, γ, otherwise $\hat{H}$ ansatz would be incompatible with w flow continuity described by eqs. (5, 20). Really, if they differ, it would introduce additional constraint on w, γ of the form $\hat{L}_x(w,\gamma)=0$, where $\hat{L}_x$ is the operator in x-derivatives. However, w, γ presumably are independent DFs, hence appearance of such constraint would result in open contradiction. Thus, $\hat{H}$ ansatz can be obtained from the term by term comparison of eqs. (19,20). In particular, $\hat{Z}g$ imaginary part includes the highest γ derivative as the term:

$$\mathrm{Im}\left(\hat{Z}g\right)=...-b_{2l}\sqrt{w}\frac{\partial^{2l}\gamma}{\partial x^{2l}},$$

however, for eq. (20) the highest γ deriavtive is proportional to $\sqrt{w}\frac{\partial^2\gamma}{\partial x^2}$. As the result, it gives: $b_2=\frac{1}{2\mu}$ and $b_{2l}=0$ for $l>1$, only in this case both expressions for $\frac{\partial\sqrt{w}}{\partial t}$ would coincide. Thus, g free evolution is described by single $\hat{H}_0$ term: $-\frac{1}{2\mu}\frac{\partial^2}{\partial x^2}$, so $\hat{H}$ is Schroedinger Hamiltonian for particle with mass m.

5. General Fuzzy Dynamics

In the previous section, 1-dimensional FM formalism was derived from CFG premises assuming that $|\zeta\rangle$ evolution is linear and its phase component $\chi(w)$ can be neglected. Now these assumptions will be dropped one by one and FM formalism extended for 3-dimensional case. The conditions of QM linearity initially were formulated by Wigner [20], however, they are extensively discussed up to now [21]. Recently, Jordan has shown that they are essentially weaker than Wigner theorem asserts [22]. Namely, if the following two conditions are fulfilled:

i) the evolution maps the set of all pure states one to one onto itself

ii) for the arbitrary mixture of orthogonal states $\rho(t)=\sum P_i(t)\rho_i(t)$ it conserves all P_i constant in time.

Then, it follows that such evolution is linear. Both these conditions are in good correspondence with geometric framework proposed by FM formalism. Really, it was shown that the pure states $|\zeta\rangle$ describe the evolution of geometric object FP m from its initial to final state. Plainly, such evolution shouldn't result in appearance of object probabilistic mixture. It's also natural to assume that such evolution is unambiguous and reversible.

Now let's return to $\eta(x)$ ansatz of (14). Analogously to eq. (16), the evolution equation for η is of the first time order, and Jordan theorem is applicable also to η evolution, hence it follows that in this case $\hat{H}$ also should be linear. However, in this case $\hat{H}$ linearity becomes incompatible with w flow continuity expressed by eq. (5), due to the appearance of additional terms in $\hat{H}$ which contain χ derivatives. Hence, to avoid the contradictions, $\eta(x,t)$ should coincide with QM wave function $g(x,t)$ considered above.

3-dimensional FM, in fact, doesn't demand any principal modification of described formalism. In that case FP a_j is described by the normalized, nonnegative function $w^j(\vec{r})$. The particle m corresponds to FP $a(t)$ characterized by its state $|\zeta\rangle$ which depends on density $w(\vec{r},t)$ and, probably, some other m DFs. Assuming that w evolution depends on local parameters only, it can be expressed as:

$$\frac{\partial w}{\partial t}(\vec{r},t) = -\Phi(\vec{r},t), \tag{21}$$

where Φ is an arbitrary local function. Then from w norm conservation

$$\int_V \Phi(\vec{r},t)d^3r = \int_V \frac{\partial w}{\partial t}(\vec{r},t)d^3r = \frac{\partial}{\partial t}\int_V w(\vec{r},t)d^3r = 0, \tag{22}$$

where V denotes the infinite volume with $|\vec{r}| \to \infty$ in all directions. Substituting $\Phi = -div\vec{J}$, w evolution described by the flow continuity equation

$$\frac{\partial w}{\partial t} = -div\vec{J}, \tag{23}$$

assuming also that $\vec{J}$ flow through infinite V surface S is zero. One can decompose formally $\vec{J} = w\vec{\upsilon}$ and consider w flow velocity $\vec{\upsilon}(\vec{r})$ as independent m DFs. m state $|\zeta\rangle$ supposedly depends on w, $\vec{\upsilon}$ DFs only, so array $\rho^0 = \{w,\vec{\upsilon}\}$ describes observational $|\zeta\rangle$ representation. Analogously to 1-dimensional case, one should find dynamical $|\zeta\rangle$ representation η which can be, for example, quaternion. However, we'll start its search from minimal ansatz in form of complex w, $\vec{\upsilon}$ functional $\eta(\vec{r}) = Y_{\vec{r}}(w,\vec{\upsilon})$. Then it follows that its minimal ansatz is $\eta = f(w)e^{i\Lambda_{\vec{r}}(w,\vec{\upsilon})}$. As in 1-dimension for m particle as the whole, its velocity is supposedly characterized by fuzzy vector $\vec{u}$ with corresponding distribution $w_u(\vec{u})$, so that

$$\langle\vec{u}\rangle = \int \vec{u}w_u(\vec{u})d^3u = \int \vec{\upsilon}(\vec{r})w(\vec{r})d^3r, \tag{24}$$

m kinematical fuzzy momentum defined as: $\vec{p} = \mu\vec{u}$. From that, analogously to calculations described by eqs. (7-14), standard QM ansatz for 3-dimensional m state obtained: $g(\vec{r}) = \sqrt{w}e^{i\gamma}$, for which $\gamma(\vec{r},t)$ obeys to the equality $\mu\vec{\upsilon} = grad(\gamma)$. To make m state $g(\vec{r})$ unambiguous, the additional constraint should be imposed on $\vec{\upsilon}(\vec{r})$ field; namely, for any closed loop l

$$\oint \vec{\upsilon}d\vec{l} = \frac{2\pi}{\mu}n_l \tag{25}$$

Normally, $n_l = 0$, but if the topological defect or nodal region located inside loop l, then $n_l = 0,\pm1,\pm2,...$, etc. [23].

Considering g linear evolution, for free m evolution its operator $\hat{H}_0$ should be the even polinom of the form

$$\hat{H}_0 = -\sum_{l=1}^{n} b_{2l}\frac{\partial^{2l}}{\partial\vec{r}^{2l}}. \tag{26}$$

If potential field influence can be described by the addition of real function $U(\vec{r},t)$ to $\hat{H}_0$, then from $i\frac{\partial g}{\partial t}$ the term $\frac{\partial\sqrt{w}}{\partial t}$ can be extracted and expressed via corresponding w, γ, $\vec{r}$-derivatives. From their comparison with corresponding $\hat{H}g$ $\vec{r}$-derivatives, the Schroedinger equation follows for m evolution. Applicability of Jordan theorem to 3-dimensional case is obvious, because the derivation of

$\hat{H}$ linearity doesn't depend on dimension number. The same is true for the proof of uniqueness of $\eta(\vec{r},t)$ ansatz, i.e. that $\chi(w) \neq 0$.

Planck constant $\hbar = 1$ in this formalism, but the same value ascribed to it in relativistic unit system in which velocity of light; $c = 1$ in FM framework, $\hbar$ only connects $\vec{r}$, $\vec{p}$ scales and doesn't have any other meaning. Note that superposition principle doesn't need to be postulated separately in such formalism. Rather, as follows from previous calculations, the sum of two physical m states $g_{1,2}$ with proper complex coefficients $a_{1,2}$ can be considered as the physical m state also. In our approach, the state space is defined by the underlying geometry and corresponding dynamics i.e. is derivable concept. For states of massive particle m it was found to be equivalent to H, but, in principle, the resulting state space can differ for other systems analogously to algebraic QM where the state space is defined by the observable algebra [1]. In this framework, the observable p_x corresponds to the operator $\hat{p}_x = -i\frac{\partial}{\partial x}$ acting on $g(\vec{r})$. Thus, x and p_x observables are described by the linear self-adjoint operators, which obey to the commutation relation $[\hat{x}, \hat{p}_x] = i$. We admit that all m projective observables $\{Q\}$ are the linear, self-adjoint operators on H with standard extension to POVM. The flow velocity $\vec{\upsilon}(\vec{r})$ isn't observable, but its value can be consistently defined as the ratio of $\vec{J}(\vec{r})$, $w(\vec{r})$ observable expectation values [24]; here $w(\vec{r})$ observable is described by the projection operator $\hat{\Pi}(\vec{r})$; observable $\vec{J}(\vec{r})$ defined in [24-25].

The particle evolution in QM in some aspects is similar to the motion of continuous media; this analogy is exploited in hydrodynamical QM model [23]. Its connection with FM will be discussed elsewhere; here it's worth to note that despite some formal similarity, their basic premises are quite different. Namely, in hydrodynamical QM model the Schroedinger equation is postulated and from it the flow continuity equation and other continuous media properties of quantum systems derived. In FM formalism, it's shown that FP density evolution described by flow continuity equation, and under simple assumptions, the corresponding particle state should obey to Schroedinger equation. In our formalism m state $|\zeta\rangle$ defined by measurable functions $(\vec{r})$, $\vec{\upsilon}(\vec{r})$, whereas in QM it characterized by $w(\vec{r})$, $a(\vec{r})$, yet, quantum phase $a(\vec{r})$, isn't observable.

In standard QM the evolution equations are or postulated *ad hoc* or derived assuming Galilean invariance of system states [1,25]. In FM the Schroedinger equation for massive particle was derived assuming only space-time shift and rotational invariance which are essentially weaker assumptions. Meanwhile, it's well known that Galilean invariance can be derived from Schroedinger equation if a reference frame is associated with free system S with mass $M_s \to \infty$ and which initial state $\Psi(t_0)$ is the wave packet with the width $\sigma_x \to 0$ [25-26].

6. Conclusion

In this paper it was shown that QFG formalism can be considered as the mathematical basis for the consistent description of quantum particle dynamics. Novel features of considered formalism can be revealed in most simple way from the comparison with Schroedinger QM formalism. From the formal side, standard QM exploits two fundamental structures of different nature: the spacetime manifold $R^3 \cdot T$ and function space H defined on R^3. In distinction, minimal FM formalism involves only one basic structure - the fuzzy manifold $\tilde{R}^3 \cdot T$, massive particle densities $w(\vec{r},t)$ are $\tilde{R}^3$ points called FPs. Their evolution induces the physical states which are equivalent to H Dirac vectors. Resulting dynamics of massive particles described by Shroedinger equation, which is the basis of universal quantum dynamics. The quantum-classical transition in such theory is essentially more simple than in standard QM, it's just the transition from $\tilde{R}^3$ manifold to R^3 one, for which the classical particles correspond to ordered points $\vec{r}(t)$. Here only the evolution of nonrelativistic particle was considered in CFG formalism, but all QM formalism is, in fact, based on its analysis, and there are no obvious obstacles, which forbid constructing universal quantum dynamics in this approach.

It seems that FM approach is based on essentially more natural axiomatic than standard QM. Its main postulate is the principal uncertainty of some system parameters, like coordinate or momentum, such uncertainty has geometric origin and isn't connected a priory with wave-like system properties. Rather, in FM the wave-like system evolution is stipulated by these geometric features. In particular, in FM the particle state is described by the fuzzy density $w(\vec{r})$ and flow velocity $\upsilon(\vec{r})$ which characterizes w change with time. Quantum phase, which in QM formalism is interpreted as the phase of de Brogile matter wave, is shown to be $\vec{\upsilon}$ functional.

Plainly, general relativity is essentially geometric theory; meanwhile, up to now the attempts to quantize gravity meet serious difficulties even at axiomatic level. Hence, if consistent geometric QM formalism will be constructed, it can help, in principle, to develop quantum theory of gravity. Currently, the main impact of QM geometrization studies is done on the exploit of Hilbert manifolds ([3] and refs. therein), however, the results obtained in this approach have quite abstract form, and their applicability to particular physical problems isn't obvious. The considered FM formalism possesses simple and logical axiomatics which origin is basically geometrical, hence it can become the appropriate part of QM geometrization program, its implications can be important also for the analysis of QM foundations.

Appendix

Here the proof of Du Bois-Reimond lemma is considered according to [27], it extended also on functions defined on R^3.

Lemma: if for continuous function $N(x)$ and arbitrary continuous, differentiable function $\tau(x)$ for which $\tau(a)=\tau(b)=0$, ($a=-\infty$, $b=\infty$ permitted) and $\tau'=\frac{\partial\tau}{\partial x}$

$$I(\tau)=\int_a^b N(x)\tau' dx=0,$$

then $N(x)$ is constant on $[a,b]$ interval. Suppose that on the opposite, $N(x)$ varies on $[a,b]$, then it should be at least two points c_1, c_2 for which N values differ, for example, $N(c_2)<N(c_1)$. Let's d_1, d_2 to be the numbers for which the following inequality holds

$$N(c_2)<d_2<d_1<N(c_1)$$

For large enough n it's always possible to construct the nonintersecting intervals $\left[x_0, x_0+\frac{\pi}{n}\right]$, $\left[x_1, x_1+\frac{\pi}{n}\right]$ confined in $[a,b]$ and such that inside $\left[x_0, x_0+\frac{\pi}{n}\right]$ the inequality $N(x)>d_1$ holds, and inside another one: $N(x)<d_2$. Let's chose $\tau'(x)$ as follows:

$$\tau'=\sin^3\left[n(x-x_0)\right] \text{ on } \left[x_0, x_0+\frac{\pi}{n}\right];$$

$$\tau'=-\sin^3\left[n(x-x_1)\right] \text{ on } \left[x_1, x_1+\frac{\pi}{n}\right];$$

$$\tau' = 0 \text{ on the rest of } [a,b].$$

In that case, function $\tau(x)$ and its derivative are continuous. In addition, lemma premises suppose that $I(\tau)=0$, but for admitted N variations in considered interval it follows that for the same integral

$$I(\tau)=\int_a^b N(x)\tau'(x)dx > (d_1 - d_2)\int_0^{\pi/n} \sin^3 nx dx > 0. \tag{27}$$

Thus $N(x)$ should be constant on $[a,b]$ interval.

For R^3 the proof is given only for functions defined on unbounded volume. For our problem its possible to substitute $\delta w(\vec{r}) = div\vec{E}(\vec{r})$. Then one should prove that if for continuous scalar $N(\vec{r})$ and any continuous vector $\vec{E}(\vec{r})$ for which $div\vec{E} \to 0$ for $|\vec{r}| \to \infty$

$$I_w(\vec{E}) = \int N div\vec{E} d^3 r = 0,$$

then $N(\vec{r})$ is constant on R^3.

Suppose that on the opposite, $N(\vec{r})$ varies on R^3, then it should be at least two points $\vec{r}_1$, $\vec{r}_2$ for which N values differ, for example, $N(\vec{r}_2) < N(\vec{r}_1)$. Let's d_1, d_2 to be the numbers for which the following inequality holds

$$N(\vec{r}_2) < d_2 < d_1 < N(\vec{r}_1).$$

For large enough n it's always possible to construct the nonintersecting spheres $S_{1,2} : |\vec{r} - \vec{r}_{1,2}| \le \frac{\pi}{2n}$, such that inside S_1 the inequality $N(\vec{r}) > d_1$ holds, and inside S_2, $N(\vec{r}) < d_2$. Let's chose $div\vec{E}$ as follows:

$$div\vec{E} = \cos^3\left(n|\vec{r} - \vec{r}_1|\right) \text{ in } S_1;$$

$$div\vec{E} = -\cos^3\left(n|\vec{r} - \vec{r}_2|\right) \text{ in } S_2;$$

$$div\vec{E} = 0 \text{ outside of } S_{1,2}.$$

Here functions $\vec{E}(\vec{r})$, $div\vec{E}(\vec{r})$ are continuous. From lemma assumption $I_w(\vec{E}) = 0$, but for admitted N difference in considered $S_{1,2}$ regions it follows

$$I_w(\vec{E}) = \int N div(\vec{E}) d^3 r > (d_1 - d_2)\int_0^{\pi/2n} \cos^3\left(n|\vec{r}|\right) d^3 r > 0. \tag{28}$$

Thus, $N(\vec{r})$ should be constant. Under these conditions the relation between δw and $\vec{E}$ described by the solution of Laplase equation well known in electrodynamics [28].

References

[1] Jauch J M 1968 *Foundations of Quantum Mechanics* (Reading: Addison-Wesly)
[2] Kibble T 1979 *Comm. Math. Phys.* **62** 189
[3] Marmo G and Volkert G 2010 *Phys. Scrip.* **82** 038117
[4] Dubois D and Prade H 1980 *Fuzzy Sets and Systems. Theory and Applications* (New-York: Academic Press)
[5] Bandemer H and Gottwald S 1995 *Fuzzy sets, Fuzzy Logics, Fuzzy Methods with Apllications* (New-York: Wiley)
[6] Dodson C T 1974 *Bull. London Math. Soc.* **6** 191

[7] Zeeman C 1961 *Topology of 3-manifolds Ed K Fort* (New Jersey: Prentice-Hall) p 240
[8] Poston T 1971 *Manifold* **10** 25
[9] Peres A 2002 *Quantum Theory: Concepts and Methods* (New-York: Kluwer)
[10] Pykaz J 2015 *Int. J. Theor. Phys.* **54** 4367
[11] Sostak A 1996 *J. Math. Sci.* **6** 662
[12] Mayburov S 2007 *J. Phys.* A **41** 164071
[13] Mayburov S 2014 *Phys. Part. Nucl. Lett* **11** 1
[14] Buckley J J and Eslami E 1997 *Fuzzy Sets Syst.* **86** 179
[15] Balachandran A P, Kurkcuoglu S and Vaidia S 2007 *Lectures on Fuzzy and Fuzzy SUSY Physics* (Singapore:World Scientific)
[16] Schroder B 2003 *Ordered Sets: An Introduction* (Boston: Birkhauser)
[17] Landau L and Lifshitz E 1976 *Mechanics of Continuous Media* (Oxford: Pergamon Press)
[18] Gelfand I and Fomin S 2000 *Calculus of Variations* ed R Silverman (New-York: Dover)
[19] Rees C S, Shah S M and Stanojevic G V 1981 *Theory and Applications of Fourier Analysis* (New York: Marcel Devker)
[20] Wigner E P 1959 *Group Theory* (New-York: Academic)
[21] Jordan T 1993 *Ann. Phys.* **225** 83
[22] Jordan T 2006 *Phys. Rev.* A **73** 022101
[23] Ghosh S K and Deb B M 1982 *Phys. Rep.* **92** 1
[24] Aharonov Y and Vaidman L 1993 *Quantum Control and Measurement* (Vol 99) eds H Ezawa and Y Murayama (Tokyo: Elsvier)
[25] Landau L and Lifshitz E 1976 *Quantum Mechanics* (Oxford: Pergamon Press)
[26] Aharonov Y and Kaufherr T 1984 *Phys. Rev.* D **30** 368
[27] Lavrentiev M and Lusternik L 1938 *Lectures on Variation Calculus* (Moscow: Gosteorizdat)
[28] Landau L and Lifshitz E 1976 *Classical Field Theory* (Oxford: Pergamon Press)

Superalgebraic structure of Lorentz transformations

V V Monakhov

Saint Petersburg State University, Saint Petersburg, Russia
E-mail: v.v.monahov@spbu.ru

Abstract. We constructed superalgebraic representation of the algebra of second quantization of spinors. It is a fermionic analog of the Bargmann-Fock representation and is based on the use of the density of Grassmann variables in impulse space and their derivatives. We showed that Dirac matrices and Lorentz transformation generators can be expressed in terms of such densities. Lorentz transformations are considered as Bogolyubov transformations of creation and annihilation operators. We constructed superalgebraic form of the Dirac equation and the vacuum state vector. We showed that in the superalgebraic form of the complex Clifford algebra generators corresponding to Dirac gamma matrices are not equivalent. Clifford vector corresponding to diagonal matrix γ^0 annihilates the vacuum, and the remaining ones give nonzero values. This means that there is asymmetric direction corresponding to the time axis.

1. Introduction

Modern relativistic theory of the second quantization of fermion and boson fields is based on the use of the mathematical apparatus of C*-algebras [1-2] and Lie superalgebras [3]. In this case, for fermions, the Lorentz transformations are considered as Bogolyubov transformations of creation and annihilation operators [3]. However, in this approach one can not obtain an explicit form of the Dirac gamma-matrices.

In [4, 5] the author proposed a superalgebraic representation of spinors and Dirac matrices, which makes it possible to give a unified algebraic interpretation to the spinors and Dirac matrices. However, in the second quantization method, two terms correspond to birth operators, and two terms correspond to the annihilation operators, and they are differently transformed under Lorentz transformations. In [4-5], this distinction was an external condition that does not follow from the algebraic nature of the spinors.

In the framework of the superalgebraic approach, when using Grassmann variables θ^α and derivatives with respect to them $\frac{\partial}{\partial\theta^\alpha}$, at rest, the birth operators are naturally considered operators of multiplication by Grassmann variables θ^α, and the annihilation operators are derivatives with respect to them $\frac{\partial}{\partial\theta^\alpha}$.

2. Transformation of global spinor variables

Let us consider infinitesimal rotation operators $e^{iG^{\mu\nu}d\omega_{\mu\nu}}$ with generators $G^{\mu\nu}$ that take one spinor state vector to another spinor state vector. With them, the state vector Φ is transformed as $\Phi' = e^{iG^{\mu\nu}d\omega_{\mu\nu}}\Phi = (1 + iG^{\mu\nu}d\omega_{\mu\nu})\Phi$.

Denote $G^{\mu\nu}\, d\omega_{\mu\nu}$ by dG. The law of transformation of an arbitrary operator T acting on the spinor state vector Φ, and in doing this, again transforming it into the spinor state vector, neglecting infinitesimal higher order, will be

$$T' = e^{idG}\, T\, e^{-idG} = T + i[dG, T] = (1 + i[dG,*])T\,,$$

where * denotes the place to which you want to substitute the operator on which the transformation operates. If it consists of the product of several operators, then each of them is transformed according to the same law:

$$(1+[i\,dG,*])T_1T_2...T_m\Phi = (T_1 + [i\,dG, T_1])(T_2 + [i\,dG, T_2])...(T_m + [i\,dG, T_m])(1 + i\,dG)\Phi. \qquad (1)$$

It follows from (1) that if there are generators of Lorentz transformations for state vectors, operators that preserve the spinors by spinors are transformed by means of commutators with these generators. Since in the second quantization the solutions of the Dirac equation are treated as operators, they must also be transformed by such commutators. The Lorentz transformation causes a simultaneous change in the basis for all factors, acting on each of them. We denote such operators $[A,*]$ by $\hat{A}$. An operator $\hat{A} = [A,*]$ is simply a Lie derivative (commutator) acting on operators.

The formula for finite angles of rotation is obtained by integrating infinitesimal transformations:

$$T' = e^{i\int dG}\, T e^{-i\int dG} = e^{iG} T_1 e^{-iG}\, e^{iG} T_2 e^{-iG} ... e^{iG} T_m e^{-iG} = T_1' T_2' ... T_m'\,, \qquad (2)$$

where $G = \int dG$ and $T_r' = e^{iG}\, T_r e^{-iG}$ for each r. Note that (2) can be written in another form:

$$T' = e^{i\int [dG,*]}\, T = (e^{i\int [dG,*]} T_1)(e^{i\int [dG,*]} T_2)...(e^{i\int [dG,*]} T_m) = T_1' T_2' ... T_m'.$$

In this case, the parentheses limit the scope of the operators $[dG,*]$. Denoting $\hat{G} = [G,*] = \int [dG,*]$, we obtain

$$T' = e^{i\hat{G}}\, T = (e^{i\hat{G}} T_1)(e^{i\hat{G}} T_2)...(e^{i\hat{G}} T_m) = T_1' T_2' ... T_m'.$$

Variables θ and derivatives $\frac{\partial}{\partial\theta}$ are operators, so for infinitesimal transformations in accordance with (1) they must be transformed as

$$\theta' = (1 + [iG^{\mu\nu}\, d\omega_{\mu\nu},*])\theta$$

and

$$(\frac{\partial}{\partial\theta})' = (1 + [iG^{\mu\nu}\, d\omega_{\mu\nu},*])\frac{\partial}{\partial\theta}.$$

It is easy to see that in the secondary quantization formalism operators $[\theta^\alpha \frac{\partial}{\partial\theta^\beta},*]$ are analogous to $\theta^\alpha \frac{\partial}{\partial\theta^\beta}$ used in [4, 5]. They correspond to matrices mixing θ^α and θ^β, as well $\frac{\partial}{\partial\theta^\alpha}$ and $\frac{\partial}{\partial\theta^\beta}$. From these operators it is possible to construct matrices corresponding to spatial rotations.

Boosts should mix the "big" and "small" components of the spinor, and therefore they must mix $\frac{\partial}{\partial\theta^\alpha}$ and θ^τ. These are the operator $[\frac{\partial}{\partial\theta^\alpha}\frac{\partial}{\partial\theta^\tau},*]$ that transforms θ^τ to $\frac{\partial}{\partial\theta^\alpha}$ and θ^α to $-\frac{\partial}{\partial\theta^\tau}$, and the operator $[\theta^\alpha\theta^\tau,*]$ that transforms $\frac{\partial}{\partial\theta^\alpha}$ to $-\theta^\tau$ and $\frac{\partial}{\partial\theta^\tau}$ to θ^α.

3. Density of Grassmann variables and their derivatives

So far we have considered Grassmann variables θ^α and their derivatives $\dfrac{\partial}{\partial\theta^\alpha}$ as global quantities that do not depend on coordinates and impulses. With this interpretation, a problem arises, since the standard commutation relations for the creation and annihilation operators

$$\begin{aligned}&\{b_\alpha(p),b_\beta{}^+(p')\}=\delta^\alpha_\beta\,\delta^3(p-p'),\\&\{d_\tau(p),d_\rho{}^+(p')\}=\delta^\tau_\rho\,\delta^3(p-p'),\\&\{b_\alpha(p),d_\tau(p')\}=\{b_\alpha(p),d_\tau{}^+(p')\}=\{b_\alpha{}^+(p),d_\tau(p')\}=\{b_\alpha{}^+(p),d_\tau{}^+(p')\}=0.\end{aligned}\qquad(3)$$

contain a three-dimensional delta function $\delta^3(p-p')=\delta(p_1-p_1')\delta(p_2-p_2')\delta(p_3-p_3')$, but the anticommutator of Grassmann variables and derivatives with respect to them does not allow us to obtain delta functions. Therefore, we introduce in the momentum space Grassmann variables $\theta^a(p)$ that depend on the impulse, and local derivatives with respect to them $\dfrac{\partial}{\partial\theta^b(p)}$, which satisfy conditions

$$\{\frac{\partial}{\partial\theta^b(p)},\theta^a(p')\}=\delta^a_b\,\delta^3(p-p'),$$

$$\left\{\frac{\partial}{\partial\theta^a(p)},\frac{\partial}{\partial\theta^b(p')}\right\}=\{\theta^a(p),\theta^b(p')\}=0.$$

They can be regarded as densities in impulse space.

4. Superalgebraic representation of matrix algebra

Consider the operator

$$[M,*]=\int d^3p\;m_a{}^b\left[\theta^a(p)\frac{\partial}{\partial\theta^b(p)},*\right].$$

When it acts on an element of the algebra of spinors

$$\chi=\int d^3p\;\chi_c(p)\theta^c(p),$$

where $\chi_c(p)$ are numerical coefficients, we obtain

$$[M,*]\int d^3p\;\chi_c(p)\theta^c(p)=\int d^3p\;m_a{}^b\,\chi_b(p)\theta^a(p),$$

but when we act on an element

$$\phi=\int d^3p\;\phi^a(p)\frac{\partial}{\partial\theta^a(p)},$$

where $\phi^a(p)$ are numerical coefficients, we obtain

$$[M,*]\phi=-\int d^3p\;\phi^a(p)m_a{}^b(p)\frac{\partial}{\partial\theta^b(p)}.$$

Similarly, $\theta^a(p)$ and $\dfrac{\partial}{\partial\theta^b(p)}$ are mixing by operators

$$[A,*]=\int d^3p\;a_{ab}(p)[\theta^a(p)\theta^b(p),*]$$

and

$$[B,*]=\int d^3p b^{ab}(p)\left[\frac{\partial}{\partial\theta^b(p)}\frac{\partial}{\partial\theta^a(p)},*\right].$$

IOP Conf. Series: Journal of Physics: Conf. Series **1051** (2018) 012023 doi:10.1088/1742-6596/1051/1/012023

We denote the matrix column-spinor, in which the upper element is 1, and the remaining elements are equal to zero, as u_1, a column in which 1 element is on the second line, and the remaining 0 as u_2, etc. We will make a mapping of the matrix spinor-columns u_k, $k=1,2,3,4$ to the operators $\theta^{\alpha}(p)$ and $\frac{\partial}{\partial\theta^{\alpha}(p)}$ corresponding to the given spatial impulse p:

$$u_1 \cong \frac{\partial}{\partial\theta^1(p)},\ u_2 \cong \frac{\partial}{\partial\theta^2(p)},\ u_3 \cong \theta^3(p),\ u_4 \cong \theta^4(p).$$

This is some simplification, since it is necessary to consider the sum of the operators corresponding to all impulses. But the transformations of the elements corresponding to different impulses occur independently, so this simplification is very convenient.
A mapping is obtained between the operators $[M,*]$, $[A,*]$, $[B,*]$ and the elements of the matrix space generated by Dirac gamma matrices. Each such operator can be associated with the same matrix, regardless of the impulse values corresponding to the spinor elements to which this operator acts.

5. Creation and annihilation operators

The Hermitian conjugation is analogous to the Hermitian conjugation of superalgebraic spinors, described by the author in [4,5]: when conjugated, the numerical factors are replaced by complex conjugates, $\theta^a(p)$ are replaced by $\frac{\partial}{\partial\theta^a(p)}$, and $\frac{\partial}{\partial\theta^a(p)}$ are replaced by $\theta^a(p)$. In this way

$$(\theta^a(p))^+ = \frac{\partial}{\partial\theta^a(p)},$$

$$(\frac{\partial}{\partial\theta^a(p)})^+ = \theta^a(p).$$

It is obvious that

$$\{\theta^a(p'),(\theta^b(p))^+\} = \{\theta^a(p'),\frac{\partial}{\partial\theta^b(p)}\} = \delta^a_b\delta^3(p-p').$$

Thus, if there are spinors

$$\Psi(p') = \psi_\alpha(p')\frac{\partial}{\partial\theta^\alpha(p')} + \psi_\tau(p')\theta^\tau(p')$$

and

$$\Phi(p) = \phi_\beta(p)\frac{\partial}{\partial\theta^\beta(p)} + \phi_\rho(p)\theta^\rho(p),$$

then

$$\Psi^+(p') = \psi_\alpha^*(p')\theta^\alpha(p') + \psi_\tau^*(p')\frac{\partial}{\partial\theta^\tau(p')},$$

and

$$\{\Psi^+(p'),\Phi(p)\} = (\psi_\alpha^*(p)\phi_\alpha(p) + \psi_\tau^*(p)\phi_\tau(p))\delta^3(p-p'). \quad (4)$$

It follows from (4) that for spinors with a given impulse value one can define a scalar product

$$(\Psi(p'),\Phi(p)) = \{\Psi^+(p'),\Phi(p)\}. \quad (5)$$

This allows us to define an orthogonal basis in which the independent generators are orthogonal. In this case, the independent operators of spinor fields must be orthogonal. In accordance with (5), if $\Psi(p')$ and $\Phi(p)$ are independent, then $\Psi^+(p')$ and $\Phi(p)$ must anticommute.

6. Superalgebraic representation of Dirac matrices

The presence of the mapping of matrix elements to superalgebraic operators allows us to introduce operators $\hat{\gamma}^\eta, \eta = 0,1,2,3,5$, corresponding to the Dirac matrices γ^η (in the Dirac representation) by the following formulas:

$$\tilde{\gamma}^0 = \int d^3p \left(-\theta^1(p)\frac{\partial}{\partial\theta^1(p)} - \theta^2(p)\frac{\partial}{\partial\theta^2(p)} - \theta^3(p)\frac{\partial}{\partial\theta^3(p)} - \theta^4(p)\frac{\partial}{\partial\theta^4(p)} \right), \tag{6}$$

$$\tilde{\gamma}^1 = \int d^3p \left(\frac{\partial}{\partial\theta^2(p)}\frac{\partial}{\partial\theta^3(p)} + \frac{\partial}{\partial\theta^1(p)}\frac{\partial}{\partial\theta^4(p)} - \theta^3(p)\theta^2(p) - \theta^4(p)\theta^1(p) \right), \tag{7}$$

$$\tilde{\gamma}^2 = i\int d^3p \left(\frac{\partial}{\partial\theta^2(p)}\frac{\partial}{\partial\theta^3(p)} - \frac{\partial}{\partial\theta^1(p)}\frac{\partial}{\partial\theta^4(p)} + \theta^3(p)\theta^2(p) - \theta^4(p)\theta^1(p) \right), \tag{8}$$

$$\tilde{\gamma}^3 = \int d^3p \left(\frac{\partial}{\partial\theta^1(p)}\frac{\partial}{\partial\theta^3(p)} - \frac{\partial}{\partial\theta^2(p)}\frac{\partial}{\partial\theta^4(p)} - \theta^3(p)\theta^1(p) + \theta^4(p)\theta^2(p) \right), \tag{9}$$

$$\tilde{\gamma}^5 = \int d^3p \left(\frac{\partial}{\partial\theta^1(p)}\frac{\partial}{\partial\theta^3(p)} + \frac{\partial}{\partial\theta^2(p)}\frac{\partial}{\partial\theta^4(p)} + \theta^3(p)\theta^1(p) + \theta^4(p)\theta^2(p) \right), \tag{10}$$

$$\hat{\gamma}^\eta = [\tilde{\gamma}^\eta, *]. \tag{11}$$

Operators $\hat{\gamma}^\eta$ convert basis elements $\frac{\partial}{\partial\theta^1(p)}, \frac{\partial}{\partial\theta^2(p)}, \theta^3(p), \theta^4(p)$ and their linear combinations in the same way as the Dirac matrices γ^η convert the columns u_k and their linear combinations. Moreover, a linear combination of products of an arbitrary number of operators $\hat{\gamma}^\eta$ corresponds to a matrix from the corresponding linear combination of products of matrices γ^η.

The situation is completely analogous to the generators of the Lorentz transformations. In the matrix form they are given by formulas

$$\sigma^{\mu\nu} = \frac{i}{2}[\gamma^\mu, \gamma^\nu],\ \sigma^{0k} = i\begin{pmatrix} 0 & \tau^k \\ \tau^k & 0 \end{pmatrix},\ \sigma^{kl} = \begin{pmatrix} \tau^m & 0 \\ 0 & \tau^m \end{pmatrix}, \tag{12}$$

where $\mu,\nu = 0,1,2,3,5$, τ^k are the Pauli matrices, and the indices k,l,m are cyclic permutation of 1, 2, 3. Similarly to (6)-(10), we can define $\tilde{\sigma}^{\mu\nu}$ by the mapping of the matrix elements (12) to operators:

$$\tilde{\sigma}^{01} = i\int d^3p \left(\frac{\partial}{\partial\theta^2(p)}\frac{\partial}{\partial\theta^3(p)} + \frac{\partial}{\partial\theta^1(p)}\frac{\partial}{\partial\theta^4(p)} + \theta^3(p)\theta^2(p) + \theta^4(p)\theta^1(p) \right)$$

and so on. The operators $\tilde{\sigma}^{\mu\nu}$ satisfy equality

$$\tilde{\sigma}^{\mu\nu} = \frac{i}{2}[\tilde{\gamma}^\mu, \tilde{\gamma}^\nu].$$

We define the operators that are generators of the Lorentz transformations

$$\hat{\sigma}^{\mu\nu} = \frac{i}{2}(\hat{\gamma}^\mu\hat{\gamma}^\nu - \hat{\gamma}^\nu\hat{\gamma}^\mu).$$

It is easy to verify that

$$\hat{\sigma}^{\mu\nu} = [\tilde{\sigma}^{\mu\nu}, *]. \tag{13}$$

Thus, operators $\hat{\sigma}^{\mu\nu}$ can match matrices in exactly the same way as we matched $\hat{\gamma}^\mu$ the matrix operators γ^μ.

IOP Conf. Series: Journal of Physics: Conf. Series **1051** (2018) 012023 doi:10.1088/1742-6596/1051/1/012023

7. Lorentz transformations

For an infinitesimal Lorentz transformation with generator $dG=\frac{1}{4}\tilde{\sigma}^{\mu\nu}d\omega_{\mu\nu}$, for the product of the spinor operators $\Psi_1\Psi_2...\Psi_m$ it follows from equation (1) that

$$\Psi'_1\Psi'_2...\Psi'_m=(\Psi_1+i[dG,\Psi_1])(\Psi_2+i[dG,\Psi_2])...(\Psi_m+i[dG,\Psi_m]). \tag{14}$$

In accordance with (13), the operators $\tilde{\sigma}^{\mu\nu}$ generate the Lorentz transformation generators $\hat{\sigma}^{\mu\nu}$, and each of the spinor operators in (14) is transformed as follows:

$$\Psi'_k=(1+\frac{i}{4}\hat{\sigma}^{\mu\nu}d\omega_{\mu\nu})\Psi_k. \tag{15}$$

Spinors Ψ_k are a linear combination of $\frac{\partial}{\partial\theta^1(p)},\frac{\partial}{\partial\theta^2(p)},\theta^3(p),\theta^4(p)$, and by construction the action of operators $\hat{\sigma}^{\mu\nu}$ on them corresponds to the action of matrices $\sigma^{\mu\nu}$ on a column. Wherein

$$(i\hat{\sigma}^{01})^2\Psi_k=\Psi_k, \tag{16}$$

$$(i\hat{\sigma}^{rl})^2\Psi_k=-\Psi_k. \tag{17}$$

Integrating (15) with allowance for (16), we find that a boost on a finite angle $\phi=-\omega_{01}$ in the plane γ^0,γ^1 corresponds a transformation $\Psi'_k=(ch\frac{\phi}{2}-i\hat{\sigma}^{01}sh\frac{\phi}{2})\Psi_k$.

Similarly, from (15) and (17) it follows that for spatial rotation by an angle α in the plane γ^r,γ^l we obtain $\Psi'_k=(\cos\frac{\alpha}{2}-i\hat{\sigma}^{rl}\sin\frac{\alpha}{2})\Psi_k$.

The operator carrying out a boost in the plane γ^0,γ^1 can be written in the form

$$\hat{T}=e^{\hat{\gamma}^0\hat{\gamma}^1\phi/2}=e^{[\hat{\gamma}^0,*][\hat{\gamma}^1,*]\phi/2}.$$

As a result of the transformation we get

$$\Psi'=e^{\hat{\gamma}^0\hat{\gamma}^1\phi/2}\Psi. \tag{18}$$

Similarly for spatial rotation

$$\Psi'=e^{\hat{\gamma}^r\hat{\gamma}^l\alpha/2}\Psi.$$

8. Hermitian conjugation of gamma-matrices

The Hermitian conjugation of matrices is given by the scalar product (5), but in the general case for arbitrary impulses it is necessary to consider the anticommutator of the $\Psi^+(x')=(\int d^3p'\,\Psi(p'))^+$ and $\Phi(x)=\int d^3p\,\Phi(p)$, that is, to specify the scalar product of the spinor operators in the form

$$(\Psi(x'),\Phi(x))=\{\Psi^+(x'),\Phi(x)\}. \tag{19}$$

By the definition of a Hermitian adjoint operator

$$(\Psi(x'),\hat{\gamma}^\mu\Phi(x))=((\hat{\gamma}^\mu)^+\Psi(x'),\Phi(x)) \tag{20}$$

for arbitrary $\Psi(x')$ and $\Phi(x)$. Therefore, to find $(\hat{\gamma}^\mu)^+$ it is necessary to find an operator A such that $(A\Psi(x'),\Phi(x))=(\Psi(x'),\hat{\gamma}^\mu\Phi(x))$ for any admissible $\Psi(x')$ and $\Phi(x)$. We will consider the expansion $\Phi(x)$ and $\Psi(x')$ with respect to impulses in the form

$$\Phi(x)=\int d^3p\,\Phi(p)=\int d^3p\left(\phi_\alpha(p)\frac{\partial}{\partial\theta^\alpha(p)}e^{-ip_\mu x^\mu}+\phi_\tau(p)\theta^\tau(p)e^{ip_\mu x^\mu}\right), \tag{21}$$

$$\Psi(x')=\int d^3p'\,\Psi(p')=\int d^3p'\left(\psi_\alpha(p')\frac{\partial}{\partial\theta^\alpha(p')}e^{-ip'_\mu x'^\mu}+\psi_\tau(p')\theta^\tau(p')e^{ip'_\mu x'^\mu}\right).$$

Consider the action of $\hat{\gamma}^0$ on these spinors:

$$\hat{\gamma}^0\Phi(x)=\int d^3p\left(\phi_\alpha(p)\frac{\partial}{\partial\theta^\alpha(p)}e^{-ip_\mu x^\mu}-\phi_\tau(p)\theta^\tau(p)e^{ip_\mu x^\mu}\right), \tag{22}$$

$$\hat{\gamma}^0\Psi(x')=\int d^3p'\left(\psi_\alpha(p')\frac{\partial}{\partial\theta^\alpha(p')}e^{-ip'_\mu x'^\mu}-\psi_\tau(p')\theta^\tau(p')e^{ip'_\mu x'^\mu}\right).$$

Hermitian adjoint to $\Psi(x')$ and $\hat{\gamma}^0\Psi(x')$ spinors:

$$\Psi^+(x')=\int d^3p'\left(\psi^*_\alpha(p')\theta^\alpha(p')e^{ip'_\mu x'^\mu}+\psi^*_\tau(p')\frac{\partial}{\partial\theta^\tau(p')}e^{-ip'_\mu x'^\mu}\right), \tag{23}$$

$$(\hat{\gamma}^0\Psi(x'))^+=\int d^3p'\left(\psi^*_\alpha(p')\theta^\alpha(p')e^{ip'_\mu x'^\mu}-\psi^*_\tau(p')\frac{\partial}{\partial\theta^\tau(p')}e^{-ip'_\mu x'^\mu}\right). \tag{24}$$

From the definition (19), and also relations (23) and (22) it follows that

$$(\Psi(x'),\hat{\gamma}^0\Phi(x))=\{\Psi^+(x'),\hat{\gamma}^0\Phi(x)\}=$$
$$=\int d^3p\left(\psi^*_\alpha(p)\phi_\alpha(p)e^{ip_\mu(x'^\mu-x^\mu)}-\psi^*_\tau(p)\phi_\tau(p)e^{-ip_\mu(x'^\mu-x^\mu)}\right).$$

Similarly, it follows from (19), (24) and (21) that

$$(\hat{\gamma}^0\Psi(x'),\Phi(x))=\{(\hat{\gamma}^0\Psi(x'))^+,\Phi(x)\}=$$
$$=\int d^3p\left(\psi^*_\alpha(p)\phi_\alpha(p)e^{ip_\mu(x'^\mu-x^\mu)}-\psi^*_\tau(p)\phi_\tau(p)e^{-ip_\mu(x'^\mu-x^\mu)}\right).$$

In this way, $(\Psi(x'),\hat{\gamma}^0\Phi(x))=(\hat{\gamma}^0\Psi(x'),\Phi(x))$, therefore, from the definition (20) we obtain that

$$(\hat{\gamma}^0)^+=\hat{\gamma}^0. \tag{25}$$

Similarly we obtain

$$(\hat{\gamma}^k)^+=-\hat{\gamma}^k,\ k=1,2,3, \tag{26}$$

$$(\hat{\gamma}^5)^+=\hat{\gamma}^5. \tag{27}$$

For $\tilde{\gamma}^\mu$ and for the spinor operator $\Psi(x)$, the operation of a Hermitian conjugation is given in the same way as for elements of the algebra of ordinary superalgebraic spinors [4,5]. The Hermitian conjugating relation $\hat{\gamma}^\mu\Psi=\tilde{\gamma}^\mu\Psi-\Psi\tilde{\gamma}^\mu$, we obtain

$$(\hat{\gamma}^\mu\Psi)^+=\Psi^+(\tilde{\gamma}^\mu)^+-(\tilde{\gamma}^\mu)^+\Psi^+=-[(\tilde{\gamma}^\mu)^+,*]\Psi^+.$$

Hermitian conjugating (6)-(10) we find that

$$(\tilde{\gamma}^0)^+=\tilde{\gamma}^0;\ (\tilde{\gamma}^k)^+=-\tilde{\gamma}^k,\ k=1,2,3;\ (\tilde{\gamma}^5)^+=\tilde{\gamma}^5.$$

Therefore

$$[(\tilde{\gamma}^0)^+,*]=[\tilde{\gamma}^0,*]=\hat{\gamma}^0=(\hat{\gamma}^0)^+,$$
$$[(\tilde{\gamma}^k)^+,*]=-[\tilde{\gamma}^k,*]=-\hat{\gamma}^k=(\hat{\gamma}^k)^+$$

and

$$[(\tilde{\gamma}^5)^+,*]=[\tilde{\gamma}^5,*]=\hat{\gamma}^5=(\hat{\gamma}^5)^+.$$

That is

$$[(\tilde{\gamma}^\mu)^+,*]=[\tilde{\gamma}^\mu,*]^+=(\hat{\gamma}^\mu)^+,$$

whereby

$$(\hat{\gamma}^\mu\Psi)^+=-(\hat{\gamma}^\mu)^+\Psi^+. \tag{28}$$

So that

$$(\hat{\gamma}^\mu\hat{\gamma}^\nu\Psi)^+=-\hat{\gamma}^{\mu+}(\hat{\gamma}^\nu\Psi)^+=\hat{\gamma}^{\mu+}\hat{\gamma}^{\nu+}\Psi^+. \tag{29}$$

It follows from (29) and (27) that

$$(\hat{\gamma}^k\hat{\gamma}^l\Psi)^+=\hat{\gamma}^{k+}\hat{\gamma}^{l+}\Psi^+=\hat{\gamma}^k\hat{\gamma}^l\Psi^+, \tag{30}$$

IOP Conf. Series: Journal of Physics: Conf. Series **1051** (2018) 012023 doi:10.1088/1742-6596/1051/1/012023

$$(\hat{\gamma}^0\hat{\gamma}^k\Psi)^+ = \hat{\gamma}^{0+}\hat{\gamma}^{k+}\Psi^+ = -\hat{\gamma}^0\hat{\gamma}^k\Psi^+, \tag{31}$$

where $k,l = 1,2,3$.

The relations (25)-(27) are completely analogous to the matrix, but (28)-(31) essentially differs from the usual matrix rules of Hermitian conjugation. This is due to the fact that in this approach there are no right-to-left operators, and all operators operate from left to right.

9. Superalgebraic form of the Dirac equation

The Dirac equation in the superalgebraic representation of the second quantization is written in the form

$$i\hat{\gamma}^\mu\partial_\mu\Psi = m\Psi. \tag{32}$$

It is easy to show that in the Dirac adjoint spinor $\overline{\Psi} = (\hat{\gamma}^0\Psi)^+ = -\hat{\gamma}^0\Psi^+$ satisfies the analogous equation:

$$i\hat{\gamma}^\mu\partial_\mu\overline{\Psi} = m\overline{\Psi}. \tag{33}$$

It must be taken into account that in the superalgebraic representation the repeated Dirac conjugation changes sign before the spinor: $\overline{\overline{\Psi}} = -\Psi$.

Solutions (32) and (33) can be written in the form

$$\Psi(t,x) = \int \frac{d^3p}{\sqrt{(2\pi)^3}} \left(\lambda_\alpha b_\alpha(p)e^{-ip_\mu x^\mu} + \lambda_\tau \overline{d}_\tau(p)e^{ip_\mu x^\mu}\right), \tag{34}$$

$$\overline{\Psi}(t,x) = \int \frac{d^3p}{\sqrt{(2\pi)^3}} \left(-\lambda_\tau^* d_\tau(p)e^{-ip_\mu x^\mu} + \lambda_\alpha^* \overline{b}_\alpha(p)e^{ip_\mu x^\mu}\right),$$

here λ_α and λ_τ are constants (possibly depending on p), the annihilation operators $b_\alpha(p)$ and $d_\tau(p)$ are obtained by Lorentz rotations of $\frac{\partial}{\partial\theta^\alpha(0)}$ and $\frac{\partial}{\partial\theta^\tau(0)}$, the creation operators $\overline{b}_\alpha(p)$ and $\overline{d}_\tau(p)$ are Lorentz rotations of $\theta^\alpha(0)$ and $\theta^\tau(0)$. In (34), instead of the notation $d_\tau^+(p)$ used in the matrix theory and, in particular, in (3), we use $\overline{d}_\tau(p)$, since $b_\alpha(p)$ and $\overline{b}_\alpha(p)$ under the Lorentz transformations are transformed covariantly, while from (18) and (31) it follows that $b_\alpha^+(p)$ transforms noncovariantly, and therefore can not be an operator acting on spinors. The same applies to $d_\tau^+(p)$.

10. Discrete impulse space and fermionic vacuum

We divide the impulse space into infinitesimal volumes $\Delta^3 p_j$. We denote

$$B_\alpha(p_j) = \frac{1}{\Delta^3 p_j}\int_{\Delta^3 p_j} d^3p\, b_\alpha(p), \tag{35}$$

$$D_\tau(p_j) = \frac{1}{\Delta^3 p_j}\int_{\Delta^3 p_j} d^3p\, d_\tau(p). \tag{36}$$

The factor $\frac{1}{\Delta^3 p_j}$ is necessary for the $B_\alpha(p_j)$ and $D_\tau(p_j)$ to be transformed covariantly. It follows from (35)-(36) that $B_\alpha(p_j)$ under Lorentz transformations it is transformed in the same way as $b_\alpha(p)$, and $D_\tau(p_j)$ in the same way as $d_\tau(p)$.

Consider an anticommutator of the $\overline{B}_\beta(p_j)$ with $B_\alpha(p_k)$:

$$\{\bar{B}_\beta(p_j), B_\alpha(p_k)\} = \frac{1}{(\Delta^3 p_j)^2} \int_{\Delta^3 p_j} d^3p \int_{\Delta^3 p_k} d^3p' \{\bar{b}_\beta(p), b_\alpha(p')\} = \frac{1}{\Delta^3 p_j} \delta_j^k \delta_\alpha^\beta \tag{37}$$

(there is no summation over j). Similarly,

$$\{\bar{D}_\rho(p_j), D_\tau(p_k)\} = \frac{1}{\Delta^3 p_j} \delta_j^k \delta_\tau^\rho . \tag{38}$$

By virtue of (35) and the anticommutativity of $b_\alpha(p)$ and $b_\alpha(p')$, and also of $\bar{b}_\alpha(p)$ and $\bar{b}_\alpha(p')$

$$(B_\alpha(p_j))^2 = (B_\alpha^{\ +}(p_j))^2 = (\bar{B}_\alpha(p_j))^2 = 0 . \tag{39}$$

Similarly we obtain

$$(D_\tau(p_j))^2 = (D_\tau^{\ +}(p_j))^2 = (\bar{D}_\tau(p_j))^2 = 0 . \tag{40}$$

Relations (37)-(40) show the realization of the transition from the continuous densities of the Grassmann variables $b_\alpha(p)$, $d_\tau(p)$, $\bar{b}_\alpha(p)$ and $\bar{d}_\tau(p)$ to discrete Grassmann variables $B_\alpha(p_j)$, $D_\tau(p_j)$ and $\bar{B}_\alpha(p_j)$, $\bar{D}_\tau(p_j)$. In this case, the integrals are replaced by sums over discrete values $\Delta^3 p_j$ of the impulse, and for the anticommutator of the conjugate quantities related to the same volume $\Delta^3 p_j$, the delta function is replaced by a large but finite value $\frac{1}{\Delta^3 p_j}$, and zero for those belonging to different volumes.

Consider the spinor $b_1(p_1)$ obtained from $b_1(0)$ by the boost (18), and the associated with it operator $\bar{b}_1(p_1)$:

$$b_1(p_1) = \cosh\frac{\phi}{2}\frac{\partial}{\partial\theta^1(p_1)} + \sinh\frac{\phi}{2}\theta^4(p_1),$$

$$\bar{b}_1(p_1) = \cosh\frac{\phi}{2}\theta^1(p_1) - \sinh\frac{\phi}{2}\frac{\partial}{\partial\theta^4(p_1)}.$$

Wherein

$$\{b_1(p_1), \bar{b}_1(p_1')\} = \delta(p_1 - p_1') .$$

Since the direction p_1 is arbitrary, these relations are valid for any spatial impulses. Similarly,

$$\{\bar{B}_\beta(p_j), B_\alpha(p_k)\} = \frac{1}{(\Delta^3 p_j)^2} \int_{\Delta^3 p_j} d^3p \int_{\Delta^3 p_k} d^3p' \{\bar{b}_\beta(p), b_\alpha(p')\} = \frac{1}{\Delta^3 p_j} \delta_j^k \delta_\alpha^\beta .$$

Operator $b_{<\alpha>}(0)\bar{b}_{<\alpha>}(0)$ (summation over indices enclosed in triangular brackets, is not present) under Lorentz rotations transferring the spinors $b_\alpha(0)$ into $b_\alpha(p)$ goes into $b_{<\alpha>}(p)\bar{b}_{<\alpha>}(p)$. Likewise, $B_{<\alpha>}(0)\,\bar{B}_{<\alpha>}(0)$ goes into $B_{<\alpha>}(p_j)\,\bar{B}_{<\alpha>}(p_j)$.

Operators $\sqrt{\Delta^3 p_j}\,B_\alpha(p_j)$, $\sqrt{\Delta^3 p_j}\,D_\tau(p_j)$, $\sqrt{\Delta^3 p_j}\,\bar{B}_\alpha(p_j)$ and $\sqrt{\Delta^3 p_j}\,\bar{D}_\tau(p_j)$ play the role of fermionic variables, from which, by analogy with the theory of algebraic spinors [6], a fermionic vacuum can be constructed. Let

$$\Psi_{B_\alpha j} = \sqrt{\Delta^3 p_j}\,B_\alpha(p_j)\sqrt{\Delta^3 p_j}\,\bar{B}_\alpha(p_j),$$

$$\Psi_{D_\tau j} = \sqrt{\Delta^3 p_j}\,D_\tau(p_j)\sqrt{\Delta^3 p_j}\,\bar{D}_\tau(p_j),$$

$$\Psi_{V_j} = \Psi_{B_1 j}\Psi_{B_2 j}\Psi_{D_3 j}\Psi_{D_4 j} .$$

So that $(\Psi_{V_j})^2 = \Psi_{V_j}$, and the operators Ψ_{V_j} can also be regarded as local fermionic vacuum in impulse space. We define the fermionic vacuum operator as

$$\Psi_V = \prod_j \Psi_{V_j},$$

where the product goes over all physically possible values of j. Obviously, Ψ_V is an idempotent, but its Hermitian conjugation gives a vacuum that differs from Ψ_V, although equivalent to it.

11. Number of particles and energy-impulse operators

We define the operators of the number of particles

$$\hat{N}_\alpha(p_j) = [\bar{B}_{<\alpha>}(p_j)B_{<\alpha>}(p_j),*] = -[B_{<\alpha>}(p_j)\bar{B}_{<\alpha>}(p_j),*],$$

$$\hat{N}_\tau(p_j) = [\bar{D}_{<\tau>}(p_j)D_{<\tau>}(p_j),*] = -[D_{<\tau>}(p_j)\bar{D}_{<\tau>}(p_j),*].$$

Their effect on the vacuum, by virtue of (39) and (40), gives zero, and on the one-particle state

$$\hat{N}_\alpha(p_j)\Delta^3 p_j \bar{B}_\beta(p_k)\Psi_V = \Delta^3 p_j[\bar{B}_{<\alpha>}(p_j)B_{<\alpha>}(p_j), \bar{B}_\beta(p_k)]\Psi_V = \delta_j^k \delta_\alpha^\beta \bar{B}_{<\alpha>}(p_j)\Psi_V.$$

That is, the operator $\hat{N}_\alpha(p_j)\Delta^3 p_j$ corresponds to the operator of number of particles in the volume $\Delta^3 p_j$.

The energy-impulse operator can be written in the form

$$\hat{P}_\mu = \sum_j \Delta^3 p_j\, p_\mu(p_j)\,[\hat{N}_1(p_j) + \hat{N}_2(p_j) + \hat{N}_3(p_j) + \hat{N}_4(p_j),*] = \sum_j \hat{P}_\mu(p_j).$$

In a similar way, we can write the substitution in (6)-(11) of integrals by a sum:

$$\hat{\gamma}^\mu = \sum_j \hat{\gamma}^\mu(p_j).$$

So that superalgebraic form of the Dirac equation in terms of the energy-impulse operator is

$$\hat{\gamma}^\mu \hat{P}_\mu \Psi = m\Psi. \tag{41}$$

We can write equivalent form of (41) by means of replacement of Dirac operator $\hat{\gamma}^\mu \hat{P}_\mu$ with modified Dirac operator $\sum_j \hat{\gamma}^\mu(p_j)\hat{P}_\mu(p_j)$. We have

$$\hat{\gamma}^\mu \hat{P}_\mu \Psi = \sum_i \sum_j \hat{\gamma}^\mu(p_i)\hat{P}_\mu(p_j)\sum_k \Psi(p_k) = \sum_k \hat{\gamma}^\mu(p_k)\hat{P}_\mu(p_k)\Psi(p_k) = \sum_k \hat{\gamma}^\mu(p_k)\hat{P}_\mu(p_k)\Psi,$$

that is why (41) leads to (42)

$$\sum_k \hat{\gamma}^\mu(p_k)\hat{P}_\mu(p_k)\Psi = m\Psi, \tag{42}$$

and vice versa, equation (42) leads to (41).

12. The Dirac equation for a state vector

However, this equivalence is proper for field operator Ψ, not for a state vector $\Psi\Psi_V$. The action of Dirac operator $\hat{\gamma}^\mu \hat{P}_\mu$ on $\Psi\Psi_V$ raises a problem. It is clear for the case of given energy-impulse $\hat{P}_\mu \Psi = p_\mu \Psi$, since

$$\hat{\gamma}^\mu \hat{P}_\mu \Psi\Psi_V = \hat{\gamma}^\mu p_\mu \Psi\Psi_V = (\hat{\gamma}^\mu p_\mu \Psi)\Psi_V + \Psi(\hat{\gamma}^\mu p_\mu \Psi_V) = m\Psi\Psi_V + \Psi(\hat{\gamma}^\mu p_\mu \Psi_V)$$

and wherein $\hat{\gamma}^\mu p_\mu \Psi_V \neq 0$. That is, the Dirac equation for the state vector is not satisfied. However, action of modified Dirac operator on the vacuum gives zero:

$$\sum_j \hat{\gamma}^\mu(p_j)p_\mu(p_j)\Psi_V = \sum_j \hat{\gamma}^\mu(p_j)p_\mu(p_j)\prod_k \Psi_{V_k} = \sum_j (\hat{\gamma}^\mu(p_j)p_\mu(p_j)\Psi_{V_j})\prod_{k\neq j}\Psi_{V_k} =$$
$$= \sum_j (\hat{\gamma}^\mu(p_j)p_\mu(p_j)B_1(p_j)\bar{B}_1(p_j)B_2(p_j)\bar{B}_2(p_j)D_3(p_j)\bar{D}_3(p_j)D_4(p_j)\bar{D}_4(p_j))\prod_{k\neq j}\Psi_{V_k} = \quad (43)$$
$$= \sum_j e^{iG_j}\left[\hat{\gamma}^0(0)m, \frac{\partial}{\partial\theta^1(0)}\theta^1(0)\frac{\partial}{\partial\theta^2(0)}\theta^2(0)\frac{\partial}{\partial\theta^3(0)}\theta^3(0)\frac{\partial}{\partial\theta^4(0)}\theta^4(0)\right]e^{-iG_j}\prod_{k\neq j}\Psi_{V_k} = 0,$$

where the impulse p_j is obtained from the state of rest by a transformation of the form (2), and $\hat{\gamma}^0 m$ then goes over into $\hat{\gamma}^\mu p_\mu$.

Therefore, we obtain

$$\sum_j \hat{\gamma}^\mu(p_j)\hat{P}_\mu(p_j)\Psi\Psi_V = m\Psi\Psi_V + \Psi\sum_j \hat{\gamma}^\mu(p_j)p_\mu(p_j)\Psi_V = m\Psi\Psi_V,$$

that is, modified Dirac equation

$$\sum_j \hat{\gamma}^\mu(p_j)\hat{P}_\mu(p_j)\,\Psi\Psi_V = m\Psi\Psi_V \quad (44)$$

is satisfied. This is meaning that (44) is proper superalgebraic form of Dirac equation for state vector. It is important to note that the fulfillment of (43) is due to the fact that $\hat{\gamma}^0(0)$ annihilates the local vacuum Ψ_{V_0} corresponding to the state of rest:

$$\Psi_{V_0} = \frac{\partial}{\partial\theta^1(0)}\theta^1(0)\frac{\partial}{\partial\theta^2(0)}\theta^2(0)\frac{\partial}{\partial\theta^3(0)}\theta^3(0)\frac{\partial}{\partial\theta^4(0)}\theta^4(0),$$
$$\hat{\gamma}^0(0)\Psi_{V_0} = -[\frac{\partial}{\partial\theta^1(0)}\theta^1(0) + \frac{\partial}{\partial\theta^2(0)}\theta^2(0) + \frac{\partial}{\partial\theta^3(0)}\theta^3(0) + \frac{\partial}{\partial\theta^4(0)}\theta^4(0), \Psi_{V_0}] = 0.$$

At the same time $\hat{\gamma}^k(0)\Psi_{V_0} \neq 0$ and $\hat{\gamma}^5(0)\Psi_{V_0} \neq 0$. Therefore, the Dirac equation can be satisfied only in the case when $\hat{\gamma}^0$ has signature +1 and corresponds to the time axis. Operators $i\hat{\gamma}^k$ and $\hat{\gamma}^5$ which has the same signature +1 can not correspond to the time axis, since they do not annihilate the local vacuum Ψ_{V_0}, and therefore three operators $\hat{\gamma}^k$ (or two of them and $i\hat{\gamma}^5$) must correspond to the spatial axes, and $i\hat{\gamma}^5$ (or the third of the operators $\hat{\gamma}^k$) corresponds to the Clifford 4-volume element. In the Clifford algebra, gamma matrices of the same signature are equivalent, however the presence of a spinor vacuum leads to a distinction of the matrix corresponding to diagonal operator $\hat{\gamma}^0$. This means that there is asymmetric direction corresponding to the time axis.

13. Conclusion

The constructed representation is a fermionic analog of the Bargmann-Fock representation for operators of creation and annihilation of bosons. Main difference between these representations is due to the fact that for fermions it is necessary to consider the dependence of all quantities, including Dirac matrices, on Grassmann variables and derivatives with respect to them.

References

[1] Bongaarts P J M 1970 The electron-positron field, coupled to external electromagnetic potentials, as an elementary C* algebra theory *Annals of Physics* **56** 1 pp 108-139
[2] Baez J C, Segal I E and Zhou Z 2014 *Introduction to algebraic and constructive quantum field theory* (Princeton University Press)
[3] Grosse H and Langmann E 1991 On current superalgebras and super-Schwinger terms *Letters in Mathematical Physics* **21** 1 pp 69-76
[4] Monakhov V V 2016 Superalgebraic representation of Dirac matrices *Theoretical and Mathematical Physics* **186** 1 pp 70-82

[5] Monakhov V V 2016 Dirac matrices as elements of superalgebraic matrix algebra *Bulletin of the Russian Academy of Sciences: Physics* **80** 8 pp 985-988

[6] Monakhov V V 2017 Construction of a fermionic vacuum and the fermionic operators of creation and annihilation in the theory of algebraic spinors *Physics of Particles and Nuclei* **48** 5 pp 836–838

IOP Conf. Series: Journal of Physics: Conf. Series **1051** (2018) 012024 doi:10.1088/1742-6596/1051/1/012024

General relativity as a unified fluid and field theory

H A Múnera

International Centre for Physics (CIF), Bogota, Colombia
E-mail: hmunera@hotmail.com

Abstract. Einstein's dream for a unified field theory of Nature is attained with a classical fluid theory founded on space, time and spin, rather than on Einstein's spacetime. An invariant quantum theory for the primordial fluid obeys the homogeneous Klein-Gordon equation, which is the same three-dimensional classical wave equation (CWE) initially tried by Schrödinger to formulate quantum mechanics (QM), but abandoned by linear superposition considerations. Primordial fluid pervades universe, obeys energy and momentum conservation, and is formed by sagions: energy-like, discrete, extended Planck-size objects of finite size, carriers of linear momentum and spin, moving in absolute 3D-curved space with speed C along straightest path. We briefly describe our novel non-harmonic and inherently quantized solutions for CWE in spherical coordinates, discovered in 1995. Solutions include a steady-state background field (possibly related to the CMB, to non-locality, and to action-at-a-distance), quantized helices, and inherently quantized functions exhibiting stable dynamic equilibrium, and isomorphism under many transformations, including the classical Doppler case, and the relativistic Lorentz, Poincaré, and Einstein transformations. Isomorphism pre-empts *ab initio* a few interpretative issues regarding relativistic and classical transformations. Mathematical fields represent the realistic physical temporal evolution of the primordial fluid in curved 3D-space.

1. Introduction: controversial questions in classical and quantum descriptions of Nature

Besides unprecedented technological progress, 20th century is characterized for abandoning cherished and fundamental notions rooted in Ancient Greece: logic, causality, space, and a careful distinction between the container and what is contained; inherently contradictory notions — as a vacuum with physical properties — are now common currency. Special relativity (SR) and quantum mechanics (QM) attributed an overemphasized role to the observer, leading to theories about how an observer sees Nature, rather than to how Nature behaves when the observer is not there as it was the case until the turn of 20th century. For Aristotle Earth was at the center of Ancient Cosmos, but a modern competitor has stronger claims: the anthropic principle places humankind at the center of universe, conveniently forgetting that tomorrow, on the geological scale, a collision with an asteroid may erase earth and humankind from the universe — event that would hardly affect the rest of cosmos.

According to some views, the main open question in physics today is compatibility of quantum and gravity theories. Einstein dreams were more ambitious: only one Nature with all interactions described by a single theory. In the recent opinion of Gerard 'tHooft, a unified theory should also address the question of "*how matter behaves*" [1]. Present paper outlines an approach that works: an objective, causal and Cartesian unified fluid and field theory, (hopefully) free of internal inconsistencies.

As general context, let us mention two controversial aspects in Newton's Principia [2]: (a) circular definition of mass, and (b) the notion of non-contact forces, in particular the notorious omission of a mechanism for generation and propagation of gravitational force. The latter omission led to the widely spread belief that Newton introduced action-at-a-distance (AAAD) in physics. Due to the high

predictive capability of Newtonian mechanics, defects (a) and (b) were pragmatically accepted or tolerated. Unfortunately, it is not generally known that in his private letters [3] Newton perseveringly and strongly argued for an ether (be it material, or immaterial) to explain gravity propagation [4].
An economic solution for issues (a) and (b) is provided by a Cartesian approach based on conservation of quantity of motion as a primitive notion [5-6] — thus downgrading force and mass from the role they have in Newtonian mechanics. A consistent kinematic definition of mass was proposed by Barré de Saint Venant around 1850 [7-8]. Without mentioning him, the method of Saint Venant is implemented in some modern textbooks [9]: stroboscopic photography to measure average velocity of two interacting bodies before and after a collision upon a pneumatic table. In contrast, the better-known operational definition of mass proposed by Mach around 1870 is based on mutually induced accelerations during the interaction of two bodies [8,10]. In criticizing the notions of instant velocity and instant acceleration fields, Russell [11] noted that to measure average velocity two observations at separate intervals of time are required, while an estimate for average acceleration requires at least three observations. From Ockham's razor, Saint Venant's more economical method is preferable to Mach's approach. Our energy-like primordial fluid (PF) is a modern version of ether, which inherently carries an indestructible quantity of motion, thereby explaining propagation of the derived notion of force — which merely is the average exchange of quantity of motion in a collision. Such PF provides a causal and physically objective basis for the mathematical notion of field [4,12-13].
Passing now to open questions in QM, the completeness issue noted by Einstein still is there. Secondly, the Copenhagen interpretation of QM accepts the existence of undefined states of Nature between being and not-being —as the iconic Schrödinger's cat, and Wigner's friend. To stress the philosophical absurdity of the latter, Einstein used to pose the rhetoric question of whether the Moon existed when an observer was not looking at her! The Einstein-Podolsky-Rosen paradox and the ensuing Bell theorem are common knowledge. This writer agrees [14] with Einstein's views on causality and objectivity. Finally, there is Born's interpretation of the QM-wave-function as a probability. Since Jacob Bernoulli's Ars Conjectandi (circa 1713), it is well-known that probability is not a physical magnitude, but a mere mathematical construct of human mind [15]. Further, present writer posits that the rest of Nature — as far as known on earth — does not conjecture about outcomes of forthcoming events. If there is little information, probability is subjective and is a mere (non-) informed guess; probability is less subjective if there is symmetry, or when it is viable to enumerate the (hopefully) complete list of physical trajectories leading to an outcome, as in Feynman's continuous path integrals, or in discrete decision trees [16]. When long-term observations or controlled experimentation are viable, objective frequency-based probability is obtained. However, in any of the three previous scenarios probability is not a physical object.
In 1952 Louis de Broglie recalled [17] a distinction he made in the mid-1920s between the continuous solutions of the linear QM-equations and its interaction with a localized (presumably material) corpuscle in such wave field, leading to an objective and causal "*dual solution*", that foreshadowed t'Hooft remarks [1] for eighty years. Since this mathematically coupled problem was difficult to solve, de Broglie settled for a simpler pilot-wave model: motion of a material corpuscle is controlled by the wave function Ψ. However, he also abandoned this proposal after realizing what we already said in previous paragraph: "*Ψ certainly is not a physical reality... it only is a representation of probabilities dependent upon the state of our knowledge*" [17] (our literal translation from French, page 266). As alternative to the Copenhagen view of QM, de Broglie suggested in the 1950s a stochastic interpretation, where Schrödinger's linear equation represents fluctuations of a fluid [18], but the fluid itself was proposed much earlier by Madelung [19-20].
The five-dimensional universe suggested by Klein in 1926, led de Broglie in 1927 [21] to an alternative for Schrödinger's linear equation: the non-linear wave equation known today as Klein-Gordon equation (KGE). De Broglie also entered the long-standing controversy between Cartesians and Newtonians [3], and pointed out that a most appealing aspect of general relativity (GR) was to eliminate "*the metaphysical concept of force from gravitational theory*" [21]. In 1962 de Broglie insisted once again on non-linearity: "*agreeing with some deep remarks by Einstein, all exchanges of*

energy and linear momentum between corpuscles probably arise from transitory non-linear processes, whose description is beyond current linear theories" [22] (our translation from French, p 425). De Broglie also used a KGE in relativistic thermodynamics [23], soliton solutions of KGE are reported in [24], and a summary of de Broglie's pilot-wave theory is in [25].

De Broglie consistently noted that all waves in his theory had physical reality, thus requiring an underlying physical fluid, often identified with Dirac's ether [26]. In 1999 present author proposed a primordial fluid (PF) [27] similar to Dirac's, but with two conceptual differences: (a) an unidirectional arrow of time from past to future, thus excluding reversibility, and science-fiction travels to the past, and (b) the fluid constituents were material 3D-extended preons. From Leibniz continuity principle stating that a change from state A to B has to go through all intermediate states, and considering the controversy on impenetrability of matter [28-30], we concluded that matter should have structure at all scales, so that the smallest bit of matter is divisible, but its components cannot possibly be material. So, instead of material preons, it is postulated here that the primordial fluid is formed by energy-like sagions, described as discrete, 3D-extended, spherical and Planck-size objects of radius R, carrying quantity of motion P, spin S, and moving with speed C along the straightest path [12]. Our fluid theory agrees with de Broglie's relativistic QM and with the notion that matter has electromagnetic origin. In our theory the simplest entity of Nature is an energy-like sagion, rather than a material object as in the Leucippus and Democritus atom.

2. Reinterpretation of empirical evidence usually cited to support Einsteinian relativity

Next question: where does the primordial fluid exist? In a Newtonian context the sagion moves with speed C in absolute space, described in 1590 by Francesco Patrizi as physical space Σ [31-32]. For Gauss, the geometrical nature of space was an empirical question, tackled by measuring the angles of a triangle defined by three mountain peaks in Germany. Gauss found a locally flat space up to his empirical accuracy. However, the observation of light-bending by Eddington in the 1919 solar eclipse suggests that, at a larger cosmological scale, space is positively curved.

2.1. On the two different principles of equivalence

In contrast, for Einstein light-bending was a strong support for GR [33]. In his correspondence during the Swiss years [34], Einstein mentions three possible methods to confirm GR: (1) examination of existing solar photographs, (2) observation of stars near the sun in full daylight, idea suggested by astronomer Erwin Freundlich-Finlay (EFF) of Berlin Observatory, and (3) observation of star light during the 1914 solar eclipse in an expedition of EFF to Russia that was abruptly ended by the outbreak of World War I [35]. Einstein mentions light-bending in a letter to EFF dated September 1/1911 from Prague (L281 = letter 281 in [34]); correspondence with EFF on this subject continued on September 21/1911 (L287), January 8/1912 (L336), October 27/1912 (L420), mid-August 1913 (L468), August 26/1913 (L472), December 7/1913 (L492), and January 20/1914 (L506). Einstein also corresponded on light-bending with astronomer George Hale on October 16/1913 (L477) and November 8/1913 (L483).

In L448 sent to Ernst Mach in June 25/1913 Einstein stated [34]: "*Next year, during the solar eclipse, we shall learn whether light rays are deflected by the sun, or in other words, whether the underlying fundamental assumption of the equivalence of the acceleration of the reference system, on the one hand, and the gravitational field, on the other hand, is really correct*". This is a clear statement of Einstein's principle of equivalence of acceleration and gravity (EP-A henceforth) mentioned by Einstein five years before in a letter to Arnold Sommerfeld dated January 5/1908 (L72 in [34]). According to Ginzburg [36-37] EP-A was formulated in 1907 [38]. In the opinion of the present writer, EP-A contains the deepest physical and philosophical implications of Einstein's GR, and has everyday manifestations easily seen in the laboratory [39]. Hence, there is no need to invoke light-bending as empirical proof for EP-A.

There is a different, although related, question: is the ratio between inertial mass and gravitational mass a constant, independent of the composition of matter? This is the equivalence principle for mass

IOP Conf. Series: Journal of Physics: Conf. Series **1051** (2018) 012024 doi:10.1088/1742-6596/1051/1/012024

EP-M. For Newton, EP-M was an empirical question that he answered with pendulum experiments described in book III, proposition VI of the Principia [2] (see p 108 in [3]). For Einstein EP-M was also an empirical question, and in July 10/1912 he asked help from Prof. Wilhelm Wien to carry out two "*quite easy*" experiments involving (a) two pendulums, and (b) a torsion balance (L413, p 319-320 in [34]). It seems that Einstein was unaware of the earlier experiment by Eötvös involving torsion balances carried out in Hungary in the 1890s [40], and published around 1910 [41]. A re-analysis of Eötvös data by the Fischbach group in 1986 [42-44] shows that gravitational mass depends on the number and distribution of protons and neutrons in the atomic nucleus. Fischbach fitted the dashed line in Figure 1 to Eötvös observations with their Yukawa-type gravity model, and the correlation was poor. The anomaly remained unexplained [41] until in 2011 present writer separately analyzed the Pt-series and the Cu-series in Eötvös experiments (Figure 1a). Taking into account orbital electrons plus the binding energy of protons and neutrons in the nucleus, an atomic-like pushing gravity model was obtained [45-46], which is consistent with Eötvös data at 92% correlation coefficient (Figure 1b). The only adjustable parameter turned out to be, as expected, the electron/proton mass ratio.

Summarizing, EP-A and EP-M are two different questions; this may clarify some controversies regarding Einstein's equivalence principle [36-37, 47-49]; also note that Einstein suggested different and separate empirical methods to verify EP-A and EP-M. In our view, EP-A is locally valid, but EP-M is not. We also depart from Einstein's interpretation of star light-bending; instead, it is curvature of absolute 3D-space in the local region from star to earth.

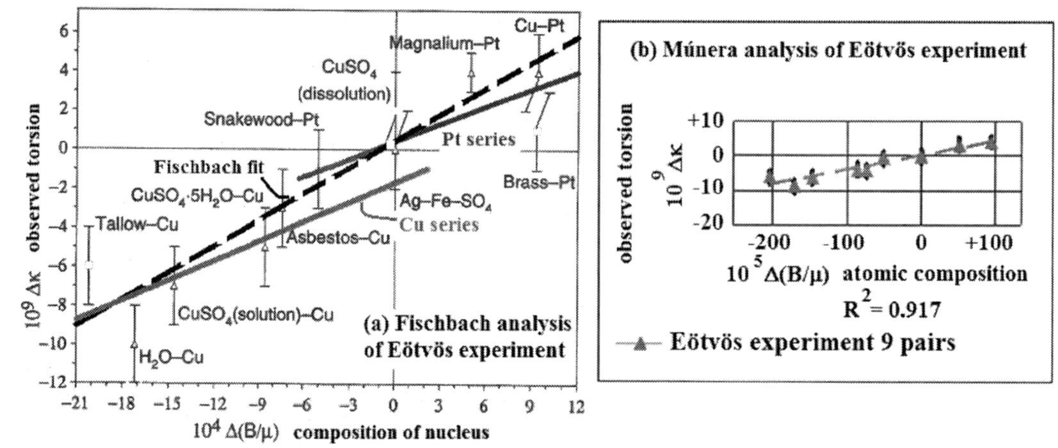

Figure 1. Relative torsion in Eötvös experiment depends on composition: (a) nuclear composition effect analysed by Fischbach [42-44], (b) atomic composition effect analysed by Múnera [45-46].

2.2. On the evidence against absolute space

Einstein rightly rejected the far-fetched properties of electromagnetic ether models in the second half of the 19th century, but he erred in his 1905 general dismissal of ether, including Newton's [2] (possibly immaterial) fluid-like ether [4]. In our view, when Einstein formulated EP-A in 1907 [38] he immediately sensed that he was a prisoner of his 1905 dismissal. Indeed, in L73 to Sommerfeld dated January 14/1908 Einstein wrote from Bern [34]: "*If the Michelson-Morley experiment had not put us in the worst predicament, no one would have perceived the relativity theory as a (half) salvation*". In L448 to Mach dated June 1913 Einstein considered the physics inside a shell rotating "*relative to the fixed stars ('Restsystem')*", which to an unaware reader sounds very similar to Newton's ideas.

According to Kostro [50], Einstein reintroduced ether around 1916, but it seems that the seeds for such reinstatement were planted around 1907. At any rate, Einstein explicitly said at Leyden in 1920: "*Newton might no less well have called his absolute space 'Ether'; what is essential is merely that besides observable objects, another thing, which is not perceptible, must be looked upon as real, to enable acceleration or rotation to be looked upon as something real*" [51]. Surprisingly, Einstein did not distinguish between the container (absolute space) and its contents (ether).

In the mid-1990s present author uncovered several weaknesses in Michelson-Morley experiment (MMX) [52]. Most significant was that only the fractional part of the fringe-shift was recorded, leading to under-estimation of fringe-shift amplitude [53], which should be of several fringes as calculated with modern values of solar velocity [54]. This prompted us to repeat the MM experiment at CIF in Bogota (Colombia) in a continuous manner during two years (2003-2005). As usual in second-order experiments, two velocities of sun relative to a preferred frame were obtained from our data: (a) CIF-S: VS = 500 km/s, R.A. = 16h-40m, Dec = -75° (southern celestial hemisphere) [55], and (b) CIF-N: VS = 365 km/s, R.A. = 5h-24m, Dec = 79° (northern celestial hemisphere) [56]. Our solar velocities are compatible with other observations supporting absolute motion [57].

Of significant interest is the compatibility of our results with two contemporary experiments: (1) de Haan [58] experiments involving standing waves and a Fabry-Perot interferometer, and (2) Lipa vertical interferometer at Stanford in 2002 [59], temperature controlled to a micro-Kelvin accuracy, that was interpreted as a "null" MMX. Applying procedure described in [54], absolute velocity of surface of earth at Stanford University (Palo Alto, California) was calculated for dates of Lipa's experiment using our CIF-S value for solar velocity. Figure 2 shows projection of our calculated absolute velocity upon Lipa's laboratory floor for May 30/2002. The observed frequency variations recorded by Lipa on same date [59] exhibit correlation coefficients at 99.8% with absolute speed, and 99.1% with absolute direction of motion, which imply that EM frequency may be affected by absolute motion, contrary to Lipa's interpretation. The paradox is easy to explain: Lipa attributed observed frequency variations to unexplained "*mechanical disturbances*", subtracted them, and obtained a noise of zero amplitude. Hence, they got the reported "null" result!

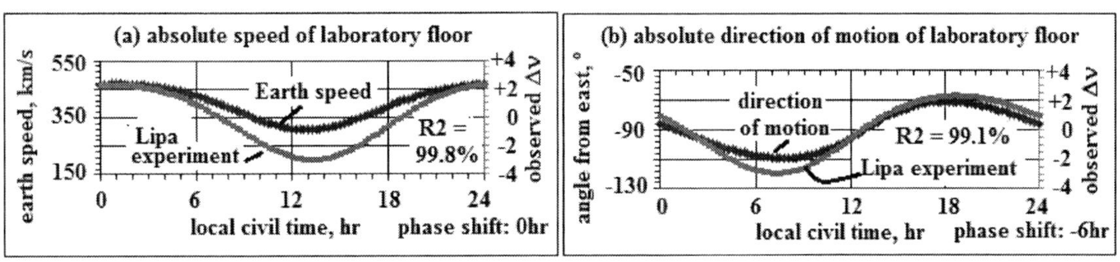

Figure 2. Projection of absolute velocity on the floor of the laboratory at Stanford University in May 30/2002. Observed frequency variations on same day [59] are highly correlated with absolute motion.

3. New solutions for the homogeneous Klein-Gordon equation or classical wave equation

Empirical evidence sketched in previous section implies absence of evidence against absolute space; in Einsteinian mood, the *Michelson-Morley experiment no longer puts us in any predicament.* It is postulated that a primordial fluid entirely pervades the curved 3D-absolute space Σ, and that it obeys conservation of total energy, and linear and angular momentum. In regions of universe devoid of matter, the fluid is locally described by the homogeneous Klein-Gordon equation (1), which is the same classical wave equation; the physical potential field $\Pi \equiv (\phi, \mathrm{A})$ contains a scalar potential ϕ (conservation of total energy) and a vector potential A (conservation of linear momentum), whose dimensions $ML^{-1}T^{-2}$ correspond to density of total energy, and to flow of linear momentum per unit area (or, equivalently, density of angular momentum flow) respectively. Equation (1) represents De Broglie's pilot wave [21-23], and also complies with classical electrodynamics [60]:

$$\Box \Pi = \left(\frac{\partial^2}{\partial^2 w} - \nabla^2 \right) \Pi = 0 \text{, where } w \equiv C^* t \text{, } \Pi \equiv (\phi, \mathrm{A}). \quad (1)$$

Time is treated as length w with C^* being the local average speed of primordial fluid. In presence of matter, right hand side is non-zero due to elastic and inelastic sagion-matter interactions; a complete

IOP Conf. Series: Journal of Physics: Conf. Series **1051** (2018) 012024 doi:10.1088/1742-6596/1051/1/012024

(3+1+1) formulation requires Dirac's equations containing spin, characterized by a positive/negative absolute sense of rotation: counterclockwise/clockwise. In isotropic absolute space all directions (θ,φ) are equivalent, and linear motion becomes one-dimensional along arbitrary ray (r,t), or (r,w). Thus, we use spherical coordinates (r,w,θ,φ) to solve equation (1). The standard separation of variables in equation (2) leads to harmonic solutions, with $F(r,w)$ being the one-dimensional travelling wave equation found by D'Alembert in the 18th century:

$$\Pi(r,w,\theta,\varphi) = F(r,w)D(\theta,\varphi),\ F(r,w) = R(r)T(w),\ D(\theta,\varphi) = Y(\theta,\varphi;\ell,m), \quad \text{where } \ell > m,\ m^2 \geq 0. \tag{2}$$

Directional component $D(\theta,\varphi)$ is the same for the linear non-relativistic Schrödinger equation and for the relativistic Klein-Gordon equation (1). Thus, the Lorentz invariance of the KGE is fully contained in the one-dimensional motion $F(r,w)$ along arbitrary ray r, which is equivalent to Einstein's special relativity motion along arbitrary X-axis (axes Y and Z being ignored in practice). In QM, directional solution $D(\theta,\varphi)$ is restricted to $m^2 \geq 0$, and reduces to usual spherical harmonics $Y(.,.)$. In the mid-1990s we studied the (usually neglected) inherently quantized helicoidal functions $H(\theta,\varphi)$ that appear for $m^2 < 0$ [12, 61-64]. There are two families of helices spiraling inward/outward on the surface of quantized cones shown in Figure 3. Inward growth has been observed in circular micro-pyramids of $YBa_2Cu_3O_7$ at the 125 nanometer scale [65], and silicon carbide crystals at the 100 micrometer scale [66]; at a larger scale, swirls, tornados and hurricanes are everyday examples. On the contrary, snail- and sea-shells display outward spiral growth. Whether spiral galaxies grow inwards or outwards is left open.

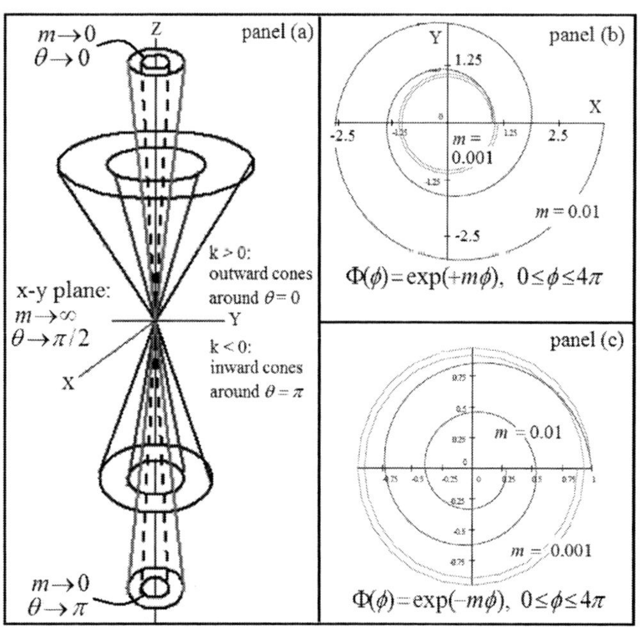

Figure 3. Helicoidal solutions for equation (1) with $m^2 < 0$ for counter-clockwise upward motion.
Panel (a): Two families of cones around Z-axis; the vertex angle is inherently quantized.
Panel (b): Outward spirals grow from $x = 1$ outwards.
Panel (c): Inward spirals grow from $x = 1$ inwards towards the Z-axis.

Searching for longitudinal non-harmonic solutions for $F(r,w)$ in electromagnetic contexts, we re-wrote the one-dimensional travelling wave as $M(r,q)$, where the novel independent variable q intermingles spatial distance (r) and time (w) in an intimate un-separable manner. But, in contrast to

SR, variable q is a mere mathematical manipulation of physical magnitudes without ontological implications regarding space and time, which remain two separate notions:

$$\Pi(r,w,\theta,\phi) = M(r,q)D(\theta,\phi) = \big(N(r)+Q(q)\big)D(\theta,\phi) = B(r,\theta,\phi) + U(q,\theta,\phi),\quad q \equiv w/r. \tag{3}$$

Separation of functions $N(r)$ and $Q(q)$ in the differential equation for $M(r,q)$ generates a new quantum number η that appears in the generic solution for $Q(q)$ in terms of Q-functions of the first, second and third kind, related to associate Legendre functions. Q-functions were first described in [61], further details appear in [12, 62-63]:

$$B(r,\theta,\varphi) = N(r)D(\theta,\varphi),\quad U(q,\theta,\varphi) = Q(q)D(\theta,\varphi),\ Q(q) = A_1Q_1(q;\ell) + A_2Q_2(q;\ell) + \eta Q_3(q;\ell). \tag{4}$$

As usual in the theory of time-dependent differential equations, the background field $B(r,\theta,\varphi)$ in equation (3) is a particular solution of equation (1). This time-independent steady-state background may provide an alternative interpretation to the origin of cosmic microwave background radiation, and yield a causal explanation for action-at-a-distance and non-locality features in QM and EM theory, along the lines discussed in [67]. New universal field $U(q,\theta,\varphi)$ depends on only three explicit dimensionless ratios of length, viewed as cosines of time (q) and spatial direction (θ,φ). Then, velocity dependent function $f(\beta)$ cancels in numerator and denominator of any transformation of moving coordinates leading to the remarkable property of isomorphism, outlined in equations (5):

$$s \to s' = f(\beta)(s-\beta w),\quad w \to w' = f(\beta)(w-\beta r),\ s = x,y,z,r,\ \text{where } \beta = V/C.$$

$$\text{Isomorph: } \left(\frac{w}{r},\frac{s}{r}\right) = (q,\cos\theta) \to (q',\cos\theta') = \left(\frac{w-\beta r}{r-\beta w},\frac{s-\beta w}{r-\beta w}\right),\ \text{where } \cos\theta = s/r, \tag{5}$$

$$\text{Doppler: } f(\beta)=1,\ \text{Lorentz: } f(\beta)=\gamma,\ \text{Pointcaré: } f(\beta)=\gamma l(\beta),\ \text{where } \gamma = \left(1-\beta^2\right)^{-1/2}.$$

The practical implication is that four competing transformations are equivalent: (1) Doppler formed by the classical Galilei transformation for distance *r*, and the local time defined by Voigt in 1887 [68], and also mentioned by Poincaré in 1900 [69], (2) Poincaré [70-71], (3) Lorentz [72], (4) Einstein [72]. There is a four-fold repetitive pattern in $Q_1(q,l)$ functions (see Figure 4) according to $l = 4n-1, 4n, 4n+1, 4n+2$, where *n* is the number of stable spherical surfaces of dynamic equilibrium [12]. The latter easily explain atomic structure [73], and Titius-Bode rule in both solar and exoplanetary systems with coefficients of correlation usually above 99% [74-75]. In far-field $(r \to \infty)$ Q_1 is monotonous and tends asymptotically to nonzero for odd-l.

The attractive non-zero far field of the $l = 4n-1$ group easily explains the flat rotation rate of galaxies, without ad hoc dark matter, while the repulsive non-zero far field of the $l = 4n+1$ group is similar to nuclear and sub-nuclear confining forces. The far field is zero for even-l, for $l = 4n$ it is attractive as in Newton and Coulomb forces, and for $l = 4n+2$ it is repulsive as in Coulomb force between equal sign charges. Q_1-functions are very similar to Boscovich unified force [76], which is inherently quantized in near field $(r \to 0)$ as Thomson [77] noted before inception of QM. Spherical surfaces of dynamic equilibrium lead to a simple method to solve the three-body problem, and to construct Trojans orbits [74].

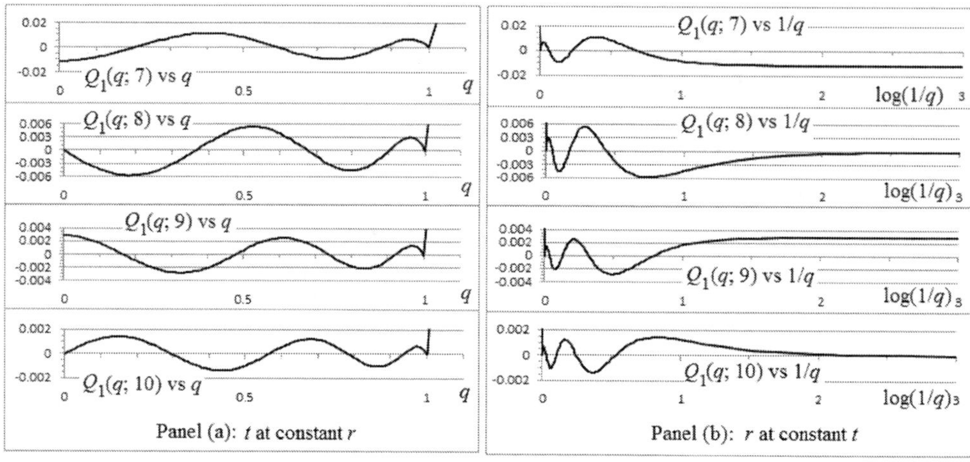

Figure 4. Function Q_1 for $n = 2$. Note stable intercepts along horizontal axis for the gravitational potential $\ell = 4n = 8$ (graphs plotted by C E Cedeño).

References

[1] 't Hooft G 2002 Determinism beneath quantum mechanics *Preprint* arXiv:0212095v1
[2] Newton I 1687 *Philosophiae Naturalis Principia Mathematica* (Madrid: Editorial Tecnos, 1987)
[3] Thayer H S 1953 *Newton's Philosophy of Nature—Selections from his Writings* (New York: Hafner Publishing) pp 183 198
[4] Múnera H A 2015 From the classical ethers of Descartes and Newton to cosmons and sagions *Apeiron* **20** 2 pp 1-67
[5] Lindsay R B 1961 *Physical Mechanics* (Princeton NJ USA: Van Nostrand)
[6] Desloge E A 1989 Empirical foundation of classical dynamics *Am. J. Phys.* **57** pp 704-706
[7] De Saint Venant B 1845 Memoire sur les sommes et les différences géometriques et sur leur usage pour simplifier la mécanique *Comptes rendus Académie des Sciences (Paris)* **21** 620
[8] Jammer M 1961 *Concepts of Mass in Classical and Modern Physics* (Harvard: Harvard University Press) (Republished in Mineola NY USA: Dover 1997) pp 90, 92
[9] Eisberg R M and Lerner L M 1981 *Physics: Foundations and Applications* (USA: McGraw-Hill) ch 4
[10] Mach E 1883 *The Science of Mechanics* translated from German ed 1908, 1912, 1921, 1933 by T J McCormack (La Salle Ill USA: Open Court) p 303
[11] Russell B 1902 *The Principles of Mathematics* 2nd (New York: W.W. Norton 1938)
[12] Múnera H A 2016 *Interconnection of all forces of nature via the energy and momentum equations for a fluid aether* ed R L Amoroso *et al* pp 247-267
[13] Amoroso R L, Kauffman L H and Rowlands P 2016 *Unified Field Mechanics: Natural Science Beyond the Veil of Spacetime* (Singapore: World Scientific)
[14] Múnera H A 2002 Bell's-type inequalities revisited: new constraints from objective reality *Momento* **25** pp 3-27 (*Preprint* http://www.revistas.unal.edu.co/index.php/momento/article/view/44239)
[15] Laplace P 1814 *A Philosophical Essay on Probabilities* translated from French by F W Truscott and F L Emory (New York: Dover 1995)
[16] Múnera H A 1992 A deterministic event tree approach to uncertainty, randomness and probability in individual chance processes *Theory and Decision* **32** pp 21-55
[17] De Broglie L 1952 Sur la possibilité d'une interprétation causale et objective de la mécanique ondulatoire *Comptes rendus Académie des Sciences* (Paris) **234** pp 265-268
[18] Bohm D and Vigier J P 1954 Model of the causal interpretation of quantum theory in terms of a fluid with irregular fluctuations *Physical Review* **96** 1 pp 208-217

[19] Madelung E 1926 Eine anshauliche Deutung der Gleichung von Schrödinger *Naturwissenschaften* **14** 45 1004
[20] Madelung E 1927 Quantentheorie in hydrodynamischer Form *Zeits. Phys.* **40** 322-6
[21] De Broglie L 1927 L'univers a cinq dimensions et la mécanique ondulatoire *Le Journal de Physique et le Radium* series VI **8** 2 pp 65-73 and 242-4
[22] De Broglie L 1962 Remarques sur l'interprétation de la dualité des ondes et des corpuscles *Cahiers de Physique* **16** 147 pp 425-45
[23] De Broglie L 1970 The reinterpretation of wave mechanics *Foundations of Physics* **1** 1 pp 5-15.
[24] Gueret P and Vigier J P 1982 Non-linear Klein-Gordon equation carrying a nondispersive soliton-like singularity *Lettere al Nuovo Cimento* **35** 8 pp 256-9
[25] Chebotarev L V 2000 The de Broglie-Bohm-Vigier approach to quantum mechanics *Jean-Pierre Vigier and the Stochastic Interpretation of Quantum Mechanics* ed S Jeffers *et al* (Montreal: Apeiron) pp 1-17
[26] Cufaro-Petroni N and Vigier J P 1983 Dirac's aether in relativistic quantum mechanics *Foundations of Physics* **13** 2 pp 253-285.
[27] Múnera H A 2000 A realistic four-dimensional hydrodynamic aether interpreted as a unified field equation *Lorentz Group, CPT and Neutrinos* ed A E Chubykalo *et al* (Singapore: World Scientific) pp 425-33
[28] Hankins T L 1965 18th-century attempts to resolve the vis viva controversy *Isis* **56** 281-97
[29] Grant E 1978 The principle of the impenetrability of bodies in the history of concepts of separate space from the Middle Ages to the seventeenth century *Isis* **69** 249 551-71
[30] Grmek M D 1996 La méthodologie de Boscovich *Revue d'Histoire de Science* **49** 4 pp 379-400
[31] Patrizi F 1593 *Nova de Universis Philosophia* Translated by Brickman (Venice, Italy)
[32] Brickman B 1943 On physical space: Francesco Patrizi *Journal for History of Ideas* **4** 224-45
[33] Einstein A 1911 Ueber der Einfluss der Schwerkraft auf die Ausbreitung des Lichtes *Annalen der Physik* **35** pp 898-908
[34] Beck A and Howard D 1995 *The Collected Papers of Albert Einstein Volume 5 The Swiss Years: Correspondence, 1902-1914* (Princeton NJ USA: Princeton University Press)
[35] Speziali P 1994 *Albert Einstein Correspondencia con Michele Besso* translated into Spanish by M Puigcerver from Correspondence (1903-1955) (Paris: Hermann 1979) (Barcelona, Spain: Tusquets Editores) p 106
[36] Ginzburg V L and Eroshenko Y N 1996 Commens on paper below *Physics-Uspekhi* **39** 1 81-2
[37] Logunov A A, Mestvirishvili M A and Chugreev Y V On incorrect formulations of the equivalence principle *Physics-Uspekhi* **39** 1 73-9
[38] Einstein A 1907 *Jahrbuch der Radioaktivität und Elektronik* **4** 411
[39] Múnera H A 2010 The Principle of Equivalence: Demonstrations of Local Effective Vertical and Horizontal *The Physics Teacher* **48** 131-3
[40] Von Eötvös R, Pekár D and Fekete E 1922 Beiträge zum Gesetze Proportionalität von Trägheit und Gravität *Annalen der Physik* series 4 **68** 9 pp 12-66.
[41] Nieto M M, Hughes R J and Goldman T 1989 Actually, Eötvös did publish his results in 1910, it's just that no one knows about it.... *Am. J. Phys.* **57** pp 397-404
[42] Fischbach E, Sudarsky D, Szafer A and Talmadge C 1986 Reanalysis of the Eötvös experiment. *Phys. Rev. Lett.* **56** pp 3-6
[43] Fischbach E, Sudarsky D, Szafer A, Talmadge C and Aronson S H 1988 Long-range forces and the Eötvös experiment *Annals of Physics* **182** pp 1-89
[44] Fischbach E and Talmadge C L 1999 *The Search for Non-Newtonian Gravity* (New York: Springer Verlag)
[45] Múnera H A 2011 A Le Sagian atomic-type model for propagation and generation of gravity *Should the Laws of Gravitation be Reconsidered?–The Scientific Legacy of Maurice Allais* ed H A Múnera (Montreal: Apeiron) pp 385-422
[46] Múnera H A 2013 The empirical basis for the equivalence principle: the Eötvös, Pekár and

Fekete experiment revisited - Once again *Proc. of Natural Philosophy Alliance* **10** pp 204-212

[47] Ohanian H C 1977 What is the principle of equivalence? *Am. J. Phys.* **45** 10 903-9

[48] Walstad A 1979 The equivalence principle *Am. J. Phys.* **47** 6 565-6

[49] Ohanian H C 1979 Reply to Professor Walstad *Am. J. Phys.* **47** 11 p 1006

[50] Kostro L 2000 *Einstein and the Ether* (Montreal: Apeiron)

[51] Einstein A 1983 *Sidelights on Relativity* (New York: Dover) p 17

[52] Múnera H A 1998 Michelson-Morley experiments revisited: systematic errors, consistency among different experiments, and compatibility with absolute space *Apeiron* **5** (1-2) 371-6

[53] Múnera H A 2006 The evidence for length contraction at the turn of the 20th century: Non-existent *Einstein and Poincaré: The Physical Vacuum* ed V V Dvoeglazov (Montreal: Apeiron Press) pp 87-102

[54] Múnera H A 2002 The effect of solar motion upon the fringe-shifts in a Michelson-Morley interferometer à la Miller *Annales de la Fondation Louis de Broglie* **27** 3 463-84

[55] Múnera H A, Hernández-Deckers D, Arenas G and Alfonso E 2007 Observation of a significant influence of earth's motion on the velocity of photons in our terrestrial laboratory *Proceedings of SPIE* **6664** 66640K.

[56] Múnera H A, Hernández-Deckers D, Arenas G, Alfonso E and López I 2009 Observation of a non-conventional influence of earth's motion on the velocity of photons, and calculation of the velocity of our galaxy *Proc. of PIERS 2009 Progress in Electromagnetics Research Symp.* Beijing (23-27 March 2009)

[57] Múnera H A 2009 Towards the reinstatement of absolute space, and possible cosmological implications *ICFAI University Journal of Physics* **2** 2-3 pp 9-24

[58] De Haan V O 2015 Experiments to test special relativity *Proc. of International Scientific Meeting PIRT-2015* Moscow (29 June-02 July) 131-9

[59] Lipa J A, Nissen J A, Wang S, Stricker D A and Avaloff D 2003 A new limit on signals of Lorentz violation in electrodynamics *Physical Review Letters* **90** 060403

[60] Malet H 1926 Sur la synthèse mathématique des lois de l'électrodynamique *Comptes rendus Académie des Sciences* (Paris) **183** pp 191-193

[61] Múnera H A, Buriticá D, Guzmán O and Vallejo J I 1995 Non-conventional solutions for the travelling wave equation (in Spanish) *Revista Colombiana de Física* **27** 215-8

[62] Múnera H A and Guzmán O 1997 New explicit nonperiodic solutions of the homogeneous wave equation *Foundations of Physics Letters* **10** 1 pp 31-41

[63] Múnera H A 2000 New closed solutions in spherical coordinates for the 3D-homogeneous wave equation (in Spanish) *Momento* **20** 1-30 (*Preprint* http://www.revistas.unal.edu.co/index.php/momento)

[64] Múnera H A 2017 Neo-Cartesian Unified Fluid Theory: From the Classical Wave Equation to De Broglie's Lorentzian Quantized Mechanics and Quantized Gravity *Proc. of Vigier X Symp.* ed R L Amoroso (Singapore: World Scientific) in the press

[65] Stäuble-Pümpin B, Mendoza G A, Guzmán O, Clavijo J, Prieto P. and Dam B 2001 Two-component model to describe the growth of physical-vapour-deposited YBa2Cu3O7 films *Physica* C **356** 161

[66] McGraw-Hill 1977 *Encyclopedia of Science and Technology* (New York: McGraw-Hill) **3** 636

[67] Chubykalo A, Espinoza A and Alvarado-Flores R 2016 Helmholtz theorems, gauge transformations, general covariance and the empirical meaning of gauge conditions *Journal of Modern Physics* **7** 1021-44 (*Preprint* http://dx.doi.org/10.4236/jmp.2016.79092)

[68] Voigt W 1887 On the Doppler's principle (in German) *Nachrichten von der Königlichen Gesellschaft der Wissenschaften und der Georg-Augusts Universität zu Göttingen* **2** pp 41-51

[69] Poincaré H 1900 Thèorie de Lorentz et principe de réaction *Archives Neerlandaises* **5** 252-78

[70] Poincaré H 1905 Sur la dynamique de l'électron *Compt. Rend. Acad. Sci.*(Paris) **140** 1504-8

[71] Poincaré H 1906 Sur la dynamique de l'électron *Rend. Circolo Matematico Palermo* **21** 129-76

[72] Pustovoit V I 2013 On the problems of experimental verification of general relativity *Proc.Int. Scientific Meeting PIRT-2013* pp 462-68
[73] Múnera H A 2016 Unified field theory from classical wave equation: application to atomic and nuclear structure *AIP Conf. Proc.* **1753** 030015
[74] Múnera H A 2016 Novel approach to gravitation from fluid theory: Titius-Bode structures, flat rotation rate of galaxies, and other predictions *Astronomy in Focus-XXIXB* ed P Benvenuti (Cambridge, UK: Cambridge University Press)
[75] Múnera H A 2017 Gravity from classical fluid theory: prediction and retrodiction of quantized planetary structures *The Galileo of Palomar: Essays in memory of Halton Arp* ed C C Fulton and M Kokus (Montreal: Apeiron)
[76] Boscovich R J 1758 *Theory of Natural Philosophy* tr J M Child (Chicago: Open Court, 1922)
[77] Thomson J J 1907 *Corpuscular theory of matter* (New York: Charles Scribner) p160

Connection between functionals of the field-theory equations and state functionals of the mathematical physics equations

L Petrova

Moscow State University, Moscow, Russia
E-mail: ptr@cs.msu.su

Abstract. It is well known that the field-theory equations are equations for such functionals as the action functional, entropy, Pointing's vector, Einstein's tensor, wave function, and others. It turns out that the equations of mathematical physics describing material systems (material media) such as the thermodynamic, gas-dynamic and cosmic systems as well as the systems of charged particles and others can be expressed in terms of the same functionals. Such a duality of functionals points out to a connection between the field-theory equations, which describe physical fields, and the mathematical physics equation, which describe material media. This discloses a connection of physical fields and material media.

1. Introduction

As it is known, the field-theory equations are equations for such functional as the action functional, entropy, Pointing's vector, Einstein's tensor, wave function, and others.

It has been shown that a relation for such functionals can be obtained from the mathematical physics equations, which describe material media.

Investigation of the relation for functionals, which is evolutionary and nonidentical, enables one to disclose specific properties and physical meaning of functionals.

This investigation showed that there exists a correspondence between evolutionary relation for functionals obtained from the mathematical physics equations and the field-theory equations. This fact points out to a connection between the field-theory equations and the mathematical physics equations.

The present investigation bases on the properties of conservation laws, which lie at the basis of the field-theory equations and the mathematical physics equations. A peculiarity consists in the fact that physical fields described by the field-theory equations and material media described by the mathematical physics equations are subject to different conservation laws. The conservation laws for physical fields mean a statement on an existence of conservative quantities or objects. Such conservation laws (which can be referred to as exact ones) are described by closed skew-symmetric forms. Whereas the conservation laws of material media are conservation laws for energy, momentum, angular momentum, and mass that are described by differential equations. It turns out that there exists a connection between conservation laws for physical fields and material media, and these points out to a connection of the field-theory equations with the mathematical physics equations.

It appears to be possible to obtain the results of present paper due to using the mathematical apparatus of skew-symmetric forms, which can describe conservation laws [1-2]. The peculiarity of such investigation consisted in the fact that, in addition to exterior forms, skew-symmetric forms, which basis, as opposed to the basis of exterior forms, is nonintegrable manifolds, were used. Such skew-symmetric forms which are obtained from differential equations and are evolutionary ones [1], possess a unique property, namely, they can generate closed exterior forms that are invariants and corresponds to conservation laws for physical fields (exact conservation laws). Such properties of skew-symmetric forms defined on nonintegrable manifolds (non of existing mathematical formalism possesses these properties) enable to describe an advent of various structures, quantum transitions and so on.

2. Peculiarities of the mathematical physics equations for material media

The equations of mathematical physics, which describe material media, consist of the equations of conservation laws for energy, linear momentum, angular momentum, and mass [3-5].
Commonly the equations of mathematical physics are applied for a description of a change of physical quantities (such as energy, pressure, density) that characterize material media.
However, it turns out that such equations possess properties that also give a possibility to describe the processes of emergence of various structures. The physical structures from which physical fields are formatted may be such structures. This discloses a connection between the field-theory equations and the equations of mathematical physics. Below it will be shown that such a connection is caused by the properties of corresponding functionals.
However, such properties of the mathematical physics equations and potentialities caused by that appear to be hidden ones. They do not directly follow from differential equations and reveal only in describing evolutionary processes. This result follows from the properties of conservation laws and reveals when the consistency of the conservation law equations, from which the mathematical physics equations are made up, are accounted for.

2.1. Investigation of the consistency of the conservation law equations. Evolutionary relation for the state functionals

For investigation of the consistency of the conservation law equations it is necessary to use two nonequivalent frames of reference.
Usually the equations of mathematical physics are written in the inertial frame of reference (the Euler frame of reference is an example of such frame).
In the case under consideration, in addition to the equations in inertial frame of reference, the equations obtained after transition from inertial frame of reference to accompanying one were used. The accompanying frame of reference is a frame of reference connected with the manifold made up by the trajectories of the material media elements (the Lagrange frame of reference is an example of such a frame).
A specific feature of present investigation is the point that the conservation law equations are transformed into equations expressed in the terms of state functionals. Such a possibility relates to the properties of desired physical quantities of material media. Since the physical quantities (like temperature, energy, pressure, density) relates to a single material medium, a connection between them should exist. Such a connection is described by state functionals. The functionals such as wave function, entropy, the action functional, the Pointing vector, the Einstein tensor and so on, which are the field-theory functionals, are also functionals of the equations describing material media [6]. (As it will be shown below, such functionals describe a material medium state, that is, they are state functionals).
Let us analyze the correlation of the equations that describe the conservation laws for energy and linear momentum.
In the inertial frame of reference the energy equation can be reduced to the following form:

$$\frac{D\psi}{Dt} = A_1, \tag{1}$$

where D/Dt is the total derivative with respect to time, A_1 is a quantity that depends on specific features of material system (material medium) and on external energy actions onto the system, $d\psi$ is the functional that specifies a material system (and which is a state functional). The action functional, entropy, wave function, Einstein's tensor, Pointing's vector and others can be regarded as examples of the functional ψ.

In the accompanying frame of reference, which is tied to the manifold made up by the trajectories of elements (particles) of material system the total derivative with respect to time is a derivative along the trajectory. For this reason, in the accompanying frame of reference the equation of energy is written in the form

$$\frac{\partial \psi}{\partial \xi^1} = A_1. \tag{2}$$

Here ξ^1 are the coordinates along the trajectory.

Thus, the equation for energy expressed in terms of the action functional S has a similar form: $\partial S / \partial \xi^1 = DS / Dt = L$. And the equation for the energy of ideal gas can be presented in the form [3]: $\partial s / \partial \xi^1 = Ds / Dt = 0$, where s is the entropy.

In a similar way, in the accompanying frame of reference the equation for linear momentum appears to be reduced to the equation of the form

$$\frac{\partial \psi}{\partial \xi^\nu} = A_\nu, \quad \nu = 2,... \tag{3}$$

where ξ^ν are the coordinates in the direction normal to the trajectory, A_ν are the quantities that depend on the specific features of material system and on force actions.

Here it should be called attention to a certain peculiarity, namely, for a single quantity (the functional ψ) on has two equations.

How to study such redefined set of equations?

Since equations (2) and (3) are expressions for derivatives along different directions, they can be convoluted into the relation

$$d\psi = A_\mu d\xi^\mu, \quad \mu = 1, \nu. \tag{4}$$

Relation (4) can be rewritten as

$$d\psi = \omega, \tag{5}$$

here $\omega = A_\mu d\xi^\mu$ is a skew-symmetric differential form of the first degree. (A summing over repeated indices is carried out.)

In the general case (for energy, linear momentum, angular momentum and mass) this relation will be the written as

$$d\psi = \omega^p, \tag{6}$$

where ω^p is the form degree p (p takes the values $p = 0,1,2,3$). (The relation for $p = 0$ is an analog to that in the differential forms, and it has been obtained from the interaction of energy and time.)

Since the conservation law equations are evolutionary ones, the relations obtained are also evolutionary relations, and the skew-symmetric form ω^p is evolutionary ones.

In this case $p = 2$ the functional is Pointing's vector [2]. The relation for Einstein's tensor is obtained when integrating the evolutionary relation for $p = 3$.

The evolutionary relation obtained from the equations of the conservation law possesses the properties which describe the connection of the field-theory equations with the equations of mathematical physics.

2.2. Specific properties of evolutionary relation

The evolutionary relation possesses a peculiarity, namely, it appears to be nonidentical. The evolutionary relation was obtained in the accompanying frame of reference, which is connected with the manifold built up by the trajectories of the material system elements. Such a manifold is a deforming nonintegrable one. The skew-symmetric form defined on nonintegrable manifold cannot be closed since the commutator of skew-symmetric form defined on such manifold includes an additional term, namely, the commutator of metric form, which is nonzero (because the metric form of nonintegrable manifold is not closed one). Since the evolutionary form is unclosed and is not a differential, the evolutionary relation (6) turns out to be nonidentical.

The evolutionary relation possesses one more peculiarity, namely, this relation is a selfvarying relation.

The evolutionary nonidentical relation is a selfvarying one, because, firstly, it is a nonidentical, namely, it contains two objects one of which appears to be unmeasurable, and, secondly, it is an evolutionary relation, that is, the variation of any object of the relation in some process leads to a variation of another object; and, in turn, the variation of the latter leads to variation of the former. Since one of the objects is an unmeasurable quantity, the other cannot be compared with the first one, and hence, the process of mutual variation cannot terminate.

The evolutionary relation obtained from the equations of mathematical physics discloses the properties of the solutions of mathematical physics equations and physical properties of these solutions that enable one to describe the processes of emergence of physical structures. And this is connected with the properties of the evolutionary relation functionals.

2.3. Properties of solutions to the mathematical physics equations

From evolutionary relation it follows that the mathematical physics equations has a double solutions, namely, the solutions that are not functions (their derivatives do not made up a differential) and the solutions that are discrete functions. These are precisely the properties of the mathematical physics equations that enable one to describe evolutionary processes and the processes of emergence of various structures.

2.4. Inexact solutions to mathematical physics equations (which are not the functions)

The nonidentical evolutionary relation $d\psi = \omega^p$ cannot be integrated directly since its right-hand side contains unclosed skew-symmetric form which is not a differential. This means that the original equations of mathematical physics prove to be nonintegrable (they cannot be convoluted into identical relation for differentials and be integrated). In this case solutions to the initial equations of mathematical physics are not functions (their derivatives do not made up a differential). These solutions will depend on the commutator of the evolutionary skew-symmetric form ω^p, which is nonzero. *If the commutator be equal to zero, the evolutionary relation would be identical and this would point out to integrability of original equation.*

However, it appears that the identical relation can be obtained from nonidentical evolutionary relation. This relation can be integrated, and this fact will point out to a realization of the solutions that are functions.

2.5. Exact solutions to equations of mathematical physics (discrete functions)

From the evolutionary nonidentical relation one can obtain an identical relation composed of differentials that can be integrated. This will point out to a realization of integrability of the initial mathematical physics equation. This can happen if a closed exterior form will be obtained from the evolutionary skew-symmetric form in the right-hand side of evolutionary relation. However, since the evolutionary form differential is nonzero (because the evolutionary form is unclosed), whereas the differential of closed exterior form equals to zero, the transition from evolutionary form to closed exterior form is possible only under *degenerate transformation, namely, under a transformation that does not preserve a differential.*

Such degenerate transformations can take place under additional conditions, which are due degrees of freedom. The vanishing of such functional expressions as determinants, Jacobians, Poisson's brackets, residues, and others corresponds to these additional conditions. These conditions can be realized (spontaneously) under a change of nonidentical evolutionary relation, which, as it was noted, appears to be a selfvarying relation. Under degenerate transformation from the unclosed evolutionary form ω^p

(see evolutionary relation (6)) with non-vanishing differential $d\omega^p \neq 0$, one can obtain a closed inexact (only on some pseudostructure) exterior form with vanishing (interior) differential.
That is, it is realized the transition

$$d\omega^p \neq 0 \rightarrow \text{(degenerate transformation)} \rightarrow d_\pi \omega^p = 0,\ \ d_\pi {}^*\omega^p = 0.$$

The realization of the conditions $d_\pi {}^*\omega^p = 0$ and $d_\pi \omega^p = 0$, means that it is realized the closed dual form ${}^*\omega^p$, which describes some structure π (this is a pseudostructure with respect to its metric properties), and the closed exterior (inexact) form ω^p_π, which basis is a pseudostructure, is obtained.
The realization of closed inexact exterior form ω^p_π leads to the fact that on a pseudostructure from evolutionary relation (6) it follows the relation

$$d\psi_\pi = \omega^p_\pi, \tag{7}$$

that occurs to be an identical one, since the form ω^p_π is a differential.
Since the identical relation can be integrated (because it contains only of differentials), this means that on the pseudostructure the equations of mathematical physics become locally integrable (only on pseudostructure). In this case the pseudostructure is an integrable structure. The solutions to the mathematical physics equations on integrable structures are generalized solutions, which are discrete functions, since they are realized only under additional conditions (on the integrable structures).
Due to the fact that generalized solutions are defined only on integral structures, the derivatives of generalized solutions have a break in the direction normal to integral structure.

3. Description of evolutionary processes in material media and the processes of physical structure emergence

It turns out that inexact solution describes nonequilibrium state of material media, whereas exact solution describes locally-equilibrium state. This follows from the evolutionary relation and is connected with the property of the functional ψ.

3.1. Nonequilibrium state of material media

The functional ψ in the left-hand side of the evolutionary relation $d\psi = \omega^p$ possesses a unique property, namely, this functional specifies a state of material media.
The differential of functional availability means that there exists a state function, and this fact points out to equilibrium state of material medium. However, in this case there is a certain delicate point related to the nonidentity of evolutionary relation. Since the evolutionary relation turns out to be not identical, from this relation one cannot get the differential $d\psi$ that could point out to the equilibrium state of material medium. The absence of the differential $d\psi$ means that the state of material medium is nonequilibrium.
And this means that an internal force acts in material medium. It is evident that the internal force originates at the expense of some quantity described by the evolutionary form commutator. (If the evolutionary form commutator be zero, the evolutionary relation would be identical, and this would point to the equilibrium state, i.e. the absence of internal forces.) Everything that gives a contribution into the evolutionary form commutator leads to emergence of internal force.
No equilibrium state of material medium is caused by the properties of conservation laws, namely, its noncommutativity. From the nonidentity of the evolutionary relation it follows that the equations of balance conservation laws turn out to be inconsistent. And this points to a noncommutativity of balance conservation laws. The reason of noncommutativity of balance conservation laws is the fact that the actions of different nature (energetic, force and others), which are inconsistent with a nature of material medium itself, act to material medium (that is, these actions are non-potential), and hence these actions cannot be directly transformed into physical quantities of material medium system itself.

IOP Conf. Series: Journal of Physics: Conf. Series **1051** (2018) 012025 doi:10.1088/1742-6596/1051/1/012025

They are accumulated as a certain nonmeasurable quantity. The evolutionary form commutator just describes such nonmeasurable quantity.

Self variation of the nonidentical evolutionary relation points to the fact that the nonequilibrium state of material medium turns out to be selfvarying. In this case the state of material medium changes but remains to be nonequilibrium since under selfvariation the evolutionary relation remains to be nonidentical.

3.2. Transition of material medium into locally equilibrium state. Advent of observable formations

From the evolutionary relation one found that inexact solutions to the mathematical physics equations describe nonequilibrium state of material medium.

It turns out that the generalized solutions, which are discrete functions, describe a locally-equilibrium state of material medium. And this follows from the physical properties of identical relation obtained.

From the identical relation $d\psi_\pi = \omega_\pi^p$ one can obtain the differential of the state functional $d\psi_\pi$, and this points out to a presence of the state function and the transition of material medium from non-equilibrium state into equilibrium one. However, such a state of material medium turns out to be realized only locally due to the fact that the differential of state functional obtained is an interior differential (only on pseudostructure). And yet the total state of material medium remains to be a non-equilibrium state because the evolutionary relation, which describes a material medium state, remains to be nonidentical one. (That is, there exists a duality. Nonidentical evolutionary relation goes on to act simultaneously with identical relation.)

It can be noted that these results point out to the fact that the functionals of evolutionary relation, i.e., functionals of the mathematical physics equations, are really state functionals, that is, they characterize a state of material medium. Under realization of the degenerate transformation condition, which can be caused by any degrees of freedom, and realization of identical relation they locally convert into state function.

The transition from non-equilibrium state to locally-equilibrium state means that unmeasurable quantity, which is described by the evolutionary form commutator and act as internal force, converts into a measurable quantity of material medium.

The transition of unmeasurable quantity into a measurable quantity of material medium reveals in emergence in material medium of some observed formations. Waves, vortices, fluctuations, turbulent pulsations and so on are examples of such formations. The intensity of such formations is controlled by a quantity accumulated by the evolutionary form commutator. (This discloses a mechanism of such processes like an origin of vortices and turbulence [7].)

Such emerged formations are described by generalized solutions to the equations of mathematical physics. The functions that correspond to generalized solutions, as it is known, are discrete functions that have breaks of functions itself or its derivatives [8].

3.3. Emergence of physical structures

It was shown that the transition of material medium from nonequilibrium state to the locally-equilibrium one is accompanied by the emergence of observable formations in material medium.

The advent of observable formations relates to emergence of physical structures.

The transition from nonidentical relation to identical one, which describes a transition of material medium from nonequilibrium state to a locally-equilibrium one, and an advent of observable formations, occurs at realization of closed inexact exterior and dual forms. This points out to emergence of a pseudostructure (closed dual form) with conservative quantity (closed inexact exterior form), that is, emergence of a physical structure (a pseudostructure on which exact conservation law fulfills). Massless particles, structures made up by eiconal surfaces and wave fronts, and so on are examples of physical structures.

Duality of physical structures and observed formations is described by identical relation (7). The left-hand side of this relation includes the differential, which specifies material medium and whose availability points to the locally-equilibrium state of material medium and an advent of observable

IOP Conf. Series: Journal of Physics: Conf. Series **1051** (2018) 012025 doi:10.1088/1742-6596/1051/1/012025

measurable formations. And the right-hand side includes a closed inexact form, which is a characteristic of physical structures.

Thus, one can see that the emergence of physical structures in evolutionary process (to which the realization of closed inexact exterior form points out) in material system reveals as an advent of certain observable formations, which develop spontaneously. It appears those physical structures and the observable formations of material media are a manifestation of the same phenomena. The light is an example of such a duality, namely, as a massless particle (photon) and as a wave.

However, physical structures and observable formations are not identical objects. Whereas the wave is an observable formation, the element of wave front made up the physical structure in the process of its motion.

The generation of physical structures by material media discloses a connection between the field-theory equations and the equations of mathematical physics.

The physical structures are structures on which exact conservation laws are fulfilled, that is, there exist conservative quantities or objects.

It is evident that physical fields, which obey exact conservation laws, are formatted by such physical structures on which exact conservation law for physical fields are fulfilled. To this it also points the fact that there exists a correspondence between the evolutionary relation, from which follows closed exterior forms describing physical structures, and the field-theory equations. Closed exterior forms also obtained from evolutionary relation are solutions to the field-theory equations. And besides, the field-theory equations, as opposed to usual differential equations, have the form of nonidentical relations (or its tensor and differential analogs).

4. State functionals of equations of mathematical physics. Connection between functionals of the field-theory equations and functionals of the mathematical physics equations

From a description of the processes of emerging physical structures and observable formations one can see a role of functionals of the mathematical physics equations in a description of these processes. It is evident that functionals of the mathematical physics equations are state functionals. They characterize a state of material medium. Such a role of functionals of the mathematical physics equations relates to a specific peculiarity of functionals. They can at once be either functionals and state functions or potentials. As functionals they describe a nonequilibrium state of material medium, and as state functions they point out to a locally-equilibrium state of material medium. The transition from functionals to state functions describe a transition of material medium from nonequilibrium state to locally-equilibrium one which is accompanied by emergence of physical structures. That is, the transition from functionals to state functions describes the mechanism of physical structure origination. This discloses physical meaning of functionals both as functionals of mathematical physics equations and as field-theory functionals. This demonstrates a connection between field-theory functionals and functionals of mathematical physics. Such a duality of functionals point out to a connection between the field-theory equations, which describe physical fields, and the mathematical physics equations, which describe material media. This discloses a connection of physical fields with material media. As it was shown, from the mathematical physics equations, which describe material media, it follows the evolutionary relation, which describes the process of emergence of physical structures, from which physical fields are formatted. This points out to the fact that physical fields are generated by material media.

References

[1] Petrova L I 2008 *Exterior and evolutionary differential forms in mathematical physics: Theory and Applications*, 157 p

[2] Petrova L I 2003 Exterior and evolutionary skew-symmetric differential forms and their role in mathematical physics *Preprint* Arxiv:0310050v1

[3] Clark J F and Machesney M 1964 *The Dynamics of Real Gases* (Butterworths, London)

[4] Tolman R C 1969 *Relativity, Thermodynamics, and Cosmology* (Clarendon Press, Oxford, UK)

[5] Fock V A 1955 *Theory of space, time and gravitation* (Moscow, Tech.~Theor.~Lit.)
[6] Petrova L I 2012 Physical meaning and a duality of concepts of wave function, action functional, entropy, the Pointing vector, the Einstein tensor *Journal of Mathematics Research* **4** 3 pp 78-88
[7] Petrova L I 2017 Features of Numerical Simulation of Euler and Navier–Stokes Equations *Journal Computational Mathematics and Modeling* **28** 1 pp 32-36
[8] Petrova L I 2009 Relationships between discontinuities of derivatives on characteristics and trajectories *J. Computational Mathematics and Modeling* **20** 4 pp 367-372

IOP Conf. Series: Journal of Physics: Conf. Series **1051** (2018) 012026 doi:10.1088/1742-6596/1051/1/012026

Some solutions for scalar models and conformal invariance

A O Shishanin

Bauman Moscow State University, Moscow, Russia
E-mail: shishandr@rambler.ru

Abstract. It is discussed solutions in scalar conformal field theories with finite action. One can observe such solutions in flat and curved spaces. As example is considered sphere S^4. Another solution with non-zero action is given. Also it is discussed a similar solution in cosmology.

1. Introduction

Conformal field theories (theories with conformal symmetry) are very important in modern physics. Note that electrodynamics and Yang-Mills theory are conformal at classical level. Conformal field theories have applications in cosmology, string theory and condensed matter physics. For example, at first physicists had believed in 60-th that the two-dimensional Ising model is conformal invariant at critical point. This statement was proved by Smirnov about 2000 year.

Conformal invariance is more complicated than scale invariance. If space-time has dimension $d > 2$ then the conformal group is $SO(2,d)$ but when $d = 2$ the conformal group is infinite-dimensional.

$2d$ Conformal Field Theory CFT describes many phase transitions and string models.

This article is devoted classical conformal scalar field in space with dimension $d > 2$. Conformal scalar field theories are studied intensively long time. There is wonderful connection between $d > 2$ conformal scalar theory and the Liouville theory in $d = 2$ [1].

2. Fubini-Lipatov instanton

Let us consider $d = 4$ conformal scalar model in Euclidian space $\mathbf{R}^4$ with the follow action

$$S = \int \left(\frac{1}{2} \partial_\mu \phi \partial_\mu \phi - \lambda \phi^4 \right) d^4 x. \tag{1}$$

The field equation here is

$$\Box \phi + 4\lambda \phi^3 = 0 . \tag{2}$$

It has the solution called Fubini-Lipatov "instanton" [2-3]

$$\phi_{FL} = \sqrt{\frac{2}{\lambda}} \frac{a}{\left(x - x_0\right)^2 + a^2} . \tag{3}$$

Here x_0 are "coordinates" of instant on and a is its "size". This solution describes decay of the false trivial vacuum $\phi = 0$. I will call a solution with finite action instanton. It is more correct to call an instanton on the solution with action is proportional to the topological charge. It is usually happen in gauge theories and sigma-models.

And action for this solution is

$$S_{FL} = \frac{2\pi^2}{3\lambda}. \tag{4}$$

There is the same solution in the Minkowski space $\mathbf{R}^{1,3}$ but this solution is singular.
Note that such an equation of motion (1) appears in some substitution for $d = 4$ Yang-Mills equations [10]. The self-duality equations are obtained when λ is zero. It is obtained for the t'Hooft ansatz. Solutions of the Laplace equation (harmonic functions) describe multi-instantons.
The action (1) has d-dimensional conformal analog except the case $d = 2$. Here the action is

$$S = \int\left(\frac{1}{2}\partial_\mu\phi\partial_\mu\phi - \lambda\phi^{\frac{2d}{d-2}}\right)d^d x. \tag{5}$$

The scale invariance is evident. This model is invariant under following transformations

$$\delta\phi = f^\alpha\partial_\alpha\phi + \frac{d-2}{2d}\phi\partial_\alpha f^\alpha,$$

where f^α is the conformal Killing vector in flat space.
Case $d = 1$ corresponds conformal quantum mechanics [4]. By the way this theory is renormalized in quantum case because of conformal invariance. In $d > 2$ the power is integer when $d = 3,4$ and 6. In general case the mass dimensional Δ is

$$\Delta = \frac{d-2}{2}.$$

And the action for this solution

$$S^d{}_{FL} = \frac{C}{\lambda^\Delta}. \tag{6}$$

Here C is some constant.
Fubini-Lipatov instanton is important in cosmology [5]. See [6] for connection of tunneling effect in cosmology and instantons in conformal theory. Instantons are described transition between two locally stable Anti-de Sitter (AdS) vacua (two stable minimum states with negative energy) [7]. It is possible to estimate transition probability using standard formula from quantum mechanics

$$p(\Lambda_- \to \Lambda_+) \sim e^{-\frac{C}{\lambda^\Delta}}.$$

This equation is justified by a holographic approach.

3. Non-flat case

Let us consider the conformal scalar theory in the Riemann space with metric $g^{\mu\nu}$

$$S = \int\left(\frac{1}{2}g^{\mu\nu}\partial_\mu\phi\partial_\nu\phi + \frac{1}{12}R\phi^2 - \lambda\phi^4\right)\sqrt{g}d^4x,$$

where R is the Ricci scalar. Here the first and second terms cancel out the other for conformal transformations. It is important that this theory has Weyl's symmetry.
For d-dimensional case the action has form

$$S = \int\left(\frac{1}{2}g^{\mu\nu}\partial_\mu\phi\partial_\nu\phi + \frac{d-2}{8(d-1)}R\phi^2 - \lambda\phi^{\frac{2d}{d-2}}\right)\sqrt{g}d^d x. \tag{7}$$

Then here the field equation is

$$\Delta_{BL}\phi - \frac{d-2}{4(d-1)}R\phi + \lambda\frac{d}{\Delta}\phi^{\frac{d+2}{d-2}} = 0, \tag{8}$$

where Δ_{BL} is Beltrami-Laplace operator

$$\Delta_{BL} = \frac{1}{\sqrt{g}}\partial_\mu\left(\sqrt{g}g^{\mu\nu}\partial_\nu\right).$$

IOP Conf. Series: Journal of Physics: Conf. Series **1051** (2018) 012026 doi:10.1088/1742-6596/1051/1/012026

Let us consider sphere S^d with unit radius, where $R = d(d-1)$. Then there is simple constant solution

$$\phi = \left(\frac{\Delta^2}{2\lambda}\right)^{\frac{\Delta}{2}}. \tag{9}$$

It is useful to note that the actions for flat case and for case the sphere coincide.
Another approach can be use. We can obtain solution of the field equation here using a conformal transformation for conformally flat manifolds [8]. For example, for sphere S^4 [9] a more general solution can be obtained than (9). Let us choice metric on a sphere with radius r in conformal form

$$g_{\mu\nu} = \frac{4r^4\delta_{\mu\nu}}{\left(r^2+x^2\right)^2}.$$

This metric is related with flat metric

$$\eta_{\mu\nu} = \Omega^2 g_{\mu\nu},\ \Omega = \frac{r^2+x^2}{2r^2}.$$

Then we can obtain the new solution

$$\phi_{S^4} = \Omega\phi_{FL} = \sqrt{\frac{1}{2\lambda}}\frac{a\left(r^2+x^2\right)}{r^2\left(\left(x-x_0\right)^2+a^2\right)}. \tag{10}$$

If $x_0 = 0$ and $a = r$ then (10) coincides with (9).
Note that in addition to Fubini-Lipatov solution of equation (2) there is another solution [10]

$$\phi_{Lop} = \frac{1}{x_\mu b^\mu + \alpha}, \tag{11}$$

where $b_\mu b^\mu = -2\lambda$. This solution has zero action. Using conformal transformation is given for

$$\phi^S{}_L = \frac{1}{2r^2}\frac{r^2+x^2}{x_\mu b^\mu + \alpha}.$$

One can consider equation (8) for sphere S^4. It is convenient to describe a point on sphere S^4 by three angles which are usual θ,φ and new angle β. Let us consider that the field ϕ depends only on β. Then equation of motion can be transform to form

$$\left(1+y^2\right)^2\phi_{yy} = 2\phi - g\phi^3,$$

where $y = ctg\beta$ and $g = 4\lambda r^2$. We know solution for case $g = 0$ which is also conformal

$$\phi = C_1\frac{\cosh\left(C_2-\beta\right)}{\sin\beta},$$

where C_1, C_2 are constant. It was shown numerically, for example, with $C_2 = \frac{\pi}{2}$ that in this case the action is finite.

4. Some cosmological solution

It is possible to use solution (3) in cosmology. Let us consider a massless scalar field (inflanton) with ϕ^4-potential in metric which has form

$$\ddot{\phi} + 3H\dot{\phi} + \lambda\phi^3 = 0. \tag{12}$$

Here H is Hubble's constant. The scale factor $a(t)$ depends on the equation of state $p = w\varepsilon$, where p is the pressure and ε is the density of energy. It well known that

$$a(t) \sim t^{\frac{2}{3(1+w)}}.$$

Put $H = \frac{1}{t}$. This corresponds $w = -\frac{1}{3}$. Environment in which both positive and negative gravity is absent. Then there is obtained the equation (2). And its solution

$$\phi = \sqrt{\frac{2}{\lambda}} \frac{d}{t^2 + d^2}.$$

It would be interesting to find solutions with a finite action like (3) for multicomponent fields with conformal invariance. For example, the field ϕ becomes complex or vector.

I would like to thank Alexei Rosly and Igor Fomin for useful discussion.

References

[1] Jackiw R 2006 Weyl symmetry and the Liouville theory *Theor. Math. Phys.* **1480** pp 941-947
[2] Fubini S 1976 A new approach to conformal invariant field theory *Nuovo Cim.* A **34** 521
[3] Lipatov L N 1977 Divergence of the perturbation theory series and the quasiclassical theory *Zh. Eksp. Teor. Fiz.* **72** 411-427
[4] de Alfaro V, Fubini S and Furlan G 1976 Conformal invariance in quantum mechanics *Nuovo Cim.* A **34** 569
[5] Linde A D 1990 Particle physics and inflationary cosmology *Contemp. Concepts Phys.***5** pp 1-362 (*Preprint* arXiv:0503203)
[6] Barbon J L F and Rabinovichi E 2011 Holography of AdS bubbles *Nucl. Phys* B **216** 121-146
[7] Coleman S and de Luccia F 1980 Gravitational effects on and of vacuum decay *Phys. Rev.* D **212** 3305
[8] Tagirov E A and Todorov I T 1979 A geometric approach to the solution of conformal invariant field equation *Acta Phys. Austriaca* **51** 135
[9] Khlebnikov S 2002 Decoherence, instantons, and cosmological horizons *Nucl. Phys.* B **631** pp 307-324 (*Preprint* arXiv:0111194v1)
[10] Lopushansky J T Relations between the classical Yang-Mills theory and massless phi*4 theories *IC* **7777** (*Preprint*)

The analysis of frequency-independent jumps of CMB according to the Planck data

V Ya Vargashkin

Senior research associate, Bauman State Technical University, Moscow, Russia
Associate professor, State University named after I.S. Tourguenev, Oryol, Russia
E-mail: finstp@bk.ru

Abstract. The temporal analysis of the changes of temperature anisotropy of CMB revealed in the Planck database at frequencies of 30, 44 and 70 *GHz* is submitted. Existence of jumps of temperature anisotropy which not depend on the frequency of the accepted radiation is established. It is shown that the found number of such jumps significantly exceeds their number which could be caused by statistical probability in a case imposing of random changes of temperature anisotropy. The assumption that emergence of such jumps can be caused by a gravitational lensing of CMB is made.

1. Introduction

In the present time experiments on detection of a gravitational lensing of CMB which, in particular, are executed by the joint groups "Planck" and "BICEP2" [1] are known. Search of the directions of the celestial sphere in which lensing events are implemented is based on the analysis of "cold" and "hot" spots on allocation maps of anisotropy of temperature of CMB.

The experimental studies of a gravitational lensing are implemented by search of couples of images of the same object possessing identical frequency spectrum. If couples of images do not manage to be separated by means of instruments, then about an event of a lensing draw conclusion by compliance of observed time diagrams of brightness of tight couples of images to theoretical dependences. One of the main differences of the phenomenon of a gravitational microlensing of starlike objects from the phenomenon of variability of stars is that the changes of luminosity of stars happen out synchronously at different frequencies of the accepted radiation. The achromatism of change of brightness can be used also for search of the phenomena of a gravitational lensing of CMB.

So far the longest researches of CMB are satellite experiments with use of the probes "WMAP" (USA) [2–3] and "Planck" (EU) [4–6]. Measurements of anisotropy of temperature of CMB by means of the first of probes continued for nine years. Measurements by means of the second probe lasted four years. The published results are integrated on nine cycles of measurements by year duration by means of the probe "*WMAP*" and also on eight cycles of measurements by means of the probe "*Planck*" duration on half a year.

The considerable duration of measuring cycles of accumulation of data and rather short general operating time of satellites do not allow to research in details temporal dependences of anisotropy of temperature of radiation in the directions of the celestial sphere for the purpose of detection of compliance of their look to the form of standard theoretical curves of a gravitational microlensing. However the available data allow to find the most significant amplitudes of jumps of temperature of CMB between adjacent cycles of measurements and to check them for existence in different frequency

domains, i.e. for independence of appearance of the jumps on frequency of the accepted radiation. The similar achromatism can be put in a basis for a conclusion about possible existence in corresponding direction of the phenomenon of a gravitational lensing of CMB.
The purpose of this research is detection and a research of properties of frequency-independent jumps of anisotropy of temperature of CMB.

2. Identification of temporary jumps of anisotropy of temperature of CMB

Multi-channel measurements of temperature anisotropy of CMB were executed by the probe "*Planck*" on two frequencies ranges. The lower range had frequencies 30, 44 and also 70 *GHz*. The high range had frequencies 100, 143, 217, 35, 545 and also 857 *GHz*. For timing analysis of jumps ΔT_{CMB} of temperature anisotropy T_{CMB} of CMB were used the data of multi-channel measurements in the low-frequency range as these data are fully submitted. At the same time the anisotropy of temperature T of radiations which coincides with temperature of absolutely black body of T_0 value (T_0=2,72548±0,00057 *K* [7]) was exposed to measurements. Deviations T_{CMB} of temperature T of radiations from T_0 value on an absolute value reached of the several tenth shares of millikelvin.
The values T_{CMB} were measured in the lower frequencies range for more than twelve million directions of the celestial sphere. It corresponds to the linear angular permission of measurements approximately in 8 angular minutes. Distributions of T_{CMB} on the celestial sphere for the listed above frequencies according to the first and second measuring accumulation cycles of data contain of Figure 1. Change ΔT_{CMB} of temperature anisotropy TCMB as the differences of these second and first measuring cycles are postponed in various directions of the celestial sphere are presented of Figure 1.
In Figure 2 the frequencies distributions of temperature anisotropy of CMB for cycles of the measurements which are available in Figure 1 are represented. The analysis of Figure 2 shows that the measured values of temperature anisotropy for the first of measuring cycles belong to ranges from the lower T_{min} to the high T_{max} borders. These borders form intervals from $-0{,}0156\cdot 10^{-1}$ to $2{,}1498\cdot 10^{-1}$ *мK* at a frequency of 30 *GHz*; from $-0{,}0148\cdot 10^{-1}$ to $1{,}2320\cdot 10^{-1}$ *мK* at a frequency of 44 *GHz* and from $-0{,}0209\cdot 10^{-1}$ to $1{,}5229\cdot 10^{-1}$ *мK* at a frequency of 70 *GHz*. At the same time the difference of T_{min} from zero arises owing to a deviation of limit of sensibility of gages from zero value at a smallness of gaged value.
Similarly, the measured values T_{CMB} for the second of measuring cycles have borders of T_{min} and T_{max}. These borders form intervals from $-0{,}0176\cdot 10^{-1}$ to $2{,}1506\cdot 10^{-1}$ *мK* at a frequency of 30 *GHz*, from $-0{,}0148\cdot 10^{-1}$ to $1{,}2323\cdot 10^{-1}$ *мK* at a frequency of 44 *GHz* and from $-0{,}0209\cdot 10^{-1}$ to $1{,}5229\cdot 10^{-1}$ *мK* at a frequency of 70 *GHz*.
Thus, temperature anisotropy T_{CMB} of CMB on average in all three frequencies of reception in the first of measuring cycles was characterized by boundary values of range from 0 to T_{max0}, and *Tmax* value is equal to $1{,}64\cdot 10^{-1}$ *мK*. Similarly, in the second measuring cycle the T_{max0} value is equal to $1{,}62\cdot 10^{-1}$ *мK*. At the same time changes ΔT_{CMB} of temperature anisotropy cannot exceed the received T_{CMB} values. The distribution of changes of temperature anisotropy on the celestial sphere presented of fragments (a), (b) and (c) of Figure 1, is characterized by statistical distributions ΔT_{CMB}, represented in Figure 3. Axes of ordinates of schedules of this drawing are represented in logarithmic scale for the purpose of identification of features of tails of these distributions which overstep the bounds ±3σ from the corresponding mean values. The analysis of the received distributions shows that their central part which is in borders ±3σ for frequencies 30, 44 and also 70 *GHz*, are characterized by values ΔT_{CMB} respectively from $-0{,}7291\cdot 10^{-3}$ to $0{,}7287\cdot 10^{-3}$ *мK*, from $-0{,}8116\cdot$10-3 to $0{,}8101\cdot 10^{-3}$ *мK* and also from $-0{,}7038\cdot$-3 to $0{,}7044\cdot 10^{-3}$ *мK*.

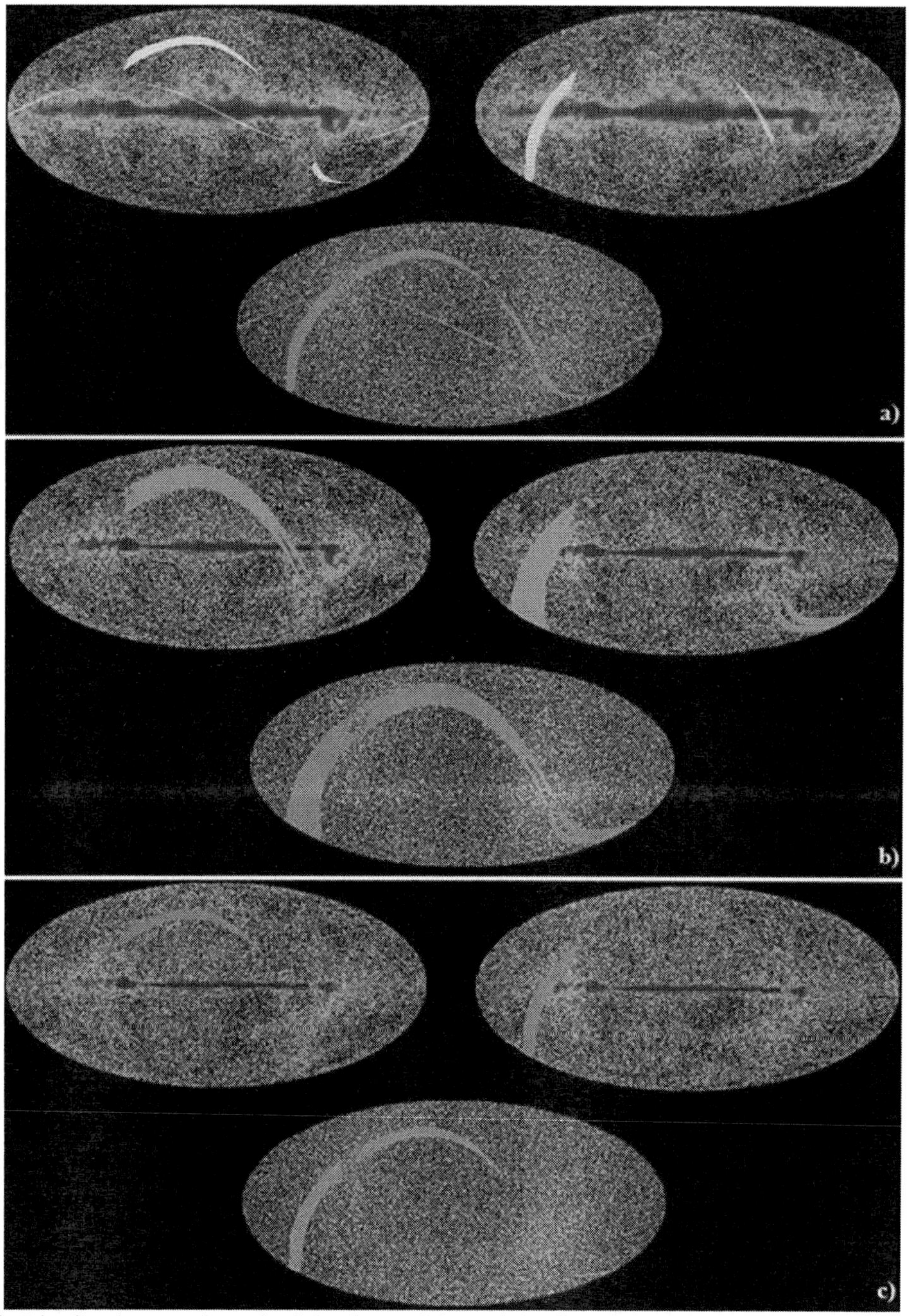

Figure 1. Distributions on the celestial sphere of data of the first (left row) and the second (right row) cycles of measurement of temperature anisotropy T_{CMB} and also changes ΔT_{CMB} of anisotropy (average row) at frequencies of 30 *GHz* (a), 44 *GHz* (b) and 70 *GHz* (c)

IOP Conf. Series: Journal of Physics: Conf. Series **1051** (2018) 012027 doi:10.1088/1742-6596/1051/1/012027

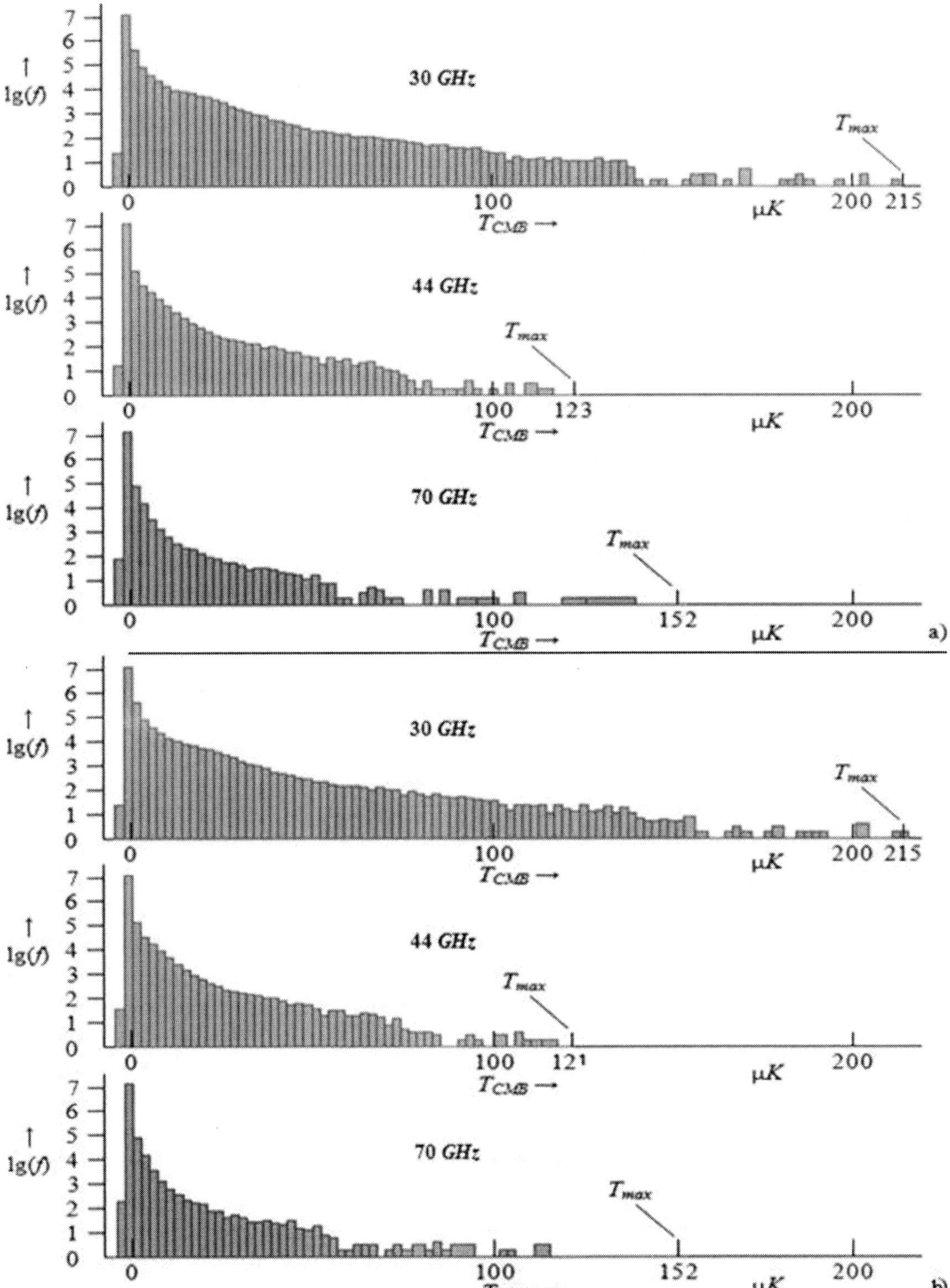

Figure 2. A statistical distribution of temperature anisotropy of T_{CMB} at frequencies 30 *GHz*, 44 *GHz* and also 70 *GHz* and also 70 *GHz* for the first (a) and second (b) measuring cycles.

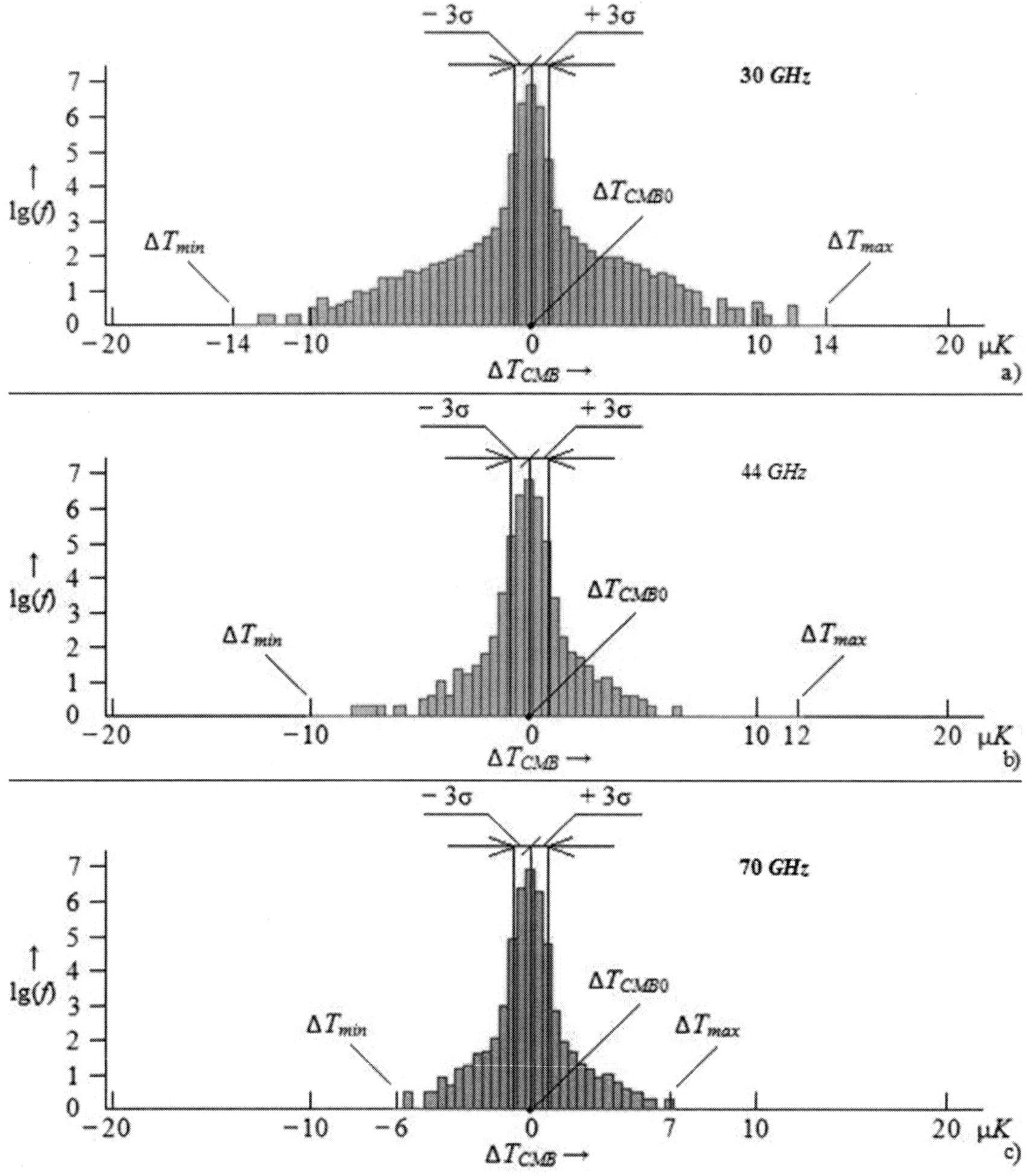

Figure 3. Histograms of distributions of changes ΔT_{CMB} of temperature anisotropy T_{CMB} between the second and first measuring cycles at frequencies 30 *GHz* (a), 44 *GHz* (b) and also 70 *GHz* (c).

Thus, these changes of temperature anisotropy do not exceed 0,5 % of T_{max0} value.
The tails of the analyzed distribution which are going beyond ±3σ from mean value for frequencies 30, 44 and also 70 *GHz*, are characterized by absolute values ΔT_{CMB} which are equal respectively $14{,}03\cdot 10^{-3}$; $9{,}74\cdot 10^{-3}$ and also $6{,}03\cdot 10^{-3}$ *мK* for the left tails of distributions and $14{,}32\cdot 10^{-3}$; $11{,}456\cdot 10^{-3}$ and also $6{,}65\cdot 10^{-3}$ *мK* for the right tails of distributions. Thus, at tails of series of satellite there is changes ΔT_{CMB} temperature anisotropy of T_{CMB} which can reach 7% of the range of T_{max0} value of temperature anisotropy of CMB.

The analysis of results of satellite measurements shows that the conclusions received for the first pair of cycles of satellite measurements are fair also for all other pair of measuring cycles. At the same time changes of temperature anisotropy can reach value of 9 % of T_{max0}.
Further it is necessary to consider a question of that in how many times the temperature anisotropy T_{CMB} of CMB can change during realization of the revealed jumps.

3. Statistical analysis of multiplication (decrease) coefficient of temperature anisotropy of CMB

Value of change of temperature anisotropy T_{CMB} of CMB can be described by means of multiplication (decrease) coefficient K. This coefficient is equal to the module of the relation of the greatest and the least values of temperature anisotropy ΔT_{CMB} implemented during its change in some direction of the celestial sphere, between two neighboring measuring cycles, for example, the second and the first. The frequencies distributions of coefficient of K between the second and first measuring cycles for frequencies 30, 44 and also 70 GHz are represented in Figure 4. Axes of ordinates of schedules of this drawing are represented in logarithmic scale for the purpose of identification of features of these distributions.
It is established that the maximal K_{max} values of coefficient of K which were implemented between the first and second measuring cycles for frequencies 30, 44 and also 70 GHz are equal respectively $4{,}7 \cdot 10^5$, $0{,}34 \cdot 10^5$ and also $15 \cdot 10^5$ times. The analysis of distributions demonstrates that the maximal values of size K can be casual leaps. They are caused by the fact that the minimum values of temperature anisotropy of a CMB which were realized in the course of its changes can be in value of limit of sensibility of a satellite gage, i.e. near zero. Using of the maximal value of temperature anisotropy to such value can cause essential uprating of result of division.
Thereof it is more expedient to characterize of change of anisotropy not by maximal value K_{max}, but K_0 averages values of coefficient of K. The K_0 values which were realized between the first and second measuring cycles for frequencies 30, 44 and also 70 GHz are equal respectively 25, 10 and also 37 times. Thus, the realized change of temperature anisotropy can correspond to multiplication (decrease) coefficient of temperature anisotropy on average to forty times. The analysis of results of satellite measurements shows that the conclusions received for the first couple of cycles of satellite measurements are fair also for all other couples of measuring cycles. It is established that on all set of the available observed data the average value K_0 of multiplication (decrease) coefficient K of temperature anisotropy of T_{CMB} during change can reach of value 45 times.
Further it is necessary to establish repeatability of change of temperature anisotropy of CMB at various frequencies of its measurement.

4. Analysis of the frequencies independence of temporary jumps of temperature anisotropy of CMB

The relation of number of changes ΔT_{CMB} temperature anisotropy of T_{CMB} which are in tails of statistical distributions of these changes of Figure 3 to total change for each of measuring cycles and on each of frequencies of measurement corresponds to probability p of emergence of the change in a tail of the corresponding distribution.
Such probability for changes of temperature anisotropy which in various directions of the celestial sphere between the second and first measuring cycles for frequencies of the radiation 30, 44 and also 70 GHz are equal respectively to $p_1 = 4{,}72 \cdot 10^{-3}$, $p_2 = 4{,}90 \cdot 10^{-3}$ and also $p_3 = 4{,}83 \cdot 10^{-3}$.

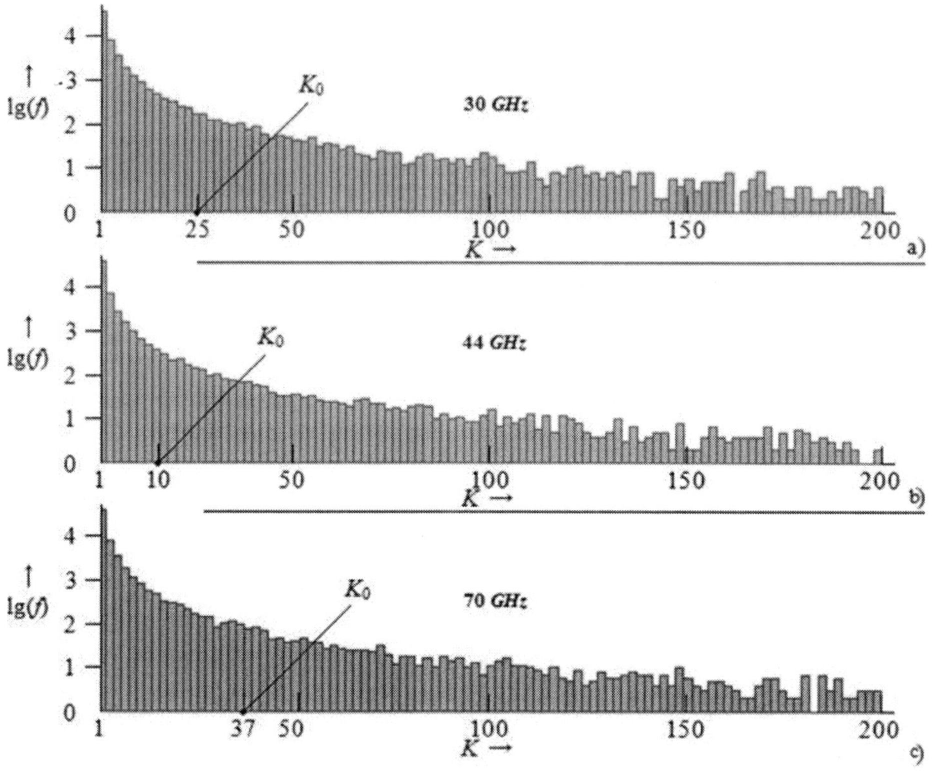

Figure 4. Histograms of distributions of the coefficient of *K* between the second and the first measuring cycles at frequencies of 30 *GHz* (a), 44 *GHz* (b) and also 70 *GHz* (c).

The probability of identification of changes at two frequencies in the same direction of the celestial sphere corresponds to the multiplication of probabilities of their identification at each frequency separately. This theoretical probability in a case with changes, revealed between the second and first measuring cycles, for a combination of the considered higher than the frequencies of radiation of 30 and 44 *GHz*, 30 and 70 *GHz* and also 44 and 70 *GHz* are equal respectively to $p_{1,2}$=2,31·10^{-5}, $p_{1,3}$=2,28·10^{-5} and also $p_{2,3}$=2,37·10^{-5}.
However use of the "*Planck*" database showed that the changes revealed by results to experiments are characterized by higher probabilities of their paired emergence of $p_{1,2}$=15,9·10^{-5}, $p_{1,3}$=10,1·10^{-5} and equal $p_{2,3}$=7,86·10^{-5}. The established excess of values of the probabilities received on the basis of the experimental data over theoretical values of the corresponding probabilities confirms possible existence of frequency-independent physical processes which do a part of changes ΔT_{CMB} to temperature anisotropy of T_{CMB} in a number of the directions of the celestial sphere. The assumption of existence of similar processes can be checked by calculation of probability of detection of jumps in various directions of the celestial sphere between the second and first measuring cycles for all three frequencies of low-frequency range of measurement of the probe. It demands the subsequent comparison of this probability with the probability established experimentally.

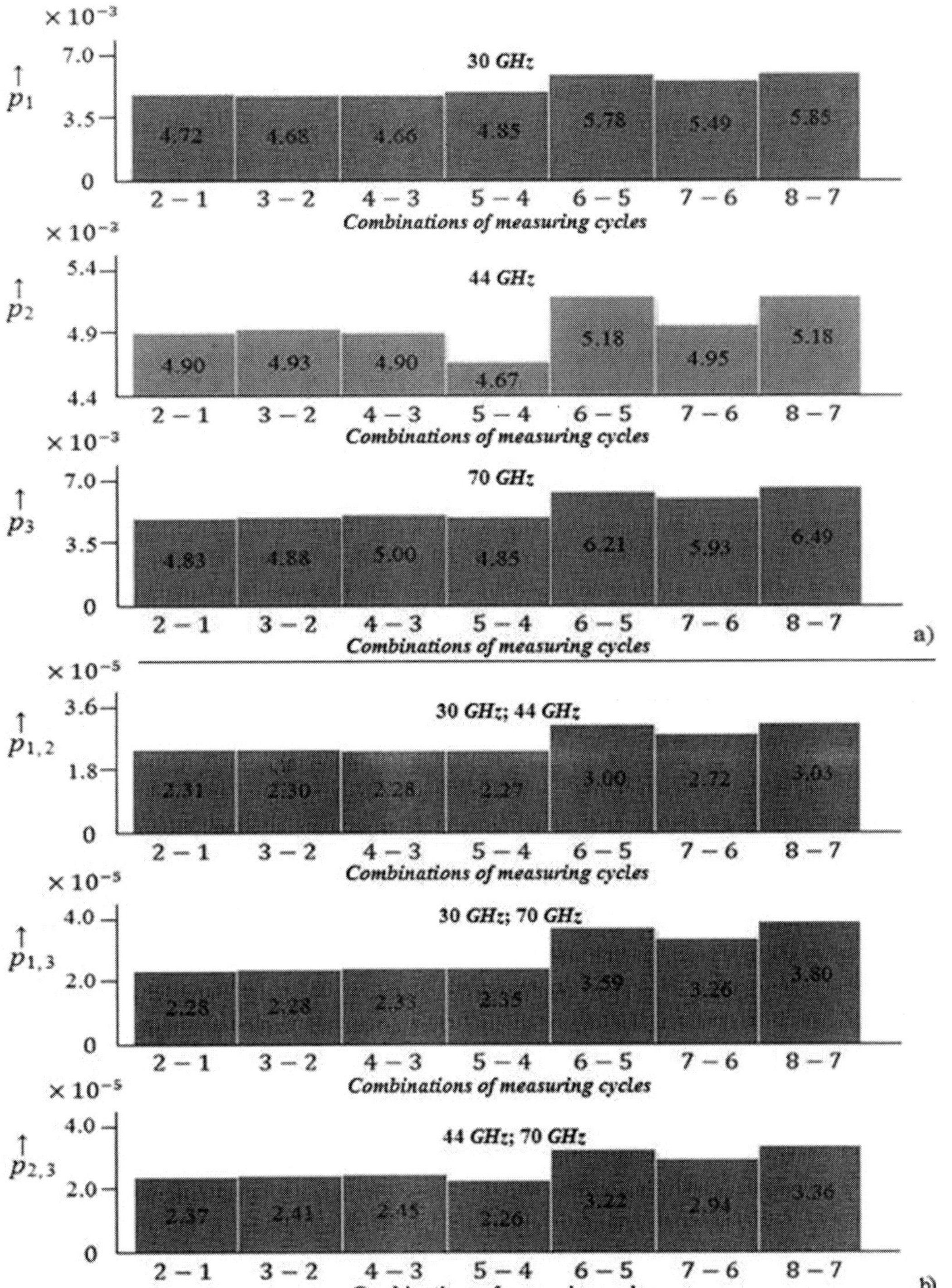

Figure 5. Diagrams of the theoretical p_1; p_2; p_3 (a); $p_{1,2}$; $p_{1,3}$; $p_{2,3}$ (b) and the experimental p_1; p_2; p_3 (a) probabilities illustrating the frequencies independence of jumps of temperature anisotropy of CMB.

IOP Conf. Series: Journal of Physics: Conf. Series **1051** (2018) 012027 doi:10.1088/1742-6596/1051/1/012027

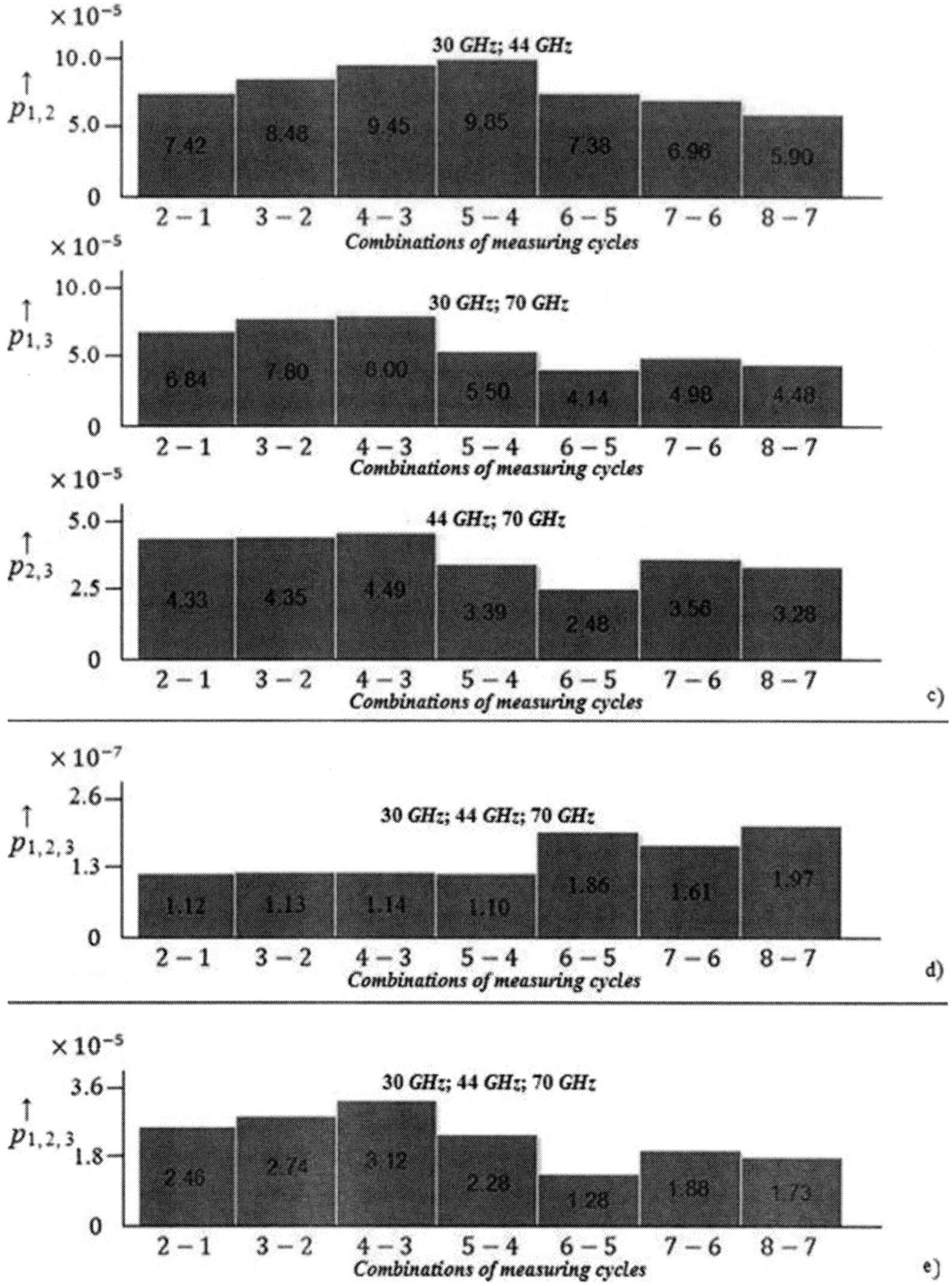

Figure 6. Diagrams of the theoretical $p_{1,2,3}$ (d) and the experimental $p_{1,2}$; $p_{1,3}$; $p_{2,3}$ (c); $p_{1,2,3}$ (e) probabilities illustrating the frequencies independence of jumps of temperature anisotropy of CMB.

The probability of identification of jumps at three frequencies in the same directions on the celestial sphere corresponds to the multiplication of probabilities of their identification on each of three frequencies separately. This probability was equal $p_{1,2,3}$=1,12·10^{-7}. However the analysis of the experimental data gave an assessment to this probability, more than by 400 times exceeding theoretical

and equal $p_{1,2,3}$=4,47·10^{-5}. The received experimental probability corresponds to identification of several hundred frequency-independent jumps of temperature anisotropy within half a year.
In Figure 5 - 6 diagrams of theoretical and experimental probabilities are shown. They demonstrate that the established excess of the probability of $p_{1,2,3}$ received experimentally over similar theoretical probability for seven neighboring pair of measuring cycles in eight years of functioning of the probe "*Planck*" is in limits from 130 up to 490 times. It confirms the received conclusion about existence of the physical processes causing frequency-independent jumps of anisotropy of temperature of CMB.

5. Conclusions

The made analysis allows making a conclusion on a possibility of searching of candidates for events of a gravitational lensing of CMB by the analysis of jumps ΔT_{CMB} its temperature anisotropy T_{CMB} of within neighboring pair of satellite measuring cycles of anisotropy.
The method of identification of similar candidates consists of the sequence of actions including four stages.
At the first stage it is necessary to calculate of changes ΔT_{CMB}, i.e. differences of values of temperature anisotropy T_{CMB} between neighboring cycles of satellite measurements.
At the second stage it is necessary to establish a statistical distribution of changes of temperature anisotropy and also to execute selection of the tails of this distribution remote from mean value, for example, more than on $\pm 3\sigma$.
At the third stage it is necessary to reveal jumps which are present at all frequencies of reception of CMB between the analyzed cycles of satellite measurements among the selected changes of temperature anisotropy.
At the fourth stage among frequency-independent jumps it is necessary to select set of those from them which amplitudes are the most close on values.
In relation to the corresponding directions on the celestial sphere it is possible to draw a conclusion on realization in them frequency-independent processes which can be considered candidates for a gravitational lensing of CMB.

References

[1] Ade P A R *et al* 2015 Joint Analysis of BICEP2/Keck Array and Planck Data *Phys. Rev. Lett.*, **114** p 101301
[2] Bennett C L *et al* 2013 Nine-Year Wilkinson Microwave Anisotropy Probe (WMAP) Observations: Final Maps and Results *ApJS.* **208** 20B
[3] Hinshaw G F *et al* 2013 Nine-Year Wilkinson Microwave Anisotropy Probe (WMAP) Observations: Cosmology Results *ApJS.* **208** 19H
[4] Planck Collaboration *et al* 2016 Planck 2015 results. I. Overview of products and scientific results *A&A.* **594A** 1P
[5] Planck Collaboration *et al* 2016 Planck 2015 results. II. Low Frequency Instrument data processing *A&A.* **594A** 2P
[6] Planck Collaboration *et al* 2016 Planck 2015 results. I. LFI mapmaking *A&A.* **594A** 6P
[7] Fixen D J 2009 The Temperature of the Cosmic Microwave Background *Astrophys. J.* **707** 916

Singularities and Phenomenological aspects of Asymptotic Safe Gravity

V Zarikas[1,2] and G Kofinas[3]

[1]Nazarbayev University, School of Engineering, Astana, Kazakhstan
[2]University of Applied Sciences of Central Greece, Lamia, Greece
[3]Research Group of Geometry, Dynamical Systems and Cosmology, University of the Aegean, Samos, Greece
E-mail: vasileios.zarikas@nu.edu.kz; gkofinas@aegean.gr

Abstract. Asymptotic Safety (AS) Program for quantum gravity keeps the same fields and symmetries with General Relativity and studies the associated gravitational action as a fundamental part of the complete theory at the nonperturbative level with the help of functional renormalization group (RG) techniques. An important phenomenological task that can test the new point of view of AS approach is the discovery of RG improved cosmologies and black holes. In this work, we analyze the properties of recently found non-singular spherically symmetric and non-singular cosmological solutions. Furthermore, we derive a novel consistent set of modified Einstein field equations, in the spirit of AS, which respects the Bianchi identities. This new set of equations completes previously published modified Einstein equations which arise by adding appropriate covariant kinetic terms to the action.

1. Introduction

The singularity theorems of Hawking and Penrose erased the hope that spacetime singularities in classical General Relativity are just a consequence of the high degree of symmetries assumed in the collapse or in the energy momentum tensor. Now we know that the appearance of spacetime singularities is a proof of the limit of validity of General Relativity. An important property of any serious attempt to quantize gravity is the ability to resolve the breakdown of physics in these singular regions of spacetime. Loop Quantum Gravity and spin foam models offer ideas to solve this problem since they assume that the spacetime is intrinsically discrete. Concerning models on continuous spacetimes, one interesting proposal is that of Asymptotic Safety Program. Asymptotic Safety (AS) framework for quantum gravity keeps the same fields and symmetries with General Relativity and studies the associated gravitational action as a fundamental part of the complete theory at the nonperturbative level (with the help of functional renormalization group (RG) techniques) [1-60].

The key idea behind the AS scenario was first proposed by Weinberg. The central requirement is the existence of a non-Gaussian fixed point (NGFP) of the RG flow for gravity. If this NGFP exists, this defines the behavior of the theory at the UV. Furthermore, in this regime all measured quantities are free from nonphysical divergences. Asymptotic safe gravity does not emerge from a direct quantization of General Relativity. The Einstein-Hilbert action is a bare action that corresponds to a non-trivial fixed point of the RG flow and is a prediction assuming asymptotic safety. The AS framework justifies someone to work with the gravitational effective average action which includes only the effect of the quantum fluctuations with momenta $p^2 > k^2$. Thus the effective average action represents an

approximate description of physics at the momentum scale $p^2 \approx k^2$. In this way, it is possible to develop, following various different paths, a phenomenological study of asymptotic safety.
The scope of this work is to investigate the structure of spacetime at high energies and in particular both in the vicinity of the center of spherically symmetric solutions and in the big bang regime of cosmological solutions. Both these type of solutions are generated from quantum modified Einstein equations according to the AS framework. The present paper has two targets. First, we analyze the properties of recently found non-singular spherically symmetric and cosmological solutions. Secondly, we present a novel improved version of modified Einstein equations according to the AS idea. The proposed new equations are compatible with a new alternative and natural scheme for energy conservation. The way we will produce these new modified Einstein equation, valid for short distance interiors of spherically symmetric configurations and early times cosmologies, follows. AS framework uses a system of truncated RG flow equations that contains two running couplings, the gravitational constant $G(k)$ and the (positive) cosmological constant $\Lambda(k)$. Near the non-Gaussian UV fixed point the coupling G approaches zero, while on the other hand, the coupling Λ behaves as the square of the momenta. Due to this scaling of G, Λ in the UV regime, we work on the belief that in the Big bang regime and in the center of black hole the Λ term plays the dominant role. Thus, we also study the modified Einstein vacuum equations where only $\Lambda(k)$ appears. More precisely, the result of the existence of the non-Gaussian fixed point at the UV is utilized. This allows us to adopt the scaling behaviour of Λ as $\Lambda \sim k^2$.
With a running cosmological constant as the source, novel uniquely defined covariant gravitational equations that modify minimally the Einstein gravity are derived. They are constructed so that the Bianchi identities are satisfied by construction. We first have to find the appropriate covariant kinetic terms that support an arbitrary source term $\Lambda(k)$ without any symmetry assumption and construct modified Einstein equations which respect the Bianchi identities. If we add a spacetime-dependent cosmological constant $\Lambda(x)$ in the 4-dimensional vacuum Einstein equations so that $G_{\mu\nu} = -\Lambda g_{\mu\nu} + 8\pi G T_{\mu\nu}$, obviously, this equation does not satisfy the Bianchi identities. Thus, we will add an energy-momentum tensor $\theta_{\mu\nu}$ to support $\Lambda(x)$, so that the Bianchi identities are satisfied. This formalism has the advance that allows the term $\Lambda(x)$ to be kept arbitrary (later we will specify the function $\Lambda(x)$).
The outline of the paper is as follows: In the second section of the present work a review of recently found solutions from quantum modified equations is presented. In the third section a novel scheme of quantum modified equations is developed with a different energy conservation scheme. The novel scheme of energy conservation includes all the matter energy-momentum tensor and the varying couplings G, Λ. Finally, in the last section various conclusions are discussed.

2. Review of recently found solutions

In this section we will try to provide an insight into recently developed solutions from quantum modified Einstein equations according to the AS framework, [61-62]. These new modified Einstein equations describe how a classical spacetime may appear in the presence of a quantum vacuum. The quantum vacuum is modeled through a cosmological constant term which is energy dependent according to the AS program.
Vacuum solutions where the cosmological constant is solely a source are of great importance. A motivation for this is that the vacuum birth of our universe from a vacuum fluctuation is an interesting proposal in various quantum gravity models. The same also holds for the central regions of black holes where we expect spacetime to be associated with antigravity properties. Thus, the vacuum modified Einstein equations have been constructed with the scope to be valid in the UV regime. In this transplanckian regime the cosmological constant plays the dominant role according to AS. This reason why in [61-62] only kinetic terms for $\Lambda(k)$ are included. A spacetime-dependent cosmological

constant $\Lambda(k)$ has to be introduced in the 4-dimensional vacuum Einstein equations so that $G_{\mu\nu} = -\Lambda g_{\mu\nu}$. However, this equation is inconsistent since the cosmological constant term is not any more constant. Thus, it does not satisfy the Bianchi identities. Therefore, it is necessary to add an energy-momentum tensor $\theta_{\mu\nu}$ to support $\Lambda(k)$, with the demand that Bianchi identities are satisfied. The new modified Einstein equations become

$$G_{\mu\nu} = -\Lambda g_{\mu\nu} + \theta_{\mu\nu}. \tag{1}$$

This equivalently gives

$$R_{\mu\nu} = \Lambda g_{\mu\nu} + \theta_{\mu\nu} - \frac{1}{2}\theta g_{\mu\nu}, \tag{2}$$

where $\theta = \theta^{\mu}{}_{\mu}$. Now the Bianchi identities are

$$\theta_{\mu\nu}{}^{;\mu} = \Lambda_{;\nu}, \tag{3}$$

where ; denotes covariant differentiation. For Λ constant, the quantity $\theta_{\mu\nu}$ should vanish since in this case the Bianchi identities are trivially satisfied and there is no need to add anything.
Assuming that the tensor $\theta_{\mu\nu}$ can be constructed by ψ and its first and second derivatives, it is possible to express it as $\theta_{\mu\nu} = A\psi_{;\mu}\psi_{;\nu} + Bg_{\mu\nu}\psi^{;\rho}\psi_{;\rho} + \tilde{N}\psi_{;\mu;\nu} + Eg_{\mu\nu}\Box\psi + Fg_{\mu\nu}$, where $\Box\psi = \psi_{;\mu}{}^{;\mu}$. Note that ψ is defined as $\Lambda = \bar{\Lambda}e^{\psi}$ where $\bar{\Lambda}$ is an arbitrary constant reference value. It is possible to solve the system of equations (3) and find finally that $\theta_{\mu\nu}$ is given by $\theta_{\mu\nu} = -\frac{1}{2}\psi_{;\mu}\psi_{;\nu} - \frac{1}{4}g_{\mu\nu}\psi^{;\rho}\psi_{;\rho} + \psi_{;\mu;\nu} - g_{\mu\nu}\Box\psi$. From this expression we obtain the self-consistent modified Einstein equations in terms of Λ. Note that ψ does not have its own equation of motion since we are working in the AS framework.
Now one crucial question is how we can consistently include an extra matter energy-momentum tensor $T_{\mu\nu}$ in the vacuum field equations. Let us write

$$G_{\mu\nu} = -\bar{\Lambda}e^{\psi}g_{\mu\nu} - \frac{1}{2}\psi_{;\mu}\psi_{;\nu} - \frac{1}{4}g_{\mu\nu}\psi^{;\rho}\psi_{;\rho} + \psi_{;\mu;\nu} - g_{\mu\nu}\Box\psi + 8\pi G T_{\mu\nu}. \tag{4}$$

Note that the gravitational constant G is in general also spacetime dependent. In order to satisfy the Bianchi identities we have to take into account that the various ψ-kinetic terms in (4) are by construction covariantly conserved. Thus in order to satisfy the Bianchi identities we have to conserve energy-momentum in combination with G, i.e.

$$\left(GT_{\mu\nu}\right)^{;\nu} = 0. \tag{5}$$

If one insists on the exact energy-momentum conservation, then either extra covariant terms constructed out of G would be necessary to be added in (4) or the approach presented in the next section is needed. In case we allow kinetic terms for both Λ and G, then this has the serious drawback that at the end the modified Einstein equations are very hard to be solved. Such an approach with starting from an action that generates equations of motion with quantum corrections of G and Λ can be found in [4]. However, in this approach the full Λ and G-dependent terms, which make the equations consistent, have not been found in general. Thus, our philosophy behind the construction of both versions of modified Einstein equations ala AS, is that we keep only the kinetic terms of Λ for both simplicity/tractability and vacuum domination at UV. A remark for equations (4) is that it is not clear if they arise from an action or not.
Finally, the RG flow provides the exact energy dependent law for $\Lambda(k)$, $G(k)$, and the system of equations (4), (5) is a consistent system of equations valid near the NGFP in the presence of matter. The

"cosmological constant" term scales at high energies as the asymptotic safety scenario suggests, i.e. $\Lambda(k) \propto k^2$. The generation of solutions needs also the connection of the momentum scale k with some length scale.

In [61], several families of novel spherically symmetric solutions have been derived in the context of AS. The analysis of all the solutions provides evidence for the spacetime structure close to the centre of spherical symmetry. The signature of a spherically symmetric metric close to this centre is not known, and this is the reason for considering both options, not only the one with the "conventional" signs for the temporal-radial metric components $(- + + +)$, but also the alternative one with signature $(+ - + +)$. It is well known that the second signature appears in the interior Schwarzschild black hole for the same spherical coordinates.

There are two categories of physically significant solutions, i.e. solutions which improve or resolve the singularity problem. The first category is represented by one solution with the "conventional" signature that originates from $r = 0$ and is associated with a curvature singularity there. The interesting feature is that no timelike geodesic can reach the origin since the geodesic of an infalling massive particle is reflected at some finite distance. The explanation of this behavior is that there is a repulsive force caused by the cosmological constant along with the other kinetic terms. Note that this feature is not met by a standard cosmological constant. However, the massless particles reach the origin in finite affine parameter.

The second category concerns several solutions of both types of signatures. These solutions are not extended to include the centre of spherical symmetry and range from a finite radius outwards. All their curvature invariants at this minimum radius are also finite. Thus, they can also be considered as non-singular physically accepted geometries.

In [62], general families of new cosmological solutions have been derived in the context of Asymptotic Safe Gravity at high energies close to the NGFP. The derived vacuum solutions were found from the same set of vacuum modified Einstein equations as in [61]. In addition an extension of the vacuum modified Einstein equations is presented in [62], capable to describe also matter. The new equations use a certain scheme for energy conservations described from Eq. (5). Thus, the work [62] can describe two different cosmic eras. First, a speculated vacuum birth from a quantum vacuum state and second a radiation-matter evolution at high energies.

In the first such cosmic era the only source is an energy-dependent cosmological constant that scales at the UV as the AS framework proposes near the NGFP. The importance of the derived physically interesting vacuum solutions is the fact that they describe an inflationary expansion with completely removed initial singularity for the scale factor, the energy density and the curvature invariants. The end of inflation happens naturally since in the AS scenario, as the energy scale becomes lower, $\Lambda(k)$ becomes insignificant and standard decelerating cosmology arises.

In the second cosmic era, an energy exchange due to Eq. (5) is realized between the matter and the varying gravitational constant $G(k) \propto k^{-2}$. An interesting discovery is the appearance of negative non-equilibrium pressure beyond the thermodynamic one. This can be interpreted as either a particle production or a mechanism of bulk viscosity. In both alternatives, there are general solutions which are non-singular and inflationary. The same holds for any spatial topology. A suspiring finding is that in the case of bulk viscosity the relation between the energy scale and the time is implied by the theory of AS. Finally, perhaps the most important feature of the matter solutions is that they suggest either particle production with entropy generation or bulk viscosity with entropy production and reheating.

3. New Field equations with a varying cosmological constant

Here, in this section we provide an alternative version of quantum modified Einstein equations following an alternative approach for encapsulating Bianchi identities. We also add a spacetime-dependent cosmological constant $\Lambda(k)$ in the 4-dimensional vacuum Einstein equations so that $G_{\mu\nu} = -\Lambda g_{\mu\nu} + 8\pi G T_{\mu\nu}$. In the framework of asymptotic safety, $\Lambda(k)$ is supposed to be determined along the RG flow. Nevertheless, we want to construct a general formulation valid for an arbitrary $\Lambda(k)$ which consequently will be able to carry any $\Lambda(k)$ specified by RG equations. As we have already explained, the previous equation is inconsistent since it does not satisfy the Bianchi identities. Therefore an energy-momentum tensor $\theta_{\mu\nu}$ will be introduced in order to support $\Lambda(k)$, so that the Bianchi $-4\pi G(Tg_{\mu\nu}-2T_{\mu\nu})$ identities are satisfied. In the new equations to be derived here, the key point is that the matter energy-momentum tensor will participate in the construction and satisfaction of the Bianchi identities.

The gravitational equations of motion are

$$G_{\mu\nu} = -\Lambda g_{\mu\nu} + \theta_{\mu\nu} + 8\pi G T_{\mu\nu} \tag{6}$$

or equivalently

$$R_{\mu\nu} = \Lambda g_{\mu\nu} + \theta_{\mu\nu} - \frac{1}{2}\theta g_{\mu\nu} - 4\pi G T g_{\mu\nu} + 8\pi G T_{\mu\nu}\,. \tag{7}$$

It is $R = 4\Lambda - \theta - 8\pi G T$, where $\theta = \theta^{\mu}{}_{\mu}$, $T = T^{\mu}{}_{\mu}$. The Bianchi identities take the form

$$\theta_{\mu\nu}{}^{;\mu} = \Lambda_{;\nu} - 8\pi\left(G T_{\mu\nu}{}^{;\mu} + T_{\mu\nu} G^{;\mu}\right), \tag{8}$$

where ";" denotes as usually covariant differentiation with respect to the Christoffel connection of $g_{\mu\nu}$. We introduce for convenience the quantity $\psi(x)$ by

$$\Lambda = \bar{\Lambda} e^{\psi}, \tag{9}$$

where $\bar{\Lambda}$ is an arbitrary constant reference value.

Let us construct such a tensor $\theta_{\mu\nu}$ from ψ and its first and second derivatives. Therefore, we work with the ansatz

$$\theta_{\mu\nu} = A(\psi)\psi_{;\mu}\psi_{;\nu} + B(\psi)g_{\mu\nu}\psi^{;\rho}\psi_{;\rho} + \tilde{N}(\psi)\psi_{;\mu;\nu} + E(\psi)g_{\mu\nu}\Box\psi + F(\psi)g_{\mu\nu}, \tag{10}$$

where $\Box\psi = \psi_{;\mu}{}^{;\mu}$. Then,

$$\theta = (A+4B)\psi^{;\mu}\psi_{;\mu} + (C+4E)\Box\psi + 4F, \tag{11}$$

$$R_{\mu\nu} = A\psi_{;\mu}\psi_{;\nu} - \left(\frac{A}{2}+B\right)g_{\mu\nu}\psi^{;\rho}\psi_{;\rho} + \tilde{N}\psi_{;\mu;\nu} - \left(\frac{C}{2}+E\right)g_{\mu\nu}\Box\psi + \left(\bar{\Lambda}e^{\psi} - F\right)g_{\mu\nu}\,. \tag{12}$$

Since $\Box(\psi_{;\nu}) = (\Box\psi)_{;\nu} + R_{\nu\sigma}\psi^{;\sigma}$, it is from (10)

$$\begin{aligned}\theta_{\mu\nu}{}^{;\mu} &= (A'+B')\psi^{;\mu}\psi_{;\mu}\psi_{;\nu} + (A+E')\Box\psi\psi_{;\nu} + (C'+A+2B)\psi^{;\mu}\psi_{;\mu;\nu} + \\ &+ (C+E)(\Box\psi)_{;\nu} + F'\psi_{;\nu} + CR_{\mu\nu}\psi^{;\mu},\end{aligned} \tag{13}$$

where a prime denotes differentiation with respect to ψ. Due to Eq. (12), Eq. (13) can be written as

$$\begin{aligned}\theta_{\mu\nu}{}^{;\mu} &= \left(A'+B'+\frac{1}{2}AC - BC\right)\psi^{;\mu}\psi_{;\mu}\psi_{;\nu} + \left(E'+A-CE-\frac{1}{2}C^2\right)\Box\psi\psi_{;\nu} + \\ &+ \left(C'+A+2B+C^2\right)\psi^{;\mu}\psi_{;\mu;\nu} + (C+E)(\Box\psi)_{;\nu} + \left(F'-CF+C\bar{\Lambda}e^{\psi} - 4\pi GCT\right)\psi_{;\nu} + \\ &+ 8\pi GCT_{\mu\nu}\psi^{;\mu}.\end{aligned} \tag{14}$$

The consistency condition Eq. (8) becomes

$$\begin{aligned}&\left(A'+B'+\frac{1}{2}AC-BC\right)\psi^{;\mu}\psi_{;\mu}\psi_{;\nu} + \left(E'+A-CE-\frac{1}{2}C^2\right)\Box\psi\psi_{;\mu} + \left(C'+A+2B+C^2\right)\psi^{;\mu}\psi_{;\mu;\nu} + \\ &+ (C+E)(\Box\psi)_{;\nu} + \left[F'-CF+(C-1)\bar{\Lambda}e^{\psi} - 4\pi GCT\right]\psi_{;\nu} = -8\pi\left(GC\psi^{;\mu} + G^{;\mu}\right)T_{\mu\nu} - 8\pi G T_{\mu\nu}.\end{aligned} \tag{15}$$

One option to proceed would be to embody the term $-8\pi GC\psi^{;\mu}T_{\mu\nu}$ of the right-hand side of

Eq. (15) with the last term of the left-hand side being proportional to $g_{\mu\nu}\psi^{;\mu}$. Since Eq. (15) must be satisfied for any ψ, the various prefactors should vanish separately and the prefactor of $\psi^{;\mu}$ would give $T_{\mu\nu}\propto g_{\mu\nu}$ with a function as a proportionality factor (the vanishing of the rest terms on the right-hand side would give an equation of motion for the matter). Such an energy-momentum tensor is of a very special form, not of particular interest, so we quit this option. The other alternative is to vanish the prefactors of the various kinetic terms on the left-hand side as they stand. The right-hand side gives the equation of motion for the matter. Therefore, we obtain the following system of equations

$$\begin{gathered} A'+B'+\frac{1}{2}AC-BC=0\,,\;\; E'+A-CE-\frac{1}{2}C^2=0 \\ C'+A+2B+C^2=0\,,\;\; C+E=0 \\ F'-CF+(C-1)\bar{\Lambda}e^{\psi}-4\pi GCT=0 \\ GT_{\mu\nu}{}^{;\mu}+\left(GC\psi^{;\mu}+G^{;\mu}\right)T_{\mu\nu}=0\,. \end{gathered} \tag{16}$$

It can be seen that the first equation is redundant and the previous system gets the form

$$\begin{gathered} A+B=-\frac{3}{4}C^2\,,\;\; E=-C\,,\;\; C'=\frac{1}{2}C^2+A \\ F'-CF+(C-1)\bar{\Lambda}e^{\psi}-4\pi GCT=0 \\ GT_{\mu\nu}{}^{;\mu}+\left(GC\psi^{;\mu}+G^{;\mu}\right)T_{\mu\nu}=0\,. \end{gathered} \tag{17}$$

The differential equation for F is linear, so its solution is

$$F=e^{\int C}\left\{\sigma+\int\left[(1-C)\bar{\Lambda}e^{\psi}+4\pi GCT\right]e^{-\int C}\right\}, \tag{18}$$

where σ is integration constant. Since for ψ constant, $\theta_{\mu\nu}$ must vanish, it arises from Eq. (10) that for ψ constant the factor F must vanish. From the form of the previous solution for F, for this to happen for any constant ψ, it needs to be $C=1, T=0$ and furthermore $\sigma=0$. Thus, although in the system of Eqs. (17) there are four equations containing the five unknowns A,B,C,E,F, the above requirement gives a unique solution. The system of Eqs. (17) is written equivalently as

$$A=-\frac{1}{2}\,,\;\; B=-\frac{1}{4}\,,\;\; C=1\,,\;\; E=-1\,,\;\; F=0 \tag{19}$$

$$T=0 \tag{20}$$

$$T_{\mu\nu}{}^{;\mu}+\left(\psi^{;\mu}+\frac{G^{;\mu}}{G}\right)T_{\mu\nu}=0. \tag{21}$$

For any given Λ, G the non-conservation equation (21) defines the equation of motion of the matter. Our final generic result is equation

$$G_{\mu\nu}=-\bar{\Lambda}e^{\psi}g_{\mu\nu}-\frac{1}{2}\psi_{;\mu}\psi_{;\nu}-\frac{1}{4}g_{\mu\nu}\psi^{;\rho}\psi_{;\rho}+\psi_{;\mu;\nu}-g_{\mu\nu}\psi+8\pi GT_{\mu\nu}\,, \tag{22}$$

where $T_{\mu\nu}$ satisfies the two equations (20), (21). The vanishing of the trace of the energy-momentum tensor, $T=0$, is a surprising outcome of the study, since it constraints the possible equations of state in an interesting way. For the case of a perfect fluid, this equation results to a radiation-dominated Universe. This is a result which is compatible with what we expect near the NGFP taking into account that the system of equations could have suggested something totally unacceptable as possible equation of state near the NGFP. It is also noticeable that although the new energy conservation scheme in the present work is totally different than the one used in [62], leading to different system of equations, still the kinetic terms of the cosmological constant coupling are the same, thus Eq. (22) is the same. Finally, the non-conservation equation (21) expresses an energy exchange between the matter fields and the

couplings $G(k)$, $\Lambda(k)$.
As in the previous section with the vacuum solutions, we also here assume a spatially homogeneous and isotropic metric for the cosmic spacetime of the form Eq. (3). Since the external fields $\Lambda(k)$, $G(k)$ carry the same symmetries, they will be of the form $\Lambda(t)$, $G(t)$. We consider a diagonal energy-momentum tensor T^{μ}_{ν}, so we take as matter content a non-perfect fluid with energy density ρ, thermodynamic pressure p and a possible non-equilibrium part π [54-57]. Due to the working symmetries of isotropy and homogeneity shear viscosity and energy fluxes are disregarded. The energy momentum tensor is

$$T^{\mu\nu} = \rho u^{\mu} u^{\nu} + (p+\pi)\left(g^{\mu\nu} + u^{\mu} u^{\nu}\right) \tag{23}$$

with u^{μ} the fluid 4-velocity. The extra pressure π can either be associated to a pressure due to particle production/destruction or to a bulk viscous pressure and could be important during the transition phase that connects the quantum vacuum stage of the universe to the subsequent era with non-zero matter density. We will consider in the following a very early high-energy universe with the assumption of AS running of the couplings.
The two independent components of (22) are

$$3\left(H^2 + \frac{\kappa}{a^2}\right) = \bar{\Lambda} e^{\psi} - 3H\frac{\dot{\psi}}{n} - \frac{3\dot{\psi}^2}{4n^2} + 8\pi G\rho \tag{24}$$

$$\frac{2}{n}\dot{H} + 3H^2 + \frac{\kappa}{a^2} = \bar{\Lambda} e^{\psi} - 2H\frac{\dot{\psi}}{n} - \frac{\dot{\psi}^2}{4n^2} - \frac{1}{n}\left(\frac{\dot{\psi}}{n}\right)^{\cdot} + 8\pi G P, \tag{25}$$

where the total effective pressure is $P = p + \pi$ and n denotes the lapse function.
In the NGFP regime of AS it is $G(k) \propto k^{-2}$, $\Lambda(k) \propto k^2$, therefore Eq. (21) implies the exact conservation $T_{\mu\nu}{}^{;\mu} = 0$. Thus, although energy can be exchanged between $G(k)$ and $\Lambda(k)$, the matter energy momentum tensor is separately conserved something that it is correct and expected in our semiclassical approach at the tree level! Alternatively, if someone demands $T_{\mu\nu}{}^{;\mu} = 0$ (again motivated by the correct tree level behavior) it can be proven generally that $G(k)\Lambda(k) = const$, which is exactly the expected behavior at the NGFP! Thus, a consistent picture arises, considering the derived modified Einstein equations associated with the energy conservation scheme of Eq. (8).
For cosmology this conservation equation gives the standard expression

$$\dot{\rho} + 3nH(\rho + P) = 0. \tag{26}$$

The system of equations (24)-(25) is satisfied by construction for any $\psi(t)$. However, the effective pressure P is constrained to be $P = \frac{1}{3}\rho$. This is different than the result obtained in [62] where the opposite sign appered in the equation if state. Indeed, here differentiating (23) with respect to t and using (26) to substitute $\dot{\rho}$ and also (24) itself we find in the UV regime

$$\left(1 + \frac{\dot{\psi}}{2nH}\right)\left[\frac{2}{n}\dot{H} + 3H^2 + \frac{\kappa}{a^2} - \bar{\Lambda} e^{\psi} + 2H\frac{\dot{\psi}}{n} + \frac{\dot{\psi}^2}{4n^2} G(k) + \frac{1}{n}\left(\frac{\dot{\psi}}{n}\right)^{\cdot} + 8\pi G P\right] = 4\pi G\frac{\dot{\psi}}{nH}\left(P - \frac{1}{3}\rho\right). \tag{27}$$

Therefore, from Eq. (25) it arises that $P = \frac{1}{3}\rho$. For vanishing π, we obtain $p = \frac{1}{3}\rho$, which is a radiation equation of state, quite natural in the early universe. Our result is equation (24) with $\rho \propto a^{-4}$ and ψ is expressed in terms of k (and the other cosmological variables) as explained above.
Far from the UV regime, the evolution of $G(k)$, $\Lambda(k)$ is not the simple power laws and Eq. (21) gets the more general form

$$\dot{\rho}+3nH\left(\rho+P\right)+\rho\left(\frac{\dot{G}}{G}+\dot{\psi}\right)=0. \tag{28}$$

This equation expresses a non-trivial energy exchange between the energy density and the time-dependent couplings and has to be solved together with Eq. (24), (25).

4. Conclusions

In this paper, we present the non singular properties of recently found spherical and cosmological solutions of quantum modified Einstein equations according to AS framework. The general solutions contain a variety of smoothing behaviors and several explicit non-singular solutions. The solutions solve the uniquely defined modified Einstein equations which arise by adding appropriate covariant kinetic terms to the equations of motion in order to ensure the satisfaction of the Bianchi identities. We assume that the cosmological constant kinetic terms dominate near NGFP, something that it is expected in the AS. The importance of the presented solutions lies on the fact that they may describe the small distance behavior of quantum corrected spacetimes. In the case of spherical solutions, we claim that these solution may describe the non-singular geometry at the center of black holes, while in the case of cosmology the non-singular solutions most probably represent the geometry of Big bang after a vacuum birth according to asymptotically safe gravity. The cosmological solutions could also be utilized towards other directions that explore high energy corrections to the expansion rate for explaining baryogenesis [63] or dark energy [64].

Although the modified Einstein equations ala AS are uniquely defined, the way we satisfy energy conservation has three alternatives. In this paper we present for the first time the remaining two alternatives of energy conservation we can have. These new two schemes of energy conservation accompany consistently the modified Einstein equations. A surprising fact is that both these schemes are associated with constrained form of equations of state that are compatible with the UV regime. The fact that the possible energy momentum tensors are constrained from the type of energy conservation is desirable and sensible. It is sensible since the form of matter and its interactions are determining which type of energy exchange is allowed and which is not. Furthermore, one of these schemes predicts a radiation equation of state in the early universe.

References

[1] 't Hooft G and Veltman M J G 1974 *Annales Poincaré Phys. Theor.* A **69**
[2] Goroff M H and Sagnotti A 1985 *Phys. Lett.* B **81**
[3] van de Ven A E M 1992 *Nucl. Phys.* B **309**
[4] Niedermaier M and Reuter M 2006 The asymptotic safety scenario in quantum gravity. *Living Reviews in Relativity* **9** 5
[5] Reuter M and Saueressig F 2010 Functional renormalization group equations, asymptotic safety and Quantum Einstein Gravity *Geometric and Topological Methods for Quantum Field Theory* eds H Ocampo, S Paycha and A Vargas (Cambridge Univ. Press, Cambridge), (*Preprint* arXiv:0708.1317)
[6] Percacci R 2009 *Approaches to Quantum Gravity: Towards a New Understanding of Space, Time and Matter* ed D Oriti (Cambridge University Press, Cambridge) (*Preprint* arXiv:0709.3851)
[7] Lauscher O and Reuter M 2007 *Quantum Gravity* eds B Fauser, J Tolksdorf and E Zeidler (Birkhäuser, Basel), (*Preprint* arXiv:0511260)
[8] Reuter M and Saueressig F 2012 *New J. Phys.* 055022 (*Preprint* arXiv:1202.2274)
[9] Koch B and Saueressig F 2014 *Int. J. Mod. Phys.* A **8** 1430011 (*Preprint* arXiv:1401.4452)
[10] Bonanno A 2011 *PoS CLAQG08* 008 (*Preprint* arXiv:0911.2727)
[11] Weinberg S 1979 *General Relativity, an Einstein Centenary Survey* eds S W Hawking and W Israel (Cambridge University Press)
[12] Weinberg S 1997 *Preprint* arXiv:9702027
[13] Weinberg S 2009 *Preprint* arXiv:0903.0568

[14] Weinberg S 2009 *PoS C D09* 001 (*Preprint* arXiv:0908.1964)
[15] Reuter M 1998 *Phys. Rev.* D **971** (*Preprint* arXiv:9605030)
[16] Wetterich C 1993 *Phys. Lett.* B **90**
[17] Morris T R 1994 *Int. J. Mod. Phys.* A **2411** (*Preprint* arXiv:9308265)
[18] Falkenberg S and Odintsov S D 1998 *Int. J. Mod. Phys.* A **607** (*Preprint* arXiv:9612019)
[19] Dou D and Percacci R 1998 *Class. Quant. Grav.* 3449 (*Preprint* arXiv:9707239)
[20] Lauscher O and Reuter M 2002 *Phys. Rev.* D 025013 (*Preprint* arXiv:0108040)
[21] Reuter M and Saueressig F 2002 *Phys. Rev.* D 065016 (*Preprint* arXiv:0110054)
[22] Lauscher O and Reuter M 2002 *Class. Quant. Grav.* 483 (*Preprint* arXiv:0110021)
[23] Lauscher O and Reuter M 2002 *Int. J. Mod. Phys.* A 993 (*Preprint* arXiv:0110054)
[24] Lauscher O and Reuter M 2002 *Class. Quant. Grav.* 483 (*Preprint* arXiv:0112089)
[25] Souma W 1999 *Prog. Theor. Phys.* 181 (*Preprint* arXiv:9907027)
[26] Reuter M and Saueressig F 2002 *Phys. Rev.* D 125001 (*Preprint* arXiv:0206145)
[27] Reuter M and Saueressig F 2004 *Fortschr. Phys.* **52** 650 (*Preprint* arXiv:0311056)
[28] Bonanno A and Reuter M 2005 *JHEP* 035 (*Preprint* arXiv:0410191)
[29] Percacci R and Perini D 2003 *Phys. Rev.* D 081503 (*Preprint* arXiv:0207033)
[30] Percacci R and Perini D 2003 *Phys. Rev.* D 044018 (*Preprint* arXiv:0304222)
[31] Percacci R and Perini D 2004 *Class. Quant. Grav.* 5035 (*Preprint* arXiv:0401071)
[32] Codello A and Percacci R 2006 *Phys. Rev. Lett.* 221301 (*Preprint* arXiv:0607128)
[33] Litim D 2004 *Phys. Rev. Lett.* 201301 (*Preprint* arXiv:0312114)
[34] Fischer P and Litim D 2006 *Phys. Lett.* B 497 (*Preprint* arXiv:0602203)
[35] Percacci R and Perini D 2004 *Class. Quant. Grav.* 5035 (*Preprint* arXiv:0401071)
[36] Codello A, Percacci R and Rahmede C 2008 *Int. J. Mod. Phys.* A 143 (*Preprint* arXiv:0705.1769)
[37] Machado P F and Saueressig F 2008 *Phys. Rev.* D 124045 (*Preprint* arXiv:0712.0445)
[38] Codello A, Percacci R and Rahmede C 2009 *Annals Phys.* 414 (*Preprint* arXiv:0805.2909)
[39] Reuter M and Weyer H 2009 *Phys. Rev.* D 105005 (*Preprint* arXiv:0801.3287)
[40] Reuter M and Weyer H 2009 *Gen. Rel. Grav.* 983 (*Preprint* arXiv:0903.2971)
[41] Reuter M and Weyer H 2009 *Phys. Rev.* D 025001 (*Preprint* arXiv:0804.1475)
[42] Machado P F and Percacci R 2009 *Phys. Rev.* D 024020 (*Preprint* arXiv:0904.2510)
[43] Daum J E and Reuter M 2009 *Adv. Sci. Lett.* 255 (*Preprint* arXiv:0806.3907)
[44] Benedetti D, Machado P F and Saueressig F 2009 *Mod. Phys. Lett.* A 2233 (*Preprint* arXiv:0901.2984)
[45] Benedetti D, Machado P F and Saueressig F 2010 *Nucl. Phys.* B 168 (*Preprint* hep-th/ arXiv:0902.4630)
[46] Benedetti D, Machado P F and Saueressig F *Preprint* arXiv:0909.3265
[47] Daum J.-E, Harst U and Reuter M 2010 *JHEP* 084 (*Preprint* arXiv:0910.4938)
[48] Harst U and Reuter M 2011 *JHEP* 119 (*Preprint* arXiv:1101.6007)
[49] Rahmede C 2011 *PoS CLAQG* 011
[50] Benedetti D, Groh K, Machado P F and Saueressig F 2011 *JHEP* 079 (*Preprint* arXiv:1012.3081)
[51] Bonanno A and Reuter M 2002 *Phys. Rev.* D 043508 (*Preprint* arXiv:0106133)
[52] Bonanno A and Reuter M 2002 *Phys. Lett.* B **9** (*Preprint* arXiv:0106468)
[53] Reuter M and Saueressig F 2013 *Lect. Notes Phys.* 185 (*Preprint* arXiv:1205.5431)
[54] Bonanno A and Reuter M 1999 *Phys. Rev.* D 084011 (*Preprint* arXiv:9811026)
[55] Bonanno A and Reuter M 2000 *Phys. Rev.* D 043008 (*Preprint* arXiv:0002196)
[56] Bonanno A and Reuter M 2006 *Phys. Rev.* D 083005 (*Preprint* arXiv:0602159)
[57] Reuter M and Tuiran E *Preprint* arXiv:0612037
[58] Reuter M and Tuiran E 2011 *Phys. Rev.* D 044041 (*Preprint* arXiv:1009.3528)
[59] Falls K and Litim D F 2014 *Phys. Rev.* D 084002 (*Preprint* arXiv:1212.1821)
[60] Cai Y.-F and Easson D A 2010 *JCAP* 002 (*Preprint* arXiv:1007.1317)
[61] Becker D and Reuter M 2012 *JHEP* 172 (*Preprint* arXiv:1205.3583)
[62] Becker D and Reuter M *Preprint* arXiv:1212.4274

[63] Emoto H 2005 Asymptotic safety of quantum gravity and improved spacetime of black hole singularity by cutoff identification *Preprint* arXiv:0511075
[64] Ward B F L 2006 *Acta Phys. Polon.* B 1967 (*Preprint* arXiv:0605054)
[65] Falls K, Litim D F and Raghuraman A 2012 *Int. J. Mod. Phys.* A 1250019 (*Preprint* arXiv: 1002.0260)
[66] Basu S and Mattingly D 2010 *Phys. Rev.* D **82** 124017 (*Preprint* arXiv:1006.0718)
[67] Casadio R, Hsu S D H and Mirza B 2011 Phys. Lett. B 317 (*Preprint* arXiv:1008.2768)
[68] Contreras C, Koch B and Rioseco P 2013 *Class. Quant. Grav.* 175009 (*Preprint* arXiv:1303.3892)
[69] Koch B and Rioseco P 2015 *Preprint* arXiv:1501.00904
[70] Lauscher O and Reuter M 2005 *JHEP* 050 (*Preprint* arXiv:0508202)
[71] Reuter M and Saueressig F 2011 JHEP 012 (*Preprint* arXiv:1110.5224)
[72] Rechenberger S and Saueressig F 2012 *Phys. Rev.* D 024018 (*Preprint* arXiv:1206.0657)
[73] Calcagni G, Eichhorn A and Saueressig F 2013 *Phys. Rev.* D 124028 (*Preprint* arXiv:1304.7247)
[74] Nink A and Reuter M 2013 *JHEP* 062 (*Preprint* arXiv:1208.0031)
[75] Kofinas G and Zarikas V 2015 *JCAP* **10** 069 (*Preprint* arXiv:1506.02965)
[76] Kofinas G and Zarikas V 2016 *Phys. Rev.* D **10** 103514 (*Preprint* arXiv:1605.02241)
[77] Aliferis G, Kofinas G and Zarikas V 2015 *Phys. Rev.* D **4** 045002 (*Preprint* arXiv:1406.6215)
[78] Kofinas G and Zarikas V 2013 *Eur. Phys. J.* C **4** 2379 (*Preprint* arXiv:1107.2602)

Acknowledgements
Vasileios Zarikas acknowledges the hospitality and support of Nazarbayev University (social policy SPG).

IOP Conf. Series: Journal of Physics: Conf. Series **1051** (2018) 012029 doi:10.1088/1742-6596/1051/1/012029

To the question of a common field theory

S O Gladkov
Professor, Moscow aviation institute (National Research University), Moscow, Russia
E-mail: sglad@newmail.ru

Abstract. The paper suggests an alternative approach devoted to the derivation of the equations of the theory of gravitation, the electromagnetic field and other fields, taking into account their interaction with each other. The theory is based on the calculation of the potential and kinetic energy of the surface of a five-dimensional hypermembrane, where the vector of its displacement $\vec{\xi}$, associated with fields of purely physical nature and it plays the role of the fifth dimension.

1. Basic functional of the theory

In this paper we will discuss one alternative approach devoted to the theory of generalized physical fields, the interaction between which, thanks to the calculation method proposed below, can be calculated exactly with a single parameter of the theory, denoted by the letter L_G and having a dimension of the length. We calculate this parameter precisely, starting from the general properties of the theory of dimensions. Note that the number of articles devoted to the General Theory of Relativity (abbreviated GTR) and the "M-theory" has a huge amount of the work. That's why we will not cite here any original works for the simple reason that simply not to offend those on whom we did not refer, and who either did not notice, or simply missed in the sea of publications on this big topic. In this connection we confine ourselves to a number of classical monographs and we will dwell in detail on our understanding of the question concerning the formal mathematical side of the possible combination of interactions between the currently known physical fields. We also emphasize that the theory presented below is not of a quantum but of a classical nature. It is worthwhile to note in this connection that the transition to the quantum limit is not difficult, since the algorithm for quantizing the amplitudes of interacting fields developed to date is rather well developed. Firstly we begin with some general provisions of GTR. The fundamental functional of the Hilbert-Einstein theory is a classical action and it has the form [1-2]

$$S\left(g^{ik}\right)=\frac{c^4}{\alpha\pi G}\int_{\Omega} R\sqrt{-g}\,d\Omega\,, \tag{1}$$

where $d\Omega = dx^0 dx^1 dx^2 dx^3, dx^0 = cdt$ - the four-dimensional volume in which dt - own time interval and the contravariant components dx^1, dx^2, dx^3 - contravariant coordinates, c - is the velocity of the light in a vacuum, G - constant of gravitation, R - curvature of the space (it will be told just below), which is the convolution of the Riemann tensor, i.e. $R = R^i_i = R^i_{mni}$, where fourth-rank tensor R^i_{klm} - Riemann tensor, $-g$ - determinant of the covariant metric tensor g_{ik}. And at last the parameter α in (1) is the coefficient which is a certain adjustable parameter that ensures a correct transition in the nonrelativistic approximation to the Newton potential, following from the Poisson equation. In [1] it is assumed equal to 16.

In calculating the variation of expression (1) with respect to the components of the contravariant metric tensor g^{ik} the remarkable theorem of Hilbert comes to the aid that the variation of the covariant Ricci tensor R_{ik}, exactly is $\delta R_{ik} = \frac{\partial R_{ik}}{\partial g_{nm}} \delta g_{nm} = 0$. This immediately leads to the Hilbert-Einstein equation in the absence of the matter: $R_{ik} - \frac{1}{2} R g_{ik} = 0$. In the event that matter is taken into account, an energy-momentum tensor T_{ik} with a multiplier appears on the right $\frac{\alpha \pi G}{4c^4}$. As it was noted in [3] that the basic equation of the general relativity theory must always be solved only in the linear approximation with respect to small corrections to the pseudo-Euclidean metric

$$g_{0ik} = \begin{pmatrix} 1 & 0 & 0 & 0 \\ 0 & -1 & 0 & 0 \\ 0 & 0 & -1 & 0 \\ 0 & 0 & 0 & -1 \end{pmatrix}.$$

We note here that this assertion follows from a comparison of the scales of the left and right sides of the basic equation GTR $R_{ik} - \frac{1}{2} R g_{ik} = \frac{k\pi G}{4c^4} T_{ik}$. At the same time, it evidently follows from this that the characteristic scale of the change of the coordinates in order of magnitude should correspond to the distances $\delta x \sim \frac{c}{\sqrt{\rho G}} \sim 10^{13}$ cm (approximately one hundred million kilometers). It is clear that if $\delta x << \frac{c}{\sqrt{\rho G}}$, then we have in the right to not take matter into account at all, and solve the equation of general relativity without the right-hand side of. The latest, however, will relate only to geometry, but not to physics. The GRT equation acquires the physical meaning only in the case when the solution is sought in the form of a small additive correction to the pseudo-Euclidean metric (which was noted in [3]). This fact turned out to be the key to the theory, which will be discussed now. Its main idea is as follows.

Let us imagine a formal dimensional $n + 1$ manifold surrounded by n dimensional hypersurface that makes small oscillations with respect to some fixed equilibrium position with some purely conventional name of the type "Beginning of the world" or "global equilibrium", but for which the main characteristic feature is the condition that the scalar curvature Riemann $R = 0$. Such a rather pathetic name, however, quite adequately characterizes the essence of the described. Indeed, if there exist small oscillations of the hypersurface of four-dimensional space-time, then they can be generated only by the physical fields known now. It is the electromagnetic (potential $A^i = (i\psi, \mathrm{A})$, where i - imaginary unit, A - the magnetic field potential, ψ - the electric field potential (potential $A^i = (i\psi, \mathrm{A})$, the gravitational (potential φ)), strong (potential φ_N), electroweak and, which has some potential φ_D while only a purely hypothetical "dark matter". This means that as soon as at least one of these fields appears, the process of oscillations of the hypersurface with the "simultaneous" appearance of one of the set of Universes of our World begins automatically. Such a picture will not contradict the theory of the "Big Bang" in according to which, at the time (if one can speak about it) bubble bursts, all known and unknown physical fields appear simultaneously, leading (the proof of this, see below) to fluctuations of the space-time Continuum, which is its inalienable property. From a formal point of view, it means that $R \neq 0$. The transition from a flat world $R = 0$ to ours $R \neq 0$ means only that, firstly, there is a kind of "geometric phase transition" and secondly, as a consequence, that a small value δR should be proportional to the deviations of the hypersurface from the position of "global equilibrium". At the same time, it becomes perfectly clear that such hypersurfaces surrounding any material objects can be as many as you like. The magnitude of the deviation, which we will denote as a vector by a letter $\vec{\xi}$ for different physical fields has a different scale of magnitude. For example, for a gravitational potential, it will be much larger than for a nuclear one. From simple analogies with vibrations of a one-dimensional elastic string in a two-dimensional (time-coordinate) case for its potential energy, one can write that $U_1 = B_1 \int\limits_l \frac{dl}{R^2}$, where B_1 - a certain constant that provides the correct dimensionality of energy, which is a rigidity, dl - is the element of a string length, R - its radius of curvature at a given point. By introducing a small displacement of the string $\xi = \xi_y = \xi(x,t)$, on a three-dimensional manifold $\xi - x - t$, the element of length on the pseudo-plane $x - t$ can be represented as with a metric tensor $\hat{g}_{ik} = \begin{pmatrix} a^2 - \xi_t^2 & i\xi_t\xi_x \\ i\xi_t\xi_x & -\left(1+\xi_x^2\right) \end{pmatrix}$, where a - is the velocity of the longitudinal vibrations of the string, $\xi_x = \frac{\partial \xi}{\partial x}$, $\xi_t = \frac{\partial \xi}{\partial t}$ $(c=1)$. In doing so, we took advantage of the presentation of $dx^1 = adt, dx^2 = idx$. As we can see, the determinant of this transformation is $g = -a^2\left(1+\xi_x^2 - \frac{\xi_t^2}{a^2}\right)$. Therefore, the total potential energy can be represented as a double integral:

$$U_1 = -\frac{B_1 a}{L} \int\limits_{\Sigma_2} \frac{\sqrt{-g}\,dxdt}{R^2}, \tag{2}$$

where $d\Sigma_2 = a\sqrt{1+\xi_x^2 - \frac{\xi_t^2}{a^2}}dxdt$ - pseudo-plane element of the hypersurface, L - the parameter having the dimension of the length. In this example this is the length of the string. From the extremum of the functional (2) with respect to the variable ξ, i.e. $\delta U_1(\xi) = 0$, in the case of small oscillations, when, $|\xi_x|$, $|\xi_t| << 1$, we are getting the well-known equation $\frac{\partial^2 \xi}{\partial t^2} = a^2 \frac{\partial^2 \xi}{\partial x^2}$ (see, for example, [4]). Its solution defines a family of extremals that provide a minimum of potential energy (2). In this example, the metric describes a two-dimensional problem on a pseudo-plane, but for a one-dimensional line. The generalization (2) to the n - dimensional case is obvious. In the framework of the problem we are considering and we must write down the potential energy of the five-dimensional hypermembrane in the following form:

$$U_3 = B_3 \int\limits_{\Sigma_4} R d\Sigma_4, \tag{3}$$

where B_3 - is the same of the constant value as in the formula (2), but of a different dimension, the element of the hypersurface $d\Sigma_4 = cdxdydzdt\sqrt{1+\xi_x^2+\xi_y^2+\xi_z^2 - \frac{\xi_t^2}{c^2}}$, but here the value R is the scalar curvature of Riemann. We note the fundamental difference between expression (3) for the potential energy from the Hilbert-Einstein action (1). When constructing the action of S functional in (1), the energy density was multiplied by $dx^0 = cdt$. As a result, the Hilbert-Einstein functional (abbreviated H-E) was obtained, the variation of which leads to the basic equation of GRT in the absence of the matter [1-2]. $\left(R_k^i - \frac{1}{2}R\delta_k^i = 0\right)$ As can be seen from (3), at our disposal there is already ready potential energy describing the displacement of points lying on the hypersurface of the space-time continuum, but not action. This means that in the five-dimensional case the membrane will be a four-dimensional hypersurface with an area element of $d\Sigma_4 = cdxdydzdt\sqrt{1+\xi_x^2+\xi_y^2+\xi_z^2-\frac{\xi_t^2}{c^2}}$. In this case, small displacements of this hypersurface will obey the usual d'Alembert equation $\Box\xi = 0$, where the d'Alembert operator is $\Box = \frac{1}{c^2}\frac{\partial^2}{\partial t^2} - \Delta$, c - the velocity on the three-dimensional hypersurface of the four-dimensional manifold, where $\vec{\xi}$ is the fourth dimension. Formally, we are talking about a five-dimensional continuum, in which time plays the role of the fifth dimension. The reasoning above leads us directly to the formulation of our problem, automatically indicating the course of its solution. Indeed if we assume that the solution of the problem should be sought in a five-dimensional (at least) space in which the formally introduced fourth vector dimension must have a purely physical nature (see below), then the role of the fifth dimension will play time. We note here that the approach under consideration has no points of intersection with the 11-dimensional "M-theory", but has a purely alternative classical character. Thus, the potential energy of oscillations of the four-dimensional hypermembrane in the five-dimensional continuum can be easily written in the obvious form:

$$U_4 = B_4 \int_{\Sigma_\Omega} R d\Sigma_\Omega , \quad (4)$$

where $d\Sigma_\Omega = \sqrt{-g}\,dx^1 dx^2 dx^3 dx^0$ - is the element of its hypersurface in the case of the pseudo-Euclidean metric. The element of the metric on it is representable in the following general form $ds_5^2 = c^2dt^2 - \frac{(dx^1)^2}{k^2} - \frac{(dx^2)^2}{k^2} - \frac{(dx^3)^2}{k^2} - d\vec{\xi}^2$, where the introduced coefficient k allows to automatically take into account the transition from macroscopic dimensions to smaller ones and is simply a scale factor, $\vec{\xi}$ - the vector characterizing a purely physical fifth dimension (see above text). It is a consideration of the influence of various physical fields under the action of which and quite definite oscillations of the hypermembrane occur relative to the position of the "Global Equilibrium". We define it as a vector with components $\vec{\xi} = (\xi_1^i, \xi_2, \xi_3, \xi_4)$, which will be denoted below as ξ^α with a Greek index of α, running four values, exactly $\alpha = 1,2,3,4$ as well as the upper Latin index ξ_1^i, where up index $i = 1,2,3,4$. This four-dimensional displacement vector ξ_1^i, is proportional to $A^i = (\psi, -i\mathrm{A})$ - (A - the potential of the electromagnetic field, ψ - magnetic field potential, electric). As for the proportionality factor, which provides the correct dimensionality for the displacement, it will be strictly calculated below. The displacement $\xi_2 = L_G\frac{\varphi}{c^2}$, where L_G - is the only parameter of the problem, having the dimension of the length, and φ - is the potential of the gravitational field. As to the displacements ξ_3, ξ_4, it is possible, for example, to mean the potential for strong interaction and the potential of the "Dark matter". At least their physical meaning will be completely transparent after the derivation of the basic equation for $\vec{\xi}$. It is worth emphasizing once again that the hypersurface under consideration is an inalienable property of any space-time continuum, and it is always present. Its oscillations are generated by all existing physical fields: gravitational, electromagnetic, nuclear, "Dark matter" (if it exists), etc. Between these fields there are interactions that can be calculated exactly within the framework of the proposed model. They will appear purely formally in terms proportional to the even powers of the displacement vector of the hypersurface ξ^{2n} for $n \geq 2$, and are found from the product of Riemann's scalar curvature R to the Jacobian $\sqrt{-g}$, where $(-g)$ - the determinant of the covariant metric tensor given on this hypersurface. Moreover, the metric on this hypersurface is non-Euclidean (formulas (5) and (6)). The representation of the displacement vector $\vec{\xi}$

of the points of the surface of a physical hypermembrane in the form $\vec{\xi} = \left(\xi_1^i, \xi_2, \xi_3, \xi_4\right)$ is only a hypothesis. However, this assumption allows us to enter into the definition of $\vec{\xi}$ all the basic physical parameters that take into account their interaction with the gravitational field. Let us note, by the way, that this will lead to all the main conclusions of the general theory of relativity and, in particular, to the gravitational waves and the Newton potential.

In addition, we arrive at the correct formulation of the theory of the electromagnetic field, which includes taking into account the gravitational field and the remaining fields (see below). Taking into account that $\vec{\xi} = \left(\xi_1^i, \xi_2, \xi_3, \xi_4\right)$ we obtain

$$ds_5^2 = c^2\left(1 - \frac{\vec{\xi}_t^2}{c^2}\right)dt^2 - dx^2\left(1 + \vec{\xi}_x^2\right) - dy^2\left(1 + \vec{\xi}_y^2\right) - dz^2\left(1 + \vec{\xi}_z^2\right)$$
$$-2dxdy\vec{\xi}_x\vec{\xi}_y - 2dxdz\vec{\xi}_x\vec{\xi}_z - 2dydz\vec{\xi}_y\vec{\xi}_z - 2idxdt\vec{\xi}_x\vec{\xi}_t - 2idydt\vec{\xi}_y\vec{\xi}_t - 2idzdt\vec{\xi}_z\vec{\xi}_t, \quad (5)$$

where i - is an imaginary unit, $dx = \frac{dx^1}{k}$, $dy = \frac{dx^2}{k}$, $dz = \frac{dx^3}{k}$ - the introduced abbreviated notations.

We note that we do not introduce here the covariant and contravariant metric tensors for a purely pseudo-Euclidean metric as for example in [1], but also use very convenient imaginary coordinates due to a simple rule $dx^i = \left(dx^0, i\frac{dx^1}{k}, i\frac{dx^2}{k}, i\frac{dx^3}{k}\right)$, where $dx^0 = cdt$. Within the framework of expression (5) it is simply pointless to speak of a pseudo-Euclidean metric, since the covariant and contravariant metric tensors on this hypersurface are 4×4 matrices. Indeed

$$\hat{g}_{ik} = \begin{pmatrix} 1 - \frac{\vec{\xi}_t^2}{c^2} & -i\frac{\vec{\xi}_x\vec{\xi}_t}{c} & -i\frac{\vec{\xi}_y\vec{\xi}_t}{c} & -i\frac{\vec{\xi}_z\vec{\xi}_t}{c} \\ -i\frac{\vec{\xi}_x\vec{\xi}_t}{c} & -\left(1+\vec{\xi}_x^2\right) & -\vec{\xi}_x\vec{\xi}_y & -\vec{\xi}_x\vec{\xi}_z \\ -i\frac{\vec{\xi}_y\vec{\xi}_t}{c} & -\vec{\xi}_x\vec{\xi}_y & -\left(1+\vec{\xi}_y^2\right) & -\vec{\xi}_y\vec{\xi}_z \\ -i\frac{\vec{\xi}_z\vec{\xi}_t}{c} & -\vec{\xi}_x\vec{\xi}_z & -\vec{\xi}_y\vec{\xi}_z & -\left(1+\vec{\xi}_z^2\right) \end{pmatrix}.$$

The determinant of this covariant tensor is given by a simple formula

$$-g = 1 + \vec{\xi}_x^2 + \vec{\xi}_y^2 + \vec{\xi}_z^2 - \frac{\vec{\xi}_t^2}{c^2}.$$

As can be seen this expression reduces to the formulas (2) and (3) with decreasing dimensionality, as it should be. Hence the sought potential energy of a hypermembrane will be

$$U_4 = \frac{c^4}{\alpha\pi G L_G}\int_{\Sigma_\Omega} R\sqrt{1 + \vec{\xi}_x^2 + \vec{\xi}_y^2 + \vec{\xi}_z^2 - \frac{\vec{\xi}_t^2}{c^2}}d\omega, \quad (6)$$

where $d\omega = dxdydzdx^0$ - a hypersurface element and $dx^0 = cdt$. To calculate the variation of the functional (6) with respect to the vector displacement function of the surface of the hypermembrane $\vec{\xi}\left(\mathrm{r}, x^0\right)$, we must first calculate the integrand in (6) in terms of the displacement function $\vec{\xi}\left(\mathrm{r}, x^0\right)$. To do this, proceed as follows. We shall not search separately for the quantities entering into (6), but we immediately use the definition of the scalar curvature of the Riemann tensor, which we write in the usual form as $R = R^i_i = R^i_{kki}$, where $R^i_{klm} = \frac{\partial \Gamma^i_{kl}}{\partial x^m} - \frac{\partial \Gamma^i_{km}}{\partial x^l} + \Gamma^i_{ln}\Gamma^n_{km} - \Gamma^i_{mn}\Gamma^n_{kl}$. Hence, the scalar is

$$R = R^i_{kki} = \frac{\partial \Gamma^i_{kk}}{\partial x^i} - \frac{\partial \Gamma^i_{ki}}{\partial x^k} + \Gamma^i_{kn}\Gamma^n_{ki} - \Gamma^i_{in}\Gamma^n_{kk}.$$

Since the Cristoffel's symbol is $\Gamma^i_{kl} = \frac{1}{2} g^{is}\left(\frac{\partial g_{sk}}{\partial x^l} + \frac{\partial g_{sl}}{\partial x^k} - \frac{\partial g_{kl}}{\partial x^s}\right)$, then

$$\Gamma^i_{kk} = \frac{1}{2} g^{is}\left(2\frac{\partial g_{sk}}{\partial x^k} - \frac{\partial g_{kk}}{\partial x^s}\right), \text{ and } \Gamma^i_{ki} = \frac{1}{2g}\nabla_k g.$$

In the first approximation, we are only interested in terms quadratic in bias $\vec{\xi}$ in the potential energy (6). The next approximation as is easy to see will be of order ξ^4, and allows us to take into account the various anharmonisms associated with the interaction of the fields, as we discussed above. To calculate the quadratic terms, we will need separate Cristoffel's symbols. Indeed for example $\Gamma^0_{00} = \frac{1}{2} g^{00}\frac{\partial g_{00}}{\partial x^0} = -\frac{\left(1 + \vec{\xi}_x^2 + \vec{\xi}_y^2 + \vec{\xi}_z^2\right)}{g}\vec{\xi}_t\vec{\xi}_{tt} \approx \vec{\xi}_t\vec{\xi}_{tt}$, where we can assume that the determinant of the metric tensor is approximately equal to -1, because the magnitude of the displacement ξ is small in comparison with the length parameter $\frac{c}{\sqrt{\rho G}}$, where ρ - is the density of the matter. Note that we assume here and further the velocity of light is unity ($c = 1$). As a result, we find

$$\Gamma^1_{11} \approx i\frac{\vec{\xi}_x\vec{\xi}_{xx}}{g} = -i\vec{\xi}_x\vec{\xi}_{xx},\ \Gamma^2_{22} \approx i\frac{\vec{\xi}_y\vec{\xi}_{yy}}{g} = -i\vec{\xi}_y\vec{\xi}_{yy},\ \Gamma^3_{33} \approx i\frac{\vec{\xi}_z\vec{\xi}_{zz}}{g} \approx -i\vec{\xi}_z\vec{\xi}_{zz}.$$

Similarly, all other components of the Cristoffel's symbol are calculated. For example, $\Gamma^1_{33} = -i\vec{\xi}_x\vec{\xi}_{zz}$, $\Gamma^1_{00} = i\vec{\xi}_x\vec{\xi}_{tt}$, $\Gamma^1_{22} = -i\vec{\xi}_x\vec{\xi}_{yy}$. As a result we have that $\Gamma^0_{kk} = i\vec{\xi}_x\square\vec{\xi}$, $\Gamma^2_{kk} = i\vec{\xi}_y\square\vec{\xi}$, $\Gamma^3_{kk} = i\vec{\xi}_z\square\vec{\xi}$, where $\square$ - is the D'Alembert operator. Therefore, in the approximation that is quadratic in the bias $\vec{\xi} = \left(\xi^i_1, \xi_2, \xi_3, \xi_4\right)$, we find the following expression

$$R \approx \frac{\partial \Gamma^i_{kk}}{\partial x^i} - \frac{\partial \Gamma^i_{ki}}{\partial x^k} = \frac{\partial}{\partial t}\left(\vec{\xi}_t\,\square\vec{\xi}\right) - \frac{\partial}{\partial x}\left(\vec{\xi}_x\,\square\vec{\xi}\right) - \frac{\partial}{\partial y}\left(\vec{\xi}_y\,\square\vec{\xi}\right) - \frac{\partial}{\partial z}\left(\vec{\xi}_z\,\square\vec{\xi}\right) + \frac{1}{2}\square g. \quad (7)$$

Hence,

$$\square g = -2\left(\vec{\xi}_x\,\square\vec{\xi}_x + \vec{\xi}_y\,\square\vec{\xi}_y + \vec{\xi}_z\,\square\vec{\xi}_z - \vec{\xi}_t\,\square\vec{\xi}_t\right) - 2\left(\xi^\alpha_{xt}\xi^\alpha_{xt} + \xi^\alpha_{yt}\xi^\alpha_{yt} + \xi^\alpha_{zt}\xi^\alpha_{zt} - \xi^\alpha_{tt}\xi^\alpha_{tt}\right)$$
$$+2\left(\nabla\xi^\alpha_x\cdot\nabla\xi^\alpha_x + \nabla\xi^\alpha_y\cdot\nabla\xi^\alpha_y + \nabla\xi^\alpha_z\cdot\nabla\xi^\alpha_z - \nabla\xi^\alpha_t\cdot\nabla\xi^\alpha_t\right).$$

Therefore, from (7), we have

$$R \approx 2(\ \square\vec{\xi}\ )^2 + \nabla\xi^\alpha_x\cdot\nabla\xi^\alpha_x + \nabla\xi^\alpha_y\cdot\nabla\xi^\alpha_y + \nabla\xi^\alpha_z\cdot\nabla\xi^\alpha_z - \nabla\xi^\alpha_t\cdot\nabla\xi^\alpha_t - \xi^\alpha_{xt}\xi^\alpha_{xt} - \xi^\alpha_{yt}\xi^\alpha_{yt} - \xi^\alpha_{zt}\xi^\alpha_{zt} + \xi^\alpha_{tt}\xi^\alpha_{tt}.$$

And, consequently, the potential energy of a hypermembrane curved by physical fields can be written in the form of the following quadratic functional:

$$U_4 \approx \frac{c^4}{\alpha\pi G L_G}\int_{\Sigma_\Omega}\Big(2(\ \square\vec{\xi}\ )^2 + \nabla\xi^\alpha_x\cdot\nabla\xi^\alpha_x + \nabla\xi^\alpha_y\cdot\nabla\xi^\alpha_y + \nabla\xi^\alpha_z\cdot\nabla\xi^\alpha_z - \nabla\xi^\alpha_t\cdot\nabla\xi^\alpha_t$$
$$-\xi^\alpha_{xt}\xi^\alpha_{xt} - \xi^\alpha_{yt}\xi^\alpha_{yt} - \xi^\alpha_{zt}\xi^\alpha_{zt} + \xi^\alpha_{tt}\xi^\alpha_{tt}\Big)d\omega. \quad (8)$$

Recall that the "nabla" operator involved here means differentiation only with respect to the spatial coordinates, that is $\nabla = \left(\frac{\partial}{\partial x}, \frac{\partial}{\partial y}, \frac{\partial}{\partial z} \right)$. Varying expression (8) by displacements ξ^{α} we easy obtain

$$\delta U_4 \approx \frac{2c^4}{\alpha \pi G L_G} \int\limits_{\Sigma_\Omega} \square^2 \vec{\xi} \delta \vec{\xi} d\omega . \tag{9}$$

To take into account the kinetic energy of the surface of the oscillating hypermembrane, let us write down its relativistic invariant expression, introducing the displacement velocity $\frac{\partial \vec{\xi}}{\partial t}$ in the following form:

$$E_k = \frac{c^2}{2L_G} \int\limits_{\Sigma_G} \rho_{\alpha\beta} \frac{\partial \xi^{\alpha}}{\partial x^i} \frac{\partial \xi^{\beta}}{\partial x^k} d\omega , \tag{10}$$

where $\rho_{\alpha\beta}$ is the matrix of the density of the corresponding matter and as a rule should have a diagonal form $\rho_{\alpha\beta} = \rho_{\alpha} \delta_{\alpha\beta}$, where $\delta_{\alpha\beta}$ - the Kronecker's symbol. In accordance with this the general formula (10) can be rewritten as:

$$E_k = \frac{c^2}{2L_G} \int\limits_{\Sigma_G} \rho_{\alpha} \frac{\partial \xi^{\alpha}}{\partial x^i} \frac{\partial \xi^{\alpha}}{\partial x^k} d\omega , \tag{11}$$

where. $dx^i = (cdt, idx, idy, idz)$. Varying expression (11) by $\vec{\xi}$, after elementary integration by parts we obtain

$$\delta E_k = \frac{c^2}{2L_G} \int\limits_{\Sigma_G} \rho_{\alpha} \frac{\partial \xi^{\alpha}}{\partial x^i} \frac{\partial \delta \xi^{\alpha}}{\partial x^i} d\omega = -\frac{c^2}{L_G} \int\limits_{\Sigma_G} \rho_{\alpha} \frac{\partial^2 \xi^{\alpha}}{\left(\partial x^i\right)^2} \delta \xi^{\alpha} d\omega = -\frac{c^2}{L_G} \int\limits_{\Sigma_G} \rho_{\alpha} \delta \vec{\xi} \,\square \vec{\xi} d\omega . \tag{12}$$

The sum of the expressions (9) and (12) must be zero, i.e

$$\frac{2c^4}{\alpha \pi G L_G} \int\limits_{\Sigma_\Omega} \square^2 \vec{\xi} \delta \vec{\xi} d\omega - \frac{c^2}{L_G} \int\limits_{\Sigma_\Omega} \rho_{\alpha} \delta \vec{\xi} \,\square \vec{\xi} d\omega = 0 .$$

Therefore, because the variation of $\delta \vec{\xi}$ are the arbitrariness we are coming to the desired equation for the displacement function

$$\square^2 \xi^{\alpha} = \kappa_{\alpha}^2 \xi^{\alpha} , \tag{13}$$

where parameter $\kappa_{\alpha}^2 = \frac{\alpha \pi G \rho_{\alpha}}{2c^2}$. On the basis of equation (13), we can make the following rather important statement. The above displacement $\vec{\xi}$ is an inherent property of any matter and thanks to this parameter, we have at our disposal the distribution of all fields in the space-time continuum, which can be mutually generated by any other physical fields (electromagnetic, strong, gravitational, dark matter, etc.). It should be noted that H-E equations do not have such an opportunity, since they contain a completely different idea, namely, that the metric properties of the space-time continuum change due to the presence of matter. This rather fundamental difference allows us, with the help of the proposed model to introduce, for example, the interaction of a gravitational field with a nuclear potential, to assert that it leads to the formation of an electromagnetic field. Similarly the EM field can interact with the nuclear potential $\varphi = \varphi\left(A^i, \psi_N\right)$, as a result of which a gravitational field with a potential is generated. It is worth emphasizing that the marked effects will appear only in the nonlinear approximation with respect to the displacement of the surface $\vec{\xi}$ and therefore with very weak interaction, which is in general quite unsurprising.

2. The basic equations

Indeed, for the displacement $\vec{\xi}$ in the absence of nonlinear terms in the metric of the hypersurface, and in the absence of currents (we shall take them into account below), we have equation (13), which, when taking into account the nonlinear terms $\vec{\xi}$ must be of a much more complicated form:

$$\Box^2 \xi^\alpha - \kappa_\alpha^2 \xi^\alpha = B_{\beta\gamma\delta}^{\alpha ijklmn} \frac{\partial}{\partial x^i}\left(\xi_{x^j}^\beta \xi_{x^k x^l}^\gamma \xi_{x^n x^m}^\delta\right), \tag{14}$$

where the down suffix members of $\vec{\xi}$ mean the partial derivatives with respect to the corresponding arguments, i.e. $\xi_{x^i}^\alpha = \frac{\partial \xi^\alpha}{\partial x^i}$, $\xi_{x^i x^k}^\alpha = \frac{\partial^2 \xi^\alpha}{\partial x^i \partial x^k}$. A dimensionless tensor of sixth rank $B_{\beta\gamma\delta}^{\alpha ijklmn}$ can be easily calculated from the general definitions for g_{ik} and g^{ik}. We also note that the length parameter L_G introduced above is also included in the definition of the tensor $B_{\beta\gamma\delta}^{\alpha ijklmn}$. As we can see the function itself $\vec{\xi}$ is not explicitly included in the definition of the metric. Its definition includes only derivatives of displacement $\vec{\xi}$. Therefore in order to find them we differentiate equation (14) with respect to x^i. Then, $\Box^2 \xi_{x^i}^\alpha = \kappa_\alpha^2 \Box \xi_{x^i}^\alpha$. The solution of this equation, like equations (14), is completely identical and in a spherically symmetric case must have purely radial dependences. In order to verify this, we return to equation (14), and find its solution. To do this, it is convenient to enter a new function $u^\alpha = -\Box \xi^\alpha$. It follows from (14) that $\Box u^\alpha - \kappa_\alpha^2 u^\alpha = 0$, where $\kappa_\beta^2 = \frac{\alpha \pi G \rho_\beta}{2c^2}$. In the simplest variant we assume that the solution of the equation can be sought in the form of an expansion into the Fourier integral with respect to the time argument, i.e. $u^\alpha(\mathrm{r},t) = \frac{1}{2\pi}\int_{-\infty}^{\infty} u_\omega^\alpha(\mathrm{r}) e^{-i\omega t} d\omega$, where $u_\omega^\alpha(\mathrm{r})$ - is a Fourier–image of the function u^α. As a result the equation for the Fourier transformations is $\Delta u_\omega^\alpha + a_\omega^2 u_\omega^\alpha = 0$, where $a_\omega = \sqrt{\kappa_\alpha^2 + \frac{\omega^2}{c^2}}$ - is the parameter. After all the simple mathematical actions the final solution can be written in the following form

$$\xi^\alpha(\mathrm{r},t) = \frac{1}{4\pi r}\int_{-\infty}^{\infty} \frac{\cos \omega r}{\omega} e^{-i\omega t} \left[\frac{C_{1\omega}^\alpha\left(e^{i(\omega + a_\omega)R} - 1\right) + C_{2\omega}^\alpha\left(e^{-i(\omega + a_\omega)R} - 1\right)}{a_\omega + \omega} - \frac{C_{1\omega}^\alpha\left(e^{i(a_\omega - \omega)R} - 1\right) + C_{2\omega}^\alpha\left(e^{-i(a_\omega - \omega)R} - 1\right)}{a_\omega - \omega} \right] d\omega.$$

If we assume that the frequency of oscillations is an monochromatic, i.e. $\xi_\omega^\alpha(\mathrm{r}) = 2\pi \xi_{\omega_0}^\alpha(\mathrm{r}) \delta(\omega - \omega_0)$, we get

$$\xi^\alpha(\mathrm{r},t) = \frac{\cos \omega r}{2r\omega} \left[\frac{C_{1\omega}^\alpha\left(e^{i(\omega + a_\omega)R} - 1\right) + C_{2\omega}^\alpha\left(e^{-i(\omega + a_\omega)R} - 1\right)}{a_\omega + \omega} - \frac{C_{1\omega}^\alpha\left(e^{i(a_\omega - \omega)R} - 1\right) + C_{2\omega}^\alpha\left(e^{-i(a_\omega - \omega)R} - 1\right)}{a_\omega - \omega} \right] e^{-i\omega t},$$

where $\omega = \omega_0 = const$. As can be seen in the stationary case for $\omega = 0$, we find $\xi^\alpha(\mathrm{r}) = -\frac{2R}{r\kappa}\cdot\frac{C_1^\alpha e^{i\kappa R} - C_2^\alpha e^{-i\kappa R}}{2i}$. Considering, that $C_1^\alpha = C_2^\alpha = -C_\alpha$, we will find

$$\xi^\alpha(\mathrm{r}) = C_\alpha \frac{2R}{r} \frac{\sin \kappa R}{\kappa}.$$

From the definition of the displacement vector of the hypermembrane $\vec{\xi}=\left(\xi_1^i,\xi_2,\xi_3,\xi_4\right)$ and because of the smallness of κ, we obtain classical solutions for the gravitational static field in the form $\xi_2(\mathrm{r})=L_G\frac{\varphi}{c^2}=C\frac{2R}{r}$. From the condition of "cross-linking" at the boundary of $-\nabla\varphi\big|_{r=R}=\mathrm{g}=G\frac{m\mathrm{R}}{R^3}$, where g - is the acceleration of free fall, and m - is the mass of the body, as it should be, we come to Newton's law $\varphi=\frac{mG}{r}$. However the most physical solution can be found after introducing the interaction of "currents" leading to inhomogeneous equations, in contrast to the homogeneous (14) (see below). Thanks to the smallness of the product κR, one can also obtain the potential of the electric field, which appears in the definition $\vec{\xi}=\left(\xi_1^i,\xi_2,\xi_3,\xi_4\right)$ in the form of a vector $\xi_1^i=L_G\left(\gamma\psi,-\gamma i\mathrm{A}\right)$, where A - is the vector potential, ψ - is the electric potential (about parameter γ we'll talk a little lower). So we have $\psi=\frac{q}{r}$, where q - is a charge. We also note that, from the above general solution, we obtain delayed potentials of the electromagnetic field, as well as other consequences of the theory of the electromagnetic field. As a result $\xi_{x^i}^{\alpha}(\mathrm{r})=\frac{D_i^{\alpha}}{r}$, where D_i^{α} - some permanents. By virtue of this solution, the covariant metric tensor is very easy to write in explicit form. Its determinant is $g=-\left(1+\frac{b_{\alpha}^2}{k^{-2}r^2}\right)$, where $b_{\alpha}^2=\left(D_1^{\alpha}\right)^2+\left(D_2^{\alpha}\right)^2+\left(D_3^{\alpha}\right)^2-\left(D_0^{\alpha}\right)^2$. As can be seen from the solution (14), all known, including purely hypothetical fields that is the gravitational, electromagnetic, nuclear and "Dark matter" should influence the metric properties of the space-time continuum. As for the D_i^{α} they enter the solution for the time being as phenomenological constants. If we assume that $D_2^{\alpha}=D_3^{\alpha}=0$, we arrive to a metric $g_{00}=1-\frac{D_0^2}{k^{-2}r^2}$, $g_{11}=-1-\frac{D_1^2}{k^{-2}r^2}$, $g_{10}=g_{01}=-i\frac{D_0D_1}{k^{-2}r^2}$, where $D_0^2=D_0^{\alpha}D_0^{\alpha}$, $D_0D_1=D_0^{\alpha}D_1^{\alpha}$, $D_1^2=D_1^{\alpha}D_1^{\alpha}$, quite strongly resembling the decisions of Schwarzschild and Kerr. In this case the metric tensor will have the following form:

$$\hat{g}_{ik}=\begin{pmatrix} 1-\frac{D_0^2}{k^{-2}r^2} & -i\frac{D_0D_1}{k^{-2}r^2} & 0 & 0 \\ -i\frac{D_0D_1}{k^{-2}r^2} & -\left(1+\frac{D_1^2}{k^{-2}r^2}\right) & 0 & 0 \\ 0 & 0 & -1 & 0 \\ 0 & 0 & 0 & -1 \end{pmatrix}. \tag{15}$$

Here it is worth noting that although the dependence of the components of the metric tensor in the Schwarzschild solution is $\frac{1}{r}$, and not $\frac{1}{r^2}$, as in (15), this is by and large the same thing since the correctness of the solution (15) is verified, as well as the Schwarzschild solution to the present moment time, unfortunately, impossible. Moreover in this connection underline that the decision (15) describing the metric on the hypersurface but not the volume one as in Schwarzschild's decision. In addition, we'd like note that in the solution (15) was obtained, the off-diagonal components of the metric tensor $g_{01}=g_{10}$ turned out to be non-zero. Perhaps some suspicion that in Schwarzschild they were simply lost are creeping in. In order for formula (15) not to look purely formal, we need to take

IOP Conf. Series: Journal of Physics: Conf. Series **1051** (2018) 012029 doi:10.1088/1742-6596/1051/1/012029

into account in the general equation (14) the currents on the surface of the hypermembrane. For this it is convenient to introduce the following quantities:

$$\varepsilon_f = L_G \int_{\Sigma_\Omega} f^a \square \xi^a d\omega\,,\; f^a = \frac{c^2}{L_G}\left(\frac{\rho_q q u_i}{L_G c^2}, -\rho, -\rho_N, -\rho_D\right),$$
$$\xi^a = L_G\left(-\frac{L_G A^i}{q}, \frac{\varphi}{c^2}, \frac{\varphi_N}{c^2}, \frac{\varphi_D}{c^2}\right), \tag{16}$$

where ε_f - is the energy associated with the displacement of the surface by the external "currents", f - is the density of the volume force, all of ρ - are the corresponding to the density (charge, matter, nuclear matter, dark matter). ξ^a - displacement of the surface of the hypermembrane, which is chosen in the only possible form, preserving the correct dimensionality for the energy ε_f. The dimensionless four-velocity

$$u^i = \frac{1}{\sqrt{1-\frac{v^2}{c^2}}}\left(1, i\frac{v}{c}\right)$$

and the contravariant four-potential of the EM field is $A^i = (\psi, -i\mathrm{A})$. When the energy (16) is varied by the displacement ξ^a after double integration by parts and using the Stokes formula for transforming the integral over the hypersurface into an integral over the hypercontour, in which the variation $\delta\xi^a = 0$, and then substituting the result in the right-hand side of equation (14) after an elementary transformation equation with allowance for nonlinear terms:

$$\square^2 \xi^a - \kappa^2 \square \xi^a = -L_G^2 \frac{4\pi G}{c^4} \square f^a + B^{aijklmn}_{\beta\gamma\delta} \frac{\partial}{\partial x^i}\left(\xi^\beta_{,x^j} \xi^\gamma_{,x^k x^l} \xi^\delta_{,x^n x^m}\right). \tag{17}$$

Let us analyze equation (17) in the case when there are no interactions that is a dimensionless tensor $\hat{B} = 0$.

3. Parameter L_G

In fact, choosing as the first example the electromagnetic potential, that is, assuming that the displacement due to it is $\xi_1^i = -L_G^2 \frac{A^i}{e}$, we are finding from equation (17)

$$\square^2 A^i - \kappa^2 \square A^i = -L_G^2 \frac{4\pi G}{c^4} \square f^1.$$

Substituting here from (16) f^1 and "removing" one D'Alembert operator we get:

$$\square A^i - \kappa^2 A^i = 4\pi\rho_q u^i \frac{e^2 G}{L_G^2 c^4}. \tag{18}$$

In order that in the limiting case of the absence of gravity that is neglecting the second term on the left-hand side of the eq. (18) it becomes the usual equation of the theory of electromagnetic potential we should do this

$$L_G = \frac{e}{c^2}\sqrt{G}. \tag{19}$$

Then instead of (18) we obtain a generalized equation of the EM potential theory with allowance for the influence of gravitational forces:

$$\Delta A^i - \frac{1}{c^2}\frac{\partial^2 A^i}{\partial t^2} + \kappa^2 A^i = -4\pi\rho_q u^i. \tag{20}$$

Substituting the numerical values of the constants in (19) leads to the following: $L_G \sim 10^{-34}$ cm. This negligible size, nevertheless, will play an important role if we need to take into account the nonlinear interactions between the fields (see below) that enter the equation (17) in the form of tensor $\hat{B}$. As we see from equation (17) in the absence of the matter and nonlinear interactions, when $\rho_\alpha = 0$, $\kappa = 0$ and $\hat{B}$, the solution should be sought in the form $\xi(\mathrm{r},t) = \xi_0 e^{i\mathrm{kr}-i\omega t}$, where ξ_0 - the amplitude of the oscillations. As a result, we immediately obtain the dispersion equation $\left(k^2c^2 - \omega^2\right)^2 = 0$, from which the dispersion laws of the gravitational and electromagnetic waves automatically follow in the usual form $\omega = ck$. If we return to eq. (17) and put $\xi_2 = L_G \dfrac{\varphi}{c^2}$ we immediately obtain from it $\dfrac{L_G}{c^2} \square^2 \varphi - \kappa^2 \dfrac{L_G}{c^2} \varphi = L_G^2 \dfrac{\alpha\pi G c^2}{c^4 L_G} \square \rho$. "Removing" here the D'Alembert operator and, making simple abbreviations we are find up to the fundamental solution $\dfrac{C}{r}$, i.e. $\square \varphi - \kappa^2 \varphi = \alpha\pi G\rho + \dfrac{C}{r}$. Setting $C = 0$, in the stationary case, we obtain the Helmholtz equation for the gravitational potential, in which the constant α should be set equal to 4 and hence

$$\Delta\varphi + \kappa^2 \varphi = -4\pi G\rho. \tag{21}$$

The solution of equation (21) is

$$\varphi(\mathrm{r}) = G \int\limits_{V_{\Sigma\Omega}} \frac{\rho(\mathrm{r}')\cos \kappa|\mathrm{r}-\mathrm{r}'|}{|\mathrm{r}-\mathrm{r}'|} dV'. \tag{22}$$

Assuming the density is homogeneous, the integral is easily calculated in a spherical coordinate system centered at its center. As a result we have the following one

$$\varphi(\mathrm{r}) = \frac{2\pi G\rho}{r} \int\limits_0^R r'dr' \int\limits_0^\pi \cos \kappa\sqrt{r^2 - 2rr'\cos\theta + r'^2}\, d\left(\sqrt{r^2 - 2rr'\cos\theta + r'^2}\right).$$

The internal integral along the azimuthal angle can be taken trivially and, as a result, the solution for the external problem will be:

$$\begin{aligned}\varphi^+(\mathrm{r}) &= \frac{2\pi G\rho}{r\kappa} \int\limits_0^R r'dr'\left[\sin \kappa(r + r') - \sin \kappa|r - r'|\right] = \frac{4\pi G\rho\cos \kappa r}{\kappa r} \int\limits_0^R r'\sin \kappa r'dr' \\ &= \frac{4\pi G\rho\cos \kappa r}{r\kappa^3}(\sin \kappa R - \kappa R\cos \kappa R).\end{aligned} \tag{23}$$

Consequently $\varphi^+(\mathrm{r}) = \dfrac{4\pi G\rho\cos \kappa r}{r\kappa^3}(\sin \kappa R - \kappa R\cos \kappa R)$. Since $\kappa R \ll 1$ then immediately follows Newton's law with an extremely small correction:

$$\varphi^+(\mathrm{r}) = \frac{MG}{r}\left(1 - \frac{(\kappa r)^2}{2} - \frac{(\kappa R)^2}{10}\right), \tag{24}$$

where M - is the mass of a spherical object. We recall that this solution was obtained from the solution of the external problem for the Helmholtz equation. It is easy to verify that for the inner problem, starting from the general solution given above, after simple computations, we obtain:

$$\varphi^-(\mathrm{r}) = 2\pi G\rho\left(R^2 - \frac{r^2}{3} - \frac{(\kappa r)^2 R^2}{6} - \frac{\kappa^2 R^4}{4} + \frac{(\kappa r)^2 r^2}{60}\right). \tag{25}$$

As can be seen from the solutions (24) and (25), they are continuously "sewn" on the boundary of the body when $r = R$, that is $\varphi^+(\mathrm{r})\big|_{r=R} = \varphi^-(\mathrm{r})\big|_{r=R}$.

4. Conclusion

To summarize, we note:

1. The functional that was used to derive the basic equation (17) is the total energy of the hypermembrane in the space-time continuum. The sum of the potential energy and work produced by all external non-conservative forces, which differs significantly from the Hilbert-Einstein method, where the approach of classical mechanics, connected with the calculation of the variation of the action functional.
2. All the physics of the problem being solved lies in the fourth spatial coordinate having vector character $\vec{\xi}$, all components of which are determined by the potentials of the physical fields.
3. The proposed approach makes it possible to find the change in the metric under the influence of different types of fields, and to calculate the interaction between fields of different physical nature, which is important in the microscopic study and evaluation of quantum contributions to various gravitational effects.

References

[1] Landau L D and Lifshitz E M 2008 *Field theory* V. 2 (Moscow: Science)
[2] Logunov A A, Mestvervishvili M A and Petrov V A 2004 How the Hilbert-Einstein equations were calculated *Preprint* (Protvino: Institute of High Energy Physics)
[3] Gladkov S O 2017 To the question of a common field theory *Int. Meeting on Physical Interpretations of Relativity Theory* (Moscow: BMSTU)
[4] El'sgolz L E 1969 *Differential equations and calculus of variations* (Moscow)
[5] Koshlyakov N S, Gliner E B and Smirnov M M 1970 *Equations in partial derivatives of mathematical physics* (Moscow: Higher school)

IOP Conf. Series: Journal of Physics: Conf. Series **1051** (2018) 012030 doi:10.1088/1742-6596/1051/1/012030

The basic principles of building an analytics platform for the search of new scientific knowledge

V O Gladyshev[1], D A Bazlev[2] and A V Kayutenko[1]
[1] Bauman Moscow State Technical University, Moscow, Russia
[2] Central Aerohydrodynamic Institute (TsAGI), Zhukovsky, Russia
E-mail: vgladyshev@mail.ru

Abstract. By analyzing the progress in the development of ideas and techniques for recording gravitational waves, the paper introduces a vector characteristic for assessing the future development of scientific knowledge $\vec{R}_i(t)$, taking into account new dynamic scientometric indicators. This characteristic is a function of time and depends on the priority assessment of the scientific problem $\vec{P}_i(t)$ of the divergence of the flow of new knowledge $\vec{Z}_i(t)$ stimulated by the solution obtained to this problem, or a new result in this area, as well as the time-varying scientometric vector potential $\vec{N}_i(t)$.

1. Introduction

The global development of information resources, the increasing speed of obtaining and accumulating information in various fields of knowledge have led to the need to analyze large amounts of data (Big Data) and to automatically single out previously unknown knowledge from these data, which can result in real benefits (Data Mining).

To a certain extent, this development is caused by the search for effective tools of managing the processes of obtaining new knowledge. Hence, bibliographic databases have been widely spread in recent years: Scopus, Ingenta, Google Scholar, Scholar.ru, Science Direct, Sci Net - Science search, Science Research Portal, Scirus, Cite Seer Publications Research Index, and others. To varying degrees these databases use scientometric indices to assess the potential of authors (researchers) and organizations (Universities, etc.). As a rule, these indices are based on identifying the citation of the articles published in scientific journals.

The use of these indices has become the norm for scientific foundations when allocating funding for projects and research teams and is a powerful tool of affecting the development of science.

If an article is published in a scientific journal or other publication, it is certain to have been reviewed and undoubtedly contains a new scientific achievement, i.e. new knowledge.

Moreover, the higher the ranking of the journal, the higher the level of the review and, consequently, the accuracy of the result is more reliable, i.e. its novelty, priority, authenticity are undeniable.

Furthermore, we often associate the ranking of the journal with the result level, i.e. its importance, significance, complexity of the research and analysis methods used. We also expect the effects the result can possibly produce on the development of scientific thought and its applications. It is a reasonable expectation, as in high-ranking journals the competition is higher as well.

However, it is obvious that even one journal publishes the articles that are distinctly important for scientific and technological progress. Thus, for example, theoretical journals can publish articles containing some innovative ideas and hypotheses that may later turn out to be erroneous, but, perhaps, they will encourage researchers to find the right solution to the problem.

In this regard, it becomes obvious that the adopted system of assessing the significance of research and the potential of the research team or the author is far from being perfect, and it is advisable to further look for the distinctive signs of scientific research and their results.

2. The problem of assessment of acquiring new scientific knowledge

The history of the development of science and scientific discoveries makes it possible through the existing examples to evolve some distinctive characteristics of scientific ideas that led to the significant development of science and technology.

For example, one of the fundamental problems of the 20th century was the problem of detecting gravitational radiation from distant space objects.

The existence of gravitational waves (GW) was theoretically predicted by Albert Einstein in his work devoted to calculating the power of gravitational radiation [1]. In [1-2] Einstein established the transversality of GW and two degrees of polarization. Later Arthur S. Eddington showed that flat GWs, as well as waves coming from a point source, propagate at the speed of light [3].

In 1988, Academician Vladimir A. Fok was the first who drew attention to the possibility of using astrophysical catastrophes as a source of powerful gravitational radiation [4].

Further calculations showed that a possible change in the distance between test masses, due to the action of a gravitational wave of cosmic origin, is very small. Therefore, for a long time the problem of detecting gravitational radiation remained the subject of theoretical research.

The existence of gravitational radiation was first confirmed in experimental studies by Joseph H. Taylor, Joel M. Weisberg and others [5-6]. These research works investigated the effect of the slowing down period of the binary stellar system PSR 1913 + 16 due to the energy losses for gravitational radiation. The obtained results accurately coincided with the calculated values, obtained according to the solution of general relativity equations.

However, many years before, different laboratories began their experimental search of methods for recording gravitational waves from space radiation sources by terrestrial gravitational antennas.

Since the general theory of relativity was built, various methods for detecting gravitational radiation have been introduced, most of which have remained unrealized either because of the lack of the method sensitivity, or because of the complexity of its technical implementation.

The article [7] contains the main ideas and gives a classification of methods for recording gravitational waves, which makes it possible to systematize the basic physical principles of recording.

In general, the most promising and technically well-provided are the projects of laser interference gravitational antennas, already built, or being constructed, with sensitivity reserves.

The idea of using a pair of free mass-mirrors and a laser interferometer to record their small oscillations caused by GWs was first proposed by Vladislav I. Pustovoit and Mikhail E. Gertsenshtein in [8].

This type of broadband gravitational antennas contains a lot of possibilities concerning the methods for recording GW, signal extraction methods, the possibility of including gravitational antennas into the network.

The main element of the laser interference GA is Fabry-Perot multi-beam free-mass resonator, on the properties of which the sensitivity and noise immunity of GA is largely dependent [9].

In scientific literature one can find the articles describing the significant progress in the use of interferometers for recording gravitational waves in the projects of VIRGO (Italy, France), LIGO (USA), TAMA (Japan), CLIO (Japan), GEO-600 (Germany) and OGran (Russia).

The first direct detection of gravitational waves by the LIGO and VIRGO collaborations was announced on February 11, 2016. The results were published in the journal Physical Review Letters [10] and brought about an extensive discussion [11].

However, long before this success in the search for gravitational waves, experimental attempts were made to record gravitational waves using the idea suggested by Joseph Weber [12]. It is on record that the first gravitational antennas were built as solid-state detectors. It took years of experiments and theoretical studies to come to the conclusion that the more promising is the idea of using displacement interferometers.

This story gives some insights into the assessment of a promising scientific idea, which is closely related to the technological progress in the field of laser, vacuum, electronic, computer technology, manufacturing of high-quality optics and solving many other problems.

3. The elements of information analysis system of scientific knowledge

When the first gravitational antennas were built, computer technology was not used yet, and it impossible to comprehensively assess the future development of the method for recording gravitational waves.

The use of computer technology in assessing scientific projects also leads to the need to introduce new vector or multivariate characteristics of the development of scientific knowledge that can analytically reflect the logic of assessing the prospects of a particular study. It would be safe to assume that such an assessment can be based on extrapolation from the available scientometric data.

At the same time, the indices used are of an integral nature and do not take into account the dynamic pattern. In particular, this leads to a young scientist or research team losing the competition, other factors being equal, although their development could be rapid.

As a result, there arises the need to develop new scientometric indices that take into account the dynamic behavior of the estimated results. The set of both adopted and new indices, can be represented by the vector potential $\vec{N}_i(t)$ of the researcher or research team.

Moreover, a drawback of the existing system for assessing scientific projects is the fact that not the scientific result itself is evaluated, but its indirect indicators (for example, citedness of the publication, or the impact factor of the journal).

To solve this problem, we need to create a system for a more accurate assessment of scientific knowledge obtained, i.e. theoretical or experimental result, technology, and give a unique digital code to it. For this purpose, it is necessary to develop a universal template of the scientific result, which would enable classification of new and known scientific knowledge in a semi-automatic mode.

The creation of such an information system for classification of scientific knowledge is an inevitable step within the conditions of the rapid growth of scientific publications throughout the world. This will prevent the results from being duplicated, as well as reduce the number of works that do not contain new scientific results. Coding scientific work will essentially mean the certification of copyright on the obtained scientific result, similar to patenting. Editorial boards of scientific journals should reject papers that contain a result duplicating already registered scientific knowledge. Literature references should contain sources with the results codes, on the basis of which a new article is based. It is these works that should receive citation indices, as the new result would not be possible without these works.

Thus, according to one scientific result, a certain number of new scientific results can be obtained at a fixed point in time, which, in turn, can become the basis for obtaining other new results. All these works represent a certain set that characterizes the power of the initial result, which became the basis for the development of scientific knowledge. This value can be identified as a number of connections between the basic work and its derivatives, i.e. made not on its basis. Because the basic work is the source of new knowledge, its significance can be assessed by the divergence of the flow of new knowledge

$$\vec{Z}_i(t) = div\,F = \lim_{V\to 0}\frac{\Phi_F}{V},\ \Phi_F = \oint_S \left(\vec{F}, d\vec{S}\right), \tag{1}$$

where Φ_F is the flow of the vector field **F** through the spherical surface of the area S, bounding the volume V. The field is a vector field. Each new scientific work is characterized by a vector in the space of scientific knowledge.

The divergence of the flow of new knowledge for each work is also a function of time. Moreover, the assessment of the priority of the scientific task $\vec{P}_i(t)$ depends on time, which is an important factor in determining whether it is reasonable or not to support a particular study.
The progress in the field of science can be defined as the non-zero vector of development $\vec{R}_i(t)$ (Figure 1). However, the development or progress does not result in a breakthrough, intermittent growth of scientific knowledge, for this purpose mutual reinforcing influence of various factors is necessary.

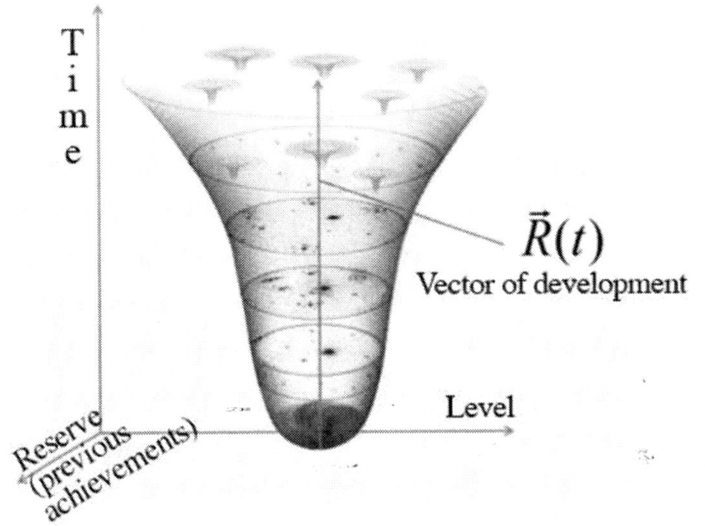

Figure 1. The vector of development $\vec{R}_i(t)$ does not necessarily end in the area where a breakthrough is possible, i. e. intermittent accumulation of knowledge and technology.

It is obvious that the expectation of a breakthrough in a particular field of knowledge should be determined not by an additive but by a multiplicative law, including the divergence of the flow of new knowledge $\vec{Z}_i(t)$, priority assessment $\vec{P}_i(t)$ and the scientometric vector potential $\vec{N}_i(t)$. It should be noted that material-and-technical, organizational and other characteristics are not taken into account here. They are commonly used when comparing scientific projects and teams. This is due to the fact that our goal is to assess the scientific potential and its ability to acquire new knowledge, i.e. other factors being equal.
In the simplest case, the extrapolation operation can be introduced as follows:

$$\vec{R}_i(t) = \vec{P}_i(t) \times \vec{Z}_i(t) \times \vec{N}_i(t)\,. \qquad (2)$$

Thus, a possible breakthrough is predicted by the superposition or intersection of three vector fields.
Let us note that currently in the field of applied and basic research, the priority assessment $\vec{P}_i(t)$ is not used explicitly, forming a list of promising, breakthrough lines of research. Therefore, this field can be formalized according to the available data.

4. Conclusion

In general, the analysis leads to the conclusion that the significant progress in the field of scientific development will occur when an information analysis system relying on the four basic elements is introduced: the extension of the standard set of scientometric indices, introduction of dynamic indices;

introduction of a universal classifier of scientific knowledge; formation of the priority task database; development of an algorithm for estimating the feasibility and forecasting a breakthrough.

References

[1] Einshtein A 1965 Approximate integration of gravitational equations *Coll. sc. pap. in 5 vol.* **1** (Moscow) pp 514–523
[2] Einshtein A 1965 On gravitational waves. *Coll. sc. pap. in 5 vol.* **1** (Moscow) pp 631–646
[3] Eddington A 1992 *Proc. Roy. Soc.* A102 pp 268–282
[4] Fok V A 1955 Theory of space, time and gravitation (Moscow, GITTL Publ.) 504 p
[5] Taylor J H and Weisberg J M 1982 A new test of general relativity: gravitational radiation and the binary pulsar PSR 1913+16 *Astrophysical Journal.* **253** 2 pt.1 pp 908-920
[6] Taylor J H 1982 Gravitational radiation and the binary pulsar *Proc. 2-th Marcel Gross. Meeting on Gen. Relativity 1979 in Amsterdam* pp 15-19
[7] Gladyshev V O and Morozov A N 2000 Classification of Gravitational-Wave Antennas by the Methods of Gravitational Radiation Detection *Measurement Techniques* **43** 9 pp 741-746
[8] Gertsenshtein M E and Pustovoit V I 1962 To the problem of detection of low frequency gravitational waves *ZhETF - Journal of Experimental and Theoretical Physics* **43** 2 (8) pp 605–607
[9] Gladyshev V O and Morozov A N 1993 Low-frequency optical resonance in a multiple-wave Fabry-Perot interferometer *Technical Physics Letters* **19** 7 pp 449-451
[10] Abbott B P *et al* (LIGO Scientific Collaboration and Virgo Collaboration) 2016 Observation of Gravitational Waves from a Binary Black Hole Merger *Physical Review Letters* **116** 6
[11] Pustovoit V I 2016 On the direct detection of gravitational waves *Phys. Usp.* **59** pp 1034–1051
[12] Weber J 1977 *Topics in Theoretical and Experimental Gravitation Physics* (New-York, London, Plenum Press)

Acknowledgements

The work was supported by Russian Foundation for Basic Research grants no. 16-02-00488 A and no. 16-08-00618 A.

Effects of moving media optics in GLONASS optical segment of new generation

V O Gladyshev, A G Strunin, V L Kauts, A V Kayutenko and D D Bazleva

Bauman Moscow State Technical University, Moscow, Russia
E-mail: vgladyshev@mail.ru

Abstract. The study introduces a system of equations describing the propagation of light rays in laser location of a moving heterogeneous medium. A recurrent system of equations is obtained for the algorithm of modeling ray trajectories in a moving space microsatellite in the form of a two-dimensional Luneberg lens. A numerical calculation and analysis of the interference optical response of the moving Luneberg lens on the examples of BLITS and BLITS-M satellites is carried out. The results obtained indicate a complex influence of the effects of the moving media optics (Doppler and Fizeau effects, Snell's law violation, dispersion phenomenon, etc.) on the spatial and temporal characteristics of the reflected interference signal on the Earth's surface, namely on the signal-to-noise ratio in the reception area.

1. Introduction

The rapid development of global satellite positioning systems is approaching the millimeter limit of measurement accuracy. A promising way to solve this problem can be the use of modern laser ranging systems, as well as the launch of reference low-orbit passive laser satellites [1].

The study [2] shows that a spherical gradient Luneberg lens can be used as a reflecting satellite. Reflectors made on the basis of gradient lenses are a good alternative to cubic angle reflectors [3]. The main property of such reflectors is the spherical aberration compensation. Calculations of miniature spherical reflectors based on the Luneberg lens [4-7] indicate a good restoring capability for a large range of horizontal viewing angles in the medium and long wavelength range of the infrared emission spectrum.

The optical response of BLITZ satellite, which is a Luneberg lens in design, is comparable with the response of a point source, and can be used to improve the accuracy of determining the coordinates in GLONASS system.

However, the optical response of a moving lens has a complex spatial structure and depends on the speed of its movement. Effects of the moving media optics affect the reflected signal of the moving lens, the accuracy of determining its orbit, which in its turn can affect the reliability of satellite laser ranging system as a whole [8].

Consequently, there arises a necessity for simulating the processes of forming the interference optical response on the earth's surface, and also for calculating the intensity distribution of the reflected coherent radiation. The mathematical model must take into account:

1) effects of the moving media optics (the Doppler effect, Fizeau effect, violation of Snell's law, dispersion phenomenon);

2) kinematic effects associated with the lens displacement during the propagation of electromagnetic radiation in it;

IOP Conf. Series: Journal of Physics: Conf. Series **1051** (2018) 012031 doi:10.1088/1742-6596/1051/1/012031

3) characteristic features of time formation of a multipath interference response.
It should be noted that in the general case the ray path bending effect in a rotating lens [9], as well as the Fermi effect [10], influences the solution of the problem. However, due to low rotation speeds, these effects can be neglected in the first approximation.
This task, apart from practical, is of fundamental importance. The reference optical satellite, orbiting in the optical GLONASS segment, provides unique opportunities for experimental verification of the equations of moving media electrodynamics.

2. Refraction on a moving media interface

The formation of the reflected interference response depends on the parameters of the laser radiation passing through the Luneberg lens at all points of intersection of the ray with the optical surfaces of the two-layer lens. Since the surfaces move relative to the Earth, it is necessary to use the solution of the problem of the fall of a flat monochromatic wave onto the moving media interface [11].

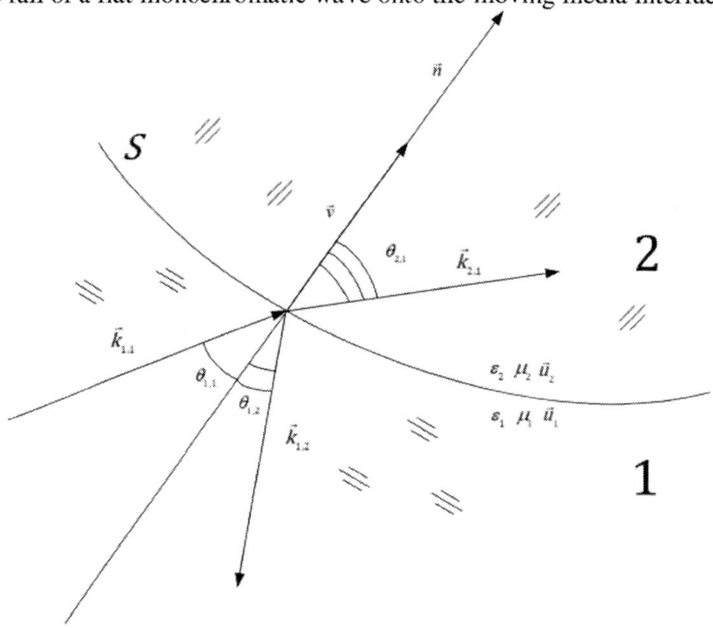

Figure 1. An electromagnetic wave with a wave vector $\vec{k}_{1,1}$ falls on the two moving media interface S, where media are moving at random velocities $\vec{u}_i$ and the interface is moving at normal velocity $\vec{v}$.

The boundary conditions for all waves in the first and second medium are the phase equalities at the two media interface. According to this requirement, phase invariants are introduced:

$$\vec{k}_{1,1t} = \vec{k}_{1,2t} = \vec{k}_{2,1t} = \vec{I}_t \quad (1)$$
$$I_t = k_0 \sin\theta_0$$

$$\omega_{1,1} - \vec{k}_{1,1n} v = \omega_{1,2} - \vec{k}_{1,2n} v = \omega_{2,2} - \vec{k}_{2,2n} v = I_1, \quad (2)$$

where $\vec{k}_{it}$ and $\vec{k}_{in}$ are the tangential and normal components of the wave vector with respect to the surface S at the point of intersection.
To find the frequencies and wave vectors in media, dispersion equations are used for each medium [2]:

IOP Conf. Series: Journal of Physics: Conf. Series **1051** (2018) 012031 doi:10.1088/1742-6596/1051/1/012031

$$\vec{k}_i^2 - \frac{\omega_i^2}{c^2} - \frac{\kappa_i}{c^2}\frac{\left(\omega - \vec{k}_i\vec{u}_i\right)^2}{1-\beta_i^2} = 0, \tag{3}$$

where $\vec{k}_i$ is the wave vector of an electromagnetic wave in the i-th medium, $\vec{u}_i$ is the velocity of the i-th layer of the medium, $\kappa_i = \varepsilon_i\mu_i - 1$, $\vec{\beta}_i = \frac{\vec{u}_i}{c}$, $i = 1,2$.

The solution of this equation for the spectral source of the incident radiation far from the medium absorption region looks as follows

$$(\omega_i)_{1,2} = I_1 \frac{1 + \kappa_i\gamma_i^2(\beta - \beta_{in})\left(\beta_{in} - \vec{d}\vec{\beta}_{it}\beta\right) \pm \beta Q_i^{1/2}}{1 - \beta^2 - \kappa_i\gamma_i^2(\beta - \beta_{in})^2} \tag{4}$$

$$Q_i = 1 + \kappa_i\gamma_i^2\left(\beta - \beta_{in}^2\right) - d^2\left[1 - \beta^2 - \kappa_i\gamma_i^2(\beta - \beta_{in})^2\right]$$
$$-\kappa_i\gamma_i^2\vec{d}\vec{\beta}_{it}\left[2(1 - \beta\beta_{in}) - (1 - \beta^2)\vec{d}\vec{\beta}_{it}\right],$$

$$\vec{d} = -\frac{c\vec{I}_t}{I_1},\ \beta = \frac{v}{c},\ \vec{\beta}_{in} = \frac{\vec{u}_{in}}{c},\ \vec{\beta}_{it} = \frac{\vec{u}_{it}}{c},\ \gamma_i^{-2} = 1 - \beta_i^2,$$

where the sign $\pm$ and indices 1,2 correspond to the refracted and reflected wave respectively, $\vec{u}_{it}$ and $\vec{u}_{in}$ are the tangential and normal components of the medium velocity with respect to the surface S at the point of intersection.

The phase invariant $\vec{I}_t$ specifies the tangential components of the wave vectors in two media. For the normal component, substituting (4) into (2), we obtain

$$(k_{in})_{1,2} = \frac{I_1}{c}\frac{\beta + \kappa_i\gamma_i^2(\beta - \beta_{in})\left(1 - \vec{d}\vec{\beta}_{it}\right) \pm Q_i^{1/2}}{1 - \beta^2 - \kappa_i\gamma_i^2(\beta - \beta_{in})^2}. \tag{5}$$

The refraction angle of an electromagnetic wave θ_i at the media interface is given by

$$\theta_i(x,y) = \mathrm{arctg}\left(\frac{\omega\sin\theta_0}{ck_{in}(x,y)}\right). \tag{6}$$

Differential equations governing the evolution of the wave vector coordinates in a medium with a complex two-dimensional motion will have the following form:

$$\frac{dx}{dt} = \frac{c}{n_i}\frac{1}{\sqrt{1 + \mathrm{tg}^2\theta_i(x,y)}},\ \frac{dy}{dt} = \frac{c}{n_i}\frac{\mathrm{tg}\theta_i(x,y)}{\sqrt{1 + \mathrm{tg}^2\theta_i(x,y)}}. \tag{7}$$

Thus, the system of equations (5) - (7) makes it possible to calculate the radiation parameters at the points of intersection of the ray and the moving media interface along the path.

3. Dispersion of an optically transparent medium

The expression (3) includes the refractive index $n = \sqrt{\mu\varepsilon}$, which in the general case is not a constant. The medium parameters can depend on the parameters of the radiation passing through it; therefore, in order to take into account the dispersion, it is necessary to recalculate the refractive index before each media interface. In the case of a moving two-layer Luneberg lens it is important, because on each surface the Doppler effect will be observed.

As the dependence of the refractive index on the wavelength of radiation, we choose Schott's approximation formula [12]:

$$n=\sqrt{A_1+A_2\lambda^2+\frac{A_3}{\lambda^2}+\frac{A_4}{\lambda^4}+\frac{A_5}{\lambda^6}+\frac{A_6}{\lambda^8}}. \tag{8}$$

Substituting (8) into (3), we obtain a 12th degree equation with respect to frequency, which can be solved by a numerical method:

$$\begin{aligned}&-\frac{1}{256c^8\left(c^2-u_1^2\right)\pi^8}\Big(A_6\omega^{12}-2A_6\omega^{11}P+\left(4A_5\pi^2c^2+A_6P^2\right)\omega^{10}-8A_5\omega^9\pi^2c^2P+\\&\left(16A_4\pi^4c^4+4A_5\pi^2c^2P^2\right)\omega^8-32A_4\omega^7\pi^4c^4P+\left(16A_4\pi^4c^4P^2+64A_3\pi^6c^6\right)\omega^6-128A_3\omega^5\pi^6c^6P+\\&\left(64A_3\pi^6c^6P^2-256\pi^8c^8u_1^2+256A_1\pi^8c^8\right)\omega^4+\left(-512A_1\pi^8c^8P+512\pi^8c^8P\right)\omega^3+\\&\left(256k^2u_1^2\pi^8c^8-256\pi^8c^8P^2+256A_1\pi^8c^8P^2-256k^2\pi^8c^{10}+1024A_2\pi^{10}c^{10}\right)\omega^2-\\&2048A_2\pi^{10}c^{10}\omega P+1024A_2\pi^{10}c^{10}P^2\Big)=0,\end{aligned} \tag{9}$$

where $P=\left(\vec{k},\vec{u_1}\right)$.

The dispersion effect was calculated in two ways. In the first case, we used a numerical solution of the algebraic equation (9) done by the step-by-step approximation method. The calculations were carried out until the left side of the equation converged to zero with the required degree of accuracy.

In the second case, the solution was found in an iterative way. At the first stage, we solved the dispersion equation (4) for a constant n and found the frequency in the medium behind the moving interface.

Then, using formula (8), we calculated the new value $n(\omega)$. For the given new value of the refractive index, we again solved the dispersion equation and repeated the process to ensure convergence (in practice, 2-3 iterations are sufficient).

4. Analytical description of the light ray propagation in an optical system of arbitrary configuration

Let us consider the propagation of coherent electromagnetic radiation in an arbitrary optical system in the geometric approximation. In this case, the flux of electromagnetic radiation can be considered as a beam of light rays characterized by spatial direction, amplitude, and phase. Imagine a light ray as a straight line p_i incident at a point $M_i\left(x_i,y_i,z_i\right)$ lying on a moving surface of the second order P_i (Figure 2). The ray comes out from the point M_{i-1} with coordinates $x_{i-1}, y_{i-1}, z_{i-1}$.

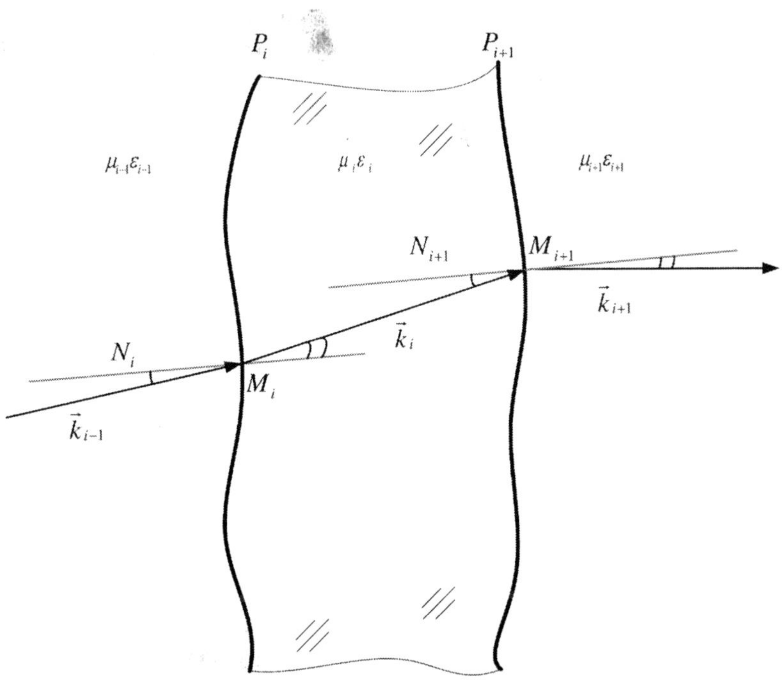

Figure 2. Propagation of an electromagnetic wave $\vec{k}_i$ in a medium considered as a set of layers, with parameters ε_i, μ_i and separated by surfaces P_i with normals N_i. Each interface between two adjacent media can have a velocity V_i. At each media interface, a refracted and a reflected waves appear.

The parametric equations of the straight line p_{i-1} have the following form:

$$x = x_{i-1} + l_{i-1}t\,,\ \ y = y_{i-1} + m_{i-1}t\,,\ \ z = z_{i-1} + n_{i-1}t\,, \tag{10}$$

where l_i, m_i, n_i are the components of the ray propagation velocity in the i-th medium.

As P_i is the second degree surface, the moving optical surface is governed by the second degree algebraic equation $F_i(x, y, z, t) = 0$:

$$\begin{aligned} &a_{i\,11}x^2 + a_{i\,22}y^2 + a_{i\,33}z^2 + a_{i\,44}t^2 \\ &+a_{i\,12}xy + a_{i\,13}xz + a_{i\,14}xt + a_{i\,23}yz + a_{i\,24}yt + a_{i\,34}zt + a_{i\,34}zt \\ &+a_{i\,15}x + a_{i\,25}y + a_{i\,35}z + a_{i\,45}t + a_{i\,55} = 0. \end{aligned} \tag{11}$$

Substituting (10) into (11), we obtain a quadratic equation with respect to time:

$$A_i t^2 + A_i t + C_i = 0\,, \tag{12}$$

where

$$\begin{aligned} A_i = {} & a_{i\,44} + 2a_{i\,14}l_{i-1} + a_{i\,11}l_{i-1}^2 + 2a_{i\,24}m_{i-1} + 2a_{i\,12}l_{i-1}m_{i-1} + a_{i\,22}m_{i-1}^2 \\ & +2a_{i\,34}n_{i-1} + 2a_{i\,13}l_{i-1}n_{i-1} + +2a_{i\,23}n_{i-1}m_{i-1} + a_{i\,33}m_{i-1}^2, \end{aligned}$$

$$\begin{aligned}B_i = &2a_{i\,45}+2a_{i\,14}x_{i-1}+2a_{i\,24}y_{i-1}+2a_{i\,34}z_{i-1}\\&+2a_{i\,15}l_{i-1}+2a_{i\,11}x_{i-1}l_{i-1}+2a_{i\,12}y_{i-1}l_{i-1}+2a_{i\,13}z_{i-1}l_{i-1}\\&+2a_{i\,25}m_{i-1}+2a_{i\,12}x_{i-1}m_{i-1}+2a_{i\,22}y_{i-1}m_{i-1}+2a_{i\,23}z_{i-1}m_{i-1}\\&+2a_{i\,35}n_{i-1}+2a_{i\,13}x_{i-1}n_{i-1}+2a_{i\,23}y_{i-1}n_{i-1}+2a_{i\,33}z_{i-1}n_{i-1},\end{aligned}$$

$$\begin{aligned}C_i = &a_{i\,55}+2a_{i\,15}x_{i-1}+a_{i\,11}x_{i-1}^2+2a_{i\,25}y_{i-1}^2+2a_{i\,12}x_{i-1}y_{i-1}+a_{i\,22}y_{i-1}^2\\&+2a_{i\,35}z_{i-1}+2a_{i\,13}x_{i-1}z_{i-1}+2a_{i\,23}y_{i-1}z_{i-1}+a_{i\,33}z_{i-1}^2.\end{aligned}$$

Then the solution of equation (12) is the time the ray travels to the i-th surface:

$$t_i = \frac{-B_i \pm \sqrt{B_i^2 - 4A_iC_i}}{2A_i}. \tag{13}$$

The sign in the expression is determined by the direction of ray propagation. Substituting t_i into equations (10) we find the coordinates of the point $M_i(x_i, y_i, z_i)$ of intersection of the line and the plane.

Thus, we obtain a recurrent system of equations (10-13), which makes it possible to describe the ray path in an arbitrary moving heterogeneous medium between adjacent media interfaces. Similarly, one can obtain expressions for the incident angle of the ray onto an arbitrary surface [13].

5. Propagation of radiation in the optical microsatellite BLITS

The study [13] describes the ray path in a two-dimensional moving Luneberg lens through the example of BLITS satellite. Fig. 3 shows the image of the central section of the lens by the plane in which the ray propagates in the limit when the velocity of the lens is zero. As the speed of the lens increases, the angle of refraction changes, the ray intersection with the surfaces shifts, the surfaces themselves move along the velocity vector of the lens.

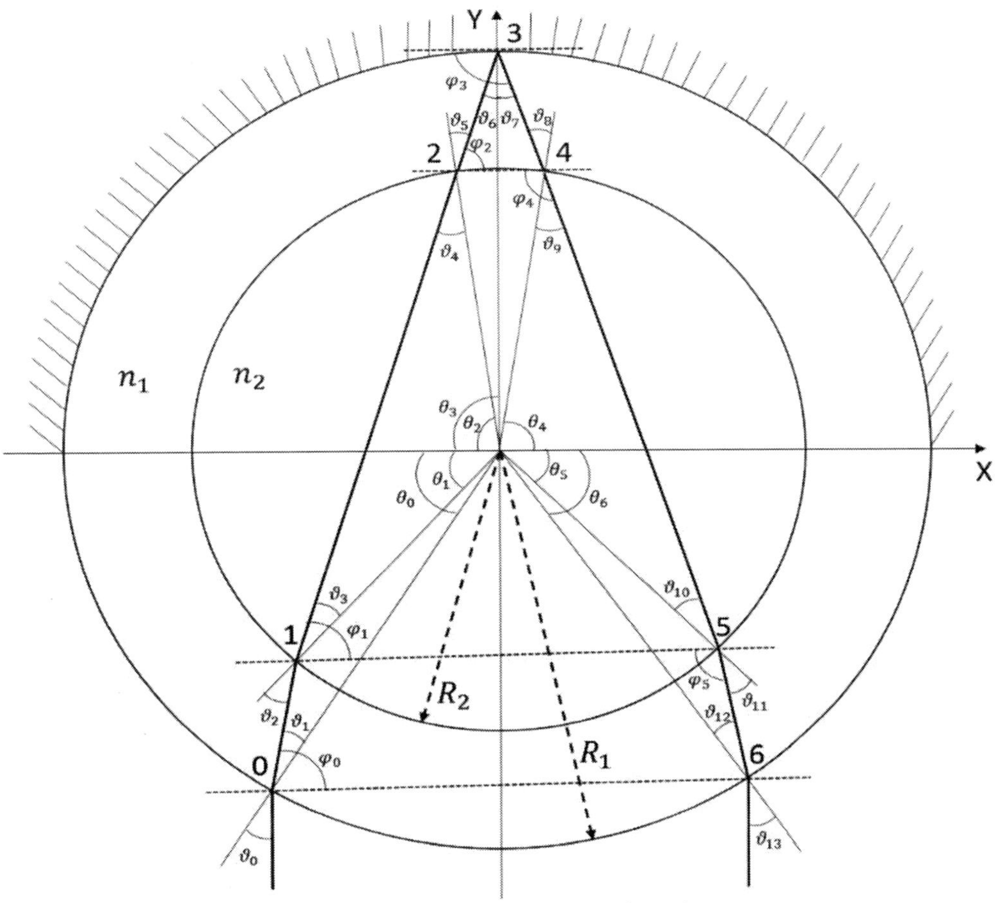

Figure 3. Ray path in the Luneberg lens.

From Figure 3 it follows that the angles θ_i are the angles of incidence and refraction, and φ_i, θ_i are the auxiliary angles.
Kinematic equations of the ray and circle are as follows:

$$\begin{cases} x_i = x_{i-1} + \dfrac{c}{n_{1,2}} t_i \cos(\varphi_{i-1}) \\ y_i = y_{i-1} \pm \dfrac{c}{n_{1,2}} t_i \sin(\varphi_{i-1}) \end{cases} \tag{14}$$

$$\left(x_k - v_x\left(t_k + T\right)\right)^2 + \left(y_k - v_y\left(t_k + T\right)\right)^2 = R_{1,2}^2, \tag{15}$$

where $k = 1 \ldots 6$, $T = \sum_{k=1}^{i} t_{k-1}$ is the total time for which the lens center shifted when considering the k-th point, $R_{1,2}$ and $n_{1,2}$ are the radii and refractive indices of the outer meniscus and inner ball. The "+" sign is chosen due to the direction of the ray along the Oy axis.
Substituting the ray equation (14) into the circle equation (15), we obtain a quadratic equation in terms of time:

$$A_k t_k^2 + B_k t_k + C_k = 0, \tag{16}$$

where

$$A_{1-3,4-6} = \left(\frac{c}{n_{1,2}}\sin(\varphi_i) \mp v_y\right)^2 + \left(\frac{c}{n_{1,2}}\cos(\varphi_i) - v_x\right)^2$$

$$B_{1-3,4-6} = 2(-Tv_x + x_{k-1})\left(\frac{c}{n_{1,2}}\cos(\varphi_i) - v_x\right) - 2(-Tv_y + y_{k-1})\left(v_y \pm \frac{c}{n_{1,2}}\sin(\varphi_i)\right)$$

$$C = (Tv_x + x_{k-1})^2 + (Tv_y + y_{k-1})^2 - R_{1,2}^2.$$

Coefficients with indices 1-3 correspond to the rays before reflection, coefficients with indices 4-6 correspond to the rays after reflection (respectively, they choose $\mp$ and $\pm$ signs), the coefficient C is not changed for both cases. Solving the equation (16), we obtain the time t_k value of the ray path between k-1 and k-points. Then we substitute the obtained value t_k into (14) and obtain the coordinates of the point where the ray intersects the surface.

To find new values of the angles φ_{i-1}, θ_i, it is necessary to know the tangential and normal components of the incident radiation wave vector with respect to the surface at points (0-6). To do this, we obtain the normal to the circle at the points (0-6) by the following formula:

$$\vec{n} = (2x_i, 2y_i). \tag{17}$$

Knowing the projection of the wave vector $\vec{k}_i$ onto the tangent and the normal to the surface at the points (0-6), we can use the phase invariants (1), (2) and the solution of the dispersion equation (5) to find the new wave vector $\vec{k}_{i+1}$. New angles φ_i, θ_{i+1} can be determined with the help of the scalar product $(\vec{k}_{i+1}, \vec{n})$ and $(\vec{k}_{i+1}, \vec{h})$, where $\vec{h}$ is the horizontal $\vec{h} = (1,0)$.

6. Calculation of the interference response in the registration plane

To obtain the ray deviation value on the earth's surface, let us find the angle θ between the refracted ray at point 6 - $\vec{k_7}$ and the horizontal $\vec{h}$. Then the ray deviation value on the earth's surface will be calculated by the formula:

$$x = y_6 \mathrm{ctg}\theta + x_6. \tag{18}$$

In this case, we assume that the center of the circles is taken as the zero along the Ox axis at the moment the ray intersects the first surface.

For the phase of the ray on the earth's surface, we can write:

$$\Phi = \sum_{i=0}^{n} w_i t_i - w_7 \Delta t - \sum_{i=0}^{n} k_i \Delta r_i, \tag{19}$$

where w_i, t_i, k_i, Δr_i are respectively, the angular frequency, the ray propagation time, the wave vector modulus and the length of the geometric path in the i section of the trajectory, $i = (0...n)$ is the number of the optical path segment in the optical system under study between the media interfaces, $\Delta t = t_{def} - \sum_{i=0}^{n} t_i$ is the difference in the time of passage of the reference (t_{def}) and calculated rays.

The intensity distribution in the observation plane of the interference pattern depends on the number of rays from the total number N falling into each coordinate interval Δx.

$$I(x) = I_0 \left(\sum_{j=1}^{N} \sin\hat{O}_j(x) \right)^2, \tag{20}$$

where I_0 is the value of the intensity of one ray, which in practice has the value of the order of 10^{-6} W/m^2.

7. The results of the numerical experiment for BLITZ and BLITZ-M satellites

The calculations were done for the parameters of BLITZ and BLITZ-M satellites of identical design (Table 1).

Table 1 - Design parameters of BLITZ and BLITZ-M

Parameters	BLITZ	BLITZ-M
Radius of the central ball $R1$, mm	53.5	63.9
Radius of the outer meniscus $R2$, mm	85	110.4
Material of the central ball	LK106	K108
Material of the external meniscus	TF105	TF105
Orbit h, km	835	1500
Linear velocity in orbit V, m/s	7500	7100

The intensity distribution graphs are plotted for the rays $N = 20000$ with incidence angles $\theta_0 = 0..7^\circ$ to the left and right of the lens axis (Figure 3). The coordinate interval $\Delta x = 1$ m was chosen for the summation of the phases (18).

The calculations used a two-dimensional approximation, so all the incident and reflected rays were in the same plane.

Figure 4 shows a graph of the intensity distribution on the earth's surface for a stationary satellite BLITZ. There are 3 intensity peaks on the graph. One is in the reception area ($X = 0$ m) and the two others are outside it ($X = \pm 74$ m).

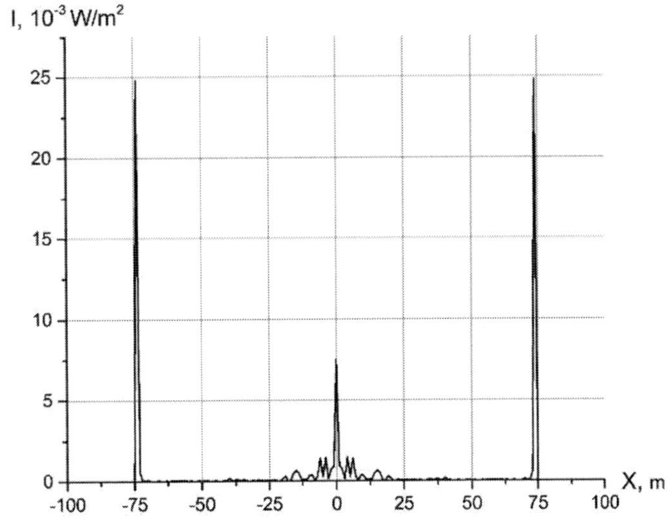

Figure 4. Dependence of the intensity $I(x)$ of the reflected radiation from the coordinate on the earth's surface at the satellite BLITZ velocity $V = 0$ m/s.

Figure 5 shows a similar graph at the satellite velocity $V = 7500$ m/s. It is noteworthy that the lateral intensity peaks shifted to the sides and markedly decreased in intensity.

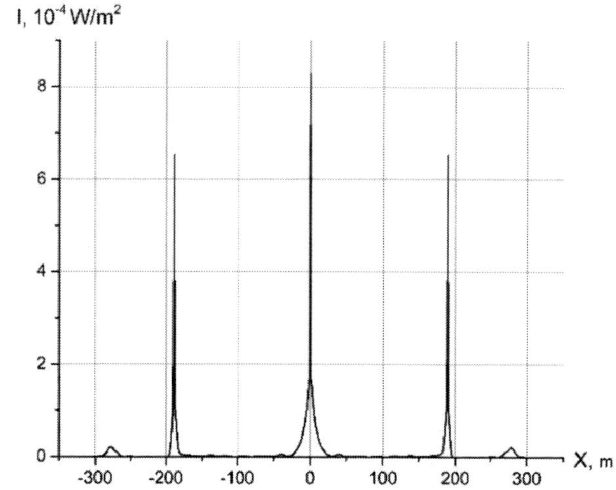

Figure 5. - Intensity distribution at the satellite BLITZ velocity $V = 7500$ m/s.

To analyze the dependence of the coordinates of the peak intensity value on the satellite velocity, we introduce the value x_m, i.e. the coordinates of the peak intensity of the reflected radiation. The dependency graph of the coordinate $x_m(V)$ on the satellite velocity V is shown in Figure 6.

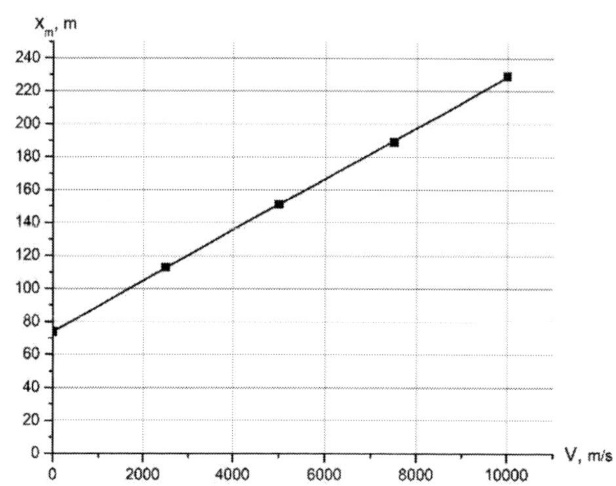

Figure 6. Dependency graph of the coordinate x_m of the peak intensity of the satellite BLITZ velocity.

To achieve the peak intensity in the reception area, one can change the geometric characteristics of the optical system. Below are the results of calculations x_m at different radii of the outer meniscus of satellite R (Figure 7).

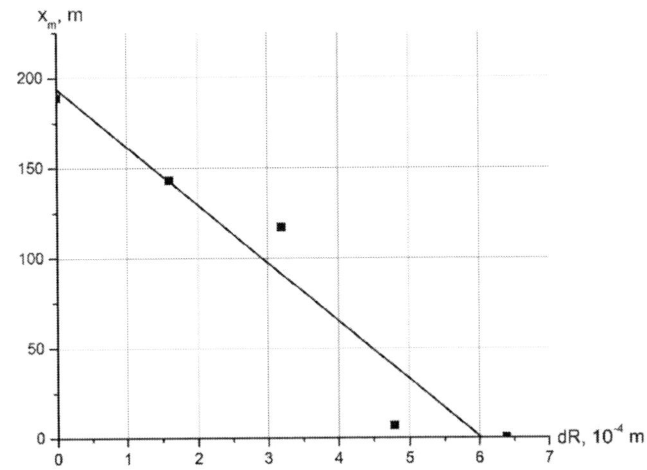

Figure 7. Dependence of the coordinate x_m of the peak intensity on the change in the radius of the meniscus dR.

As a result, for $x_m = 0$ we obtain $dR \approx 0{,}66$ mm. The radiation intensity distribution graph for the radius of the external meniscus 85,66 mm is shown in Figure 8.

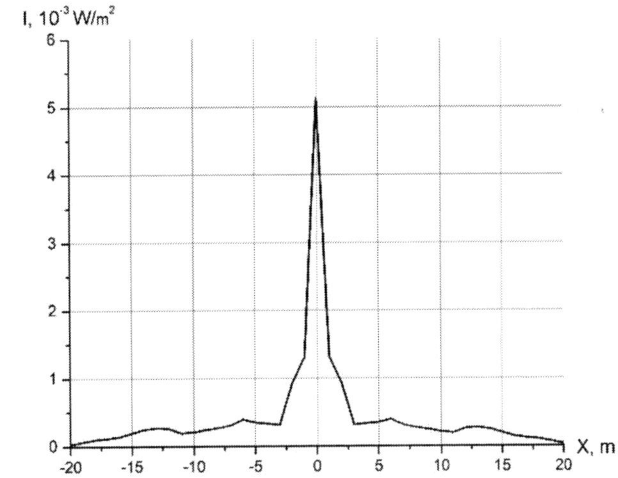

Figure 8. Intensity distribution at the satellite velocity $V = 7500$ m/s and the radius of the outer meniscus $R_1 = 85{,}66$ mm.

The change in the geometric characteristics of the satellite significantly affects the intensity distribution. From the above graph it follows that a change in the radius of the outer meniscus allows us to optimize the distribution and amplify the signal in the reception area. Intensity in $X = 0$ increased by more than 6 times.
Similar results were found for BLITS-M, which has better design parameters. Therefore, calculations showed that without taking into account the velocity, i.e. at $V = 0$ m/s, the main intensity falls on a relatively small area (Figure 9).

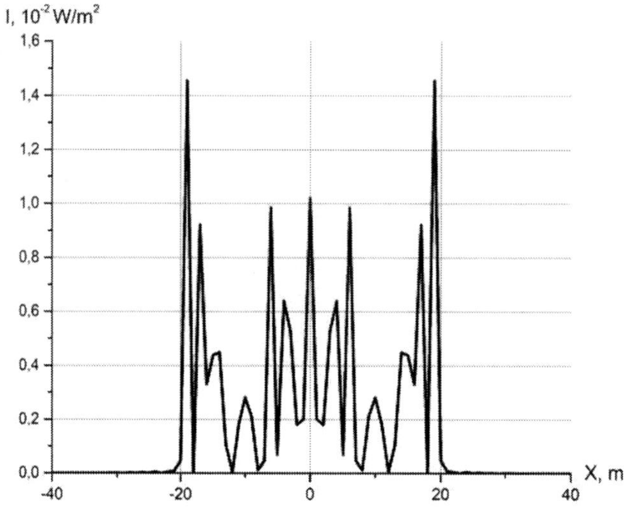

Figure 9. Dependence of radiation intensity $I(X)$ on the coordinate on the earth's surface at the satellite velocity $V = 0$ m/s.

However, with the motion of satellite BLITZ-M, the intensity distribution also undergoes a change (Figure 10).

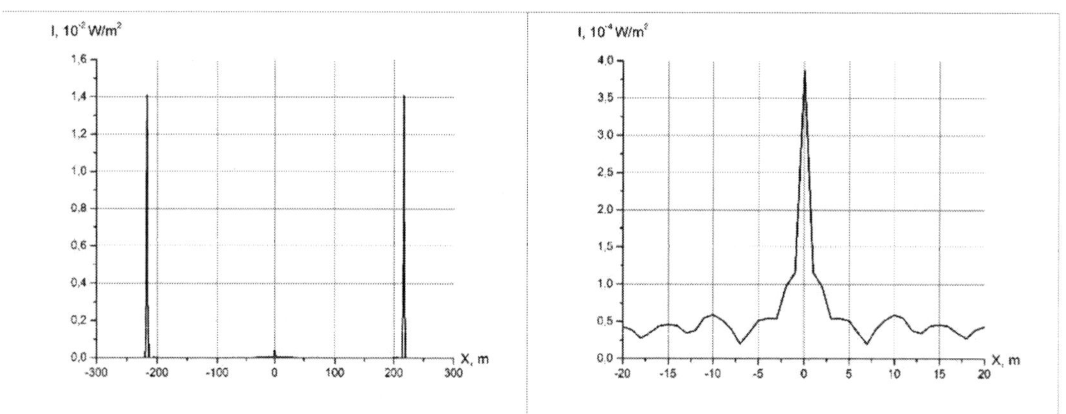

a) b)

Figure 10. Intensity distribution at the satellite velocity $V = 7100$ m/s a) in the interval $[-300, 300]$ m b) in the interval [-20,20] m.

Calculations show that a change in the radius of the outer meniscus makes it possible to correct the interference pattern and to achieve an increase in intensity in the reception area by 2.5 times.

To control the accuracy of numerical experiments, calculations were performed with a different subinterval of the reception area.

When the subinterval was changed from $\Delta x = 1$ m to $\Delta x = 0{,}1$ m to calculate the interference response, the intensity in the peaks changed by no more than $\Delta I = 3 \cdot 10^{-5}$ W/m^2, which is much less than the measured values of the intensities. Moreover, the results of calculations were verified using the Zemax software package at the satellites velocity $V = 0$ m/s.

8. Conclusion

The study investigated the effect of the moving media optics on the moving Luneberg lenslike microsatellite.

A recurrent system of equations is obtained that allows constructing an algorithm for simulating the motion of light rays in a moving Luneberg lens with allowance for dispersion.

As a result of numerical simulation, we obtained the interference response intensity distributions for BLITS and BLITS-M satellites in the registration plane on the earth's surface at the satellites velocity $V = 7500$ m/s and $V = 7100$ m/s, respectively.

The study shows that taking into account the optical satellite velocity at simulation of the interference response leads to the appearance of lateral peaks, which in intensity can exceed the central peak.

This effect is the result of the complex influence of violation of Snell's law, the longitudinal and transverse Fizeau effect, the Doppler effect, the kinematic effect of the satellite offset during the propagation of radiation in it, and the dispersion phenomenon.

Findings of the research show that the phenomena under investigation can have a significant effect on the signal-to-noise ratio in the new GLONASS optical segment. At the same time, the results can be applied to adjust the geometric characteristics of the satellite in order to obtain peak intensity in the reception area.

In general, it can be concluded that the Luneberg cosmic lens is an exceptional experimental laboratory for studying the effects of moving media electrodynamics in the optical spectrum.

References

[1] Shargorodskiy V D and Sadovnikov M A 2013 Lazerny GLONASS *Scientific Journal of Science and Technology* (Bulletin of Siberian State University of Science and Technologies) **6** 52

[2] Kucharski D, Kirchner G, Hyung-Chul Lim and Koidl F 2011 *Adv. Space Res. V.* **48** pp 1335–1340

[3] Ancheier N C, Bernacki B E and Krishnaswami K 2008 *FY 2007 Miniature Spherical Retroreflectors – Final Report* (Pacific Northwest National Laboratory) pp 1–34

[4] Handerek V, McArdle H, Williams T, Psaila N and Laycock L 2005 *Proc. SPIE. V.* 5986

[5] Ancheier N C, Bernacki B E, Klymyshyn N A, Krishnaswami K and Rodriguez C P 2009 *FY 2008 Miniature Spherical Retroreflectors – Final Report* (Pacific Northwest National Laboratory) pp 1–32

[6] Ancheier N C, Bernacki B E and Qiao H A 2011 *FY 2011 Miniature Spherical Retroreflectors – Final Report* (Pacific Northwest National Laboratory) pp 1–26

[7] Bernacki B E, Anheier N C, Krishnaswami K and Cannon B D 2008 *Proc. SPIE V.* 6940 1–7

[8] Guixin Li, Zentgraf T and Shuang Zhang 2015 *Nature Physics* **12** pp 736–740

[9] Gladyshev V O, Tereshin A A and Bazleva D D 2016 Electromagnetic Radiation in a Medium with a Velocity Gradient *JETP Letters* **103** 7 pp 440–442

[10] Gladyshev V O and Portnov D I 2016 Dependence of Laser Radiation Intensity on the Elastic Deformation of a Revolving Optical Disk with a Reflective Coating *Optics and Spectroscopy* **121** 6 pp 901–906

[11] Bolotovskii B M and Stolyarov S N 1989 Reflection of light from a moving mirror and related problems *Sov. Phys. Usp.* **32** pp 813–827

[12] Ternov I M, Mikhaylin V V and Khalilov V R 1985 *Synchrotron radiation and its applications* (Moscow. Lomonosov MSU Publ.)

[13] Gladyshev V O and Tereshin A A 2016 A Luneburg Lens in moving Coordinates *Optics and Spectroscopy* **120** 5 pp 773–780

Acknowledgements

The work was supported by Russian Foundation for Basic Research grants no. 16-02-00488 A and no. 16-08-00618 A.

Laser excitation of the dark bound photonic states during parametric processes in dielectric media

V S Gorelik

Professor, Lebedev Physical Institute of Russian Academy of Sciences, Moscow, Russia
Professor, Bauman Moscow State Technical University, Moscow, Russia
E-mail: gorelik@sci.lebedev.ru

Abstract. Dispersion laws of darktons (non radiative quasiparticles) in vacuum and in dielectric media were analyzed. The theory of dark bound photonic states generation as the result of different parametric processes is developed. Fermi resonance condition for dark bound photonic states forming is discussed. The experimental setups for dark bound photonic states detection are proposed. The generation of coherent darktons emission by using of powerful pulsed laser as the pumping source for parametric processes excitation in dielectric media is predicted. The synchronism conditions in photon-darkton conversion processes are proved on the base of unitary polaritons (refractive index close to unity) exciting.

1. Introduction

In linear approximation photons in vacuum are non interacting between itself. However, including the nonlinear (parametric) terms in Hamiltonian, results in the photon-photon interactions. Photon-photon scattering may be described if third and fourth order parametric processes of Hamiltonian are taking into account. The interactions between photons become essential only for very intensive light emission, achievable inside of stars or with the help of powerful lasers. It is known, that in solids the strong phonon-phonon scattering due to the fourth or third order anharmonicity results in the existing of scalar or pseudoscalar dark bound states of two phonon (biphonon), non interacting with polar vibration. Accordingly as the result of the strong photon-photon scattering the scalar or pseudoscalar (nonradiative) dark bound states of two photons in vacuum may be created. Dark states of photons (hidden photons and axions) in vacuum may be called as darkton states. Darktons are massive Bose-type particles, satisfying to the known relativistic law:

$$E^2 = m_0^2 c_0^4 + p^2 c_0^2 . \tag{1}$$

Here $E = \hbar\omega$, $p = \hbar k$ are the energy and momentum of darktons, m_0 is the rest mass of darktons and c_0 - the velocity of light in vacuum.

According to the modern scenarios of Universe evolution the phase transition, closed to the first order type, in isotropic media, corresponding to initial physical vacuum state (praphase), has been occurred. The initial vacuum phase state of Universe (praphase) existed at very high temperature and pressure. At these conditions darktons may be formed as some soft modes, responsible for the lowering of vacuum symmetry as the result of the phase transition during temperature lowering. The so-called standard model, describing such phase transition, is based on local (gauge) symmetry with symmetry group $\mathrm{SU} \times \mathrm{U}(1)$. As the result of the phase transition, at lower temperatures the soft mode, corresponding to the density of matter fluctuations, was freezing and some space ordering of vacuum

IOP Conf. Series: Journal of Physics: Conf. Series **1051** (2018) 012032 doi:10.1088/1742-6596/1051/1/012032

particles (maximons) was realized. Accordingly in the spectrum of elementary excitations of vacuum, the scalar type quasiparticles, known as Higgs bosons, have appeared. Such quasiparticles, corresponding to density of vacuum fluctuations, are formed as the result of bounding of two photons.
In recent years it became clear that the standard model, despite of great success in the description of the known elementary particles spectrum, requires further clarification, by taking into account the effects of time (T) breaking symmetry at small distances. In this regard the theory predicts the existence another soft modes, known as Goldstone-Nambu boson. The known theory of incommensurate phase transition in solids predicts the existence of two types of soft modes, corresponding to so-called "amplitudon" and "phason". Accordingly near the phase transition in vacuum should exist darkton (amplitudon) and darkton (phason) quasiparticles.
Thus the conclusion [1-3] was made about the existence in vacuum of Bose-type particles with the very small but finite rest mass. It is so called hidden-photons (scalar-type particles) and axions (pseudoscalar type one), corresponding to the soft mode excitations, responsible for the phase transition in vacuum. It is expected that the rest mass of these particles is very small. Its rest energy is at the range $0{,}001-1{,}0\ \mathrm{meV}$, i.e. less than the rest mass of all known elementary particles. Hidden photons and axions may be classified as some types of darktons. Such elemental excitations have appeared in the earliest stage of Universe as the result of vacuum symmetry lowering during Universe phase transition. Now there is the preposition [4-7], that hidden photons and axions are possible elemental particles of dark matter. The problems of the presence of dark matter (total share 0,23) and dark energy (the total share of 0,73) in the Universe are actual task of the modern physics. Hidden photons and axions are predicted to be some candidates on elementary particles of dark matter. "Cold" (slow) hidden photons and axions are nonrelativistic (Newtonian) particles. "Hot" (fast) hidden photons and axions are relativistic particles and move with the speed, close to the speed of light. Important property of these particles is its super weak interaction with media. According to the estimates obtained from astrophysical data, the equilibrium concentration of hidden photons and axions in our part of the galaxy is about $10^{-24}\ \mathrm{g/cm^3}$. So, owing to the extremely small rest mass, these particles should be in Bose-Einstein condensate state even at room temperature.
Now the task of detection and generation of darktons in the laboratory is discussed [3-7]. The theory predicts photon – hidden photon conversion in vacuum by using of intense laser emission as exciting light. Beside that, the feasibility of photon-axion conversion and inverse processes in the presence of strong external magnetic field in vacuum with using of powerful lasers is predicted. In works [3-5,7] it was suggested the possibility of "hot" axions creating with energy $2-3\ \mathrm{eV}$ in the laboratory as the result of photon-axion conversion in experiments as "Light shining through wall" (so-called Primakov-effect). Schematic diagram for the generation and detection of axions on the base of Primakov effect using is shown in Figure 1.

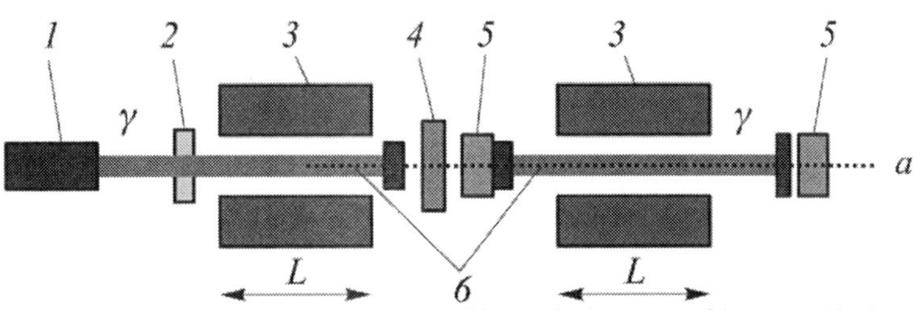

Figure 1. Schematic setup for observation of the effect of photons of laser radiation conversion into pseudoscalar bosons $\gamma \rightarrow a$ (axions) and the reverse process — the reconversion $a \rightarrow \gamma$; 1 — the source of laser radiation; 2 —mirror; 3 — solenoids; 4 — opaque wall; 5 — receivers of the secondary radiation; 6 — Fabry–Pero resonator, generating the light quanta (γ).

Taking into account the strong magnetic field presence (see Figure 1), the photon-axion ($\gamma - a$) conversion was waited. Axions, emerging at left side of setup, should penetrate through the opaque

wall 4. Then, at right part of setup, reverse processes should take place resulting axion-photon conversion. So the reconverted photons may be detected after opaque wall. As a result of the secondary photons detection the probability of these processes and effectiveness of the photon production from "hot" for axions in the laboratory may be evaluated. According to the selection rules, photon-axion conversion processes are permitted only when magnetic field, the induction of which is perpendicular to the direction of the beam of exciting radiation, is applied to the area of laser radiation. In the absence of an external magnetic field photon-hidden photon conversion processes are allowed, the probability of which is very small. As it turned out from the experiments [3-5,7], the useful signal of the secondary radiation produced as the result of conversion-reconversion processes seemed to be extremely small and the useful signal is below the sensitivity threshold of the modern detectors.
So far the attempts for laboratory hopeful observation of photon-hidden photon and photon–axion conversion in vacuum were not successful because of very small photon-photon interaction efficiency in vacuum. We consider that such experiments should be continued, but not only in vacuum and also in dielectric media, where the efficiency of photon-photon parametric processes is essentially higher with comparing to vacuum. It is known, that if photons penetrate from vacuum into dielectric media some hybrid (photon-phonon) particles are formed, named as polaritons. Accordingly, when vacuum darkton penetrate into media, hybrid vacuum darkton-media scalar or pseudoscalar excitations should take place.
In this paper the properties of different parametric processes with generating of darkton-type excitations in dielectric media and in vacuum are discussed. The laboratory experimental schemes for photon-darkton conversion processes observation under intense laser excitation of parametric processes in dielectric media are proposed.

2. Polaritons as photons in dielectric media

Consider the case of cubic dielectric crystal with one polar (dipole-active) optical branch as CsCl-type. For such case the Maxwell equations are:

$$\begin{aligned} &rot\vec{E} = -\frac{\partial \vec{B}}{\partial t};\ rot\vec{H} = \frac{\partial \vec{D}}{\partial t}; \vec{D} = \varepsilon_0 \varepsilon \vec{E} = \varepsilon_0 \vec{E} + \vec{P}; \\ &div\vec{D} = 0; \quad div\vec{B} = 0;\ \vec{B} = \mu_0 \mu \vec{H} = \mu_0 \vec{H} + \mu_0 \vec{M}. \end{aligned} \tag{2}$$

Thus, for electric field $\vec{E}$ of plane transverse monochromatic waves we obtain the wave equation:

$$\left(\nabla^2 - \frac{\varepsilon\mu}{c_0^2} \frac{\partial^2}{\partial t^2} \right) \vec{E}(\vec{r},t) = 0; \vec{E}(\vec{r},t) = \vec{E}_0 \exp\left[i\left(\vec{k}\vec{r} - \omega t \right) \right]. \tag{3}$$

Correspondingly for dispersion law $\omega(k)$ of transverse electromagnetic waves in such media we have:

$$\omega^2 = \frac{c_0^2 k^2}{\varepsilon(\omega)\mu(\omega)};\ \mu(\omega) = 1;\ \omega^2 = \frac{c_0^2 k^2}{\varepsilon(\omega)}. \tag{4}$$

Here $\varepsilon(\omega)$ and $\mu(\omega)$ are the corresponding dispersion dependences of dielectric and magnetic permittivity. At the first stage, we will not take into account the contribution to the dielectric permittivity of valence electrons, but analyze the dielectric properties of the medium taking into account only one polar vibration. The equation of motion of the polar vibrations has the form:

$$\ddot{\vec{u}} = -\omega_0^2 \vec{u} + \frac{e\sqrt{F}}{m} \vec{E},\ \vec{u} = \vec{u}_0 \exp\left(i\left(\vec{k}\vec{r} - \omega t \right) \right) \tag{5}$$

Here m is the oscillator effective mass, e is the electron charge, F is so called oscillator strength $(F \sim 1)$ and u is the magnitude of deviation of oscillating particle from equilibrium position, ω_0 - is the frequency of polar oscillator. According to Eq. (5), the corresponding amplitude of the oscillating charged particle is:

$$\vec{u}_0 = \frac{e\sqrt{F}}{m\left(\omega_0^2 - \omega^2\right)}\vec{E}_0. \tag{6}$$

Correspondingly, for the amplitudes of the dipole momentum and the polarization (V_0 - is the volume of elemental cell) take place:

$$\vec{p}_0 = \frac{e^2 F}{m\left(\omega_0^2 - \omega^2\right)}\vec{E}_0;\ \vec{P}_0 = \frac{e^2 F}{mV_0\left(\omega_0^2 - \omega^2\right)}\vec{E}_0. \tag{7}$$

Equations for polarization are written as:

$$\ddot{\vec{P}} = -\omega_0^2 \vec{P} + \frac{e^2 F}{mV_0}\vec{E};\ \vec{P} = \vec{P}_0 \exp\left[i\left(\vec{k}\vec{r} - \omega t\right)\right]. \tag{8}$$

Introducing frequency $\omega_p = \sqrt{\dfrac{e^2 F}{mV_0}}$ from Eq. (7) we arrive to the relation

$$\ddot{\vec{P}} = -\omega_0^2 \vec{P} + \omega_p^2 \vec{E}. \tag{9}$$

For the electric field induction we get the following equation:

$$\vec{D}_0 = \varepsilon_0 \vec{E}_0 + \vec{P}_0 = \varepsilon_0\left[1 + \frac{e^2 F}{mV_0(\omega_0^2 - \omega^2)}\right]\vec{E}_0 = \varepsilon_0 \varepsilon(\omega)\vec{E}_0. \tag{10}$$

Thus, the dielectric function can be presented as:

$$\varepsilon(\omega) = \frac{c_0^2 k^2}{\omega^2} = 1 + \frac{e^2 F}{mV_0(\omega_0^2 - \omega^2)} = 1 + \frac{\omega_p^2}{\omega_0^2 - \omega^2}. = \frac{\omega_l^2 - \omega^2}{\omega_0^2 - \omega^2}. \tag{11a}$$

Here $\omega_l^2 = \omega_0^2 + \omega_p^2$. Accounting for the electronic oscillator in the elementary volume V_0 results in the factorization of the dielectric function:

$$\varepsilon(\omega) = \varepsilon_\infty\left(1 + \frac{\omega_p^2}{\omega_0^2 - \omega^2}\right) = \varepsilon_\infty \frac{\omega_l^2 - \omega^2}{\omega_0^2 - \omega^2}. \tag{11b}$$

Here $\varepsilon_\infty = n_\infty^2$ dielectric constant at high frequency with comparing to ω_0. Then, the dispersion relation for photon-phonon hybrid states (polaritons [8-9]) in the considered medium takes the form:

$$\omega^2 = \frac{c_0^2 k^2}{\varepsilon(\omega)\mu(\omega)} = \frac{c_0^2 k^2\left(\omega_0^2 - \omega^2\right)}{\varepsilon_\infty\left(\omega_l^2 - \omega^2\right)} = \frac{c^2 k^2\left(\omega_0^2 - \omega^2\right)}{\left(\omega_l^2 - \omega^2\right)}. \tag{12}$$

Here $\mu(\omega) = 1$ and $c^2 = \dfrac{c_0^2}{\varepsilon_\infty}$. From Eq. (12) we arrive to biquadratic equation

$$\omega^4 - \omega^2\left(\omega_l^2 + c^2 k^2\right) + \omega_0^2 c^2 k^2 = 0. \tag{13}$$

The exact solution of Eq. (13) defines two polariton branches:

$$\omega_\pm^2 = \frac{\omega_l^2 + c^2 k^2}{2}\left[1 \pm \sqrt{1 - \frac{4\omega_0^2 c^2 k^2}{\left(\omega_l^2 + c^2 k^2\right)^2}}\right]. \tag{14}$$

For small wave vectors $(k \to 0)$ we get:

$$\omega_+^2 = \omega_l^2 + c^2\left(1 - \frac{\omega_0^2}{\omega_l^2}\right)k^2;\ \omega_-^2 = \frac{c^2 \omega_0^2}{\omega_l^2}k^2. \tag{15}$$

If we have several $(j=1,2,3..n)$ polar vibrations we come to the known following expression of dielectric function and dispersion law of polaritons:

$$\varepsilon(\omega)=\varepsilon_\infty\prod_{j=1}^{n}\frac{\omega_{lj}^2-\omega^2}{\omega_{0j}^2-\omega^2};\quad \omega^2=\frac{\tilde{n}_0^2k^2}{\varepsilon(\omega)}. \tag{16}$$

Figure 2 illustrates the shape of polariton dispersion law at the presence of one polar phonon and one polar exciton branches in dielectric crystals.

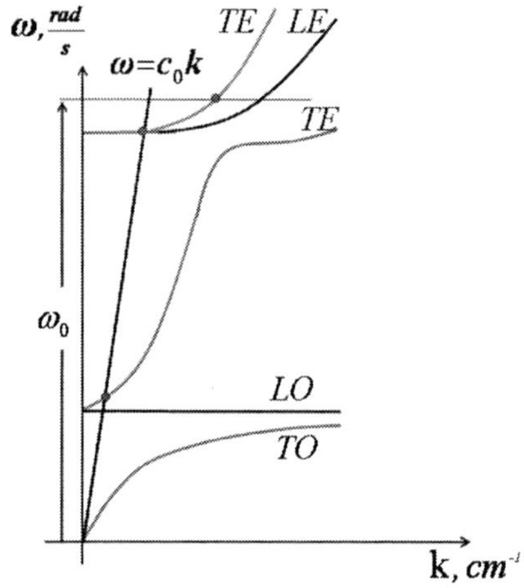

Figure 2. Dispersion law of polariton in dielectric media. The lowest curve corresponds to polar phonons, the upper- to polar excitons; $\omega=c_0k$ – dispersion law of photons in vacuum; the points of intersections between polariton branches and $\omega=c_0k$ line correspond to unitary polaritons, corresponding to refractive index equal to unity: $n=1$.

3. Darktons in dielectric medium

Instead of the three-dimensional Maxwell's equations using, we consider the dispersion relation for polaritons for one-dimensional case taking into account the hybridization of polar oscillations with the electromagnetic field that penetrates into the dielectric medium. Then the equations of motion for polar oscillations and related electromagnetic field given by the function $\xi(x,t)$ take the form [10]:

$$\begin{aligned}&\ddot{u}=-\omega_0^2u+\omega_p^2\xi,\ u=u_0\exp\left[i(kx-\omega t)\right];\\&\ddot{\xi}=-\frac{c_0^2}{\varepsilon_\infty}k^2\xi-\ddot{u},\ \xi=\xi_0\exp\left[i(kx-\omega t)\right].\end{aligned} \tag{17}$$

Substitution in Eq. (17) the solutions in the form of plane monochromatic waves leads to the dispersion relation that coincides with Eq. (14). The resemble situation takes for polar excitons with the own frequency $\omega_{ex}>>\omega_{ex}$:

$$\omega_\pm^2=\frac{\omega_l^2+c^2k^2}{2}\left[1\pm\sqrt{1-\frac{4\omega_{ex}^2c^2k^2}{\left(\omega_l^2+c^2k^2\right)^2}}\right]. \tag{18}$$

The equations of motion for scalar type lattice oscillators, interacting with hidden photons in dielectric medium, can be written as:

$$\begin{aligned}&\ddot{s}=-\omega_{0s}^2s+\omega_i^2\varsigma; && s(x,t)=s_0\exp\left[i(kx-\omega t)\right];\\&\ddot{\varsigma}=-\omega_h^2\varsigma-c_0^2k^2\varsigma-\ddot{s}; && \varsigma(x,t)=\varsigma_0\exp\left[i(kx-\omega t)\right].\end{aligned} \tag{19}$$

In these equations, the functions $\varsigma(x,t)$ and $s(x,t)$ describe the waves that correspond to hidden photons and scalar (total symmetrical type) optical phonons or excitons, respectively; the quantity ω_i characterizes the interaction of scalar excitations of media with hidden photons; ω_h is the frequency, corresponding to the rest mass of hidden photon; ω_{0s} is the frequency of fully symmetrical oscillation (optical phonons or excitons). The solutions of Eq.(19) for plane monochromatic wave are:

$$-\omega^2 s = -\omega_{0s}^2 s + \omega_i^2 \xi; \quad s = \frac{\omega_i^2 \xi}{\omega_{0s}^2 - \omega^2}; \qquad (20)$$
$$-\omega^2 \xi = -\omega_h^2 \xi - c_0^2 k^2 \xi + \omega^2 s.$$

Solving Eq. (20) we obtain the dispersion relations of the waves under consideration:

$$\omega^2 = \omega_h^2 + c_0^2 k^2 + \frac{\omega^2 \omega_i^2}{\omega_{0s}^2 - \omega^2}; \qquad (21)$$
$$k^2 = \frac{\omega^2}{c_0^2}\left(1 + \frac{\omega_i^2}{\omega_{0s}^2 - \omega^2}\right) - \frac{\omega_h^2}{c_0^2}.$$

From these relations we have function $g(\omega)$, resemble to the dielectric function $\varepsilon(\omega)$:

$$g(\omega) = \frac{c_0^2 k^2}{\omega^2} = (1 + \frac{\omega_i^2}{\omega_{0s}^2 - \omega^2} - \frac{\omega_h^2}{\omega^2}). \qquad (22)$$

In common case of several scalar phonons (excitons) we have function $g(\omega)$ as:

$$g(\omega) = \frac{c_0^2 k^2}{\omega^2} = \left(1 - \frac{\omega_h^2}{\omega^2}\right) + \prod_{j=1}^{n} \frac{\omega_{ij}^2}{\omega_{0sj}^2 - \omega^2}. \qquad (23)$$

From (22) we obtain the biquadratic equation:

$$\omega^4 - \omega^2\left(c_0^2 k^2 + \omega_{fs}^2 + \omega_h^2\right) + \omega_h^2 \omega_{0s}^2 + c_0^2 k^2 \omega_{0s}^2 = 0: \ \omega_{fs}^2 = \omega_{0s}^2 + \omega_i^2. \qquad (24)$$

Solution of (24) results in two branches for hybrid quasiparticles: scalar phonons (excitons) and hidden photons:

$$\omega_{\pm}^2 = \frac{\omega_{fs}^2 + \omega_h^2 + c_0^2 k^2}{2}\left[1 \pm \sqrt{1 - \frac{4\left(\omega_{0s}^2 \omega_h^2 + c_0^2 \omega_{0s}^2 k^2\right)}{\left(\omega_{fs}^2 + \omega_h^2 + c_0^2 k^2\right)^2}}\right]. \qquad (25)$$

For small wave vectors $(k \to 0)$ we obtain dispersion relation:

$$\omega_+^2 = \omega_{fs}^2 + \omega_h^2 - \frac{\omega_h^2 \omega_{0s}^2}{\omega_{fs}^2 + \omega_h^2} + c_0^2 k^2 \left(1 + \frac{\omega_h^2 \omega_{0s}^2}{\left(\omega_{fs}^2 + \omega_h^2\right)^2} - \frac{\omega_{0s}^2}{\omega_{fs}^2 + \omega_h^2}\right); \qquad (26)$$
$$\omega_-^2 = \frac{\omega_{0s}^2 \omega_h^2}{\omega_{fs}^2 + \omega_h^2} + c_0^2 k^2 \left[\frac{\omega_{0s}^2}{\omega_{fs}^2 + \omega_h^2} - \frac{\omega_{0s}^2 \omega_h^2}{\left(\omega_{fs}^2 + \omega_h^2\right)^2}\right].$$

For calculation of corresponding dispersion law in common case we should use the relation

$$\omega^2 = \frac{c_0^2 k^2}{g(\omega)}; g(\omega) = \frac{c_0^2 k^2}{\omega^2} = (1 - \frac{\omega_h^2}{\omega^2}) + \prod_{j=1}^{j=n} \frac{\omega_{ij}^2}{\omega_{0sj}^2 - \omega^2}. \qquad (27)$$

4. Raman scattering on polaritons in dielectric crystals

It is well known the experimental setups for parametric scattering with polariton revealing in dielectric crystals by means of laser spectroscopy technique [8-9,11-15]. In these experiments the frequency-angular $\nu(\theta)$ dependencies in Raman spectra of noncentrosymmetrical crystals were investigated. Figure 3 (a,b) illustrates such kind dependencies for GaP (Figure 3a) and IV-phase of ammonium chloride monocrystals (Figure 3b).

We can see from Figure 3(a,b), that at different scattering angles Raman spectrum have continuum nature and contain some sharp breaks (gaps).The continual curves correspond to polaritons, dispersion law of which is described by relations (12,16). In cubic GaP crystal there is only one polar mode, resulting two fundamental Raman peaks at 388 cm^{-1} (TO-mode) and 402 cm^{-1} (LO-mode) (see figure 3a). Beside fundamental (first order) Raman peaks there are a number of overtones as a result of phonon combining from the different points of Brillouin zone. Besides, there is the sharp peaks at 784 cm^{-1} (more detail picture is in [16]), corresponding to dark (A_1-type of symmetry) bound state of two photons. The frequency of corresponding fundamental mode at Γ-point is 402 cm^{-1} [16].

The gap on polariton curve at Figure 3b corresponds to resonance intersections of polaritons with two photon states in NH_4Cl-crystal. We can also see (Figure 3b) the sharp peaks, corresponding to dark (scalar-type) bound states at 2805 cm^{-1} (more detail picture is in [15]). Such dark bound state attitude to A_1 (total symmetrical) irreducible representation and can not be explained as fundamental mode. This dark bound photonic states is enhanced due to Fermi resonance of overtones with fundamental total symmetrical mode at 3042 cm^{-1}. Thus scalar A_1-type dark bound photonic state is formed as the result of bounding of two photons, frequencies of which close to $\nu_{4l}(F_2)=1418$ cm^{-1}, presented in the fundamental spectrum on ammonium chloride (IV –phase) crystal, and Fermi-resonance conditions realizing for fundamental $\nu_1(A_1)=3042$ cm^{-1}. Thus, if we use the powerful infrared laser, the frequency of which is close to the unitary polariton frequency for polar vibration $\nu_{4l}(F_2)=1418$ cm^{-1} (see Figure 3b)), dark bound photons wave generation is waited with frequency value 2805 cm^{-1}.

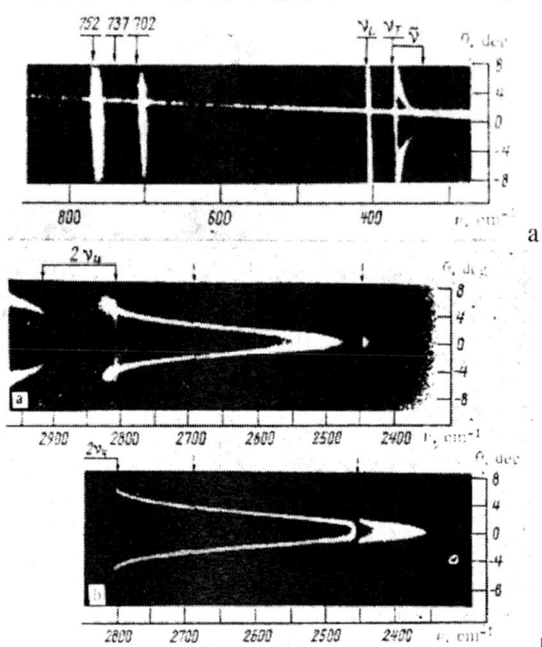

Figure 3. Frequency-angular dependence $\nu(\theta)$ of polariton Raman scattering; a – the polariton Raman spectrum near the fundamental polar mode on GaP monocrystal under excitation of copper (578,2 nm) laser; b-polariton Raman spectra at the region of vibrational excitations in NH_4Cl by using of green (510,6 nm; up) and yellow (578,2 nm) copper vapor laser.

5. To the theory of third- and fourth-order parametric processes in dielectric media

As a result of the interaction between polaritons in dielectrics, the bound states of two polaritons may be created. Such bound states of two polaritons may be classified by some irreducible representations, presenting in vector symmetric squared presentations. For example the scalar type (A_1-type) of the bound states of two polaritons in dielectrics may be formed. Before the bound states of two phonons, magnons, plasmons, excitons, and other elementary excitations of condensed matter have been discussed. In this paper we have analyzed the opportunity of two photons (polaritons) bound states existing in dielectric media, including the polar crystals and photonic crystals.

At the first step, in dielectrics the interactions between polaritons, corresponding to different frequencies and momenta, may be not taking into account. At the second step we should consider such interactions. Thus, the Hamiltonian of the photon (polariton) system in dielectrics may be presented as follows [17-19]:

$$H=\sum_k E_k a_k^+ a_k + g_3 \sum_{k_1,k_2,k_3} a_{k_1}^+ a_{k_2} a_{k_3} + g_4 \sum_{k_1,k_2,k_3,k_4} a_{k_1}^+ a_{k_2}^+ a_{k_3} a_{k_4} + \ldots.. \tag{28}$$

Here E_k is the energy of photons (polaritons), a_k^+, a_k are creation and annihilation operators of photons (polaritons), and g_3 and g_4 are the corresponding coupling parameters. Photon-photon scattering in dielectrics may be illustrated by diagrams, taking into account also photon-phonon (Raman) processes (see Figure 4).

In view of the approximations used, the Bethe–Salpeter equation for the multiple scattering of two photons (polaritons) with momentum Q has the solution ($\hbar = 1$):

$$G_2(\omega)=\frac{G_{02}(Q,\omega)}{1-g_4 G_{02}(Q,\omega)};G_{02}(\omega)=\int\frac{\rho_{02}(Q,\omega)}{\omega-E+2i\gamma\omega}. \tag{29}$$

Here $\rho_{02}(Q,\omega)$ is the density of two-photonic states. Bound photonic states appear providing that $G_2(\omega)$ function has the pole. We can see that the bound state is indeed may be created for large enough g_4 constant. Such situation may be realized, for example, if Stimulated Raman Scattering (STRS) observed at high powerful laser excitation. The spectrum of bound states of two photons (polaritons) is described by the relation:

$$I(Q,\omega)=\frac{\operatorname{Im} G_2(Q,\omega)}{\pi}. \tag{30}$$

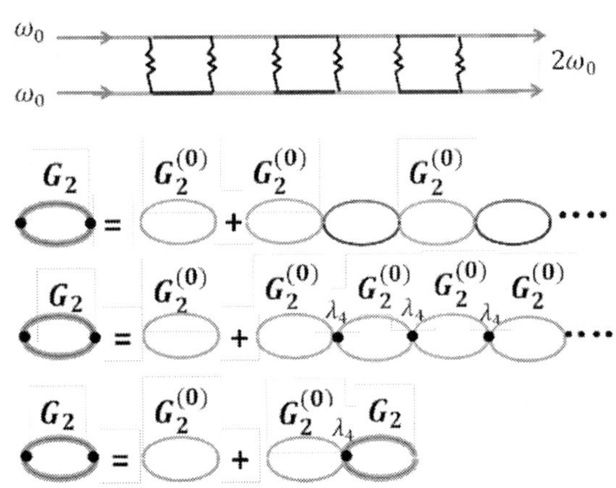

Figure 4. The illustration of photon-photon scattering processes, taking into account Stokes (red lines) and antiStokes (blue lines) Raman processes.

Taking into account the Dyson equation for interacting photons at $g_4 = 0$ and $g_3 \neq 0$, we arrive the equation:

$$G_1(\omega) = \frac{G_1^0}{1 - g_3^2 G_1^0 G_2^0(Q,\omega)}. \tag{31}$$

Thus, the bound state is formed if $G_1(\omega)$ function has the pole at high enough g_3 constant.

6. Experimental setup for dark bound states of photons generation and detection during laser excitation in dielectric media

Under high intensity laser excitation in dielectric media some photon-phonon and photon-photon processes may be realized. The illustration of elemental Spontaneous Raman Scattering (SRS) processes is given at Figure 5. The example [20] of Spontaneous Raman spectrum in $NaNO_2$ crystal is given in Figure 6. We can see, that in this spectrum the most intensive Raman satellite, corresponding to pseudoscalar mode of A_2–type symmetry, is observed.

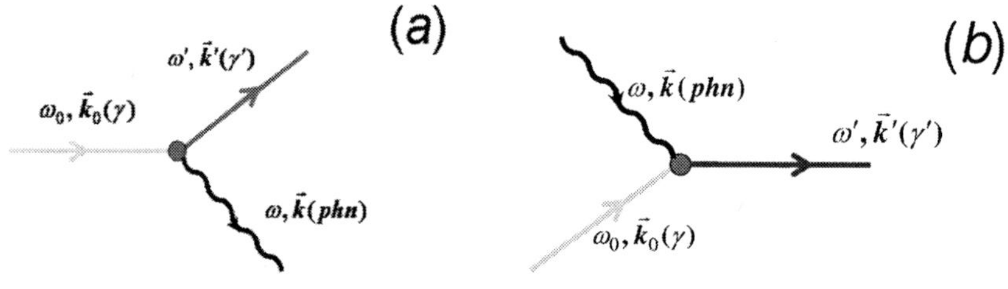

Figure 5. Elemental spontaneous Stokes and anti-Stokes Raman scattering processes.

If the very intensive laser emission for Raman scattering excitation is used the so-called Stimulated Raman Scattering (STRS) may be observed. In this case, the intensity of one Stokes satellite becomes very large and the generation of coherent corresponding vibrational mode takes place. The most intensive mode in SRS mode has the lowest STRS threshold and may be revealed as stimulated process under powerful laser excitation. In $NaNO_2$ crystal such mode corresponds A_2 (pseudoscalar–type symmetry) lattice vibration (see Figure 6).

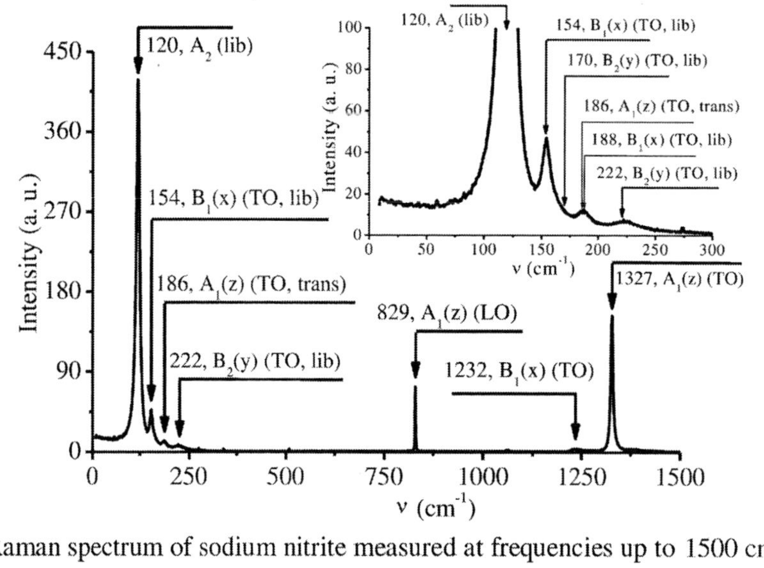

Figure 6. Raman spectrum of sodium nitrite measured at frequencies up to 1500 cm^{-1} at room temperature. The inset shows the low-frequency part of the spectrum.

Experimental setup for STRS in crystals or powders observations is illustrated by Figure 7. Thus realizing of STRS processes in $NaNO_2$ crystal opens the opportunity to generate the pseudoscalar type darkton coherent waves in dielectric media.

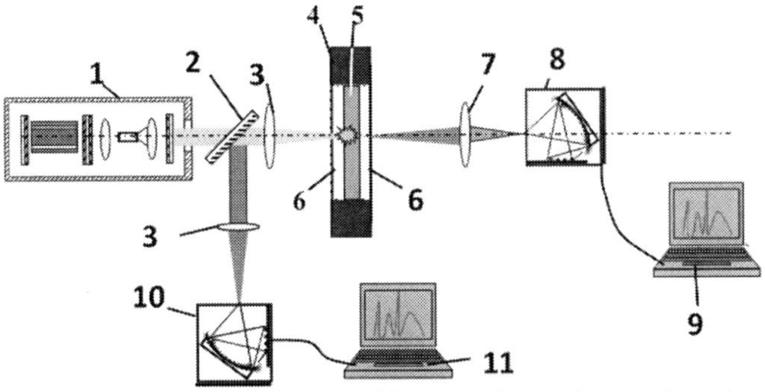

Figure 7. Experimental equipments for Stimulated Raman Scattering registration; 1 - YAG:Nd^{3+} laser (532 nm); 2,6 - glass plates; 3,7 - lens;4 - sample holder;5 - sample; 8,10 - minispectrometers; 9,11 - computers.

Usually the STRS threshold in molecular liquids or crystals is about $10^7 - 10^9$ W/cm^2 .Note, that at these intensities of exciting emission others parametric processes have been observed, for example, two-photon excited luminescence (TPEL) [13,21-22].

The shape of multifrequency parametric STRS spectra, obtained by us for $NaBrO_3$ monocrystals, is given at Figure 8. We can see a number of equidistant Stokes and anti-Stokes satellites of scalar type (A_1) symmetry, with frequency shift about 780 cm^{-1}. Such excitations are not interacting with photons, having F-type of symmetry in $NaBrO_3$ crystal, but can strong interact with scalar darktons.

Thus, under intensive laser emission in dielectric media some darkton-type vibrational or exciton modes may be excited. Several of these modes may be scalar or pseudoscalar type. Accordingly, such modes should strong interact with hidden photons or axions. Such type modes should be excited during STRS or other parametric processes [23-26]. Inside of the discussed dielectric media the coherent wave of darktons may be generated.

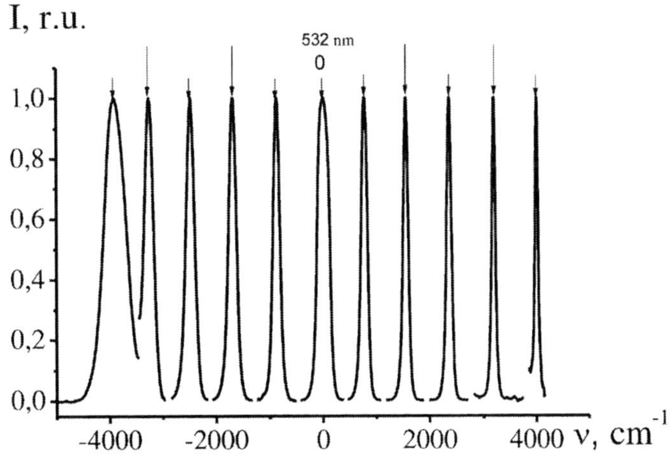

Figure 8. Stimulated Raman Scattering spectrum in $NaBrO_3$ crystal under powerful YAG:Nd^{3+} laser($\lambda = 532$ nm) excitation.

For detection of darkton waves the same crystal ($NaBrO_3$) should be used. If the bound states with frequency close to $2\omega_0$ of two photons with frequency ω_0 is generated under STRS processes the experimental setup, resemble to "Light shining through wall" in vacuum may be used (see Figure 9). The using of photonic crystals, filled by different dielectrics, opens the additional opportunities for bound state of photons observing. In this case the proximity of exciting laser emission frequency to the edges of photonic crystal band gaps results in the giant photon density of states and, correspondingly, increase of photon – dark bound state of two photons conversion efficiency [16-19, 27].

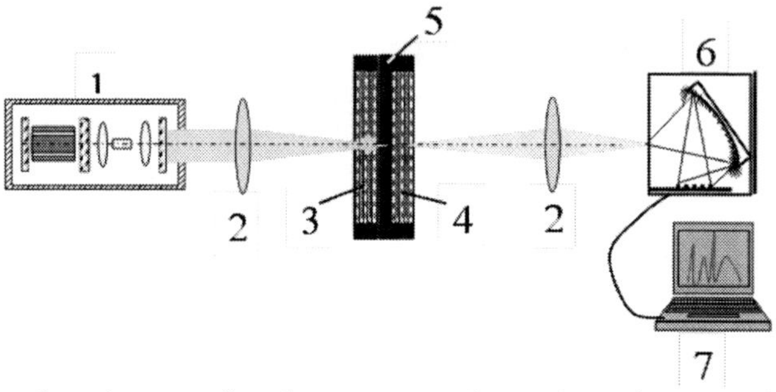

Figure 9. Experimental equipments for photon – two photon bound state – photon conversion; 1- YAG:Nd^{3+} laser(532 nm); 2 - lens; 3, 4 - samples; 5 - samples holder; 6 - minispectrometers; 7 - computer.

7. Summary

Thus it is proposed the optimization of experimental installations for generation and detection of dark bound photonic states by means of parametric active dielectric media using and application of intense laser emission for parametric processes excitation. For the generation of "hot" dark bound photonic states it is proposed to use the powerful laser with high spectral intensity in the visible or ultraviolet ranges. This ensures that the transition from the spontaneous dark bound photon states conversion to stimulated one should take place. It was shown that photon-darkton conversion efficiency may be enhanced if the frequency of exciting laser correspond to the frequency of unitary polaritons (refractive index is close to unity). In this case, momentum conservation law is not disturbed and group velocity of exciting laser wave should be very small. New opportunities opens the using of photonic crystals, proving the essential enhance of the spontaneous and stimulated parametric processes efficiency due to the known Purcell effect.

References

[1] Okun L B 1982 *Sov. Phys. JETP* **56** 502
[2] Hoffmann S 1986 *Phys. Lett.* B **193** 117
[3] Van Bibber K, Dagdeviren N R, Koonin S E, Kerman A K and Nelson H N 1987 *Phys. Rev. Lett.* **59** 759
[4] Ruoso G *et al* 1992 *Z. Phys.* C **56** 505
[5] Cameron R *et al* 1993 *Phys. Rev.* D **47** 3707
[6] Duffy L D *et al* 2006 *Phys. Rev.* D **74** 012006
[7] Sikivie P, Tanner D B and Van Bibber K 2007 *Phys. Rev. Lett.* **98** 172002
[8] Henry C H and Hopfield J J 1965 *Phys. Rev. Lett.* **15** 964
[9] Scott J F, Cheesman L E and Porto S P 1967 *Phys. Rev.* **162** 834
[10] Gorelik V S, Pyatyshev A Yu, Belyanchikov M A and Gorshunov B P 2016 *Physics of Wave Phenomena* **24** 87
[11] Gorelik V S, Mitin G G and Sushchinskii M M 1975 *Sov. Phys. JETP* **42** 419
[12] Mitin G G, Gorelik V S, Kulevskii L A, Polivanov Yu N and Sushchinskii M M 1975 *Sov. Phys.-JETP* **41** 882
[13] Gorelik V S 2010 *Eur. Phys. J. Appl. Phys* **49** 33007

[14] Gorelik V S, Maximov O P, Mitin G G and Sushchinskii M M 1977 *Solid State Communications* **21** 615
[15] Sushchinskii M M and Gorelik V S 1983 *Zhurnal Prycladnoi Spektroscopii* **38** 95
[16] Sushchinskii M M, Gorelik V S and Maximov O P 1978 *Journal of Raman Spectroscopy* **7** 26
[17] Gorelik V S 2006 *Acta Phys. Hung.* A **26** 37
[18] Gorelik V S 2006 *J. of Russ. Laser Research* **27** 400
[19] Gorelik V S 2008 *Laser Phys.* **18** 1479
[20] Gorelik V S, Pyatyshev A Yu and Krylov A S 2016 *Physics of the Solid State* **58** 170
[21] Vodchits A I, Voinov Yu P, Gorelik V S, Kudryavtseva A D, Orlovich V A and Tcherniega N V 2013 *Bulletin of the Lebedev Physics Institute* **40** 329
[22] Gorelik V S, Kudryavtseva A D, Tareeva M V and Tcherniega N V 2014 *Inorg. Mat.* **50** 1217
[23] Gorelik V S 2015 *EPJ Web of Conferences* **103** 04003
[24] Gorelik V S 2015 *Bulletin of the Russian Academy of Sciences: Physics* **80** 779
[25] Gorelik V S 2014 *Herald of the Bauman Moscow State Tech. Univ., Nat. Sci.* 6 p 3
[26] Gorelik V S 2015 *Herald of the Bauman Moscow State Tech. Univ., Nat. Sci.* 4 p 23
[27] Gorelik V S 2009 *Phys. Scr.* T **135** 014039

Acknowledgements
This work was supported in part by RFBR grant 16-52-00026 (BRFFR F16R-063).

Instability states and Ostrogradsky formalism

T F Kamalov

Moscow Institute of Physics and Technology, Dolgoprudny, Russia
E-mail: timkamalov@gmail.com

Abstract. Contemporary physics, both Classical and Quantum, requires a notion of inertial reference frames. However, how to find a physical inertial frame in reality where there always exist random weak forces? We suggest a description of the motion in instability frames by means of inclusion of higher time derivatives. They may play a role of non-local hidden variables in a more general description complementing both classical and quantum mechanics.

1. Introduction. Can Classical Description of Physical Reality be Considered Complete?

Quantum non-locality and quantum correlations are experimentally observable phenomena of quantum physics. At the same time, these experimental facts of quantum experiments have no explanation in classical physics and contradict in some cases to its axiomatic. Should we consider physics not a unified consistent science, but rather, regard classical and quantum physics distinct, independent and unlinked sciences, and then there would be no problem. But actually, this is not the case. Physics must be a unified science, with classical physics describing macroscopic objects and quantum physics, microscopic ones. And depending on the situation we apply either former or latter ones. So, their non-contradiction is required for consistency, and vice versa.

A most successful solution in providing physics consistency would be a unified axiomatic of the unified physics underlying both classical and quantum physics. The search for such an axiomatic was for the last century being carried out through the search of suitable quantum axioms. Since 1935 the issue of completeness of the quantum-mechanical description of the reality has been one of the most extensively discussed in physics. Usually the classical axiomatic was considered completed. The psychological inhibition of the extension of classical axiomatic was proliferated by extensive evidence of Newton's and Einstein's fundamental studies validity. However, a problem of incompleteness of the classical description of the physical reality has never been raised. May we pose such a question? Is the classical description of the physical reality complete?

According to Godel's theorem, there exist provisions in any theory that cannot be proved within this theory. It can also be added that no theory is complete. Axioms of any theory are not to be proved, but rather, they are conjectured, so any system of axioms may be replaced with another one. Newton's laws pertain to such improvable provisions. And Classical Physics can be extended to other field of an application.

In short, Newton's laws postulate the description of mechanical system dynamics with second order differential equations. Are there any cases of the reality description with higher order differential equations? The answer is positive, it is known that yes, but it is not Newtonian mechanics of Inertial Reference Frames.

As an example of body dynamics description with higher order differential equations let us consider body dynamics in randomly Instability Reference Frames

$$Q = f\left(t, q, \dot{q}, \ddot{q}, \ldots, q^{(n)}\right),$$
$$T = t.$$

here τ is a time interval for averaging coordinates $Q = \langle q(t) \rangle = \frac{1}{2}\left[q(t+\tau) + q(t-\tau)\right]$.

The second example for Classical Physics extension is considering the Quantum Reference Frames.

1.1. Definition

Quantum Reference Frame is a Reference frames with the transformation of coordinates and time as

$$Q = \langle q(t) \rangle = \int_{-\tau}^{+\tau} \psi^* q \psi dV,$$
$$T = t.$$

Here

$$\psi = \psi_0 \exp\left[\frac{i}{\hbar}(\Delta p \Delta q + Et)\right] = \psi_0 \exp\left[i \frac{\frac{\Delta p}{\Delta t} - F}{\frac{\hbar}{\Delta q \Delta t}}\right] = \psi_0 \exp\left[i \frac{f_0}{f_Q}\right] \quad (1)$$

is wave function with the inertial force f_0 depend of high-order derivatives coordinates on time and f_Q corresponds of inertial forces and constant force.

An Instability Reference Frames is a method to describe influences of random fields onto both the particle to be described itself and to an observer. Changing from an Instability Reference Frame to an inertial one makes a free particle to randomly oscillate correlating with other free particles oscillations. If we consider single frequency oscillations, then these vibrations look coherent. Inertial frames transformations are prescribed as Galilean transforms (in a relativistic case, Lorentz transforms). Transformations of Instability Reference Frames differ from Galilean-Lorentz transforms by remainder terms in Taylor's expansion. Then free particles in Inertial Reference Frames shall be featured with uncertainty in coordinate and momentum, time and energy equal to remainder terms of Taylor expansion.

If a transformation of an Instability Reference Frame to another one described as a Taylor expansion contains a remainder term with index N, then we may say that this free particle conserves its N-th order time derivative. Such free particle is described by N derivatives and conserves this state until interactions with other bodies (forces) perturb this state.

If such a particle interacts with other bodies (a force acts onto it), then the dynamics of such a particle is described by $(N+1)$-th order differential equations. In other words, an influence of a force adds one more derivative to the description of particle oscillations.

Considering a particle in an Inertial Reference Frame instead of Real ones, one shall either introduce inertial forces, that is, change from higher derivative description to the description without higher derivatives, but with inertia forces, or take into account remainder terms of Taylor expansion.

Modern physics (both classical and quantum) is physics of Inertial Reference Frames. The case of a Non-Inertial Reference Frame usually comes down to introducing inertia forces into an Inertial Reference Frame. Application of inertia forces enables reducing problems of physical system dynamics in a Non-Inertial Reference Frame to those in an Inertial Reference Frame through artificial introduction of inertial forces or through application of the d'Alembert's principle. At the same time, an Inertial Reference Frame doesn't exist in reality, as any reference frame is always influenced by small perturbing fields or forces. In the present study, we suggest considering only non-Inertial Instability Reference Frames as real ones. We shall call such non-Inertial Instability Reference Frames as Instability ones. Since the incipience of d'Alembert's principle and up to now realness of inertia forces is a debatable issue. We believe that the question of inertia forces realness may be reduced to the question of Inertial Reference Frame realness.

In view of the above, there is a question to be answered, how could be physical systems described in non-

Inertial Instability Reference Frames without introduction of inertia forces?
Commonly Reference Frames are called Inertial provided Newton's laws hold there; actually, Newton's laws constitute the axiomatic of Classical Physics. They postulate the description of physical systems behavior by second order differential equations. Abandonment of higher time derivatives of coordinates is related to the issue inertia forces in Inertial Reference Frames. So, to answer the above question we have to consider a more general case of higher orders differential equations and to extension Classical Physics with a description employing higher time derivatives of coordinate.
Changing from Inertial Reference Frame to Instability ones without inertia forces introduction means changing from second-order differential equations description of physical systems to their description with higher-order differential equations. Abandonment from employing higher order time derivatives of coordinate in the classical Newtonian physics does not mean they do not exist. They do exist in certain problems. But this is not the Newtonian Physics of Inertial Reference Frames.
Conversion of coordinates of a point particle between Instability Reference Frames provided τ is a time interval for averaging, shall be expressed as

$$Q = q(t) + \dot{q}(t)\tau + \Delta q(t),$$

$$\Delta q(t) = \sum_{k=2}^{N} (-1)^k \frac{1}{k!} \tau^k q^{(k)}(t)$$

same holds for momentum

$$P = p(t) + \dot{p}(t)\tau + \Delta p(t),$$

$$\Delta p(t) = \sum_{k=2}^{N} (-1)^k \frac{1}{k!} \tau^k p^{(k)}(t). \tag{2}$$

Here, $\Delta q(t)$ and $\Delta p(t)$ are remainder terms of the Taylor expansion. The remainder terms $\Delta q(t)$ and $\Delta p(t)$ in the Instability Reference Frame may be interpreted as uncertainties of coordinate and momentum of a point particle in this reference frame. In quantum mechanics, uncertainties of coordinate and momentum of a micro particle obey to the rule

$$\Delta q(t) \Delta p(t) \geq \frac{\hbar}{2}. \tag{3}$$

In the classical physics can be introduced an uncertainty relation, as there always exist random small fields and forces influencing either the very system to be described or an observer, that is

$$\left[\sum_{k=2}^{N} (-1)^k \frac{1}{k!} \tau^k q^{(k)}(t)\right] \left[\sum_{k=2}^{N} (-1)^k \frac{1}{k!} \tau^k p^{(k)}(t)\right] \geq \frac{H}{2}$$

inertial one.
The supremum of the difference of the action function in Instability Reference Frames (with higher time derivatives of the generalized coordinate) from the classical mechanics action functions (without higher derivatives) is

$$\sup \left| S\left(q, \dot{q}, \ddot{q}, \dddot{q}, \ldots, q^{(n)}, \ldots\right) - S(q, \dot{q}) \right| = H. \tag{4}$$

In this case, higher derivatives are non-local additional variables and disclose the sense of the classical analog H of the Planck's constant. The H constant defines the supremum of the influence of random fields onto the reference frame and the observer. We shall analyze this case in terms of Instability Reference Frames. In this case, H defines the supremum of the difference between an Instability Reference Frame and an Inertial Reference Frame.
Action functions with higher-derivative variables describe instability trajectories of physical systems dynamics and are accounted for via Instability Reference Frame.
In our case, the classical space is featured by infinite number of variables, same as Hilbertian one. In the search for a unified axiomatics the classical constant H shall coincide with the quantum one, i.e. the Planck constant $\hbar$. In this approach, the estimate of the Planck constant may be determined by higher derivatives, playing the role of non-local hidden variables.

In this case the state of quantum object can be describe

$$\psi(t)=\left|q,\dot{q},\ddot{q},\dddot{q},...,q^{(n)}\right\rangle=\left|Q(t)\right\rangle.$$

The transfer object from point 1 to point 2 can be describe by Lagrange function $L(Q)$

$$\left\langle Q_1,t_1\left|Q_2,t_2\right.\right\rangle=\int_{t_1}^{t_2}DQ\exp\left(\frac{i}{\hbar}L(Q)\right)dt.$$

And action function $A_{Q(R)}$ can be represent by Hamiltonian $\hat{H}$

$$A_{Q(R)}=\left\langle q(t),\dot{q}(t),\ddot{q}(t),\dddot{q}(t),...q^{(n)}(t)\right|\left[i\hbar\frac{\partial}{\partial t}-\hat{H}\right]\left|q(t),\dot{q}(t),\ddot{q}(t),\dddot{q}(t),...q^{(n)}(t)\right\rangle.$$

2. Quantum Correction of Second Newton Law

Ostrogradsky formalism uses Lagrange function

$$L=L\left(q,\dot{q},\ddot{q},\dddot{q},...,q^{(n)},...\right),$$

but not

$$L=L(q,\dot{q}).$$

Euler-Lagrange equation in this case is follow from least action principal [1-5]

$$\delta S=\delta\int_{t_1}^{t_1}L\left(q,\dot{q},\ddot{q},\dddot{q},...,q^{(n)},...\right)dt=\delta\int_{t_1}^{t_1}\sum_{n=0}^{\infty}(-1)^n\frac{d^n}{dt^n}\frac{\partial L}{\partial q^{(n)}}dt=0$$

or

$$\frac{\partial L}{\partial q}-\frac{d}{dt}\frac{\partial L}{\partial\dot{q}}+\frac{d^2}{dt^2}\frac{\partial L}{\partial\ddot{q}}-\frac{d^3}{dt^3}\frac{\partial L}{\partial\dddot{q}}+...+(-1)^{(n)}\frac{d^n}{dt^n}\frac{\partial L}{\partial q^{(n)}}+...=0.$$

This equation can be write in the form of corrected Newton Second Law of motion in Instability Reference Frames

$$F-ma+f_0=0.$$

Here

$$f_0=mw=mw+m\dot{w}\tau+m\sum_{k=2}^{\infty}(-1)^k\frac{1}{k!}\tau^k w^{(k)}$$

is a random inertial force in (1) which can be represent by Taylor expansion with high-order derivatives coordinates on time. The corrected law has the form

$$F-ma+\tau m\dot{a}-\frac{1}{2}\tau^2 m\ddot{a}+..+\frac{1}{n!}\tau^n ma^{(n)}+...=0.$$

From this formula follows the Second Newton Law in the Inertial Reference Frames because random acceleration w in this case equal to zero.

3. Verification of High-Order Derivatives as Non Local Hidden Variables

The role of High-Order Derivatives as Hidden Variables can be verified by using the Equivalence Principal when acceleration is equal to the gravitational field. Then the correlation factor for entangled photons polarization measurements may be presented as

$$|M|=\left|\left\langle\left(\lambda^i A^k g_{ik}\right)\left(\lambda^m B^n g_{mn}\right)\right\rangle\right|.$$

Here, the random variables distribution function may be considered uniform, with the photon polarization varying from 0 to π:

$$\frac{1}{\pi}\int_0^{\pi}\rho(\varphi)d\varphi=1.$$

According to the definition,

$$\cos\varphi = \frac{\lambda^i A^k g_{ik}}{\sqrt{\lambda^i \lambda_i}\sqrt{A^k A_k}},$$

$$\cos\left(\varphi+\theta\right) = \frac{\lambda^m B^n g_{mn}}{\sqrt{\lambda^m \lambda_m}\sqrt{B^n B_n}}.$$

Hence, the correlation factor with the photon polarization varying from 0 to π is

$$|M| = \left|\frac{1}{\pi}\int_0^{\pi}\rho(\phi)\cos\varphi\,\cos(\varphi+\theta)\,d\varphi + \frac{1}{\pi}\int_{\pi}^{2\pi}\rho(\phi)\cos\varphi\,\cos(\varphi+\theta)\,d\varphi\right|,$$

$$|M| = |\cos\theta|.$$

The Bell's observable in our case differs from that calculated by Bell and does not contradict to experimental data. Bell's inequalities are not violated in either classical or quantum cases of accounting for random fields, forces and waves.

4. Conclusion

Additional terms in the form of higher derivatives may play the role of hidden variables complementing both quantum and classic mechanics. Additional terms have non-local character, which enables their employment for description of nonlocal effects of quantum mechanics. The effect of quantum correlations and non-locality of quantum states can be explained by the random character of interactions if extended particles. Such an approach can be realized within the scope of Ostrogradsky higher derivatives formalism.

References

[1] Ostrogradsky M V 1850 *Mem. Acad. St. Peterburg* **6** 4 p 385
[2] Kamalov T F 2013 *J. Phys.: Conf. Series* **442** 012051
[3] Kamalov T F 2010 Physics of Non-Inertial Reference Frames *Preprint* arXiv: 0708.1584v6
[4] Kamalov T F 2011 *Quantum Computers and Computing* **11** 52 (*Preprint* arXiv: 1011.4256)
[5] El-Nabulsi R A 2014 *Qual. Theory Dyn. Sys.* **13** 149

"Ghosts" and the big bang theory

V M Koryukin[1] and A V Koryukin[2]

[1] Consultant of Physics and Mathematics Department, Mari State University, Yoshkar-Ola, Russia
[2] Engineer of Quantum Optics and Nanophotonics Department, Kazan Federal University, Kazan, Russia
E-mail: vmkoryukin@gmail.com

Abstract. We propose to use the second principle of thermodynamics and the quatum chromodynamics for the description of processes going in the Universe. Besides we give the large significance to matter properties in the degenerate state and specifically to the spontaneous breaking of symmetry.

1. Introduction

In our works we follow for principles proposed by Blokhintsev [1] for an explanation of a random character of physical laws. It allows to consider laws as approximates and to construct the physical theory as the most plausible one. We also follow for Pauli, which introduced an unobservable particle for the production of conservation laws by beta decay. Fermi named it as the neutrino and constructed the theory of the beta decay. We propose to use "ghosts" (unobservable zero-spin fermions) introduced in the quantum chromodynamics for the renormalizability of the theory [2] in the cosmology. The "ghosts" utilization necessity in the Universe is connected with the possibility for an explanation of processes with an immense energy release. We shall rely on the Landau theory of the Fermi liquid in which it is offered to replace the consideration of strong interacting original particles on the consideration of the almost ideal gas of repeat quasi-particles. We shall consider "ghosts" (which's are characterized by the $SU(3)$ symmetry) as original particles of the Universe Fermi liquid and all Universe observable particles as repeat quasi-particles of the given Fermi liquid. If observable particles are absent in the Universe, then we shall consider that the Fermi liquid entropy is minimal, so as to use the second principle of thermodynamics.

2. The space of "ghosts" and the mathematical formalism

We shall use the equivalence class $\{\Psi(x)\}$ for the description of "ghosts" in the kind of Universe, where a point $x \in M_n$, $\Psi(x)$ is the section of the vector fiber bundle M_{n+N} with the base M_n. In the given particular case $n = 8$ is the count of the freedom degree of the function space, having only the color and characterized by the symmetry of the $SU(3)$ group. Here and further the equivalence relation is determined by the infinitesimal transformation having the form:

$$\Psi(x) \to \Psi(x) + \delta\Psi(x) = \Psi(x) + \delta\omega^a T_a(x,\Psi), \tag{1}$$

where $a,b,c,d,e = 1,2,\ldots,n+r$; $\delta\omega^a(x)$ are infinitesimal parameters; $T_a(x,\Psi)$ are transition operators, generated by the collision integral. By this

$$x^i \to x^i + \delta x^i = x^i + \delta\omega^a(x)\xi_a^i(x), \tag{2}$$

where x^i are coordinates of a point $x \in M_n$; $i,j,k,l,... = 1,2,...,n$. Considering M_n as the smooth manifold it might be worthwhile to rewrite the changes $\delta\Psi(x)$ of smooth fields $\Psi(x)$ in the following form:

$$\delta\Psi(x) = \delta_0\Psi(x) + \delta\omega^a \xi^i_a \partial_i \Psi. \tag{3}$$

$\partial_i\Psi$ are the partial derivatives of fields $\Psi(x)$), where $\delta_0\Psi(x)$ are the changes of fields $\Psi(x)$ in the point x. Rewriting $\delta_0\Psi(x)$ in the form

$$\delta_0\Psi(x) = \delta\omega^a X_a(\Psi) = \delta\omega^a \left(T_a(x,\Psi) - \xi^i_a \partial_i \Psi\right), \tag{4}$$

we shall regard $X_a(\Psi)$ as the generators of the Lie local loop $G_r(x)$ [3].
Let

$$A = \int_{\Omega_n} \Lambda d_n V = \int_{\Omega_n} \kappa \overline{X}(\Psi) \rho X(\Psi) d_n V \tag{5}$$

is the action, where Λ is the Lagrangian (we shall use the system of units $\frac{h}{2\pi} = c = 1$, where h is the Planck constant and c is the velocity of light). Here and further κ is a constant; $\rho = \rho(x)$ is the density matrix ($\mathrm{tr}\rho = 1$, $\rho^+ = \rho$, the top index "+" is the symbol of the Hermitian conjugation) and the bar means the generalized Dirac conjugation which must coincide with the standard one in particular case that is to be the superposition of Hermitian conjugation and the spatial inversion of the space-time M_4. We shall name solutions $\Psi(x)$ of differential equations, which are being produced by the requirement of the minimality of the integral (5), as the maximum plausible realizations of Lie local loops [3] and we shall use for the construction of the all set of functions $\{\Psi(x)\}$ (generated by the help of the transition operators $T_a(x,\Psi)$).
Of course for this purpose we can use the analog of the maximum likelihood method employing for the probability amplitude, but not for the probability as in the mathematical statistics. As is known, according to the Feynman's hypothesis the probability amplitude of the system transition from the state $\Psi(x)$ in the state $\Psi'(x')$ equal to the following integral

$$\int_{\Omega(\Psi,\Psi')} \exp(iA) D\Psi = \lim_{N\to\infty} I_N \int d\Psi_1 ... \int d\Psi_k ... \int d\Psi_{N-1} \exp\left(i \sum_{k=1}^{N-1} \Lambda\left(\Psi\left(x^k\right)\right) \Delta_n V_k\right) \tag{6}$$

($i^2 = -1$; the constant I_N is chosen so that the limit is existing). So, the formula (6) allows describe the physical process in the quantum theory the most adequately. At the same time the functions $\Psi(x)$, being the solutions of differential equations and obtaining from the requirement of the minimum of the action A, may be the maximum likelihood ones only, but then they allow to describe the same physical system in condensed (short) form. In this approach the Lagrangian Λ plays the more fundamental role than differential equations which are generated by it. As the transition operators are constructed on the base of experimental data, then the differential equations, obtained in a result of the Lagrangian special choice in the action (5), can name as the differential equations of the root-mean-square regression Ψ on x.

3. Quarks, gluons and neutrinos

The given symmetrical state of the Universe matter is unlikely can consider a stable state if only on account of fluctuations. Boltzmann considered that the Universe formation is connected with the gigantic fluctuation. So in the first phase as a result of fluctuations are appeared particles in the excited state (quarks and antiquarks, having the spin), among "ghosts", existing in the ground state. As a result of an increase of freedom degrees (the appearance of quarks and antiquarks) the Universe temperature must be decreased so, that must goes the spontaneous breaking of symmetry with the freezing-out of freedom degrees for quarks and antiquarks (the appearance of neutrinos and antineutrinos).

IOP Conf. Series: Journal of Physics: Conf. Series **1051** (2018) 012034 doi:10.1088/1742-6596/1051/1/012034

Considering "ghosts" as original particles of a Universe matter the transition to leptons (neutrinos and antineutrinos) may be interpreted as the phase transition. At present quarks and antiquarks attend only in hadrons and black holes [4] (inside quark bags). The entropy of the all Universe matter increased, that is the indication of the given process irreversibility.

Quarks and antiquarks interact by gluons and have baryon charges, which are characterized by the potential of the magnetic type [5]. For the gluon description we shall use the sections $B(x)$ of the vector fiber bundle M_{n+r} with the base M_n. The sections $B(x)$ are named the gauge fields because the transformation law for them in a point x has the form

$$\delta_0 B = \delta\omega^a Y_a(x,B) + \partial_i \delta\omega^a Z_a^i(x,B). \tag{7}$$

The gauge fields $B(x)$ are introduced in the standard manner

$$BB^+ = \rho \operatorname{tr}\left(BB^+\right) \tag{8}$$

with the exploitation of the density matrix $\rho(x)$. The description of the Universe matter in this form is subject the quantum chromodynamics completely.

We connect the cutoff parameter of quantum chromodynamics with the Fermi energy of University background fermions. We obtain the natural physical interpretation of the cutoff parameter and also we obtain the physical justification of the lattice model. Note, that the transition to the infinity of cutoff parameter in the quantum chromodynamics is the natural transition for the general collection in the mathematical statistics (for the simplification of calculations). The given transition is the base for the use of Hausdorff spaces in the physical theory and in particular the use of Euclidean and pseudo-Euclidean spaces. The search of the estimation for the cutoff parameter can become the next task due for LHC (the large hadron collider) in CERN, because the estimation ($10^6 \div 10^{10}$ GeV) is a very crude one, received by the observation of cosmic rays of ultrahigh energies (in consequence of a modern theory such cosmic rays must not be observed for 10^{10} GeV).

4. Cooper pairs of noncharged leptons

In our opinion further it must take place Cooper pairing of neutrinos with a formation of a Bose liquid, which must cause to the density reduction of original leptons and to the appearance of photons, fundamental massive vector bosons, charged leptons (lepton bags or lepton bubbles) and hadrons (bags with quarks having electric charges). It can use the Bardeen-Cooper-Schrieffer mechanism of the microscopic superconductivity theory for the explanation of the neutrino Cooper pairs formation the more so that there were particles (quarks and antiquarks) for this in the Universe. In the Universe standard model the given process, having an explosive behavior (Big Bang) was accompanied an immense energy emission.

As a result of a Bose condensation it was showed the asymmetry of neutrinos relative to a polarization. Sterile neutrinos and sterile antineutrinos (nonobservable particles in processes of beta decay) turned out to be bigger than observable particles with the "regular" polarization. The main role goes to neutrinos and antineutrinos. We shall consider these particles as background particles. The physical properties of sterile neutrinos and sterile antineutrinos must define geometrical properties of the space-time just therefore. Note that the quantization of observed particle charges (including the spin of particles) is the direct verification of the existence of the Universe background fermions, because $\Psi(x)$ are not univalent functions (M_n is not the connected space [6]).

Note that at present in the Universe it is possible goes the reverse process of the Bose liquid decay with a density increase for neutrinos and antineutrinos and consequently with the increase of distances between galaxies (physical properties of sterile neutrinos and sterile antineutrinos define space-time geometrical properties), which is interpreted as the inflation (the effect of the Universe dark energy).

Taking into consideration the Ginsburg-Landau theory of superconductivity, the scalar field in the standard model of electroweak interactions ought to interpret as the order parameter of neutrino Cooper pairs. The necessity for the existence of fundamental scalar particles is absent in such theory. Photons and massive vector bosons (generating weak interactions) will perform their role. Masses of massive vector bosons must be an outcome of their interaction with neutrino Cooper pairs, at the same time as photons (which's do not interact with this pairs) will stay massless particles. In consequence of this the detected 125 GeV resonance is impossible to interpret as the scalar Higgs boson.

5. The application of the theory

The appearance of photons (playing the role of standards in the relativity theory) caused to the domination of electromagnetic interactions in experimental data and to the scope for division of the all Universe matter into two subsystems (slow and rapid). The matter of the slow subsystem (a thermostat [1]) does not participate in electromagnetic interactions and plays the catalyst role of stochastic processes. Using the Landau theory of the Fermi liquid for the condensed (brief) description of slow subsystem repeat particles (including neutrino Cooper pairs) we shall apply fields $\Pi(x)$ [4]. For the definition of fields $\Pi(x)$ will be used the differential of the fiber bundle M_{4+r} projection in the base M_4, components of which are $\xi^i_a(x)$. As a result it can consider that the slow subsystem of the Universe matter is described by the space-time M_4, having nontrivial properties. By this Einstein equations define the connection of geometrical parameters with parameters, characterizing matter physical properties. It is naturally, that the space-time of the Einstein theory can describe only the simplified model of the slow subsystem.

The mass existence of observed fermions can be explained by their interaction with background fermions and the numerical value must be connected with the cross-section for background fermions scattering. For illustration let us consider the lepton mass appearance. For this it amply consider the zero approximation, using the Lagrangian Λ in the form

$$\Lambda = \kappa\eta^{ij}\left(\partial_i\overline{\Psi} - \overline{T_i\Psi}\right)\left(\partial_j\Psi - T_j\Psi\right), \tag{9}$$

where η^{ij} are the contravariant components of the Minkowski space metric tensor. Writing fields $\Psi(x)$ and matrices T_i in the form:

$$\Psi(x) = \begin{pmatrix} \psi_L(x) \\ \psi_R(x) \end{pmatrix},\ \psi_L = \frac{1}{2}(I-\gamma_5)\psi,\ \ \psi_R = \frac{1}{2}(I+\gamma_5)\psi,\ \psi = \begin{pmatrix} \psi_1 \\ \psi_2 \\ \psi_3 \\ \psi_4 \end{pmatrix}, \tag{10}$$

$$T_A = -T_A^+ = \frac{i}{2\kappa}\begin{pmatrix} I-\gamma_4 & 0 \\ 0 & I+\gamma_4 \end{pmatrix} \otimes \Sigma_A,\ \ \Sigma_A = \begin{pmatrix} \sigma_A & 0 \\ 0 & \sigma_A \end{pmatrix}, \tag{11}$$

$$T_4 = -T_4^+ = \frac{i}{2\kappa}\begin{pmatrix} I-3\gamma_4 & -2m\kappa I \\ -2m\kappa I & I+3\gamma_4 \end{pmatrix} \otimes \gamma_4 \tag{12}$$

(I is the unit matrix; $\gamma_5 = -i\gamma_1\gamma_2\gamma_3\gamma_4; \gamma_1; \gamma_2; \gamma_3; \gamma_4$ are Dirac matrices; σ_A are Pauli matrices; $A = 1,2,3$) it can received the electron field Lagrangian, applied in the quantum electrodynamics. The given form of matrices T_i (characterizing zero-point quantum oscillations) shows, that the electron density matrix dimensionality of the pure state coherent mixture twice as many as for the neutrino, although the given dimensionality will stay smaller as the density matrix dimensionality of color fermions.

It may observe the matter of the slow subsystem by the matter of the rapid subsystem. As a result the particles of the rapid subsystem play the role of Brownian particles. They allow studying the properties of slow subsystem particles. Consider Heisenberg uncertainty relation:

$$\Delta x \cdot \Delta p \geq \frac{h}{4\pi}, \tag{13}$$

where Δx is the measurement accuracy of the elementary particle coordinate and Δp is the measurement accuracy of the its momentum. Writing down Δp as:

$$\Delta\left(m\frac{dx}{dt}\right)=m\frac{d(\Delta x)}{dt} \tag{14}$$

(m is the particle mass, $\frac{d}{dt}$ is the derivative with respect to a time) and integrating, we produce the law of Brown motion (λ is the wave-length of the elementary particle):

$$(\Delta x)^2 \geq \frac{h}{2\pi}\frac{\Delta t}{m}=\lambda c\Delta t \tag{15}$$

(the square of the deviation from an initial position is the directly proportional to the first degree of the time interval). Here (15) the viscosity of Fermi liquid depends on the kind of elementary particle and thereby it is assumed that the Fermi liquid is the polycomponent liquid. This statement is the generalization of the Dirac hypothesis on the particle sea in the vacuum. The given relation (15) expresses the statistical dependence of the particle coordinate value on its wave nature.

References

[1] Blokhintsev D I 1987 *Printsipinal'nye voprosy kvantovoj mekhaniki* (Moscow, Nauka)
[2] Yndurain F J 1983 *Quantum Chromodynamics: An Introduction to the theory of Quarks and Gluons* (New York, Berlin, Heidelberg, Tokyo, Springer-Verlag)
[3] Koryukin V M 1990 *Sov. J. Nucl. Phys.* **52** 573
[4] Koryukin V M and Koryukin A V 2016 *Proc. Int. Meeting on Physical Interpretations of Relativity Theory "PIRT-2015"* (Moscow) p 220
[5] Koryukin V M 2011 *Proc. XX Int. Baldin Seminar on High Energy Physics Problems: Relativistic Nuclear Physics and Quantum Chromodynamics* 34
[6] March N H and Parrinello M 1982 *Collective Effects in Solids and Liquids* (Bristol, Adam Hilger Ltd)

Objective doubts about the authenticity of the event GW150914

A Lukanenkov

Professor, Moscow State Linguistic University, Russia
E-mail: a_v_luk@mail.ru

Abstract. The paper presents the results of processing the registration data of LIGO 14.09.2015. Impulse signals were detected using the theory of optimal detection. Soliton-like signals were highlighted with the help of sub-optimal (two-step) causal filtering of the LIGO data. These signals have a different polarity. It is also found absence of two chirp-signals, which announced February 11, 2016. Doubts about the reliability of the detection of chirp-signals, which announced February 11, 2016, are justified by the results the conducted analysis.

1. Introduction

In 1916 Albert Einstein predicted the existence of gravitational waves [1].
The discovery of the binary pulsar system PSR B1913+16 by Hulse and Taylor [2] and subsequent observations of its energy loss by Taylor and Weisberg [3] demonstrated the existence of gravitational waves.
Experiments to detect gravitational waves began with Weber [4].
It was announced the observation of gravitational waves from black holes merger thanks to the two signals registered according to observatories LIGO Hanford (***H*1**) and Livingston (***L*1**) in [5].This paper presents the results of processing of registration data posted on the website LIGO (https://losc.ligo.org/events/GW150914) [6].

2. Registration data of two observatories

The time of arrival of gravitational waves:
- $T_{detect} = 1126259462.39$ =September 14, 2015, 09: 50: 45.39 UTC,
- signals were detected in the frequency band [35÷350 Hz],
- the events have a combined signal-to-noise ratio $SNR_{combined} \approx 24$ [5-6].

These data were presented in February - May, 2016 on the website [6], later, in July 2016 the time of the gravitational wave arrival was corrected and been removed 0.39 seconds, this is roughening of accuracy the time of arrival.
The detected signals are contained in two strain time series (fragments) $h_L(t)$, $h_H(t)$.
Fragments $h_L(t)$, $h_H(t)$ have a duration of 32 seconds(Figure 1), strain time series centered at GPS 1126259462=September 14 2015, 09:50:45 UTC.
The corresponding random processes have a bandwidth [10÷2048 Hz].
Sampling frequency $f_d = 4096$ Hz.
Start time of the fragment $T_{start} = 1126259446$ = September 14 2015,09:50:29 UTC, it is

16 seconds before the time of signal arrival $\mathrm{T_{detect}} = 16.39$ sec .
Further seconds and its share are relative $\mathrm{T_0} = 09:50:45$ or $\mathrm{T_{start\,0}} = 09:50:29$.
Useful signals have an amplitude of 10^{-21}.
These signals are under the noise (*SNR* <10^{-3}) (Figure 1), therefore it is necessary to apply optimal (suboptimal) detection methods.
On the page "Data release for event GW150914" (https://losc.ligo.org/events/GW150914/) are given:
- results of processing carried out by the members of LIGO Collaboration (Figure 2, *top part*);
- signals $s_H(t)$, $s_L(t)$ (Figure 2, *bottom part*) according to observation 14.09.2015.

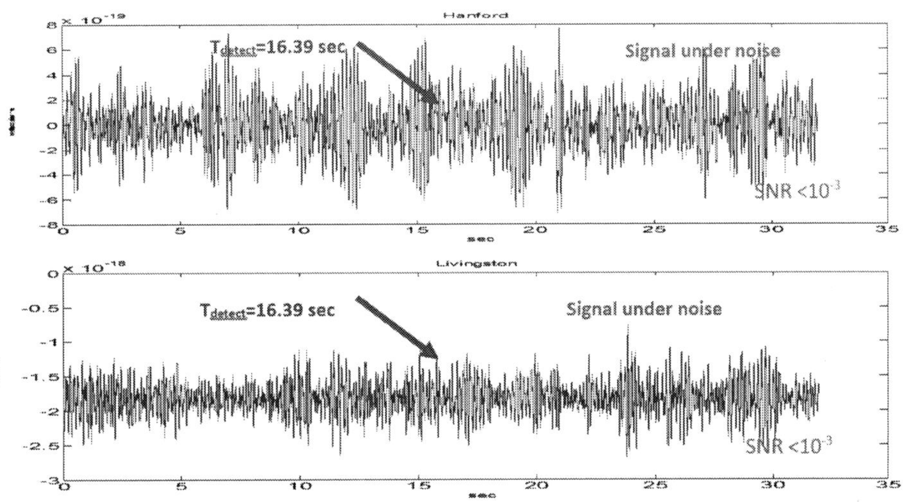

Figure 1. Fragments $h_L(t)$, $h_H(t)$. The arrows indicate the time interval of the detected signals.

Times are shown relative to September 14, 2015 at 09:50:45 UTC (Figure 2). For visualization, all time series are filtered with a 35÷350 Hz band-pass filter to suppress large fluctuations outside the detectors' most sensitive frequency band, and band-reject filters to remove the strong instrumental spectral lines [5-6].
Declared time of the entry signal (0,39 s) indicated by blue circles [5-6]. Numbers and frequencies of wave phases (Figure 2, left top part). The true signals arrive into 0.33 s (green circle).
Experimental templates, shown in Figure 2 (bottom part), can be described as a first approximation by the relation:

$$S(t) = A(t)\cos\left(\phi_0 + 2\pi\left(f_0 t + \frac{\beta t^2}{2}\right)\right), \qquad (1)$$

where $S(t)$ - the chirp-signal ; $A(t)$ - the chirp-signal amplitude; f_0 - the carrier frequency; ϕ_0 - the initial phase; β - the speed of change of frequency; $f_0 = \frac{1}{2}(F_{\max} + F_{\min})$; $\beta = \frac{F_{\max} - F_{\min}}{Ts}$; $\beta(t)$ - in the general case; $F_{\min}$, $F_{\max}$ - minimum and maximum value of the signal frequency; Ts - signal duration.

If β=constant and $\beta = \frac{F_{\max} - F_{\min}}{Ts}$, then this signal is called a linearly frequency modulated (LFM) signal = chirp-signal (linear).

3. Invariants of signals before and after filtering

1. Chirp-signal stays the chirp-signal after passing through the band filter.

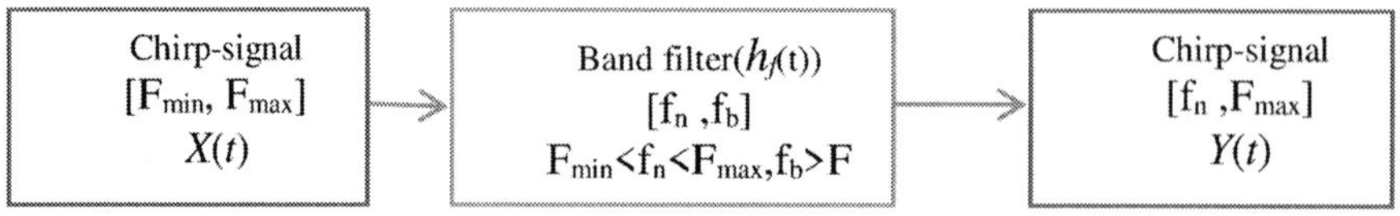

Filtering is defined by the convolution:

$$Y(t) = X(t) \otimes h_f(t), \tag{2}$$

where $h_f(t)$ - impulse response of a filter; $X(t)$ - input process; $Y(t)$ - the filtered process; $\otimes$ is the sign of the convolution.

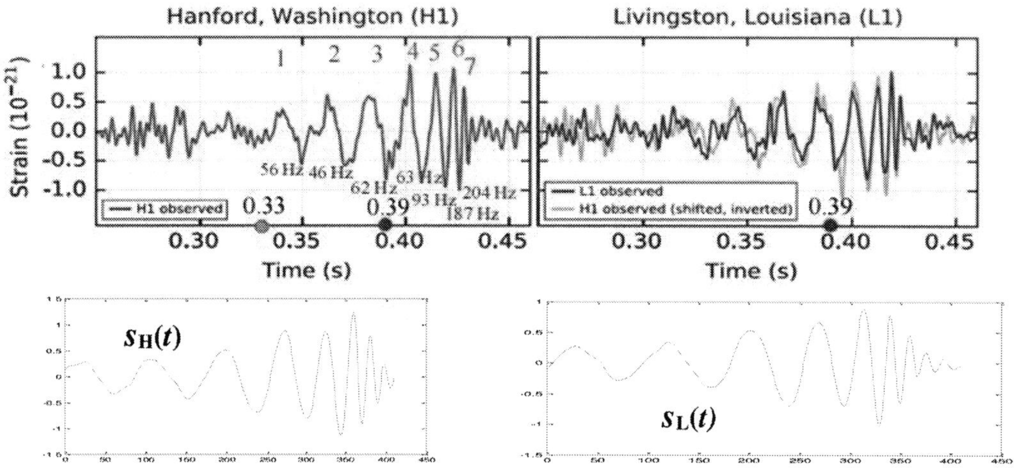

Figure 2. *Top part.* The gravitational-wave event GW150914 observed by the LIGO Hanford (***H*1**, left column panels) and Livingston (***L*1**, right column panels) detectors.
Bottom part. Experimental templates on observatories $s_H(t)$ (Hanford), $s_L(t)$ (Livingston) [6]. The horizontal axis - number in the data array, on the ordinate axis scaling factor $k = 10^{-21}$ is sometimes omitted.

2. The integral of the square of the signal remains practically unchanged if the signal in the filter band. This is a consequence of Parseval's theorem.

$$F_Y(f) = F_X(f) \cdot H(f). \tag{3}$$

$$|H(f)| \approx 1 \Rightarrow$$

$$\int_{t-T/2}^{t+T/2} X(\tau)^2 d\tau \approx \int_{F_{max}-\Delta f}^{F_{max}+\Delta f} |F_X(f)|^2 \cdot 2df \approx \int_{F_{max}-\Delta f}^{F_{max}+\Delta f} |F_Y(f)|^2 \cdot 2df \approx \int_{t-T/2}^{t+T/2} Y(\tau)^2 d\tau, \tag{4}$$

where $H(f)$ - the transfer function of the filter, $H(f) = F(h_f)$ - the spectrum of the impulse response of the filter, $|H(f)|$ - amplitude-frequency characteristic *(AFC)* of the filter .

3. The spectrum of the signal after filtering remains practically unchanged if the signal is filtered in band, where $|H(f)| \approx 1$.

4. If the registered signals $s_H(t)$ and $s_L(t)$ coincide with the delay and inversion $s_H(t) = -s_L(t+\tau)$, as stated in [5], their spectra $\left(|Fs_{Hf}(f)| \approx |Fs_{Lf}(f)|\right)$ are almost equal after filtering:

$$s_{Lf}(t) = s_L(t) \otimes h_f(t), \; s_{Hf}(t) = s_H(t) \otimes h_f(t). \tag{5}$$

IOP Conf. Series: Journal of Physics: Conf. Series **1051** (2018) 012035 doi:10.1088/1742-6596/1051/1/012035

4. The procedure of the detection of known signals

Detection of gravitational chirp-signal is based on the selection of one from two alternative hypotheses:

$$H_0: \begin{matrix} x_H(t) = n_H(t) \\ x_L(t) = n_L(t) \end{matrix} \qquad H_1: \begin{matrix} x_H(t) = n_H(t) + s_H(t) \\ x_L(t) = n_L(t) + s_L(t) \end{matrix}$$

$H_0(H_1)$ - hypothesis that ***chirp-signal is absent*** (***chirp-signal is present***) respectively.

From the mathematical theory of statistics follows [7] that the suboptimal receiver represents matched filtering procedure and then comparing to a threshold.

The matched filter has an impulse response equal to the inverted templates (Figure 2, bottom part):

$$T_L(t) = s_L(T_{template} - t),\ T_H(t) = s_H(T_{template} - t), \tag{6}$$

$T_{template}$ - template duration (duration of impulse response).

The matched filter of input records in the band [35÷350 Hz] with the signals, represented in Figure 2 (bottom part), is written by the relations:

$$h_{1f}(t) = h_{Lf}(t) \otimes T_L(t),\ h_{2f}(t) = h_{Hf}(t) \otimes T_H(t); \tag{7}$$

where $h_{Lf}(t) = h_L(t) \otimes h_f(t)$, $h_{Hf}(t) = h_H(t) \otimes T_f(t)$ - filtered processes after bandpass filtering in the band [35÷350 Hz].

The procedure for testing the hypothesis of the presence of chirp-signals in the records:

1. Produced filtering $h_H(t), h_L(t)$, then find $h_{Hf}(t), h_{Lf}(t)$.
2. The convolution is produced in accordance with (7) and find $h_{1f}(t)$ and $h_{2f}(t)$.
3. Modules of $h_{1f}(t)$ and $h_{2f}(t)$ are compared with thresholds $(\mathrm{P}_1, \mathrm{P}_2)$.

$$|h_{1f}(t)| > \boldsymbol{P_1} \Rightarrow \text{true } H_1 \text{ else } H_0,\ |h_{2f}(t)| > \boldsymbol{P_2} \Rightarrow \text{true } H_1 \text{ else } H_0$$

The signal/noise ratio is determined by the formula:

$$SNR = \frac{A_{\max}}{\sigma_n}. \tag{8}$$

For example, mixtures of signal and noise were formed to demonstrate the matched filtering efficiency

$$x_L(t) = h_L(t) \otimes h_f(t) + s_L(t - \tau_s)\frac{\sigma_L}{\sigma_s}k_{SNR},\ x_H(t) = h_H(t) \otimes h_f(t) + s_H(t - \tau_s)\frac{\sigma_H}{\sigma_s}k_{SNR}, \tag{9}$$

where $h_f(t)$ - impulse response of a band filter [35÷350 Гц]; $h_L(t)$, $h_H(t)$ - initial deformation; σ_s - rms of signal $s(t)$; $\sigma_L(\sigma_H)$ - rms of noise $n_L(t)(n_H(t))$; $\sigma = (Disp(h(t)))^{1/2}$, k_{SNR} - factor depending on *SNR*; τ_s - time of signal arrival.

Mixtures of signal and noise $x_L(t)$, $x_H(t)$ are calculated for the following parameters: time of model signal arrival $\tau_s = 15$ sec, $k_{SNR} = 1.0$.

Matched filtering results $x_L(t)$, $x_H(t)$ with model signals $s_H(t)$ and $s_L(t)$ shown in Figure 3.

The signal/noise ratio at the output SNR_{out} is determined by the SNR_{in} and gain at the expense of processing (B_{SNR}):

$$SNR_{out} = B_{SNR} \cdot SNR_{in}, \tag{10}$$

For formed processes $x_L(t), x_H(t)$ signal/noise ratio at the input in the simulation near 15th second :

$$SNR_{in\ mod} \approx 2.323$$

It is clearly distinguished model signals to the 15th second after matched filtering on Figure 3, $SNR_{out} \approx 8.9 - 10.7$ when the input signal/noise ratio ≈ 2.323.

Input signal/noise ratio for gravitational signals, if the real signals are in Figure 2:

$$SNR_{inL} \approx 5.5, \qquad SNR_{inH} \approx 5.25.$$

At such values SNR_{in} by the formula (10)

$$SNR_{out} > \frac{8.9 \cdot 5.25}{2.323} = 20, \tag{11}$$

that is more than twice as many compared with model signals at the 15th second.

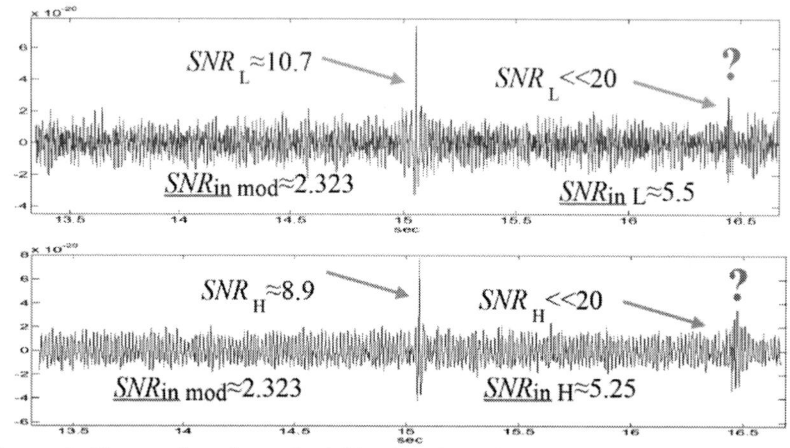

Figure 3. The results of matched filtering the mixture of signals and noises. Signals are at the 15th (test model) and 16th (supposed) seconds.

If hypothesis H_1 is true then SNR_{out} should be more than 20.

A small rise seen near 16th second ($SNR_{real} \approx 3$) as a result of matched filtering (Figure 3), which is much smaller than it should be (equation (11)).

Thus, the hypothesis H_1 is not confirmed ($SNR_{real} << 20$).

This means: **the chirp-signals are absent in the original records** (Figures 1, 2).

5. Determination of the signals and their parameters

If the chirp-signals are absent in registration data (Figure 1), it is necessary to determine the type of signals that are in LIGO records 14.09.2015.

Below are the results of processing of registration data for this purpose.

5.1. Evaluation of the times of signal arrivals

For this purpose, the optimum detector is used, which calculates the likelihood ratio:

$$L(h_H \mid s) = \ln \frac{P(h_H \mid s)}{P(h_H \mid n)}, \qquad L(h_L \mid s) = \ln \frac{P(h_L \mid s)}{P(h_L \mid n)} \quad . \tag{12}$$

From equation (12) it follows that the unknown signal detection in a first approximation, based on the calculation of the functional [7-8]:

$$U(t) = \int_{f_l}^{f_h} \frac{G_{t+T}(f)}{G_{t-T}(f)} df , \tag{13}$$

where $G_{t+T}(f)$, $G_{t-T}(f)$ - energy process spectra before and after the current time t, f_l, f_h - lower and upper frequency range of the signal.

Looking at Figure 4, you can see:

- confident detection of signals on the observatories **L1**, **H1**;
- signal arrived first at **L1** and ≤ 10 ms later at **H1**;
- $SNR \geq 7$;

- arrival time at L1 earlier 16.4 sec. $T_{arrival} \approx 16.395$ sec , $T_{arrival} > 16.33$ sec.

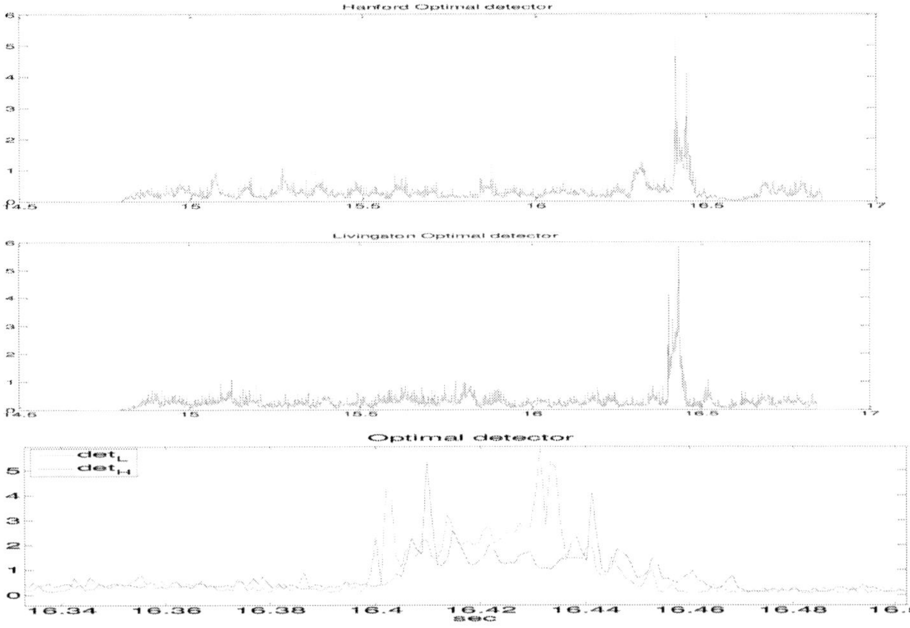

Figure 4. The results of the optimal detector **Hanford (*H*1)** - blue line, **Livingston (*L*1)**-red line.

5.2. Definition of signal waveforms using filtration

This is done using a two-stage causal filtering:

1st step. Butterworth filter, 35÷350 Hz band, the filter order = 8.

2nd step. Butterworth filter, 60÷450 Hz band, the filter order = 4 (to notch components at frequencies near 32 Hz and 60 Hz).

According to [5-6], the amplitude of the wave phases №№4-7 $A \approx 4 \cdot 10^{-21}$ (Figure 2), and their frequency is greater than 80 Hz and below 250 Hz,so after a two-step filtration (AFC $|H(f)| \approx 1$), these wave phases have not practically change.

However, these wave phases are not observed in the time interval 16.39 seconds -16.45 seconds in Figure 5 and Figure 8, and there are pulse signals (amplitude signals from $1 \cdot 10^{-21}$ to $2.5 \cdot 10^{-21}$).

This is a clear contradiction, and so the property of invariant № 1 is violated.

Similarly, one can see in violation of their of invariant № 2.

In more detail the detected signals are represented in Figure 5 (right part, Regime "Magnifier").

Signals (Figure 5, right part) are similar to the wavelets "Mexican hat".

They have the property of solitons: the higher the frequency the greater the amplitude,

$$\max A_L > \max A_H\,,\ f_L > f_H\,.$$

Spectra of these signals are shown in Figure 6.

Spectrum of signal $L1$ is a higher frequency than spectrum of signal $H1$ and invariants (№ 3,4) are not performed.

It follows from Figure 6:

$$\max S_H(t) \approx 165\,\text{Hz},\ \max S_L(t) \approx 205\,\text{Hz};$$

$$\Delta F = F_{\max L} F_{\max H} = 40\,\text{Hz} = 205 - 165\,\text{Hz}.$$

This is a clear contradiction to the declared data about detected signals [5,6], and contradicts the invariance properties of the spectra of signals at filtering (Invariants №№ 3,4).

It means: $s_H(t) \neq -s_L(t+\tau)$ **!**

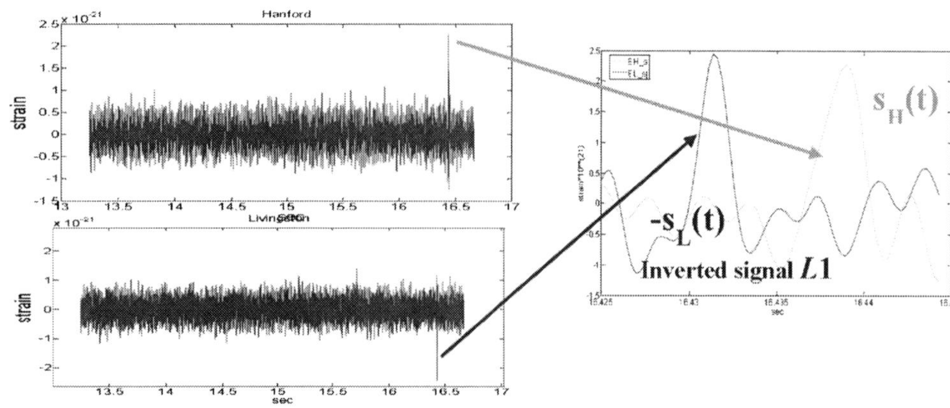

Figure 5. *Left part.* Recordings of filtered information, the time interval 13.2 sec - 16.8 sec. *Right part.* Recordings of filtered information, the time interval 16.425 sec - 16.445 sec. Regime "Magnifier". For a visual comparison the ***L*1** data are inverted.

6. Discussion of processing results

Different types of filtration are used in detection of signal with a priori unknown form: bandpass, whitening, etc. [7-8].

Physically realizable filter whitening is based on the use of optimal filtering of Wiener-Kolmogorov theory [9-10] and Levinson-Durbin procedure [11-12].

This approbation way of noise whitening was refused in [5-6].

Whitening (rectify) is proposed to conduct in the frequency domain in [5-6], using the equation:

$$Y(t) = F^{-1}\left(F_X(f)F_h(f)\right), \tag{14}$$

where $F_X(f)$ - Fourier transform of $X(t)$; $F_h(f)$ - transfer function of the whitening filter; F^{-1} - inverse Fourier transform.

The transfer function of the whitening filter is in [5]:

$$H(f) = R_{whitening}(f) + i\,\mathrm{Im}_{whitening}(f),$$
$$R_{whitening}(f) = \frac{1}{G_n(f)^{1/2}}, \quad \mathrm{Im}_{whitening}(f) = 0. \tag{15}$$

This function has a zero phase.

The transfer function of the filter (equation (15)) is a real even function. The impulse response is the inverse Fourier transform(F^{-1}) of the function (equation (15)). From the properties of the Fourier transform it follows that the impulse response will also be a real even function.

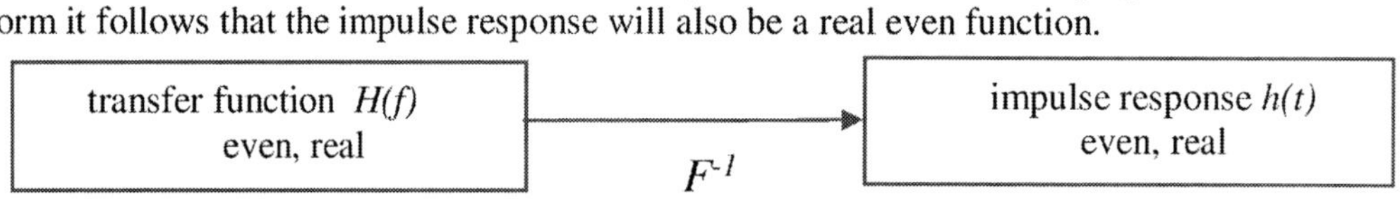

This function $h(t)$ is symmetric with respect to $t = 0$ and therefore there will be non-zero values of the impulse response in negative time $t < 0$ (Figure 7).

If there are non-zero values in the impulse response for $t < 0$, the causality principle is violated: the filter response occurs earlier than the input effect.

This is the proof that filter (equations (14), (15)) is a physically unrealizable (not causal).

The application of the corresponding filter leads to false detection of signals, shifts of the time of their arrival and distortion of their forms [13-15].

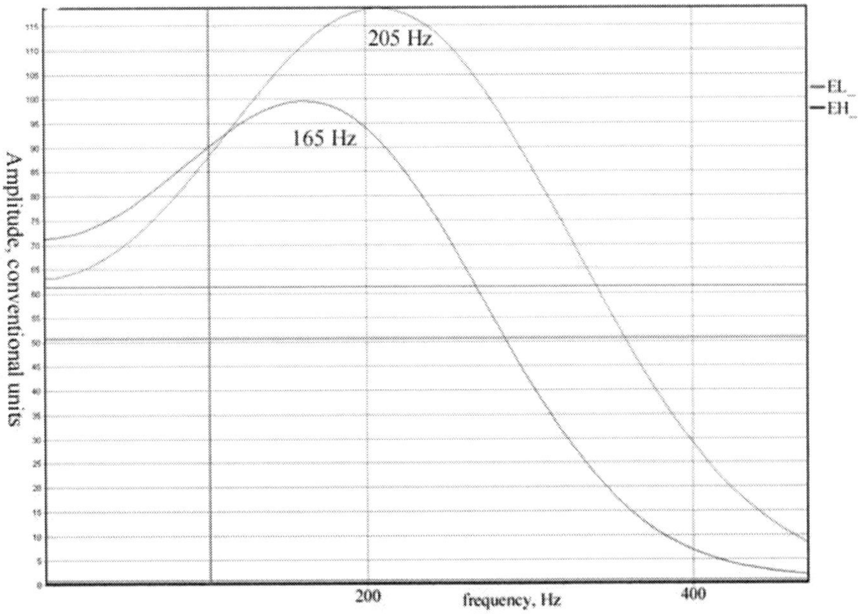

Figure 6. Signal spectra on observatories *L*1 (red line) and *H*1 (blue line).

A clear example of the impulse response of a zero-phase filter is given below (Figure 7, right part).
In this figure $h_1(t)$ is the impulse response of the Butterworth filter, 35÷350 Hz, the order is 6, the transfer function is:

$$H_1(f)= F(h_1(t))= H_{\text{batter}}(f);$$

the transfer function $H_2(f)$ is equal to the modulus $H_1(f)$:

$$H_2(f)=|H_{\text{batter}}(f)|.$$

The function $H_2(f)$ is real and even, the impulse response $h_2(t)$ is also real and even (Figure 7, right part) and non-zero values are clearly visible for $t<0$.
These two filters have the same amplitude-frequency characteristics ($|H_2(f)|=|H_1(f)|$,), and different phase characteristics ($\varphi_2(f)$ is not equal $\varphi_1(f)$), $\varphi_2(f) = 0$),
A visual comparison of the results of two different types of processing of registration data is presented in Figure 8.
Chirp-signals are not visible after causal filtering (Figure 5, 8), therefore the signals (Figure 2) are false, obtained by the not causal filters based on equations (14) and (15).

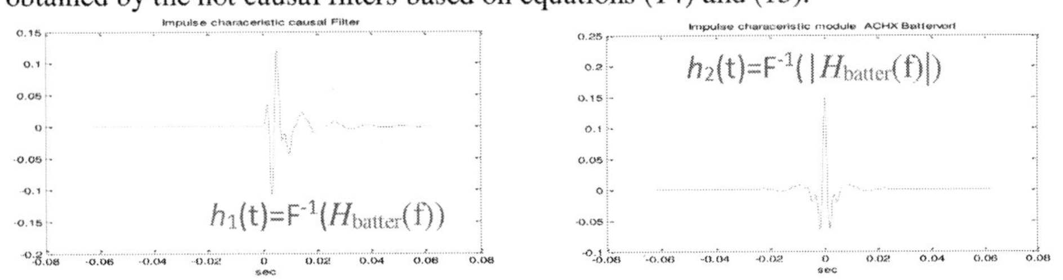

Figure 7. *Left part.* Impulse response of the Butterworth filter, 35-350 Hz, order = 6
Right part. Impulse response $h_2(t)$, transfer function $H_2(f)=|H_{batter}(f)|$.

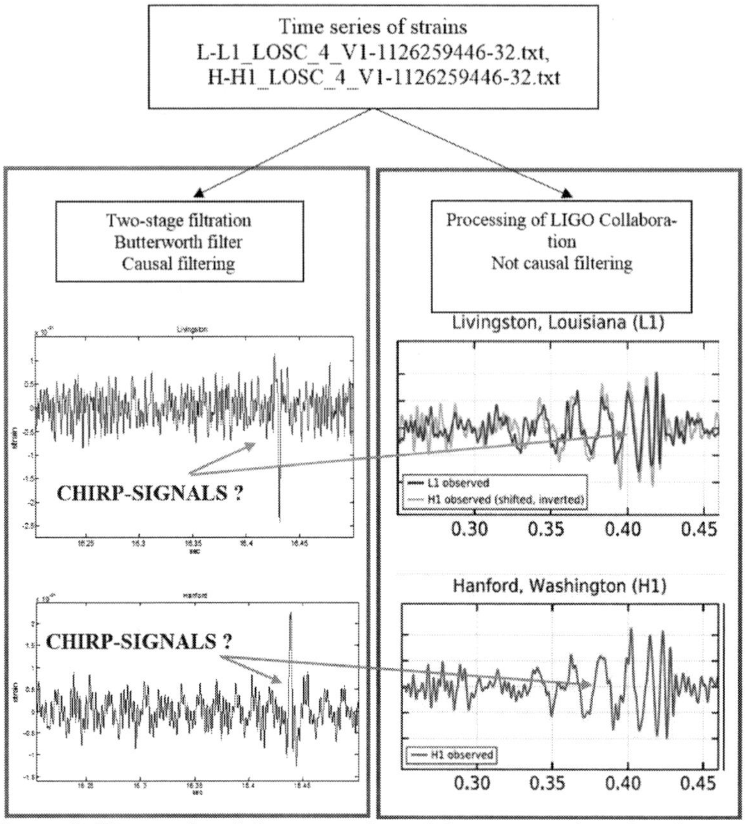

Figure 8. The results of the processing of the same registration data.
Left part - causal filtering, *right part* - not causal filtering.

In accordance with the LIGO scenario, the input useful signals (perturbations of the metric of the space at the input of the recorders of LIGO) are chirp-signals, which are obtained as solutions to the equations of rotation and merger of two black holes [16-17] along elliptical trajectories.
Above, the absence of black hole merger signals was experimentally confirmed by on the same records and signals of a different type are found.
It can be explained as follows:
parabolic or hyperbolic trajectories without the merger are more likely[18-19].
For such trajectories, the most probable signals are signals of several half-periods (example is shown at Figure 5, right part).
In this case, the body makes several turns around the black hole and once again flies off in space.

7. Conclusions

1. Chirp-signals are absent in LIGO data 14.09.2015.
This is established by means a matched filter and two-steps filtration, therefore signals of the merger of black holes are absent in the these data.
2. Signals of type "soliton" were highlighted by means of bandpass filtering (two steps). Two impulse signals of different polarity were found in these records. The corresponding wavelet "Mexican hat".
3. The spectra of the detected impulse signals have maxima:

$$\max S_H(t) \approx 165 \text{ Hz}, \ \max S_L(t) \approx 205 \text{ Hz};$$

4. The detected signals at $L1$ и $H1$: $s_H(t) \neq -s_L(t+\tau)$.

Appendix. You can filter data using standard procedures of MATLAB
Filtering program (MATLAB) for LIGO data and its description
(https://cloud.mail.ru/public/74ZC/nQUn6LryQ)
With its help everyone can make sure, that there are no signals from black hole merger in the LIGO records 14.09.2015.

References

[1] Einstein A 1916 Approximative Integration of the Field Equations of Gravitation (*Prussian Academy of Sciences, Sitzungsber Preuss. Akad. Wiss. Berlin (Math. Phys.)*) pp 688-696
[2] Hulse R A and Taylor J H 1975 Discovery of a pulsar in a binary system *Astrophysical Journal* **195** pp L51-L53
[3] Taylor J H and Weisberg J M 1982 A new test of general relativity - Gravitational radiation and the binary pulsar PSR 1913+16 *Astrophysical Journal* **253** pp 908-920
[4] Weber J 1960 Detection and Generation of Gravitational Waves *Physical Review* **117** pp.306-313
[5] Abbott B P *et al* 2016 Observation of Gravitational Waves from a Binary Black Hole Merger *Phys. Rev. Lett.* **116** 061102
[6] https://losc.ligo.org/s/events/GW150914/GW150914_tutorial.html
[7] Levin B R 1989 *Theoretical Foundations of Statistical Radio Engineering* (Moscow: Publishing house "Radio and connection") p 656
[8] Repin V G and Tartakovskiy G P 1977 *Statistical synthesis under a priori uncertainty and adapting information systems* (Moscow: Publishing house "Sov. Radio") p 432
[9] Kolmogorov A N 1941 Interpolation and extrapolation of stationary sequences *Proceedings of the Academy of Sciences of the USSR. Ser. Math* **5** pp 3-14
[10] Wiener N 1949 Extrapolation, interpolation and smoothing of stationary time *Series with engineering applications* ed N Y Wiley p 162
[11] Levinson N 1947 The Wiener RMS error criterion in filter design and prediction *J. Math. Phys.* **25** pp 261-278
[12] Durbin J 1960 The fitting of time series models *Rev. Inst. Int. Stat.* **28** pp.233-243
[13] Max J 1980 *Methods and techniques for signal processing and applications to physical measurements* (Paris Masson)
[14] Lukanenkov A V 2016 What registered LIGO 14.09.2015? *Engineering Physics* **8** pp 64-73
[15] Lukanenkov A V 2017 Experimental Confirmation of the Doubts about Authenticity of Event GW150914 *Journal of Applied Mathematics and Physics* **5** pp 538-550
[16] Shapiro C and Tyukolski S 1983 *Black holes, white dwarfs and neutron stars* N Y Wiley p 653
[17] Zeldovich B Ya and Novikov I D 1971 *The theory of gravity and evolution of stars* (Moscow: Publishing house "Science") p 484.
[18] Novikov I D and Frolov V P 1986 *The physics of black holes* (Moscow: Publishing house "Science") p 328
[19] Lukanenkov A V 2017 Gravitational experiments. Interpretation issues *Engineering Physics* **2** pp 57-66

Braiding, Majorana Fermions and the Dirac Equation

L H Kauffman

Department of Mathematics, Statistics and Computer Science University of Illinois, Chicago, United States
Department of Mechanics and Mathematics, Novosibirsk State University, Novosibirsk, Russia
E-mail: kauffman@uic.edu

Abstract. This paper reviews a construction called *iterant algebra* that reconstructs Clifford algebra and matrix algebra from the point of view of the symmetric group acting on vectors as ordered lists of their components. This algebraic background is then used to discuss the braiding of Majorana Fermions, the structure of the Dirac equation and the form of the the version of the Dirac equation due to Majorana.

1. Introduction

The simplest discrete system corresponds to the square root of minus one, when the square root of minus one is seen as an oscillation between plus and minus one. Thinking about the square root of minus one in this way, as an *iterant*, is explained below. Discrete systems can be embedded in non-commutative systems where discrete derivatives are replaced by commutators. This observation genralizes iterants and is explained at the end of Section 2 of the present paper. Iterant algebra generalizes matrix algebra and can be used to formulate the Clifford algebra for Majorana Fermions. This paper is a sequel to [8] and [5-8, 11-17] and it uses material from these papers. In this paper we give a very concise formulation of the basic concepts for our approach to iterants, Majorana Fermion operators and the Dirac equation. We have taken formulations in our previous papers and written them in the most condensed possible manner. These results in overlap with these papers, but the formulations are original to the present work. A further background of previous work of the author is given in the following references [1-4, 10, 19].

Section 2 is an introduction to the process algebra of iterants and how the square root of minus one arises from an alternating process. Section 3 discusses how Clifford algebras are fundamental to the structure of Fermions and how braiding is related to the Majorana Femions. Section 4 discusses the structure of the Dirac equation and how the nilpotent and the Majorana operators arise naturally in this context. This section provides a link between our work and the work on nilpotent structures and the Dirac equation of Peter Rowlands [22]. We end this section with an expression in split quaternions for the Majorana Dirac equation in one dimension of time and three dimensions of space. The Majorana Dirac equation can be written as follows:

$$\left(\frac{\partial}{\partial t}+\hat{\eta}\eta\frac{\partial}{\partial x}+\varepsilon\frac{\partial}{\partial y}+\hat{\varepsilon}\eta\frac{\partial}{\partial z}-\hat{\varepsilon}\hat{\eta}\eta m\right)\psi=0$$

where η and ε are the simplest generators of iterant algebra with $\eta^2=\varepsilon^2=1$ and $\eta\varepsilon+\varepsilon\eta=0$, and $\hat{\varepsilon}$, $\hat{\eta}$ form a copy of this algebra that commutes with it. This combination of the simplest Clifford algebra with itself is the underlying structure of Majorana Fermions, forming indeed the underlying structure of all Fermions.

2. Iterants

An iterant is a sum of elements of the form

$$[a_1, a_2, ..., a_n]\sigma$$

where $[a_1, a_2, ..., a_n]$ is a vector of scalars (usually real or complex numbers) and σ is a permutation on n letters. Such elements are themselves sums of elements of the form

$$[0,0,...,0,1,0,...,0]\sigma = e_i\sigma$$

where the 1 is in the i-th place. The vectors e_i are the basic idempotents that generate the iterants with the help of the permutations.

If $a = [a_1, a_2, ..., a_n]$, then let a^σ denote the vector with its elements permuted by the action of σ:

$$a^\sigma = \left[a_{\sigma(1)}, a_{\sigma(2)}, ..., a_{\sigma(n)}\right].$$

If a and b are vectors then their *elementary product* ab denotes the vector where $(ab)_i = a_i b_i$, and $a+b$ denotes the vector where $(a+b)_i = a_i + b_i$. Then products and sums are defined by the formulas below:

$$(a\sigma)(b\tau) = (ab^\sigma)\sigma\tau,$$
$$(ka)\sigma = k(a\sigma)$$

for a scalar k, and

$$(a+b)\sigma = a\sigma + b\sigma$$

where vectors are multiplied as above and we take the usual product of the permutations. All of matrix algebra and more is naturally represented in the iterant framework. See [17].

If η is the order two permutation of two elements, then $[a,b]^\eta = [b,a]$. We can define

$$i = [1,-1]\eta$$

and then

$$i^2 = [1,-1]\eta[1,-1]\eta = [1,-1][1,-1]^\eta\eta^2 = [1,-1][-1,1] = [-1,-1] = -1.$$

In this way the complex numbers arise naturally from iterants. One can interpret $[1,-1]$ as an oscillation between $+1$ and -1 and η as denoting a temporal shift operator. The $i = [1,-1]\eta$ is a time sensitive element and its self-interaction has square minus one. In this way iterants can be interpreted as a formalization of elementary discrete processes.

A more general approach to discrete processes [18] includes this interpretation of iterants and the square root of negative unity. The more general approach is worth reprising in this context. Given a sequence of discrete algebraic elements $X_t\,(t = 0,1,...)$ (we take them to be associative but not necessarily commutative for this discussion), we define an *invertible shift operator* J so that

$$X_{t+1} = J^{-1}X_tJ$$

Then one can define a discrete derivative ∇ by the equation

$$\nabla X_t = \frac{J(X_{t+1} - X_t)}{\Delta t}.$$

Note that, since $JX_{t+1} = X_t J$, this discrete derivative satisfies

$$\nabla X_t = \frac{X_t J - JX_t}{\Delta t} = \frac{[X_t, J]}{\Delta t}.$$

Thus ∇ is represented as a commutator and satisfies the Leibniz rule

$$\nabla(AB) = \nabla(A)B + A\nabla(B).$$

This means that we can shift the analysis of discrete systems to non-commutative worlds where all derivatives are represented by commutators. This is the subject of [18] and a number of our earlier papers.

3. Clifford Algebra, Majorana Fermions and Braiding

In Fermion algebra one has annihiliation operators ψ and their conjugate creation operators $\psi^\dagger$. One has $\psi^2 = 0 = (\psi^\dagger)^2$ and a fundamental commutation relation

$$\psi\psi^\dagger + \psi^\dagger\psi = 1.$$

Pairs of such operators such as ψ and φ, anti-commute:

$$\psi\varphi = -\varphi\psi.$$

Majorana Fermion operators c that satisfy $c^\dagger = c$ so that they are their own anti-particles. There is a lot of interest in such operators in relation to quasi-particles. They are related to braiding and to topological quantum computing. See [21] where researchers found quasiparticle Majorana Fermions in edge effects in nano-wires. The Fibonacci model that we discuss in [6] is also based on Majorana particles, related to collective electronic excitations. If P is a Majorana Fermion particle, then P can interact with itself to either produce itself or to annihilate itself. This is the simple "fusion algebra" for this particle. One can write $P^2 = P + 1$ to denote the two possible self-interactions the particle P. The patterns of interaction and braiding of such a particle P give rise to the Fibonacci model.
Majorana operators are related to standard Fermion operators. The algebra for Majorana operators is $c^\dagger = c$ and $cc' = -c'c$ if c and c' are distinct Majorana Fermion operators with $c^2 = 1$ and $c'^2 = 1$. A standard Fermion operator is composed from two Majorana operators via

$$\psi = (c + ic')/2,$$
$$\psi^\dagger = (c - ic')/2.$$

Similarly one can make two Majorana operators from any single Fermion operator. If you take a set of Majoranas

$$\{c_1, c_2, c_3, \ldots, c_n\}$$

then there are braiding operators that act on the vector space with these c_k as the basis. The operators are mediated by algebra elements

$$\tau_k = (1 + c_{k+1}c_k)/\sqrt{2},$$
$$\tau_k^{-1} = (1 - c_{k+1}c_k)/\sqrt{2}.$$

The braiding operators are

$$T_k : \ Span\{c_1, c_2, \ldots, c_n\} \to Span\{c_1, c_2, \ldots, c_n\}$$

via

$$T_k(x) = \tau_k x \tau_k^{-1}.$$

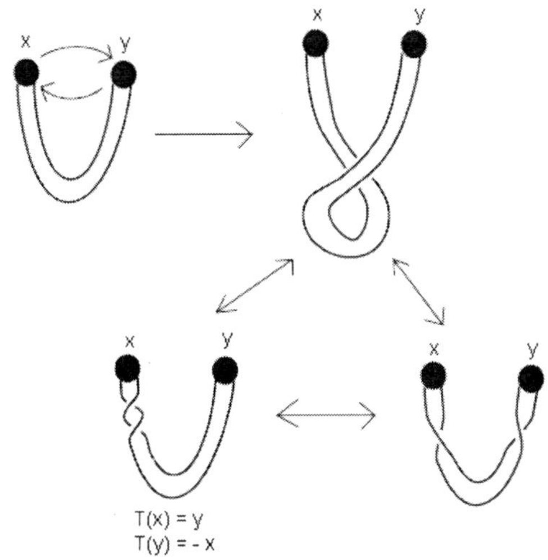

Figure 1: Braiding Action on a Pair of Fermions

The braiding is summarized by these formulas:

$$T_k(c_k) = c_{k+1},$$
$$T_k(c_{k+1}) = -c_k,$$

and T_k is the identity otherwise. This gives a very unitary representaton of the Artin braid group, and it can be used to produce universal quantum computing in the presence of key local unitary operators. See Figure 1 for a depiction of this in relation to the topology of a belt that connects the Fermion particles. The relationship with the belt depends upon the fact that in quantum mechanics one must represent rotations of three dimensional space as unitary transformations. See [9]. In the Figure, the belt does not know which of the two Fermions to endow with the phase change, but the clever algebra above makes this decision. It is remarkable that fundamental topology occurs at the level of the Majorana Fermion operators.

Now we make contact between the iterants and the algebra of the Majorana Fermion operators. Let $e = [1,-1]$. Then $e^2 = [1,1] = 1$ and $e\eta = [1,-1]\eta = [-1,1]\eta = -e\eta$. Thus

$$e^2 = 1,$$
$$\eta^2 = 1,$$

and

$$e\eta = -\eta e.$$

We regard the basic iterants e and η as a fundamental pair of Majorana Fermions.

4. The Dirac Equation and Majorana Fermions

We now construct the Dirac equation. If the speed of light is equal to 1 (by convention), then energy E, momentum p and rest mass m are related by the (Einstein) equation

$$E^2 = p^2 + m^2.$$

Dirac constructed his equation by looking for an algebraic square root of $p^2 + m^2$ so that he could have a linear operator for E that would take the same role as the Hamiltonian in the Schroedinger equation. We will get to this operator by first taking the case where p is a scalar (we use one dimension of space and one dimension of time.). Let $E = \alpha p + \beta m$ where α and β are elements of a possibly non-commutative, associative algebra. Then

$$E^2 = \alpha^2 p^2 + \beta^2 m^2 + pm(\alpha\beta + \beta\alpha).$$

Hence we satisfiy $E^2 = p^2 + m^2$ if $\alpha^2 = \beta^2 = 1$ and $\alpha\beta + \beta\alpha = 0$. This is the Clifford algebra pattern and we can apply the iterant algebra generated by e and η. Then, with the quantum operator for momentum is $-i\frac{\partial}{\partial x}$ and the operator for energy is $i\frac{\partial}{\partial t}$, we have the Dirac equation

$$i\frac{\partial\psi}{\partial t} = -i\alpha\frac{\partial\psi}{\partial x} + \beta m\psi.$$

Let

$$O = i\frac{\partial}{\partial t} + i\alpha\frac{\partial}{\partial x} - \beta m$$

so that the Dirac equation takes the form

$$O\psi(x,t) = 0.$$

Now note that

$$Oe^{i(px-Et)} = (E - \alpha p - \beta m)e^{i(px-Et)}.$$

We let

$$\Delta = (E - \alpha p - \beta m)$$

and let

$$U = \Delta\beta\alpha = (E - \alpha p - \beta m)\beta\alpha = \beta\alpha E + \beta p - \alpha m,$$

then

$$U^2 = -E^2 + p^2 + m^2 = 0.$$

This nilpotent element leads to a (plane wave) solution to the Dirac equation. We have that

$$O\psi = \Delta\psi$$

for $\psi = e^{i(px-Et)}$. It follows that

$$O(\beta\alpha\Delta\beta\alpha\psi) = \Delta\beta\alpha\Delta\beta\alpha\psi = U^2\psi = 0,$$

from which it follows that

$$\psi = \beta\alpha U e^{i(px-Et)}$$

is a (plane wave) solution to the Dirac equation.

This calculation suggests that we multiply the operator O by $\beta\alpha$ on the right, obtaining the operator

$$D = O\beta\alpha = i\beta\alpha\frac{\partial}{\partial t} + i\beta\frac{\partial}{\partial x} - \alpha m,$$

and the equivalent Dirac equation

$$D\psi = 0.$$

For the specific ψ above we have $D\left(Ue^{i(px-Et)}\right) = U^2 e^{i(px-Et)} = 0$. Rowlands [22] does this trick in the context of quaternion algebra. Here we use the split quaternions. We regard this method of construction via the split quaternions as fundamental since it connects the Dirac equation to the process algebra and iterants by which we have begun this development. Note that the solution to the Dirac equation that we have found is expressed in Clifford algebra or iterant algebra form. It can be articulated into specific vector solutions by using an iterant or matrix representation of the algebra.

We see that the algebra element $U = \beta\alpha E + \beta p - \alpha m$ with $U^2 = 0$ is the core of this plane wave solution to the Dirac equation. A natural non-commutative algebra arises directly from articulation of

discrete process and can be regarded as essential information in a Fermion. It is natural to compare this algebra structure with algebra of creation and annihilation operators that occur in quantum field theory.
If we let

$$\tilde{\psi} = e^{i(px+Et)}$$

(reversing time), then

$$D\tilde{\psi} = (-\beta\alpha E + \beta p - \alpha m)\psi = U^{\dagger}\tilde{\psi},$$

This gives a definition of $U^{\dagger}$ corresponding to the anti-particle for $U\psi$.

We have

$$U = \beta\alpha E + \beta p - \alpha m$$

and

$$U^{\dagger} = -\beta\alpha E + \beta p - \alpha m$$

Note that

$$(U + U^{\dagger})^2 = (2\beta p + \alpha m)^2 = 4(p^2 + m^2) = 4E^2,$$

and

$$(U - U^{\dagger})^2 = -(2\beta\alpha E)^2 = -4E^2.$$

Thus

$$U^2 = (U^{\dagger})^2 = 0$$

and

$$UU^{\dagger} + U^{\dagger}U = 4E^2.$$

This is a direct appearance of the Fermion algebra corresponding to the Fermion plane wave solutions to the Dirac equation. Furthermore, the decomposition of U and $U^{\dagger}$ into the Majorana Fermion operators corresponds to $E^2 = p^2 + m^2$. Normalizing by dividing by $2E$ we have

$$A = \frac{\beta p + \alpha m}{E}$$

and

$$B = i\beta\alpha.$$

so that

$$A^2 = B^2 = 1$$

and

$$AB + BA = 0.$$

then

$$U = (A + Bi)E$$

and

$$U^{\dagger} = (A - Bi)E,$$

articulating the Fermion operators in terms of the simpler Clifford algebra of Majorana operators (split quaternions once again).

4.1 Writing in the Full Dirac Algebra

We have written the Dirac equation in one dimension of space and one dimension of time. We here translate the formalism to three dimensions of space. Take an independent Clifford algebra generated by σ_1, σ_2, σ_3 with $\sigma_i^2 = 1$ for $i = 1,2,3$ and $\sigma_i\sigma_j = -\sigma_j\sigma_i$ for $i \neq j$. Assume that α and β as used above generate an independent Clifford algebra commuting with the algebra of the σ_i. Replace scalar momentum p by a 3-vector momentum $p=(p_1, p_2, p_3)$ and let $p \bullet \sigma = p_1\sigma_1 + p_2\sigma_2 + p_3\sigma_3$. Replace $\frac{\partial}{\partial x}$ with $\nabla = \left(\frac{\partial}{\partial x_1}, \frac{\partial}{\partial x_2}, \frac{\partial}{\partial x_3}\right)$ and $\frac{\partial p}{\partial x}$ with $\nabla \bullet p$.

The Dirac equation takes the form:

$$i\frac{\partial \psi}{\partial t} = -i\alpha\nabla \bullet \sigma\psi + \beta m\psi .$$

Let

$$O = i\frac{\partial}{\partial t} + i\alpha\nabla \bullet \sigma - \beta m$$

so the Dirac equation takes the form

$$O\psi(x,t) = 0 .$$

By our previous discussion we let

$$\psi(x,t) = e^{i(p\bullet x - Et)}$$

and construct solutions by applying the Dirac operator to this ψ. The two Clifford algebras interact to generalize the nilpotent solutions and Fermion algebra that we have detailed for one spatial dimension to this three dimensional case. The modified Dirac operator is

$$D = i\beta\alpha\frac{\partial}{\partial t} + \beta\nabla \bullet \sigma - \alpha m .$$

We have that

$$D\psi = U\psi$$

where

$$U = \beta\alpha E + \beta p \bullet \sigma - \alpha m .$$

Then $U^2 = 0$ and $U\psi$ is a solution to the modified Dirac Equation, as before. We articulate the structure of the Fermion operators and locate the corresponding Majorana Fermion operators. Details are left to the reader.

4.2 Majorana Fermions

We end with a brief discussion making a Dirac algebra that is distinct from the one generated by α, β, σ_1, σ_2, σ_3. From the new algebra we obtain an equation that can have real solutions. This was the strategy Majorana [20] followed to construct his Majorana Fermions. A real equation can have solutions that are invariant under complex conjugation. The solutions can correspond to particles that are their own anti-particles. We describe this Majorana algebra in terms of the split quaternions ε and η. For convenience we use the matrix representation given below. The reader of this paper can substitute the corresponding iterants.

$$\varepsilon = \begin{pmatrix} -1 & 0 \\ 0 & 1 \end{pmatrix}, \ \eta = \begin{pmatrix} 0 & 1 \\ 1 & 0 \end{pmatrix}.$$

Let $\hat{\varepsilon}$ and $\hat{\eta}$ generate another, independent algebra of split quaternions, commuting with the first algebra generated by ε and η. A totally real Majorana Dirac equation can be written as follows:

$$\left(\frac{\partial}{\partial t} + \hat{\eta}\eta\frac{\partial}{\partial x} + \varepsilon\frac{\partial}{\partial y} + \hat{\varepsilon}\eta\frac{\partial}{\partial z} - \hat{\varepsilon}\hat{\eta}\eta m\right)\psi = 0 .$$

To see that this is a correct Dirac equation, note that

$$\hat{E} = \alpha_x\hat{p}_x + \alpha_y\hat{p}_y + \alpha_z\hat{p}_z + \beta m$$

(Here the "hats" denote the quantum differential operators corresponding to the energy and momentum.) will satisfy

$$\hat{E}^2 = \hat{p}_x^2 + \hat{p}_y^2 + \hat{p}_z^2 + m^2$$

if the algebra generated by α_x, α_y, α_z, β has each generator of square one and each distinct pair of generators anti-commuting. We obtain the general Dirac equation by replacing $\hat{E}$ by $i\frac{\partial}{\partial t}$, and $\hat{p}_x$ with $-i\frac{\partial}{\partial x}$ (and same for y, z).

$$\left(i\frac{\partial}{\partial t} + i\alpha_x\frac{\partial}{\partial x} + i\alpha_y\frac{\partial}{\partial y} + i\alpha_z\frac{\partial}{\partial z} - \beta m \right)\psi = 0 .$$

This is equivalent to

$$\left(\frac{\partial}{\partial t} + \alpha_x\frac{\partial}{\partial x} + \alpha_y\frac{\partial}{\partial y} + \alpha_z\frac{\partial}{\partial z} + i\beta m \right)\psi = 0 .$$

We take

$$\alpha_x = \hat{\eta}\eta,\ \alpha_y = \varepsilon,\ \alpha_y = \hat{\varepsilon}\eta,\ \beta = i\hat{\varepsilon}\hat{\eta}\eta ,$$

These elements satisfy the requirements for the Dirac algebra. Note how we have a significant interaction between the commuting square root of minus one (i) and the element $\hat{\varepsilon}\hat{\eta}$ of square minus one in the split quaternions. This return to our original considerations about the source of the square root of minus one. Both viewpoints combine in the element $\beta = i\hat{\varepsilon}\hat{\eta}\eta$ that makes this Majorana algebra work. Since the algebra appearing in the Majorana Dirac operator is constructed entirely from two commuting copies of the split quaternions, there is no appearance of the complex numbers, and when written out in 2×2 matrices we obtain coupled real differential equations to be solved. This ending is actually a beginning of a new study of Majorana Fermions. That will commence in a sequel to the present paper.

References

[1] Kauffman L 1985 Sign and Space, In Religious Experience and Scientific Paradigms *Proceedings of the 1982 IASWR Conference, (Stony Brook, New York: Institute of Advanced Study of World Religions)* pp 118-164

[2] Kauffman L 1987 Self-reference and recursive forms *Journal of Social and Biological Structures* pp 53-72

[3] Kauffman L 1987 Special relativity and a calculus of distinctions *Proceedings of the 9th Annual Intl. Meeting of ANPA, Cambridge, England (Pub. by ANPA West)* pp 290-311

[4] Kauffman L 1987 Imaginary values in mathematical logic *Proceedings of the Seventeenth International Conference on Multiple Valued Logic, May 26-28 Boston MA (IEEE Computer Society Press)* pp 282-289

[5] Kauffman L H 1994 *Knot Logic In Knots and Applications* ed L Kauffman (World Scientific Pub. Co.) pp 1-110

[6] Kauffman L H 2016 *Knot logic and topological quantum computing with Majorana Fermions. In "Logic and algebraic structures in quantum computing and information"* (Lecture Notes in Logic) eds J Chubb, J Chubb, Ali Eskandarian and V Harizanov (Cambridge University Press)

[7] Kauffman L H and Lomonaco S J 2018 Braiding, Majorana Fermions and Topological Quantum Computing *Special Issue of QIP on Topological Quantum Computing* (to appear)

[8] Kauffman L H 2015 Iterants, Fermions and Majorana Operators *Unified Field Mechanics - Natural Science Beyond the Veil of Spacetime* eds R Amoroso, L H Kauffman and P Rowlands (World Scientific Pub. Co.) pp 1-32

[9] Kauffman L H 1991, 1994, 2001, 2012 *Knots and Physics* (World Scientific Pub)

IOP Conf. Series: Journal of Physics: Conf. Series **1051** (2018) 012036 doi:10.1088/1742-6596/1051/1/012036

[10] Kauffman L H 2002 Time imaginary value, paradox sign and space *Computing Anticipatory Systems CASYS AIP Conference Proceedings Fifth International Conference (Liege, Belgium) – 2001* ed D Dubois **627**
[11] Kauffman L H and H Noyes Pierre 1996 Discrete Physics and the Derivation of Electromagnetism from the formalism of Quantum Mechanics *Proc. of the Royal Soc. Lond.* A **452** pp 81-95
[12] Kauffman L H and H Noyes Pierre 1996 Discrete Physics and the Dirac Equation *Physics Letters* A **218** pp 139-146
[13] Kauffman L H 1998 Noncommutativity and discrete physics *Physica* D **120** pp 125-138
[14] Kauffman L H 1998 Space and time in discrete physics *Intl. J. Gen. Syst.* **27** pp 1-3, 241-273
[15] Kauffman L H 2004 Non-commutative worlds *New Journal of Physics* **6** pp 2-46
[16] Kauffman L H 2007 Differential geometry in non-commutative worlds *Quantum Gravity - Mathematical Models and Experimental Bounds* eds B Fauser, J Tolksdorf and E Zeidler, (Birkhauser) pp 61-75
[17] Kauffman L H 2017 Iterant algebra. *Entropy* **19** 7 pp 94A17
[18] Kauffman L H 2018 Non-Commutative Worlds and Classical Constraints *Entropy* **20** 483
[19] Kauffman L H and Samuel J. Lomonaco Jr 2007 q-deformed spin networks, knot polynomials and anyonic topological quantum computation. *J. Knot Theory Ramifications* **16** 3 pp 267–332
[20] Majorana E 1937 A symmetric theory of electrons and positrons *I Nuovo Cimento* **14** pp 171-184
[21] Mourik V, Zuo K, Frolov S M, Plissard S R, Bakkers E P A M and Kouwenhuven L P 2012 Signatures of Majorana Fermions in hybred superconductor-semiconductor devices (*Preprint* arXiv:1204.2792)
[22] Rowlands P 2007 Zero to Infinity - The Foundations of Physics *Series on Knots and Everything* (World Scientific Publishing Co.) **41**

Acknowledgements
It gives the author pleasure to thank G. Spencer-Brown, James Flagg, Alex Comfort, David Finkelstein, Pierre Noyes, Peter Rowlands, Sam Lomonaco, Bernd Schmeikal and Rukhsan Ul Haq for conversations related to the considerations in this paper. Nothing here is their fault, yet Nothing would have happened without them.